Nanostructured Multifunctional Materials
Synthesis, Characterization, Applications and Computational Simulation

Editor

Esteban A. Franceschini

Instituto de Investigaciones en Fisicoquimica de Córdoba
Universidad Nacional de Córdoba - CONICET
Depto. de Fisicoquimica - Facultad de Ciencias Químicas
Córdoba, Argentina

CRC Press
Taylor & Francis Group
Boca Raton   London   New York

CRC Press is an imprint of the
Taylor & Francis Group, an **informa** business

A SCIENCE PUBLISHERS BOOK

Cover credit: Cover illustrations reproduced by kind courtesy of Dr. Alejandro Fracaroli (author of Chapter 5) and Dr. Gaetano Caramazza, Departamento de Quimica Organica, Facultad de Ciencias Quımicas, Universidad Nacional de Cordoba, Cordoba, Argentina; Instituto de Investigaciones en Fisicoquımica de Cordoba INFIQC, Consejo Nacional de Investigaciones Cientificas y Tecnicas, CONICET.

First edition published 2021
by CRC Press
6000 Broken Sound Parkway NW, Suite 300, Boca Raton, FL 33487-2742

and by CRC Press
2 Park Square, Milton Park, Abingdon, Oxon, OX14 4RN

© 2021 Taylor & Francis Group, LLC

CRC Press is an imprint of Taylor & Francis Group, LLC

Library of Congress Cataloging-in-Publication Data
Names: Franceschini, Esteban A., 1985- editor.
Title: Nanostructured multifunctional materials : synthesis,
    characterization, applications and computational simulation / editor,
    Esteban A. Franceschini, Instituto de Investigaciones en Fisicoquímica
    de Córdoba, Universidad Nacional de Córdoba--CONICET, Depto. de
    Fisicoquímica--Facultad de Ciencias Químicas, Córdoba, Argentina.
Description: First edition. | Boca Raton : CRC Press, Taylor & Francis
    Group, 2021. | Includes bibliographical references and index.
Identifiers: LCCN 2020057693 | ISBN 9780367420697 (hbk)
Subjects: LCSH: Nanostructured materials.
Classification: LCC TA418.9.N35 N35424 2021 | DDC 620.1/15--dc23
LC record available at https://lccn.loc.gov/2020057693

ISBN: 978-0-367-42069-7 (hbk)
ISBN: 978-0-367-76349-7 (pbk)
ISBN: 978-0-367-82219-4 (ebk)

Typeset in Times New Roman
by Radiant Productions

Preface

With the advance in materials science, the methods to develop systems with different functionalities went from being simple trial and error procedures to rational methods designed *ad hoc* to obtain defined structures and functionalities, seeking to synergistically accompany the demands of industrial/technological developments.

In this case, the synthesis of nanomaterials plays a fundamental role in current and future technological applications, particularly nanomaterials that have multiple functionalities.

Nanostructured Multifunctional Materials: Synthesis, Characterization, Applications, and Computational Simulation provide a broad overview of the effect of nanostructuring in the multifunctionalities of different widely studied nanomaterials.

This book is divided into four sections constituting a road map that group materials sharing certain types of nanostructuring, including:

- Nanoporous materials.
- Nanoparticulate materials.
- 2D laminar materials.
- Computational methods and characterizations of the nanostructures are addressed.

This structured approach in nanomaterials research will work as valuable reference material for a broad number of chemists, (bio)engineers, physicists, nanotechnologists, undergraduates and professors.

The book also has an introductory chapter that aims to provide a useful theoretical framework for all subsequent chapters, thus avoiding repeated explanations of necessary basic concepts.

Each part is divided into chapters that explore the state of the art of different nanomaterials including analysis of the different synthesis methods, characterization techniques, approaches to a variety of applications and finally the computational studies that allow the rational design and the study of the properties that give rise to the multifunctions of nanomaterials.

The reader will find in this book a wide and very complete approach to the rational design of materials for multiple applications from an appropriate perspective for a broad audience.

The idea of this book project was initiated by the publisher. It is my honour to be invited to serve as editor for this book and allow me to invite respected colleagues as chapter contributions.

The authors would like to thank the readers for considering this book and we hope it fulfils their expectations. I would like to express my sincere thanks to all the contributing authors for their commitment and work. This would not have been possible without your knowledge and experience.

Moreover, I would especially like to thank the Argentinean public education system, *alma mater* of most of the authors of this book, working now in many universities and institutes around the globe. Without a high-quality public education, the work we do would not have been possible.

Finally, I also thank the Editorial Department of Science Publishers (CRC Press, Taylor and Francis group).

07/15/2020

E. Franceschini
Cordoba, Argentina

Contents

CHAPTER 1

Introduction to Nanostructured Multifunctional Materials

Esteban A. Franceschini

Introduction

In the last decades, great progress has been made with a focus on the rational design of materials, particularly nanomaterials, as an approach to ad-hoc improving their properties for different applications. The characteristics of these nanomaterials can be tuned by a carefully controlled selection of the synthetic parameters varying the nucleation conditions, growth, crystallization, use of structuring agents, post-synthesis treatments, among many others. The physicochemical and structural characteristics of a nanomaterial or the different combination of them, give the materials particular properties (Liu et al. 2011).

There are several classifications for nanomaterials, the most concrete are those that are based on their composition. Normally, materials can be classified as metals and non-metals, while non-metals can be divided into inorganic and organic materials. The simplest categories used are metals, ceramics, polymers and the different mixtures of these materials are called composites or hybrids, depending on the constituent parts. There are also subcategories, some of which are semiconductors, biomaterials, elastomers, glasses, etc.

More importantly, the dimensionality in the development of nanomaterials is one of the determining factors for the properties of the materials. This effect is mainly because dimensionality affects the movement of electrons, holes, excitons, plasmons and phonons in their structures, modifying their intrinsic properties, such as electrical, magnetic, and optical, among others (Hu et al. 1999), making them much different than bulk materials properties (Tran et al. 2009, Sajanlal et al. 2011).

Pokropivny and Skorokhod (2007) established a classification of nanomaterials according to their morphology, which exceeds the previous classifications established by Gleiter (2000) and Skorokhod et al. (2001), which allows including 0-D, 1-D, 2-D and 3-D structures (Fig. 1.1). This classification based on the dimensionality of the nanostructures helps to understand and even predict the properties of nanomaterials leading to develop material systems with defined functionalities and even multifunctionality (Wu et al. 2016).

Figure 1.2 shows an example of systems where the confinement effect given by the different morphologies generates electronic effects. In this example, a semiconductor material with different types of confinement (0-D, 1-D, 2-D and 3-D) is surrounded by a material of a wider band gap. Thus, the electrons in each case are confined in different dimensions giving place to different types

INFIQC, Facultad de Ciencias Químicas, Universidad Nacional de Córdoba-CONICET. 5000 Córdoba, Argentina.
Email: esteban.franceschini@mi.unc.edu.ar

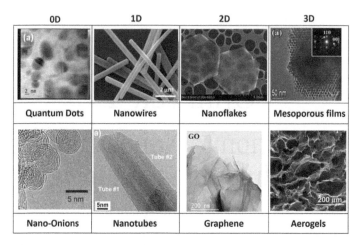

Figure 1.1: Different heterogeneous nanostructured materials based on structural complexity. 0-D, zero-dimensional; 1-D, one-dimensional; 2-D, two-dimensional; 3-D, three-dimensional. Adapted from (Nideep et al. 2020, Tomita et al. 2001, Lehman et al. 2011, Gupta et al. 2017) with permission from Elsevier and (Adapted from (Shen et al. 2017, Saathish and Miyazawa 2007, Yue et al. 2009, Hu et al. 2014) with permission from American Chemical Society).

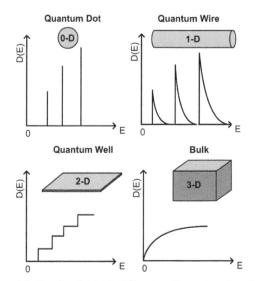

Figure 1.2: Density of states in different confinement configurations.

of band structures. The conduction and valence bands split into overlapping sub bands that become successively narrower as the electron motion is restricted in a greater number of dimensions (Saleh and Teich 2007)

Some of the definitions of each one of these categories will be reviewed.

Zero-Dimensional (0-D) Nanomaterials

In 0-D nanomaterials, the electronic confinement is found in three dimensions. Quantum dots, uniformly sized nanoparticles, that can be isolated or in arrays, similar to the core-shell, multilayer core-shell (onions), hollow spheres and nano lenses are some of the examples of 0-D nanomaterials. In these cases, the dimensional confinement of the electrons is responsible for the electronic and optical properties of 0-D materials. For instance, in Quantum Dots (QDs), there is no possibility for delocalization of electrons due to the confinement and are located in an 'infinitely deep potential well'.

Most Quantum Dots (QDs) are made up of semiconductor nanoparticles. When these particles interact with a photon of the appropriate energy, the transference of one electron from the valence bands to the conduction bands is carried out, and an electron-hole pair (exciton) is produced. So there exists an exciton Bohr radius (defined as rb), which is stated as the physical separation between the electron and the hole produced by the interaction with the photon. This Bohr radius varies not only by the composition of the semiconductors QDs but also because of the crystalline structure, coatings and interaction between nanoparticles.

Thus, the discrete energy (E_n) corresponding to the confinement occurred for different 0-D nanomaterials can be obtained by:

$$For \ 0-D: E_n = \left[\frac{\pi^2 h^2}{2mL^2}\right](n_x^2 + n_y^2 + n_z^2) \tag{1}$$

where h is the Planck's constant, m is the mass of an electron, L is the orbital perimeter (related to the particle diameter), n is the dimensional coordinates (dimensionless unit).

Therefore, the quantum confinement theory allows one to predict and easily fit the band gap values for semiconductor crystals by simply modifying the QDs diameter in comparison with rb (Bashir and Liu 2015). For instance, a decrease in size below 10 nm can increase up to 2 eV the electronic transition energy of the same material (relative to the bulk material) (Brichkin and Razumov 2016).

By definition, the length and width of 0-D materials are predominantly similar. However, the crystalline structure can change dramatically and amorphous or crystalline structures can be found. In the case of crystalline QDs, the diameter of the nanoparticle is usually the same as the crystallite size and is in the order with the rb responsible for the exciton quantum confinement effect raised by the discrete energy levels. Hence, any change in the dimensions of the nanocrystal means a significant change in the band gap and consequently, in the physicochemical properties of the material.

This extensive knowledge of the relationship between size and electronic properties of 0-D nanomaterials has led to a huge development in recent decades oriented to its varied applications.

It is important to note that although most of the classic QDs are made up of inorganic semiconductors, Carbon Quantum Dots (CQDs) have attracted great attention in the last two decades. Particularly, fluorescent CQDs that were first discovered by Xu et al. (2004) in an arc-discharge synthesis of single-walled CNTs (Xu et al. 2004). From there, CQDs become some of the most studied new 0-D materials because it is a carbon-based material with very particular physicochemical, electronic and optical properties; which made it applicable to a wide variety of technological fields such as drug delivery, biological imaging, optoelectronics, photovoltaics and photocatalysis. Particularly, one of the most interesting properties of these types of materials is the easy photoluminescence tunability which affects the photo-induced electron transfer making it exceptionally capable of light-harvesting, which makes them excellent candidates for solar energy conversion (Wang et al. 2017a).

One-Dimensional (1-D) Nanomaterials

Nanotubes, nanorods and fibres are some examples of one-dimensional nanomaterials. In these types of structures, there are two nanometric dimensions while the third one is outside the nanoscale. Thus, the electron confinement occurs in the two nanoscale dimensions, producing a restricted electrons movement (Bashir and Liu 2015).

The crystalline structure of these types of materials is more complex and interesting than that of 0-D type materials since in this case, the systems can be amorphous or crystalline, although the systems that point to crystallinity can be mono or polycrystalline, a phenomenon that it is usually not found in nanocrystals with a 0-D type structure.

This leads to the fact that along the nanomaterial and due to the presence of different exposed crystalline faces, edges, and defects, surface-dependent properties, such as catalytic activity in the case of nanocatalysts or surface reactivity, in general, could cause variations, in some cases of great magnitude.

Other examples of 1-D nanomaterials include nanowires, nanoribbons, nanobelts and nanofilaments. In these materials, the discrete energy (E_n) is given by:

$$For\ 1-D: E_n = \left[\frac{\pi^2 h^2}{2mL^2}\right](n_x^2 + n_y^2) \tag{2}$$

In this case, the incorporation of a dimension outside the nanoscale allows surface modification of these materials more easily than in the case of 0-D materials, where space is usually restricted; so that 1-D structured materials are highly versatile candidates for many applications, and, in many cases, as supports for other systems (QDs, carbon-based brushes, etc.). Some of the physicochemical properties that these types of materials acquire when gaining a dimension, unlike the 0-D nanomaterials, are related to the change in the confinement of electrons, for instance, fast electrical transport, short ion diffusion distances and variations in the volume expansion (Mai et al. 2014).

The growth in one dimension also allows other types of nanostructuring, such as the addition of structuring agents that permit the incorporation of pores in these nanostructures, increasing the area and surface energy (Taguchi and Schuth 2005, Chen and Zhang 2019). Thus, the possibility of making modifications in the structure of these nanomaterials, the incorporation of other nanostructures on their surface and inside the pores and all the new combinations of possible materials, and even the generation of tubular 1-D nanostructures generates several possibilities of using these materials for multiple applications, such as conversion and storage of energy, light adsorption, (photo) catalysis, and many other areas. Moreover, the short ion diffusion length allows these materials the possibility of using them for rapid absorption of small molecules, maximizing the contact area and reducing the time needed for applications such as water remediation, gas sensing and storage, among others (Wei et al. 2017). The tubular structures, though the carbon-based ones known as carbon nanotubes (CNT) are well known (single layer, multilayer, pristine or functionalized, etc.) can also be obtained from a wide variety of metals and mainly semiconductors, where structural stability and the relatively low restructuring rate allow their integration into various applications with very promising perspectives (Yu et al. 2019).

Two-Dimensional (2-D) Nanomaterials

Alongwith the logic described above, nanomaterials with a 2-D type structure have an electronic confinement in one dimension, while the other two dimensions are outside the nanoscale. Some of the clearer examples of these types of nanostructures are nanoplates, nanocoatings, nanosheets and nanofilms. In the same way as 1-D nanomaterials, 2-D nanomaterials can be obtained in crystalline or amorphous structures, although in this case, the extension of the dimension favours the incorporation of a greater number of inhomogeneities in the crystalline structure and in their composition, particularly in multilayer systems, where it is possible to tune the physicochemical properties of each of the ad-hoc layers for different applications.

As expected, the electrons of these materials are confined in only one dimension allowing the system to have particular properties concerning the electrical phonons and plasmon conduction, with well-defined interlayer distances, as is the case of g-C_3N_4 and other structures such as graphene, tungsten disulphide and molybdenum disulphide (Yu et al. 2019), permitting in many cases the delocalization of the electrons (Bashir and Liu 2015).

In the case of 2–D nanomaterials, the quantum confinement theory permits one to obtain the discrete energy (E_n) by the equation:

$$For\ 2-D: E_n = \left[\frac{\pi^2 h^2}{2mL^2}\right](n_x^2) \tag{3}$$

The confinement of electrons and holes in 2-D materials can lead to a wide variety of physical phenomenon in a limit of the monolayer that can be combined taking advantage of possible synergistic effects between the layers (Li et al. 2017a). The use of an atomically thin-layered 2-D structure presents a series of intrinsic structural advantages such as lightweight, flexibility, high adhesion and transparency (Li et al. 2015), added to the chemical properties related to the exposed surface area, the presence of defects, etc.

After the development of a robust method for production and characterization of graphene by Novoselov et al. in 2004, which is the quintessential 2-D material for various applications, a revival of these types of nanostructures has been seen and much of the literature has been devoted to the categorization and study of these nanomaterials, although many of them were already known before (Shih et al. 1976).

Thus, the basic and applied studies carried out in recent years in graphene and graphene-based materials and the discovery of its unique properties as such the rapid mobility of charge carriers at room temperature that reaches values of 2×10^5 cm^2 V^{-1} s^{-1}, with an exceptional conductivity (10^6 S cm^{-1}), large theoretical specific surface area (2,630 m^2 g^{-1}), and excellent optical transmittance ($\sim$ 97.7%) (Geim and Novoselov 2007), served as triggers to resume the study of an entire area of nanostructures that had not been developed until that time (Ji et al. 2016), despite the great advances that had been made in polymer and metal coatings during the 80s and 90s (Pathania et al. 2017, Beck 1988, Biallozor and Kupniewska 2005, Presuel-Moreno et al. 2008, Zheludkevich and Tedim 2012), for many applications, particularly electronics.

Hence, the development of 2-D nanomaterials, with a controlled crystalline structure and composition, and particularly its use as building blocks for multilayer conductor and semiconductor structures have enormous potential for applications in electronics, energy conversion (light-harvesting), photocatalysis, etc.

At the same time, other 2-D nanostructured materials that are presented as promising for multiple applications are boron nitride (h-BN) (Dean et al. 2010), black phosphors, and transition metal dichalcogenides (TMDC) such as MoS$_2$, WS$_2$, WSe$_2$, among others, are promising candidates for many superior quantum efficiency optoelectronic applications (Radisavljevic et al. 2011), electromagnetic shielding, energy applications, and remediation (Jacoby 2017).

In this way, a large number of materials that were developed in recent years led to a second classification for these materials:

- *TMDCs (e.g., MoS$_2$, WS$_2$)*

  Unlike graphene where all carbon atoms are in the same dimension, this type of material has a structure similar to that of multilayers with a metal atom between two chalcogenide atoms. Thus, different configurations can be found due to stacking among which we find one tetragonal (1T), two hexagonal (2H) and three rhombohedral (3R) symmetries (Shi et al. 2015) and in-between the layers, a weak Van der Waals force holds each TMDC layer mutually, although there is recent computational evidence that indicates that once the layers are separated there are repulsive forces that prevent repackaging.

- *Nitrides (e.g., VN, BN, MoN, TiN)*

  The arrangement of atoms is similar to that of graphene, with a hexagonal layered structure intercalating nitrogen atoms and a second atom, normally a non metal or a metalloid, presenting semiconductor or insulating properties.(Bhimanapati et al. 2016, Theerthagiri et al. 2018).

- *MXenes (e.g., Ti$_2$C, Ti$_3$C$_2$, Nb$_4$C$_3$ Ta$_4$C$_3$)*

  The structure of these materials consists of a few atoms of thick layers composed of carbides, nitrides or carbonitrides. To date, approximately 30 M Xenes based on Ti, Sc, Mo, Nb, Zr, V, Hf, Ta and Cr were presented in the literature (Anasori et al. 2017).

- *Xenes (e.g., borophene, gallenene, silicene, germanene)*

  Similarly, this 2-D honeycomb arrangement of atoms presents a wide variety of properties and were exhibited as promising agents for biosensors, bioimaging, therapeutic delivery, and theranostics, as well as in several other new bio-applications (Tai et al. 2015). Particularly borophene and phosphorene, with an excellent performance in nanotechnology applications, were the first materials developed in this area and showed the enormous potential of such structures.

- *Organic Materials (e.g., 2-D polymers, covalent organic frameworks)*

  These materials are a sheet-like macromolecule with laterally connected monomers stacked as molecular sheets (Cai et al. 2015, Bojdys 2016). There are numerous types of 2-D organic polymeric nanomaterials including covalently-linked polymers, coordination polymers, supramolecular polymers with many applications in surface coatings, patterning, gas separation, etc.

Three-Dimensional (3-D) Nanomaterials

3-D materials do not show confinement or the consequent electronic quantization. In these types of materials, nanostructuring generally comes from porosity and interaction between building blocks, that are put together to form hierarchical 3-D structures. Thus, the porosity in these types of materials allows obtaining a large surface area as well as the surface modification with organic brushes, anchoring of nanoparticles, generation of porous multilayers (Angelome et al. 2006, Gent et al. 2019), selective functionalization of the different surfaces present (Franceschini et al. 2016), generation of nanowires, nanotubes, etc.

In recent decades, the study of porous materials, particularly meso and microporous materials, has been one of the fundamental pillars in the development of nanomaterials for multiple applications, due to the high surface area, excellent mechanical properties, high reproducibility, excellent durability and quick mass and electron-transfer kinetics (Costa et al. 2020, Eftekhari 2017, Zhao et al. 2012). This rapid mass transfer kinetics is due to the interconnection between the pores or the use of materials with hierarchical pore structure, which allows the rational design of nanomaterials for applications in microfluidics, flow fields for fuel cells, water remediation, etc. (Bruno et al. 2012, Fuentes-Quezada et al. 2019).

Something similar happens with the 3-D porous graphene-based materials for energy applications where the synergy between the characteristics of graphene and 3-D nanomaterials are used (Cao et al. 2014, Mao et al. 2015).

Moreover, Metal-Organic Frameworks (MOFs) consisting of metallic ions or clusters interconnected having a 3-D porous structure with a long-range order, appears as promising nanomaterials for many purposes, such as molecules encapsulation, filtration or support nanoparticles (Chen et al. 2017, Wang et al. 2017b). Thus, sulphonates, phosphonates, carboxylates and heterocyclic rings work as organic linkers for inorganic building units. In this way, the combination of these two kinds of materials makes possible the tuning of the functional properties of the MOFs.

Furthermore, another type of 3-D nanomaterials called nanocrystalline materials is composed of many nanocrystals in multiple arrangements and different orientations. These materials present many interesting properties, for instance, have been reported to be considerably stronger than macro-sized crystals, although these materials were expected to have high ductility and superplasticity, they normally do not have these characteristics for different reasons (Koch 2003).

Effect of Nanostructuration on Surface Properties

The properties of nanomaterials come from both the effect of quantization and confinement because they have high surface atoms proportion per unit volume. The area/volume ratio of the material suffers substantial changes when a macroscopic object is successively divided into smaller parts. For example, for a 1 cm^3 iron cube, the surface atoms percentage would be only 10^{-5}%, while for a 1 nm^3 cube 100% of the iron atoms are on the surface. In Fig. 1.3a a scheme of the dependence of the surface atoms percentage with the nanoparticle diameter is presented. Here a quick increase

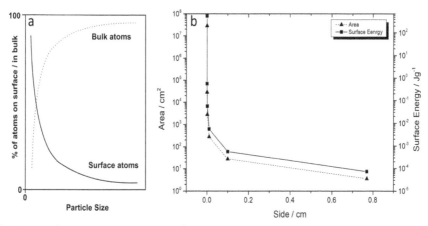

Figure 1.3: (a) Dependence of the percentage of surface atoms with cluster diameter. (b) Variation of surface energy with blocksize (Data from Adamson 1976. Adapted with the permission from John Wiley & Sons).

in the area/volume ratio occurs in particles below 10 nm, leading to huge modifications in the physicochemical properties of the nanomaterials (Nutzenadel et al. 2000). However, the effects of confinement and quantization of energy below 100 nm are also observed, particularly in the mesoscale (from 2 to 50 nm).

As expected, the total surface energy increases with the total surface area (Fig. 1.3b), which depends on the material dimension. This dependence on surface energy with the exposed area is shown in Fig. 1.3b for 1 g of sodium chloride, assuming a surface energy and edge energy 2×10^{-5} Jcm^{-2} and 3×10^{-13} Jcm^{-1}, respectively; and considering that the original cube was divided into smaller cubes (Adamson and Gast 1997).

Therefore, it is important to note that the surface area which is directly related to the total surface energy is insignificant when the particles are big but become important when the area/volume ratio is large.

High surface energy is responsible for many of the particular properties of nanomaterials; although it is also important to emphasize that this high surface energy make them thermodynamically unstable or, in the best case, metastable. Because of that, during the synthesis, it is very important to limit and prevent the growing size of the nanostructures, driven by the total surface energy minimization. Thus, having a correct understanding of the physicochemistry of solid surfaces and surface energy is basic for the stabilization of nanomaterials to reduce surface energy, avoiding agglomeration, coarsening and growth, whatever be its dimension.

Surface Energy

Superficial atoms have lower coordination numbers, and therefore have dangling bonds exposed. Due to the dangling bonds, the surface atoms are subjected to an inward force and thus, the distance between the surface and the bulk atoms, is lower than that existing between the inner atoms, causing an increase in the gravimetric density of nanoparticles compared to massive materials when particles are small. This decrease in bond length for very small particles are because the lattice parameters show a significant reduction. This extra energy located in surface atoms is called surface energy (or surface free energy) and is defined as the energy (γ) required to create a unitary surface area:

$$\gamma = \left(\frac{\partial G}{\partial A}\right) n_i, T, P \tag{4}$$

where A is the surface area. Hence, when a solid material such as that presented in Fig. 1.4a is separated into two pieces (Fig. 1.4b), the surfaces created have atoms that decrease the coordination because they are in an asymmetrical environment, decreasing their distance between first neighbours

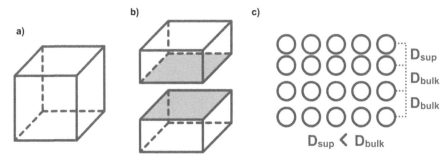

Figure 1.4: (a) and (b) Schematic representation of the change in the exposed surface area when a cube is sliced in two pieces. (c) The ideal situation of monoatomic surface contraction.

as shown in Fig. 1.4c. As expected, the surface relaxation becomes more important in crystals where the area/volume ratio is greater, and a noticeable reduction of the lattice parameter can be observed in nanoparticles (Goldstein et al. 1992). This process of surface stabilization means that extra energy is required to return the surface atoms to their original position and the required energy will be equal to the number of broken bonds, N_b, multiplied by half of the bond strength, ε.

Therefore, the surface energy is given by the equation:

$$\gamma = \frac{1}{2} N_b \varepsilon \rho_a \tag{5}$$

where ρ_a is the number of atoms per unit area on the analyzed surface. This simple model only takes into account the interactions between the first neighbours, assuming that the value of ε is the same for all atoms (whether surface or bulk), and does not consider other parameters such as entropic or volume/pressure contributions, so it is only applicable to solids with a rigid structure where superficial relaxation does not occur.

In consequence, when there is a superficial relaxation due to a rearrangement, this model tends to overestimate the value of surface energy.

Despite the simplicity of Equation (5), it serves to exemplify some general notions. Let us consider a mono elemental crystal with Face Cubic Centred structure (FCC). In this type of structure, each bulk atom has a coordination number of 12. However, each atom belonging to a crystalline surface with structure (1 0 0) has four broken chemical bonds, and the surface energy can be calculated as:

$$\gamma_{\{100\}} = \frac{1}{2} \frac{2}{a^2} 4\varepsilon = \frac{4\varepsilon}{a^2} \tag{6}$$

In the same way, each atom on (1 1 0) crystal plane presents five broken chemical bonds and each atom on a (1 1 1) has three. Thus, the surface energies for these surfaces are given by:

$$\gamma_{\{110\}} = \frac{5}{\sqrt{2}} \frac{\varepsilon}{a^2} \tag{7}$$

$$\gamma_{\{111\}} = 2\sqrt{3} \frac{\varepsilon}{a^2} \tag{8}$$

In this way, it is easy to notice that the low index planes have low surface free energy making them more thermodynamically stable than the planes with more broken bonds. Hence, these planes show a lower tendency for restructuring the reduction of surface energy.

On the one hand, amorphous solids have an isotropic microstructure and isotropic surface energy (spherical shape). On the other hand, for crystalline structures, different crystal planes can be combined in certain patterns to minimize the surface energy. Then, it is possible to estimate for a given crystal, which crystalline planes will be exposed in a thermodynamic equilibrium situation, and which planes will grow at the expense of which planes with high surface energy.

With this objective, the Wulff plot (Fig. 1.5) is commonly used to try to establish the shape (and therefore the exposed crystalline planes) of a crystal in equilibrium, that is, where the surface energy is minimal (Herring 1952, Mullins 1963).

For a crystal in the equilibrium there exists a point in the interior of the crystal such that its perpendicular distance, h_i, from the ith face is directly proportional to the surface free energy, γ_i:

$$\gamma_i = C h_i \tag{9}$$

where C is a constant and has the same value for all surfaces of a given crystal.

As a result, monocrystals with two different crystalline structures can be combined schematically as presented in Fig. 1.5 to evaluate the most stable form.

Figure 1.5 represents the lowest energy conformation for a two-dimensional hypothetical crystal using the Wulff construction (Adamsom and Gast 1997). It is important to note that this represents an ideal situation and does not consider kinetic factors, that are dependent on crystal growth conditions, which explains various anomalies found in nanoparticles, such as the formation of core-shell structures, which are not thermodynamically favoured (Matijevi 1985, Spitale et al. 2015).

The mechanisms for reducing surface energy can be grouped depending on whether they affect individual atoms or the entire structure.

The mechanisms that affect superficial atoms are varied. Some of these processes include: (i) Surface relaxation while maintaining the crystalline plane (inward shift, Fig. 1.4c) (Van Hove et al. 1986, Finnis and Heine 1974, Landman et al. 1980, Adams et al. 1982), (ii) Surface restructuring, increasing the coordination of surface atoms, which usually generates surface tensions (Chan et al. 1980, Van Hove et al. 1981, Robinson et al. 1984, Tromp et al. 1986), (iii) chemi or physisorption of chemical species (Davisson and Germer 1927, Christmann et al. 1979, Shih et al. 1976, MacLaren et al. 1987), and (iv) Segregation of the surface composition by solid-state diffusion.

In bulk solids, segregation of the composition is not significant, however, in nanomaterials, phase segregation can play an important role in reducing surface energy, considering the large area/volume ratio and the short distance of diffusion. Despite this, the mechanism is far from being the most important for the reduction of surface energy.

On the other hand, mechanisms for the reduction of surface energy at the overall system level include (i) Combining individual nanostructures to form large structures to reduce the area of the general surface by forming a new structure with a nanostructure different from that of the original particles, and (ii) Agglomeration of individual nanostructures without altering the individual nanostructures.

The mechanisms for these processes to occur include (i) Sintering, in which the individual structures fuse and (ii) Maturation of Ostwald, in which structures grow at the expense of smaller particles.

Figure 1.6 represents the two processes, sintering and Ostwald ripening. Although both processes have as a driving force the reduction of surface free energy, the resulting materials are very different (Cao 2004).

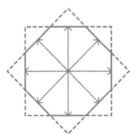

Figure 1.5: Shape given by the Wulff construction conformed for a hypothetical two-dimensional crystal; (10) plane and (11) plane.

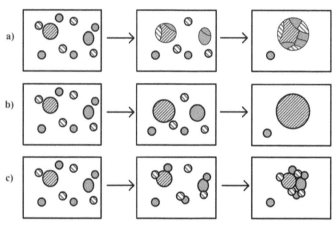

Figure 1.6: Schematic representation of (a) Sintering with the combining of individual particles to bulk with solid interphases to connect each other, (b) Ostwald ripening by merging small particles into a large particle and (c) formation of agglomerates.

In general, at room temperature (and lower) sintering is insignificant, and usually occurs when materials are heated to temperatures close to the melting point (it is usually necessary to reach at least 70% of the melting point).

Thus, taking into account the high surface energy of nanomaterials, sintering can become a problem when they are brought to even moderate temperatures, either during the manufacture, processing and use of nanomaterials, for example in PEM fuel cells, where reaching temperatures of 343–353 K allows the cells to considerably increase its efficiency (Franceschini et al. 2012, Wong et al. 2019).

This clear dependence on the sintering process with temperature is because it implies a series of processes with high activation energy such as; solid-state diffusion or dissolution-precipitation (Kingery et al. 1976, Reed 1988, DeGarmo et al. 1988), and the final product of sintering is normally a polycrystalline material.

As a rule, sintering is favoured when the distance between nanostructures, that is the diffusion distance, is short, so it is important to have a good and homogeneous distribution of the nanoparticles to avoid agglomerations (Ferreira et al. 2005).

On the other hand, Ostwald maturation is a radically different process, in which two individual nanostructures become one. First, Ostwald maturation can occur even at low temperatures as long as the nanostructures have appreciable solubility in the solvent in which they are found.

Second, a nanostructure grows at the expense of the smaller ones resulting in a single uniform structure.

A third mechanism for the reduction of the surface energy of nanostructures, in addition to sintering and Ostwald maturation, is agglomeration. In agglomerates, nanostructures are associated with each other through chemical bonds and/or forces of physical attraction. One of the main problems during the synthesis of nanostructures is to avoid agglomeration. Once formed, the agglomerates are very difficult to re-disperse and normally as in the case of graphene, the separation methods also attack the nanostructure causing damage and defects. The greater the area/volume ratio of the individual nanostructures, the stronger they associate with each other and are more difficult to separate.

Effect of Surface Curvature on the Chemical Potential

Different types of nanostructures have different surface curvatures, which give them different properties. Particularly, concave surfaces have much lower surface energy than convex surfaces generating changes in the equilibrium vapour pressure solubility and stability, between others.

To illustrate the relationship between chemical potential and surface curvature, a simple example could be considered, the transfer of material from an infinite flat ideal surface to a spherical particle.

Ideally, when transferring mass from the surface to a particle of radius R, the change in volume of the particle (dV) is equal to the atomic volume of the transferred atoms Ω multiplied by the number of particles transferred (dn), as:

$$dV = 4\pi R^2 dR = \Omega \, dn \qquad (10)$$

Then the change in chemical potential $(\Delta\mu)$ is given by the Young-Laplace equation:

$$\Delta\mu = \mu_c - \mu_\infty = \gamma \frac{dA}{dn} = \gamma 8\pi R \, dR \frac{\Omega}{dV} \qquad (11)$$

$$\Delta\mu = 2\gamma \frac{\Omega}{R} \qquad (12)$$

where μ_c is the chemical potential on the particle surface and μ_∞ is the chemical potential on the flat surface. This equation can be generalized for any type of curved surfaces considering two radii of curvature, R_1 and R_2, so one gets (Adamson 1976):

$$\Delta\mu = \gamma\Omega\left(\frac{1}{R_1} + \frac{1}{R_2}\right) \qquad (13)$$

The case of a concave surface could also be reviewed. In this system, the curvature is 'negative' and the chemical potential of an atom on that surface is lower than that of a flat surface. Under these conditions, the transfer of mass from the flat surface to a concave surface results in a decrease in the chemical potential of the surface. The opposite occurs on convex surfaces where when transferring mass from a flat surface, the chemical potential increases.

These differences in the chemical potentials of curved surfaces explain the differences in vapour pressure and the solubility of these structures. Assuming that the vapour of the solid phase complies with the law of ideal gases, for the flat surface it can be obtained with the equation:

$$\mu_v - \mu_\infty = -kT \ln P_\infty \qquad (14)$$

where μ_v is the chemical potential of a vapour atom, k the Boltzmann constant, P_∞ the equilibrium vapour pressure of an infinite flat solid surface, and T the temperature.

The same equation can be adapted for use on curved surfaces:

$$\mu_v - \mu_\infty = -kT \ln P_c \qquad (15)$$

where P_c is the equilibrium vapour pressure of the curved surface.

Thus, by combining Equations 14, 15 with the Young-Laplace equation one can obtain the Kelvin equation for a spherical particle (Fisher and Israelachvili 1981, Melrose 1989):

$$\ln\left(\frac{P_c}{P_\infty}\right) = \frac{2\gamma\Omega}{kRT} \qquad (16)$$

It is important to note that this equation can also be used to obtain the Gibbs-Thompson relation that relates the solubility with the curvature of a spherical particle such as (Vook 1982):

$$\ln\left(\frac{S_c}{S_\infty}\right) = \gamma\Omega \frac{R_1^{-1} + R_2^{-1}}{kT} \qquad (17)$$

where S_c is the solubility of a curved solid surface, S_∞ is the solubility of a flat surface.

According to Equation (17), when two particles with radii R_1 and R_2 (with $R_1 \gg R_2$) are placed in a solvent, a solubility equilibrium will be established. The solubility of the smallest particle will be greater than that of the largest particle. For this reason, a net diffusion of atoms is established from the particle with a smaller radius to the large particle, which has lower solubility. To maintain the solubility balance, the solute will be deposited in the large particle, increasing its radius and decreasing its solubility even more. Thus, the smaller radius particle, which decreases its size and increases its solubility, must continue to dissolve to compensate the amount of solute diffused.

Figure 1.7 illustrates the dependence of silica solubility as a function of surface curvature (Iler 1979), whether it is concave or convex curvature. It can be seen how the solubility increases considerably as the diameter of the convex curvature decreases (Sambles et al. 1971, Lisgarten et al. 1971, La Mer and Gruen 1952, Piuz and Borel 1972). The same behaviour is found for vapour pressure (La Mer and Gruen 1952).

Therefore, as can be seen in Fig. 1.7, Ostwald maturation can have a positive or negative influence on nanomaterials. This effect on the stability of nanostructures is one of the main reasons why the development of mesoporous materials for various applications has attracted the interest of researchers in recent decades since these materials show exceptional chemical stability under various conditions.

The case of nanostructures 0-D and 1-D is very different from that of 3-D. In the case of 0-D and 1-D, the Ostwald ripening can expand or reduce the size distribution, a process that can be controlled with a careful study of conditions. However, in most cases, the Ostwald ripening is undesirable, since it reduces the exposed area of the nanostructures.

Another interesting case is that of the synthesis of polycrystalline materials. In this case, the Ostwald ripening produces an abnormal growth of the grain, which leads to an inhomogeneous microstructure and inferior mechanical properties. Usually, a few grains grow at the expense of several small grains, which results in a non-homogeneous microstructure often stressed.

As one can see, during the manufacture and processing of nanostructures, the main barrier to overcome is the enormous total free energy of the surface to create the desired nanostructures, avoiding unwanted agglomerations and restructuring. As the dimension of the nanostructured material is reduced, the Van der Waals force between nanomaterials becomes increasingly important.

The reduction of total surface free energy is the driving force for the restructuring of nanostructures. If the stabilization mechanisms are ignored during the synthesis and processing of nanostructures, the materials are prone to form agglomerates. The main mechanisms for stabilization are electrostatic stabilization and steric stabilization. A system that uses electrostatic stabilization is kinetically stable, while steric stabilization makes the system thermodynamically stable.

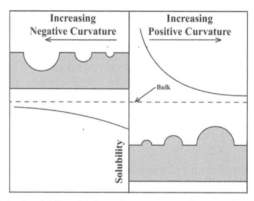

Figure 1.7: Scheme of the variation in solubility with the surface curvature radius. Negative curvature is shown as depressions or holes in the surface, and the crevices between particles. Positive curvature is shown in particles and projections from a planar surface (Adapted from Iler 1979. Adapted with the permission from John Wiley & Sons).

Effect of Porosity and Confinement

Very often for applications in catalysis, ion exchange and sensors, one tries to maximize the exposed area using porous materials with pore diameters as small as possible. However, there are other parameters to take into account that modify the diffusion of species within pores, which affects the response of these materials, particularly the electrochemical response (Calvo et al. 2009). The most significant are the tortuosity, the confinement effect, the capillary pressure inside the pores, among many other parameters.

Velasco et al. (2017) analyzed the structure of water within porous TiO_2 particles with a pore diameter of approximately 3.5 nm using Nuclear Magnetic Resonance (NMR) and Molecular Dynamics (MD) employing different water loadings within the pores. They found by ^{1}H static spectroscopy, Solid-State NMR Magic Angle Spinning (SSNMR MAS), and Double-Quantum (DQ-NMR) experiments the existence of three different water layers. The first one, a strongly adsorbed layer of water molecules which appears independently of the filling degree due to the high hydrophilicity of the TiO_2 surface.

The second, a more mobile water layer (characterized through MAS spectra), while the third inner liquid layer was detected by static spectra. As a consequence, the accessible diameter of the TiO_2 nanopores is around 1 nm smaller than the physical diameter due to water layers with different degrees of stiffness. This means that for very small pores the effective diameter of the pores decreases and the diffusion of species within the pores becomes difficult.

Takahashi et al. (2002) analyzed the diffusion coefficient of nickel nitrate in porous silica discs with different pore diameters. They found a clear dependence between the pore diameter and the diffusion coefficient of the species within those mesopores, and particularly the diffusion coefficient falls fast for pores of 2 nm diameter where the mobile water layer must be very small and the effect of the walls very important. Figure 1.8 shows the data obtained that clearly show the effect of porosity on the diffusion coefficient of a nickel nitrate solution as an example of the effect on the physicochemical properties of the nanomaterials.

Another important effect within nanopores is related to the enormous capillary pressure they present, which affects both the filling of these pores with water, as well as the formation of bubbles and the diffusion of gases in them. From the Young-Laplace equation, it is possible to calculate the pressure difference between the bulk and the pressure inside the pore. Franceschini et al. (2013) presented the particular case of a nanoporous catalyst for the oxidation of methanol in an acidic medium where the possibility of formation of CO_2 bubbles could occlude the pores preventing the arrival of methanol.

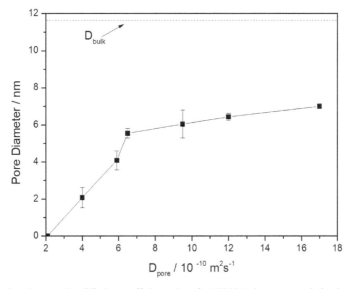

Figure 1.8: Comparison between the diffusion coefficient values for $Ni(NO_3)_2$ in aqueous solution in different conditions. (Constructed using data from Takahashi et al. 2002 with permission from Royal Society of Chemistry).

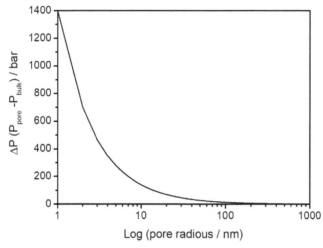

Figure 1.9: Dependence of the pressure difference (ΔP) with the pore radius according to the Young-Laplace equation.

A calculation of the CO_2 supersaturation in pores of 10 nm indicates that ΔP could be as high as 123 bar, considering a Liquid-Vapour interfacial tension of 70 mN/m. That is, CO_2 would not form bubbles inside the pores, since to form a bubble the gas pressure must be greater than the liquid pressure in the channel. In fact, in Fig. 1.9 it is evident that for pores with diameters greater than 1,000 nm the internal pressure begins to resemble the pressure of the bulk.

The Rise of Hybrid Multifunctional Materials

The combination of different materials to form complex architectures is commonly found in nature to produce amazing multifunctional structures. A clear example is living organisms, where inorganic phases often appear in conjunction with organic phases comprising of tissues with excellent mechanical resistance, such as bones and exoskeletons. These combined, organic/inorganic systems are called 'hybrids'. Usually, in these types of complex materials, the inorganic phases add mechanical strength and toughness, while the organic phases are excellent binders. However, organic phases may have many other interesting properties, such as biological functions, electrical superconductivity, corrosion resistance, thermal resistance, etc. Thus, a rational choice of the components that will make up the hybrid material is fundamental to obtain an outstanding variety of multifunctional materials or develop them ad-hoc for specific applications (Yang et al. 2017, Pandey and Mishra 2011, Tatemichi et al. 2007).

These systems with varied combinations between the organic/inorganic components are usually used in sensors, optics, photo and electrocatalysis, electronics, biomaterials, in energy conversion and storage devices, optoelectronics, photonics, etc. (Pardo et al. 2011, Heveline et al. 2017, Li et al. 2017b, McAlpine et al. 2008, Gómez et al. 2019, Franceschini and Lacconi 2018).

The absence of simple materials capable of remaining stable under critical conditions or that can present multifunctionality has been a motivation for the growing demand of hybrid materials. These interesting materials show some of the characteristics of both organic (such as their elasticity, low weight, low reactivity and resistance to corrosion, flexibility, among others) and inorganic phases (such as chemical, thermal and mechanical resistance, magnetic properties, optical properties, etc.) taking advantage of possible synergistic effects between the different materials used.

Additionally, in hybrid materials, the spatial distribution of the phases is a fundamental factor for a set of observed properties. Achieving synergistic effect, and even more, controlling it depends on the physicochemical structure of the phases and the size, morphology and distribution of the domains. As the domains of the phases become smaller, the interface influence becomes important and the synergy between the materials becomes dependent on the interactions between the phases.

It is at this point that the use of nanomaterials with different dimensions appears as an alternative of enormous technological interest for the use of nanomaterials in various applications, since it allows their stabilization, and maximizing the synergistic effect obtained in hybrid materials.

As stated above, the properties observed in hybrid materials largely depend on the interaction observed between interfaces. Hence, the nature of these interactions is normally used to classify hybrid materials into two classes (José and Prado 2005, Tomer et al. 2016, Mammeri et al. 2005, Unterlass et al. 2016).

Class I hybrid materials are materials where the interaction between the phases that compose it are mainly physical interactions, such as H-bonds, weak electrostatic forces and Van der Waals forces. Class II materials are hybrid materials where the phases interact mainly through chemical bonds. Usually, Class II materials have greater synergistic effects due to the stronger nature of the interactions than those observed in Class I materials. The interaction in Class II hybrid materials often generates a hybridization process in the reactive surface groups in the dispersed phases, which react with the matrix (continuous phase) creating covalent bonds.

Thus, the correct control of the synthesis method allows the generation of hybrids with different properties, starting from the combination of the same phases. Therefore, control of the synthesis process is crucial in the rational design of a hybrid nanomaterial. This has led to an enormous effort being made in the analysis of the integration between phases and the effect on their physicochemical properties.

As a result, the macroscopic segregation of phases that usually occurs when synthetic conditions are not controlled correctly is one of the main problems encountered in hybrid materials development. Thus, thermodynamic incompatibility between the different component phases is a difficult problem to solve in many cases. This thermodynamic incompatibility controls the segregation process and makes it difficult to control the shape and size of the domains dispersed within the continuous phase.

A strategy that can be used to improve the integration between phases is the superficial modification with organic functional groups that direct the integration of phases increasing their compatibility and reducing the effect of phase segregation (Mammeri et al. 2005).

The need for biocompatible materials is one of the fundamental goals in the development of hybrid nanomaterials. In this area, hybrids should be able to interact in the desired way with natural structures of enormous complexity. One example of these materials is those used for regeneration of bone tissue. These systems should have various properties, such as flexibility, mechanical strength and high biocompatibility. These properties depend on the phases that constitute it, but also on its spatial distribution. Huang et al. (2015) presented a hybrid material designed ad-hoc for bone regeneration with controlled phase distribution, composed of polydimethylsiloxane (PDMS), Bioactive Glass (BG) and poly (caprolactone) (PCL) (Huang et al. 2015). A correct method of synthesis, combined with a very careful choice of the constituent phases makes this monolithic hybrid material flexible and mechanically resistant, while the spatial arrangement of the phases provides the appropriate framework for the regeneration of bone tissue (Fragal et al. 2016, Tomaz et al. 2016).

Although the aforementioned is a concrete example, the development of hybrid nanomaterials (and nanocomposites) limits the development of new materials where it is not only necessary to know and control the synthesis parameters and the properties of the phases separately, but also the interaction and distribution between the constituent phases, allowing to obtain materials with an enormous capacity towards multifunctionality (Rieger et al. 2013, Ninan et al. 2015, Lee et al. 2017, Triantafillidis et al. 2013, Sun et al. 2016, Werber et al. 2016, Shao et al. 2016, Gawande et al. 2016, Seh et al. 2017, Follmann et al. 2017).

Classification and Nomenclature

The current boom related to the development of multifunctional nanomaterials, and the increasing number of materials, synthetic routes and applications need to be systematically point out differences between the various systems analyzed. Thus, it is seen in literature that an increasing number of classifications and terms that have different meanings and often are used indistinctly.

A clear example is the well-known term 'intelligent material'. This classification is quite old and great although its meaning varies significantly from one author to another.

As the number of multifunctional materials is expanded and the synthetic methods become more robust, it will be necessary to develop a defined nomenclature. To do this, in various articles some classifications have been presented that can be very useful in deepening the analysis of these complex materials.

In a recent work, Duarte Ferreira et al. (2016) proposed to name the Multifunctional Material Systems (MFMS) category as Multifunctional composites (MFC), Multifunctional materials (MFM) and another less used as multifunctional structures (MFS). Thus, the definition of Material Systems (MS) would include the common use of the terms composites, structures and materials. However, not all materials that have these structures will be MFMS, for this, they must have more than one defined functionality. Just having the structure is not enough.

In this way, it is also important to define what functionality is, and to define multifunctionality. Multifunctional materials are those that can fulfil more than one function, and that function may be simultaneous or not. The conflict appears when one considers the structural properties. These structural (or mechanical) properties may or may not be a functionality, but they are always necessary to enable the application of the nanomaterial. To do this ons needs to understand the difference between functionality and property. There are many structural properties, such as toughness, mechanical resistance, damping, stiffness, strength, ductility and fatigue resistance (Gibson 2010, Pinto 2013).

A clear example of that is flexibility. Flexibility is an intrinsic property of many materials, for example, graphene, but it can also be a specific function of material just as in the case of graphene are its electrical conductivity and optical transmittance. When this structural property is responsible for the material being able to be used for applications that would be impossible without it, one can conclude that this property allows to increase the functionality of the material and is, therefore, the MS is a multifunctional material, in the same way, that one would do it with many other properties of the materials, such as the biological, magnetic and/or electrical properties (Gupta and Srivastava 2010).

On the other hand, the nomenclature related to nanomaterials is already well established, although many authors misuse the use of the word 'nano'. There is a marked consensus that the nanoscale corresponds to the interval between 1–100 nm, while some authors call 'high nanoscale' to material systems between 100–1,000 nm, and consequently, 'low nanoscale' to the scale between 1–100 nm (Devasahayam 2019).

Then, in multifunctional materials such as composites or hybrids where there are different materials with different scales (the matrix and the dispersed phase) it is normally considered a nanostructured material (or nanocomposite) because one of its constituent parts is in the nanoscale (usually the dispersed phase), and nanostructuring affects the properties. Thus, as it was mentioned earlier, a dispersed phase with a high area/volume ratio (or to put it another way, in the nanoscale) allows greater interaction with the matrix and maximizes the synergistic effect between the constitutive phases (Kotomin 2011, Zakaria et al. 2019, Mikhalchan and Vilatela 2019, Radhamania et al. 2018).

Another important term in these kinds of materials is that of material systems with hierarchical structures. In these types of systems, the structuring changes systematically and each level fulfils a defined function. A well known example is that of systems with hierarchical pore structure, where one find pores of diameters smaller than 2 nm with low accessibility, pores with diameters in the order of 2–100 nm with good accessibility and macropores macroscopically connecting the smaller pores and serving as a feeder (Soten and Ozin 1999, Bruno et al. 2010, Fuentes-Quezada et al. 2019, Bruno et al. 2012).

Other relevant classification, at least for composite materials, is the integration of materials, that is, how intimately the materials are integrated.

As it was shown earlier, in the case of composite materials, the interaction between the dispersed phase and the matrix greatly affects the properties of the composite. There are two methods to obtain multifunctionality, integrating several functions in multi-material systems or integrating individual materials. The first method is based mainly on the addition of nanoscale or microscale fillers, as in (nano) composites or multilayer constructions. The second method allows obtaining multifunctionality through integration at the molecular level, as is the case of functionalized polymers.

These two integration methods give rise to three levels of integration that apply to most MFMS. In Type I MFMS, the materials can be segregated or not integrated; in those of Type II the materials are integrated but different phases can be distinguished (composites and hybrids) and finally in Type III materials the integration is carried out at the molecular level (Matic 2003, Asp and Greenhalgh 2012).

A group of materials of great interest are those that are mainly developed to be integrated into MFMS are Stimulus-Responsive Materials (SRM). As the name implies, these MFMS respond to a stimulus by modifying some of their physical and/or chemical properties and, in the case of being integrated into MFMS, they usually modify the properties of the whole set, so they open a huge range towards the rational development of materials with different levels of autonomy.

Some of the properties of SRMs include thermo-responsive materials (thermochromic, thermoelectric, memory shape materials), pressure/stress-responsive (piezoelectric), electro-responsive (electrochromic, electro-rheological fluids, electrostrictive), magneto-responsive (magnet or heological fluids, magnetocaloric), light-responsive (photochromic, photomechanical), pH-responsive (molecular switches).

Effect of the Application of MFMS on Production

The use of MFMS diminishes the number of pieces, reducing the necessary synthetic steps and, consequently the cost of the industrial application. An efficient integration could eliminate electrical connectors, boards and electronic circuits reducing the weight and volume of the system. On the other hand, the high adaptability of the MFMS allows that with small variations during the synthesis the properties of these materials change drastically adapting them for different applications.

Duarte Ferreira et al. (2016) exemplify the advantages of using MFMS in Unmanned Aerial Vehicles powered by electricity (UAV).

In the particular case of UAV, the factors that affect flight efficiency are; the weight stored battery energy and the battery efficiency factor. Equation (18) (Thomas et al. 2002) shows how these parameters are related. It can be seen that the decrease in the aircraft weight increases flight time by 1.5 times, while the increase in battery capacity increases the autonomy by a factor of one So, for the development of this type of aircraft, weight reduction of both, the aircraft and the battery, is a priority. However, there is an alternative that implies a greater degree of integration which could be much more efficient to increase the flight time: Combining the battery with the structural parts using MFMS.

$$\frac{\Delta t_E}{t_E} = 1\frac{\Delta(E_B \eta_B)}{E_B \eta_B} - 1.5\frac{(\Delta W_S + \Delta W_B)}{W_S + W_B} \tag{18}$$

where t_E is the flight endurance time, E_B is the nominal stored energy in the battery, η_B represents the battery efficiency factor which explains the influence of the current extraction rate, temperature, etc., on the amount of energy extracted, W_S is the aircraft weight and W_B is the battery weight. Although there are numerous examples in the literature on the integration of batteries and capacitors with structural functions (Chan et al. 2018, Shi et al. 2016), one of great interest is that presented by Moyer et al. (2020). They design ad-hoc a structural battery for a 1U CubeSat reducing considerably the weight of the system, due to the refunctionalization of part of the lithium-ion battery packs which occupy a significant volume in these systems. Specifically, in the case of four panels of structural batteries assembled in the 1U CubeSat, each with an energy density of 35 Wh/kg, it stores total energy of ~ 10 Wh, which decreases the total required mass of external batteries in ~ 30% in this configuration and creates free volume in the CubeSat chassis, approaching the operational requirements of NASA. This system was designed and tested with favourable results.

Finally, it is important to consider that, as can be seen in Fig. 1.10, currently the cost and scalability in the manufacture of MFMS is one of the biggest challenges to overcome. Many of the methods used for the synthesis are expensive and cannot be transferred to the industry because they are not scalable. It is expected that with time and the development of new methods of continuous production these problems will be overcome. Moreover, with the maturity of 3D printing technology, some of these difficulties are being surpassed rapidly. This type of technology takes full advantage of the main property of MFMS, which is that of assembly. The possibility of a layer by layer assembly can significantly reduce the number of monofunctional materials used for a device, substantially reducing the cost, weight, volume and steps required for the construction of a device.

Some of the basic concepts necessary to understand the advantages and limitations of nanostructured multifunctional materials, as well as the different possible functions and structures have been presented in this chapter. In the following chapters, the tools and definitions represented here will be used to apply them to different cases of MFMS. Methods of synthesis, characterization, applications and state of the art will be studied in materials with different dimensions of nanostructure where the nanometric dimension plays a fundamental role in functionality.

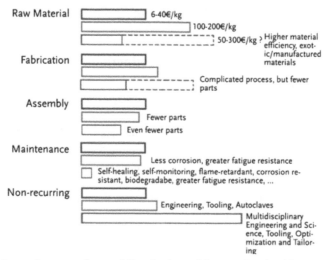

Figure 1.10: Cost study: metal vs composite vs multifunctional material system (Reprinted from Duarte Ferreira et al. 2016 with permission from Elsevier).

Acknowledgements

The authors thank financial support from Agencia Nacional de Promoción Científica y Tecnológica (PICT 2017-0250), SECyT-UNC and CONICET (PUE2017). EAF is permanent research fellows of CONICET. The author thanks to Dr. Ana Spitale for her assistance in translation.

References

Adams, D. L., H. B. Nielsen, J. N. Andersen, I. Stengsgaard, R. Friedenhans'l and J. E. Sorensen. 1982. Oscillatory relaxation of the Cu(110) surface. Phys. Rev. Lett. 49: 669.
Adamson, A. W. 1976. Physical Chemistry of Surfaces. Wiley. New York.

Adamson, A. W. and A. P. Gast. 1997. Physical Chemistry of Surfaces. 6th edition. John Wiley & Sons. New York.

Anasori, B., M. R. Lukatskaya and Y. Gogotsi. 2017. 2D metal carbides and nitrides (MXenes) for energy storage. Nat. Rev. Mater. 2: 16098.

Angelomé, P. C., M. C. Fuertes and G. J. Soler-Illia. 2006. Multifunctional, multilayer, multiscale: Integrative synthesis of complex macro and mesoporous thin films with spatial separation of porosity and function. Adv. Mater. 18(18): 2397–2402.

Asp, L. E. and E. S. Greenhalgh. 2012. Multifunctional composite materials for energy storage in structural load paths. [Slide Presentation].

Bashir, S. and J. Liu. 2015. Overviews of Synthesis of Nanomaterials, Advanced Nanomaterials and their Applications in Renewable Energy. pp. 51–115. Elsevier, Amsterdam.

Beck, F. 1988. Electrodeposition of polymer coatings. Electrochim. Acta 33: 839–850.

Bhimanapati, G. R., N. R. Glavin and J. A. Robinson. 2016. Chapter three – 2D boron nitride: synthesis and applications. Semiconduct. Semimet. 95: 101–147.

Biallozor, S. and A. Kupniewska. 2005. Conducting polymers electrodeposited on active metals. Synth. Met. 155: 443–449.

Bojdys, M. J. 2016. 2D or not 2D-layered functional (C, N) materials "beyond silicon and graphene". Macromol. Chem. Phys. 217: 232–241.

Brichkin, S. B. and V. F. Razumov. 2016. Colloidal quantum dots: synthesis, properties, and applications. Russ. Chem. Rev. 85(12): 1297–1312.

Bruno, M. M., E. A. Franceschini, G. A. Planes and H. R. Corti. 2010. Electrodeposited platinum catalysts over hierarchical carbon monolithic support. J. Appl. Electrochem. 40: 257–263.

Bruno, M. M., E. A. Franceschini, F. A. Viva, Y. R. J. Thomas and H. R. Corti. 2012. Electrodeposited mesoporous platinum catalysts over hierarchical carbon monolithic support as anode in small PEM fuel cells. Int. J. Hydrog. Energy. 37: 14911–14919.

Cai, S. L., W. G. Zhang, R. N. Zuckermann, Z. T. Li, X. Zhao and Y. Liu. 2015. The organic flatland-recent advances in synthetic 2D organic layers. Adv. Mater. 27: 5762–5770.

Calvo, A., B. Yameen, F. J. Williams, G. J. A. A. Soler-Illia and O. Azzaroni. 2009. Mesoporous films and polymer brushes helping each other to modulate ionic transport in nanoconfined environments. An interesting example of synergism in functional hybrid assemblies. J. Am. Chem. Soc. 131: 10866–10868.

Cao, G. 2004. Nanostructures and Nanomaterials, Synthesis, Properties, and Applications. Imperial College Press. London.

Cao, X., Z. Yin and H. Zhang. 2014. Three-dimensional graphene materials: preparation, structures, and application in supercapacitors. Energy Environ. Sci. 7(6): 1850–1865.

Chan, C. M., M. A. Van Hove, W. H. Weinberg and E. D. Williams. 1980. An R-factor analysis of several models of the reconstructed Ir(110)-(1 × 2) surface. Surf Sci. 91: 440–448.

Chan, K. Y., B. Jia, H. Lin, B. Zhu and K. T. Lau. 2018. Design of a structural power composite using graphene oxide as a dielectric material layer. Mater. Lett. 216: 162–165.

Chen, L., R. Luque and Y. Li. 2017. The controllable design of tunable nanostructures inside metal-organic frameworks. Chem. Soc. Rev. 46: 4614–4630.

Chen, X. and Q. Zhang. 2019. Recent advances in mesoporous metal-organic frameworks. Particuology 45: 20–34.

Christmann, K., R. J. Behm, G. Ertl, M. A. Van Hove and W. H. Weinberg. 1979. Chemisorption geometry of hydrogen on Ni(111): Order and disorder. J. Chem. Phys. 70: 4168.

Costa, J. A. S., R. A. de Jesus, D. O. Santos, J. F. Mano, L. P. C. Romão and C. M. Paranhos. 2020. Recent progresses in the adsorption of organic, inorganic, and gas compounds by MCM-41-based mesoporous materials. Micropor. Mesopor. Mat. 291: 109698.

Davisson, C. J. and L. H. Germer. 1927. Diffraction of electrons by a crystal of nickel. Phys. Rev. 29: 908.

Dean, C. R., A. F. Young, I. Meric, C. Lee, L. Wang, S. Sorgenfrei et al. 2010. Boron nitride substrates for high-quality graphene electronics. Nat. Nanotechnol. 5(10): 722–726.

DeGarmo, E. P., J. T. Black and R. A. Kohner. 1988. Materials and processes in manufacturing. Mac Millan. New York.

Devasahayam, S. 2019. Characterization and biology of nanomaterials for drug delivery. pp. 477–522. *In*: S. S. Mohapatra, S. Ranjan, N. Dasgupta, R. K. Mishra and S. Thomas [eds.]. Characterization and Biology of Nanomaterials for Drug Delivery, Nanoscience and Nanotechnology in Drug Delivery. Elsevier.

Duarte Ferreira, A. B. L., P. R. O. Nóvoa and A. T. Marques. 2016. Multifunctional material systems: A state-of-the-art review. Compos. Struct. 151: 3–35.

Eftekhari, A. 2017. Ordered mesoporous materials for lithium-ion batteries. Micropor. Mesopor. Mat. 243: 355–369.

Ferreira, P. J., G. J. la O', Y. Shao-Horn, D. Morgan, R. Makharia, S. Kocha et al. 2005. Instability of Pt/C electrocatalysts in proton exchange membrane fuel cells. J. Electrochem. Soc. 152(11): A2256–A2271.

Finnis, M. W. and V. Heine. 1974. Theory of lattice contraction at aluminium surfaces. J. Phys. F4: L37–L41.

Fisher, L. R. and J. N. Israelachvili. 1981. Experimental studies on the applicability of the Kelvin equation to highly curved concave menisci. J. Colloid Interface Sci. 80: 528–541.

Follmann, H. D. M., A. F. Naves, R. A. Araujo, V. Dubovoy, X. X. Huang, T. Asefa et al. 2017. Hybrid materials and nanocomposites as multifunctional biomaterials. Curr. Pharm. Design. 23(26): 3794–3813.

Fragal, V. H., T. S. P. Cellet, E. H. Fragal, G. M. Pereira, F. P. Garcia, C. V. Nakamura et al. 2016. Controlling cell growth with tailorable 2D nanoholes arrays. J. Colloid Interface Sci. 466: 150–161.

Franceschini, E. A., M. M. Bruno, F. A. Viva, F. J. Williams, M. Jobbágy and H. R. Corti. 2012. Mesoporous Pt electrocatalyst for methanol tolerant cathodes of DMFC. Electrochim. Acta 71: 173–180.

Franceschini, E. A., M. M. Bruno, F. J. Williams, F. A. Viva and H. R. Corti. 2013. High-activity mesoporous Pt/Ru catalysts for methanol oxidation. ACS Appl. Mater. Interfaces 5: 10437–10444.

Franceschini, E. A., E. de la Llave, F. J. Williams and G. J. A. A. Soler-Illia. 2016. A simple three-step method for selective placement of organic groups in mesoporous silica thin films. Mat. Chem. Phys. 169: 82–88.

Franceschini, E. A. and G. I. Lacconi. 2018. Synthesis and performance of nickel/reduced graphene oxide hybrid for hydrogen evolution reaction. Electrocatalysis 9: 47–58.

Fuentes-Quezada, E., E. de la Llave, E. Halac, M. Jobbágy, F. A. Viva, M. M. Bruno et al. 2019. Bimodal mesoporous hard carbons from stabilized resorcinol-formaldehyde resin and silica template with enhanced adsorption capacity. Chem. Eng. J. 360: 631–644.

Gawande, M. B., A. Goswami, F. X. Felpin, T. Asefa, X. Huang, R. Silva et al. 2016. Cu and Cu-based nanoparticles: Synthesis and applications in catalysis. Chem. Rev. 116(6): 3722–3811.

Geim, A. K. and K. S. Novoselov. 2007. The rise of graphene. Nat. Mater. 6: 183–191.

Gent, E., D. H. Taffa and M. Wark. 2019. Multi-layered mesoporous TiO$_2$ thin films: Photoelectrodes with improved activity and stability. Coatings 9: 625.

Gibson, R. F. 2010. A review of recent research on mechanics of multifunctional composite materials and structures. Compos. Struct. 2(11): 2793–810.

Gleiter, H. 2000. Nanostructured materials: basic concepts and microstructure. Acta Mater. 48: 1–29.

Goldstein, A. N., C. M. Echer and A. P. Alivisatos. 1992. Melting in semiconductor nanocrystals. Science 256: 1425–1427.

Gomez, M. J., A. Loiácono, L. A. Pérez, E. A. Franceschini and G. I. Lacconi. 2019. Highly efficient hybrid Ni/ nitrogenated graphene electrocatalysts for hydrogen evolution reaction. ACS Omega 4: 2206–2216.

Gupta, P. and R. K. M. Srivastava. 2010. Overview of multifunctional materials. Intech.

Gupta, R. K., M. Malviya, C. Verma and M. A. Quraishi. 2017. Aminoazobenzene and diaminoazobenzene functionalized graphene oxides as novel class of corrosion inhibitors for mild steel: Experimental and DFT studies. Mater. Chem. Phys. 198: 360–373.

Herring, C. 1952. Structure and Properties of Solid Surfaces. University of Chicago. Chicago, IL.

Heveline, D. M. F., F. N. Alliny, A. A. Rafael, D. Viktor, H. Xiaoxi, A. Tewodros et al. 2017. Hybrid materials and nanocomposites as multifunctional biomaterials. Curr. Pharm. Des. 23(26): 3794–3813.

Hu, J., T. W. Odom and C. M. Lieber. 1999. Chemistry and physics in one dimension: synthesis and properties of nanowires and nanotubes. Acc. Chem. Res. 32(5): 435–445.

Hu, H., Z. Zhao, Y. Gogotsi and J. Qiu. 2014. Compressible carbon nanotube–graphene hybrid aerogels with superhydrophobicity and superoleophilicity for oil sorption. Environ. Sci. Technol. Lett. 1: 214–220.

Huang, X., Z. Zhao, L. Cao, Y. Chen, E. Zhu, Z. Lin et al. 2015. High-performance transition metal-doped Pt3Ni octahedra for oxygen reduction reaction. Science 348(6240): 1230–1234.

Iler, R. K. 1979. The Chemistry of Silica: Solubility, Polymerization, Colloid and Surface Properties and Biochemistry of Silica. John Wiley and Sons, Hoboken, New Jersey.

Jacoby, M. 2017. 2-D materials go beyond graphene. Chem. Eng. News 95(22): 36–40.

Ji, L., P. Meduri, V. Agubra, X. Xiao and M. Alcoutlabi. 2016. Graphene-based nanocomposites for energy storage. Adv. Energy Mater. 6: 1502159.

José, N. M. and L. A. S. d. A. Prado. 2005. Materiais híbridos orgânico-inorgânicos: preparação e algumas aplicações. Quim. Nova 28: 281–288.

Kingery, W. D., H. W. Bowen and D. R. Uhlmann. 1976. Introduction to Ceramics. Second edition. Wiley. New York.

Koch, C. 2003. Optimization of strength and ductility in nanocrystalline and ultrafine grained metals. Scr. Mater. 49: 657–662.

Kotomin, S. 2011. Polymer molecular composites—new history. J. Thermoplast. Compos. Mater. 26: 91–108.

La Mer, V. K. and R. Gruen. 1952. A direct test of Kelvin's equation connecting vapour pressure and radius of curvature. Trans. Faraday SOC. 48: 410–416.

Landman, U., R. N. Hill and M. Mosteller. 1980. Lattice relaxation at metal surfaces: An electrostatic model. Phys. Rev. B 21: 448.

Lee, E. J., B. K. Huh, S. N. Kim, J. Y. Lee, C. G. Park, A. G. Mikos et al. 2017. Application of materials as medical devices with localized drug delivery capabilities for enhanced wound repair. PProg. Mater. Sci. 89: 392–410.

Lehman, J. H., M. Terrones, E. Mansfield, K. E. Hurst and V. Meunier. 2011. Evaluating the characteristics of multiwall carbon nanotubes. Carbon 49: 2581–2602.

Li, H. N., Y. M. Shi, M. H. Chiu and L. J. Li. 2015. Emerging energy applications of two-dimensional layered transition metal dichalcogenides. Nano Energy 18: 293–305.

Li, D., T. Liu, X. Yu, D. Wu and Z. Su. 2017a. Fabrication of graphene-biomacromolecule hybrid materials for tissue engineering application. Polym. Chem. 8(30): 4309–4321.

Li, H., Y. Li, A. Aljarb, Y. Shi and L. J. Li. 2017b. Epitaxial growth of two-dimensional layered transition-metal dichalcogenides: growth mechanism, controllability, and scalability. Chem. Rev. 118(13): 6134–6150.

Lisgarten, N. D., J. R. Sambles and L. M. Skinner. 1971. Vapour pressure over curved surfaces-the Kelvin equation. Contemp. Phys. 12: 575–593.

Liu, R., J. Duay and S. B. Lee. 2011. Heterogeneous nanostructured electrode materials for electrochemical energy storage. Chem. Commun. 47: 1384–1404.

MacLaren, J. M., J. B. Pendry, P. J. Rous, D. K. Saldin, G. A. Somorjai, M. A. Van Hove et al. 1987. Surface Crystallography Information Service. Reidel Publishing, Dordrecht.

Mai, L., X. Tian, X. Xu, L. Chang and L. Xu. 2014. Nanowire electrodes for electrochemical energy storage devices. Chem. Rev. 114: 11828.

Mammeri, F., E. L. Bourhis, L. Rozes and C. Sanchez. 2005. Mechanical properties of hybrid organic inorganic materials. J. Mater. Chem. 15(35-36): 3787–3811.

Mao, S., G. Lu and J. Chen. 2015. Three-dimensional graphene-based composites for energy applications. Nanoscale 7(16): 6924–6943.

Matic, P. 2003. Overview of multifunctional materials. pp. 61–69. 5053. Bellingham, Wash: SPIE.

Matijevi, E. 1985. Production of monodispersed colloidal particles. Annu. Rev. Muter: Sci. 15: 483–516.

McAlpine, M. C., H. D. Agnew, R. D. Rohde, M. Blanco, H. Ahmad, A. D. Stuparu et al. 2008. Peptide–nanowire hybrid materials for selective sensing of small molecules. J. Am. Chem. Soc. 130(29): 9583–9589.

Melrose, J. C. 1989. Applicability of the Kelvin equation to vapor/liquid systems in porous media. Langmuir 5: 290–293.

Mikhalchan, A. and J. J. Vilatela. 2019. A perspective on high-performance CNT fibres for structural composites. Carbon 150: 191–215.

Moyer, K., C. Meng, B. Marshall, O. Assal, J. Eaves, D. Perez et al. 2020. Carbon fiber reinforced structural lithium-ion battery composite: Multifunctional power integration for CubeSats. Energy Storage Mater. 24: 676–681.

Mullins, W. W. 1963. Metal Surfaces: Structure Energetics and Kinetics. The American Society for Metals, Metals Park, OH.

Nideep, T. K., M. Ramya and M. Kailasnath. 2020. An investigation on the photovoltaic performance of quantum dot solar cells sensitized by CdTe, CdSe and CdS having comparable size. Superlattices Microstruct. 141: 106477.

Ninan, N., M. Muthiah, I. -K. Park, T. W. Wong, S. Thomas and Y. Grohens. 2015. Natural polymer/inorganic material based hybrid scaffolds for skin wound healing. Polym. Rev. 55(3): 453–490.

Novoselov, K. S., A. K. Geim, S. V. Morozov, D. Jiang, Y. Zhang, S. V. Dubonos et al. 2004. Electric field effect in atomically thin carbon films. Science 306: 666–669.

Nutzenadel, C., A. Zuttel, D. Chartouni, G. Schmid and L. Schlapbach. 2000. Critical size and surface effect of the hydrogen interaction of palladium clusters. Eur. Phys. J. D 8: 245–250.

Pandey, S. and S. B. Mishra. 2011. Organic–inorganic hybrid of chitosan/organoclay bionanocomposites for hexavalent chromium uptake. J. Colloid Interface Sci. 361(2): 509–520.

Pardo, R., M. Zayat and D. Levy. 2011. Photochromic organic-inorganic hybrid materials. Chem. Soc. Rev. 40: 672–687.

Pathania, A., R. K. Arya and S. Ahuja. 2017. Crosslinked polymeric coatings: Preparation, characterization, and diffusion studies. Prog. Org. Coat. 105: 149–162.

Pinto, F. 2013. Smart multifunctional composite materials for improvement of structural and non-structural properties U616913 PhD. University of Bath (United Kingdom), Ann. Arbor.

Piuz, F. and J. -F. Borel. 1972. Thermodynamic size effect in small particles of silver. Phys. Status Solid. Λ 14: 129 133.

Pokropivny, V. V. and V. V. Skorokhod. 2007. Classification of nanostructures by dimensionality and concept of surface forms engineering in nanomaterial science. Mater. Sci. Eng. C 27: 990–993.

Presuel-Moreno, F., M. A. Jakab, N. Tailleart, M. Goldman and J. R. Scully. 2008. Corrosion-resistant metallic coatings. Mater. Today 11: 14–23.

Radhamania, A. V., H. C. Laua and S. Ramakrishna. 2018. CNT-reinforced metal and steel nanocomposites: A comprehensive assessment of progress and future directions. Compos. Part A Appl. Sci. Manuf. 114: 170–187.

Radisavljevic, B., A. Radenovic, J. Brivio, V. Giacometti and A. Kis. 2011. Single-layer MoS$_2$ transistors. Nat. Nanotechnol. 6(3): 147–150.

Reed, J. S. 1988. Introduction to Principles of Ceramic Processing. Wiley. New York.

Rieger, K. A., N. P. Birch and J. D. Schiffman. 2013. Designing electrospun nanofiber mats to promote wound healing—a review. J. Mater. Chem. B 1(36): 4531–4541.

Robinson, I. K., Y. Kuk and L. C. Feldman. 1984. Domain structure of the clean reconstructed Au(110) surface. Phys. Rev. B29: 4762.

Sajanlal, P. R., T. S. Sreeprasad, A. K. Samal and T. Pradeep. 2011. Anisotropic nanomaterials: structure, growth, assembly, and functions. Nano Rev. 2: 5883.

Saleh, B. E. A. and M. C. Teich. 2007. Fundamentals of photonics. Wiley Series in Pure and Applied Optics. Hoboken, New Jersey.

Sambles, J. R. 1971. An electron microscope study of evaporating gold particles: the Kelvin equation for liquid gold and the lowering of the melting point of solid gold particles. Proc. R. SOC. A324: 339–351.

Sathish, M. and K. Miyazawa. 2007. Size-tunable hexagonal fullerene (C60) nanosheets at the liquid-liquid interface. J. Am. Chem. Soc. 129: 13816–13817.

Seh, Z. W., J. Kibsgaard, C. F. Dickens, I. Chorkendorff, J. K. Nørskov and T. F. Jaramillo. 2017. Combining theory and experiment in electrocatalysis: Insights into materials design. Science 355(6321).

Shao, M., Q. Chang, J. P. Dodelet and R. Chenitz. 2016. Recent advances in electrocatalysts for oxygen reduction reaction. Chem. Rev. 116(6): 3594–3657.

Shen, W., L. Zhang, S. Zheng, Y. P. Xie and X. Lu. 2017. Lu2@C82 nanorods with enhanced photoluminescence and photoelectrochemical properties. ACS Appl. Mater. Interfaces 9: 28838–28843.

Shi, Y., H. Zhang, W. H. Chang, H. S. Shin and L. J. Li. 2015. Synthesis and structure of two-dimensional transitionmetal dichalcogenides. MRS Bull. 40: 566–576.

Shi, Y., S. R. Hallett and M. Zhu. 2017. Energy harvesting behaviour for aircraft composites structures using MacroFibre. Composite: Part I – Integration and Experiment. Compos. Struct. 160: 1279–1286.

Shih, H. D., F. Jona, D. W. Jepsen and P. M. Marcus. 1976. Atomic underlayer formation during the reaction of Ti{0001} with nitrogen. Surf. Sci. 60: 445–465.

Skorokhod, V., A. Ragulya and I. Uvarova. 2001. Physico-chemical Kinetics in Nanostructured Systems. Academperiodica. p. 180.

Soten, I. and G. A. Ozin. 1999. New directions in self assembly: materials synthesis over "all" length scales. Curr. Op. Colloid Interf. Sci. 4: 325.

Spitale, A., M. A. Perez, S. Mejía-Rosales, M. J. Yacamán and M. M. Mariscal. 2015. Gold–palladium core@shell nanoalloys: experiments and simulations. Phys. Chem. Chem. Phys. 17: 28060–28067.

Sun, M. H., S. Z. Huang, L. H. Chen, Y. Li, X. Y. Yang, Z. -Y. Yuan et al. 2016. Applications of hierarchically structured porous materials from energy storage and conversion, catalysis, photocatalysis, adsorption, separation, and sensing to biomedicine. Chem. Soc. Rev. 45(12): 3479–3563.

Taguchi, A. and F. Schuth. 2005. Ordered mesoporous materials in catalysis. Micropor. Mesopor. Mat. 77: 1–45.

Tai, G., T. Hu, Y. Zhou, X. Wang, J. Kong, T. Zeng et al. 2015. Synthesis of atomically thin boron films on copper foils. Angew. Chem. Int. Ed. 54: 15473–15477.

Takahashi, R., S. Sato, T. Sodesawa and H. Nishida. 2002. Effect of pore size on the liquid-phase pore diffusion of nickel nitrate. Phys. Chem. Chem. Phys. 4: 3800–3805.

Tatemichi, M., M. A. Sakamoto, M. Mizuhata, S. Deki and T. Takeuchi. 2007. Protein-templated organic/inorganic hybrid materials prepared by liquid-phase deposition. J. Am. Chem. Soc. 129(35): 10906–10910.

Theerthagiri, J., G. Durai, K. Karuppasamy, P. Arunachalam, V. Elakkiya, P. Kuppusami et al. 2018. Recent advances in 2-D nanostructured metal nitrides, carbides, and phosphides electrodes for electrochemical supercapacitors—A brief review. J. Ind. Eng. Chem. 67: 12–27.

Thomas, J., M. Qidwai, P. Matic, R. Everett, A. Gozdz and M. Keennon. 2002. Multifunctional Approaches for Structure-Plus-Power Concepts. 43rd AIAA/ASME/ASCE/AHS/ASC Structures, Structural Dynamics, and Materials Conference, American Institute of Aeronautics and Astronautics.

Tomaz, V. A., A. F. Rubira and R. Silva. 2016. Solid-state polymerization of EDTA and ethylenediamine as one-step approach to monodisperse hyperbranched polyamides. Rsc Adv. 6(47): 40717–40723.

Tomer, V. K., S. Devi, R. Malik and S. Duhan. 2016. Mesoporous Materials and Their Nanocomposites, Nanomaterials and Nanocomposites. pp. 223–254. Wiley-VCH Verlag GmbH & Co.

Tomita, S., T. Sakurai, H. Ohta, M. Fujii and S. Hayashi. 2001. Structure and electronic properties of carbon onions. J. Chem. Phys. 114: 7477.

Tran, H. D., D. Li and R. B. Kaner. 2009. One-dimensional conducting polymer nanostructures: bulk synthesis and applications. Adv. Mater. 21(14-15): 1487–1499.

Triantafillidis, C., M. S. Elsaesser and N. Husing. 2013. Chemical phase separation strategies towards silica monoliths with hierarchical porosity. Chem. Soc. Rev. 42(9): 3833–3846.

Tromp, R. M., R. J. Hamers and J. E. Demuth. 1986. Scanning tunneling microscopy of Si(001). Phys. Rev. B34: 5343.

Unterlass, M. M. 2016. Green synthesis of inorganic–organic hybrid materials: State of the art and future perspectives. Eur. J. Inorg. Chem. (8): 1135–1156.

Van Hove, M. A., R. J. Koestner, P. C. Stair, J. P. Birberian, L. L. Kesmodell, I. Bartos et al. 1981. The surface reconstructions of the (100) crystal faces of iridium, platinum and gold: I. Experimental observations and possible structural models. Surf. Sci. 103: 189–217.

Van Hove, M. A., W. H. Weinberg and C. M. Chan. 1986. Low-Energy Electron Diffraction. Springer-Verlag, Berlin.

Velasco, M. I., M. B. Franzoni, E. A. Franceschini, E. G. Solveyra, D. Scherlis, R. H. Acosta et al. 2017. Water Confined in mesoporous TiO_2 aerosols: Insights from NMR experiments and molecular dynamics simulations. J. Phys. Chem. C 121: 7533–7541.

Vook, R. W. 1982. Structure and growth of thin films. Int. Metals Rev. 21: 209–245.

Wang, R., D. Jin, Y. Zhang, S. Wang, J. Lang, X. Yan et al. 2017a. Engineering metal-organic framework derived 3D nanostructures for high-performance hybrid supercapacitors. Mater. Chem. A 5: 292.

Wang, R., K. Q. Lu, Z. R. Tang and Y. J. Xu. 2017b. Recent progress in carbon quantum dots: synthesis, properties, and applications in photocatalysis. Mater. Chem. 5: 3717.

Wei, Q., F. Xiong, S. Tan, L. Huang, E. H. Lan, B. Dunn et al. 2017. Porous one-dimensional nanomaterials: design, fabrication and applications in electrochemical energy storage. Adv. Mater. 29: 1602300.

Werber, J. R., C. O. Osuji and M. Elimelech. 2016. Materials for next-generation desalination and water purification membranes. Nat. Rev. Mater. 1: 16018.

Wong, C. Y., W. Y. Wong, K. Ramya, M. Khalid, K. S. Loh, W. R. W. Daud et al. 2019. Additives in proton exchange membranes for low- and high-temperature fuel cell applications: A review. Int. J. Hydrog. Energy 44: 6116–6135.

Wu, Z., S. Yang and W. Wu. 2016. Shape control of inorganic nanoparticles from solution. Nanoscale 8: 1237–1259.

Xu, X., R. Ray, Y. Gu, H. J. Ploehn, L. Gearheart, K. Raker et al. 2004. Electrophoretic analysis and purification of fluorescent single-walled carbon nanotube fragments. J. Am. Chem. Soc. 126: 12736.

Yang, H. K., L. -L. Liu, X. Yuan and S. -M. Wu. 2017. Using a facile experimental manipulation to fabricate and tune a polyoxometalate-cholesterol hybrid material. J. Colloid Interface Sci. 496(Supplement C): 150–157.

Yu, P., F. Wang, T. A. Shifa, X. Zhan, X. Lou, F. Xia et al. 2019. Earth abundant materials beyond transition metal dichalcogenides: A focus on electrocatalyzing hydrogen evolution reaction. Nano Energy 58: 244–276.

Yue, W., X. Xu, J. T. S. Irvine, P. S. Attidekou, C. Liu, H. He et al. 2009. Mesoporous monocrystalline TiO_2 and its solid-state electrochemical properties. Chem. Mater. 21: 2540–2546.

Zakaria, M. R., H. M. Akil, M. H. A. Kudus, F. Ullah, F. Javed and N. Nosbi. 2019. Hybrid carbon fiber-carbon nanotubes reinforced polymer composites: A review. Compos. B Eng. 176: 107313.

Zhao, L., H. Qin, R. Wu and H. Zou. 2012. Recent advances of mesoporous materials in sample preparation. J. Chromatogr. A 1228: 193–204.

Zheludkevich, M. L., J. Tedim and M. G. S. Ferreira. 2012. "Smart" coatings for active corrosion protection based on multi-functional micro and nanocontainers. Electrochim. Acta 82: 314–323.

CHAPTER 2

Mesoporous Particles by Combination of Aerosol Route and Sol-Gel Process

María Verónica Lombardo,* Andrea Verónica Bordoni and *Alejandro Wolosiuk*

Introduction

During the late 1800s, light-scattering optical experiments introduced by Tyndall (1869) and Lord Rayleigh's studies (1871) on colloidal sulphur solutions (*hydrosols*) demonstrated the existence of highly stable phases composed of submicroscopic objects homogeneously dispersed in a continuous media. As a consequence, colloidal science experienced an important boost but mostly restricted to the study of dispersed solid particles in liquid solutions. In this context, it was not until by the end of World War I that the term *aerosol*, introduced by Irish chemist Frederick G. Donnan while studying warfare chemical smokes, described the dispersion of clouds composed of fine particles and matter in the air. This concept further evolved in 1920 when August Schmauß, a German meteorologist, pointed the resemblance between liquid colloidal solutions or hydrosols and the stability of atmospheric clouds due to electrical forces. Since then, a suspension of small solid particles scattered in a gaseous phase was called an aerosol and paved the way to the development of new branches of chemistry with a tremendous social and ecological impact in the form of atmospheric and environmental chemistries. The aerosol dispersion of finely divided matter is a characteristic of several processes that occur in nature: pollution, forest fires, acid rains and volcanic eruptions. As the particles can encompass a wide range size, from a few angstroms to several microns, it is only necessary that the suspension remains stable while various chemical reactions are taking place and overcome gravitational settling, just as it occurs during a liquid phase synthesis of colloidal solutions. Besides, as atmospheric chemistry has taught one that chemical reactions happen in the Earth's atmosphere, it is evident that one can recreate a new chemical environment where finely dispersed materials are synthesized within a gas phase just as it can be done it in a chemical beaker. In this case, the dispersed phase can be either in solid or liquid form (*droplets*) and these species will respond under new reaction conditions.

Spray drying techniques related to aerosol processes have a long history in food manufacture (e.g., powdered milk, dried juices, lipids, carotenoids) (Gharsallaoui et al. 2007) and the pharmaceutical industry (Ziaee et al. 2019). From the aforementioned examples, it is evident that the operational conditions of spray drying are benign, as they preserve the nutritive value of food or the molecular structure of drug components. In the case of high throughput particle oxide production, research can be traced back to 1930–1940 where flame hydrolysis techniques produced fumed

Gerencia Química – Centro Atómico Constituyentes, Comisión Nacional de Energía, Atómica, CONICET, Av. Gral. Paz 1499, (B1650KNA) San Martín, Buenos Aires, Argentina.
* Corresponding author: marialombardo@cnea.gov.ar

silica, also known as pyrogenic silica (e.g., Aerosil® and Cab-O-Sil®) (Bergna and Roberts 2005). Milder chemistries, such as sol-gel hydrolysis-condensation reactions, eventually managed to be a promising approach for obtaining colloidal particles in considerable amounts, an important requisite from an industrial perspective (Visca and Matijević 1979).

A simple description of the equipment used for aerosol high throughput of mesoporous oxide particles will be given: first, a solution or suspension is transformed into very small drops using an atomizer, where these drops can be considered as individual reactors. Then, transported by a carrier gas these micro reactors come into contact with a hot gas (spray drying) or pass through an oven where a chemical reaction and pyrolysis occurs (spray pyrolysis); a collector at the end of the line holds the synthesized product (See Fig. 2.1). Both spray drying/pyrolysis setups are continuous production methods, that minimize the use of precursors and considerably reduce the waste generated during materials synthesis.

There is a great similarity in the *spray drying* and *spray pyrolysis* processes. The steps of the spray pyrolysis method include: (1) generation of an aerosol; (2) initial evaporation of the solvent; (3) thermolysis of the precipitate at high temperatures; (4) formation of solid particles and (5) sintering of the particles (Gavrilović et al. 2018). When no chemical reaction occurs within the droplet (the drops simply dry out), the method is called *spray drying*. Given the lower operation temperatures, this method is widely used in industries such as pharmaceuticals (Bürki et al. 2011, Wendel and Celik 2005) and the food industry (Anandharamakrishnan 2014, Jafari et al. 2008, Murugesan and Orsat 2012). As can be will seen, *spray drying* techniques have been extended to systems with chemical reactions that require mild temperatures conditions (sol-gel chemistry). Sometimes, a spray dryer is generally used in the synthetic first steps, but a subsequent chemical reaction at higher temperature leads to the final products. It is important to note that synthesis temperatures used in a spray dryer equipment are low (< 400°C), while in the *spray pyrolysis* method the working temperature can be higher (Pal et al. 2016, Waldron et al. 2014).

In this chapter, the focus will be on the synthesis and applications of mesoporous oxide colloidal particles obtained from aerosol-assisted sol-gel processes based on spray drying and spray pyrolysis. In addition, critical parameters concerning the chemical and physical processes required for a successful synthesis will be presented.

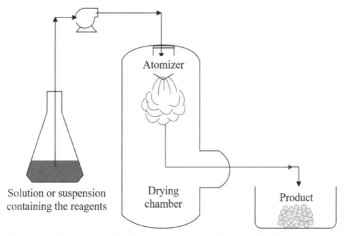

Figure 2.1: Simplified scheme of the steps involved in an aerosol synthesis. A chemical reaction may or may not be involved.

Sol-Gel Chemistry

General Aspects

The synthesis of novel materials depends on 'building blocks' that control both the connectivity and the physicochemical properties of the final product. These building blocks can be small molecules, nanoparticles, biomolecules, macromolecules, polymers, colloids, etc. However, the chemistry for binding these components should keep the original properties of these blocks.

In this context, sol-gel chemistry has emerged in the last decades as a synthetic approach that facilitates the integration of various molecular building blocks with precise control of both their physical and chemical properties (Livage et al. 1988, Yang et al. 2017). In essence, this chemistry is the 'weapon of choice' for preparing metal oxide and mixed oxide composites. Starting from an inorganic precursor, based on a metal or metalloid element surrounded by oxo-ligands, a series of hydrolysis and condensation reactions generates hydroxylated compounds which further connect through oxo-bridges (Scheme 2.1). This 'inorganic polymerization' results in a highly interconnected framework, in closely resembling the polymerization chemistry of organic compounds (Brinker and Scherer 1990).

Following the hydrolysis-condensation steps, gelation, sintering and/or structural condensation/ expansion of the synthesized framework results in different products (i.e., colloidal particles, films, powders, xerogels, etc.). Moreover, this chemistry can be carried out under mild conditions (pH, solvents, low temperatures) as it is often called '*chimie douce*' (Livage 1987).

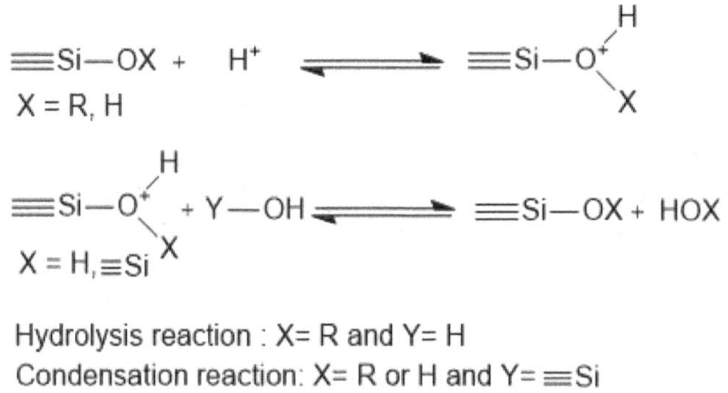

$$\equiv Si-OX \; + \; H^+ \; \rightleftharpoons \; \equiv Si-O^{+\diagup}_{\diagdown X}^{H}$$
$$X = R, H$$

$$\equiv Si-O^{+\diagup}_{\diagdown X}^{H} \; + \; Y-OH \; \rightleftharpoons \; \equiv Si-OX + HOX$$
$$X = H, \equiv Si$$

Hydrolysis reaction : X= R and Y= H
Condensation reaction: X= R or H and Y= $\equiv$Si

Scheme 2.1: Sol-Gel reactions (Hydrolysis and condensation).

Chemical Precursors for Silica Sol-Gel Chemistry: Organoalkoxysilanes

The success of the sol-gel approach for materials synthesis is due to a diverse library of organoalkoxysilanes precursors, having the general formula $R'_n Si(OR)_{4-n}$ (Pagliaro and Chemistry 2009, Gómez-Romero and Sanchez 2006). The hydrolysis and condensation of the alkoxy group (R = -OCH$_3$, -OCH$_2$CH$_3$, etc.) result in an inorganic polycondensed matrix of Si-O-Si covalent bonds. On the other hand, the non-hydrolysable R' (i.e., amino, cyano, methyl or vinyl groups) ends integrated within the Si-O-Si framework, either on its surface or within the transition metal oxide framework; besides, they are easily integrated into transition metal oxide frameworks or along with the condensation of other precursors (i.e., Zr/Ti) (Angelomé and Soler-Illia 2005). Alkoxysilanes shine because they allow easy surface functionalization, keeping both the texture of the substrate (particles, powders, plain surfaces) and their bulk properties (density, refractive index, magnetism) and they can be bought from standard chemical suppliers (Bordoni et al. 2016). These features add for the quest to find simple and trouble-free coupling chemistries, a hot subject for anyone looking to tailor a certain material with simple organic synthetic steps (Wei et al. 2015, Sheldon and Van Pelt 2013, Li et al. 2012, Arcos and Vallet-Regí 2010, Lebeau and Innocenzi 2011, Ciriminna et al. 2013, Cattoën et al. 2014).

Some Kinetical Aspects in the Hydrolysis and Condensation of SiO$_2$ and MOx Based Materials

From a chemical kinetics perspective, the hydrolysis and condensation sol-gel reactions depicted in Scheme 2.1 are competitive reactions. Moreover, the condensation can involve more than one molecular species: cluster–monomer aggregation, the formation of oligomeric species by cluster–cluster aggregation or more complex intermolecular structures. This scenario complicates the kinetic analysis as each entity will have a different reaction rate leading to different intermediates formed as hydrolysis and condensation reactions proceed. As the kinetics of these two important reactions are modulated by numerous variables, from a practical point of view, it is easier to analyze the relative rates of the hydrolysis and condensation reactions only. In this analysis, similar to what is done in organic polymerizations, molecular species present in the sol-gel systems are usually oversimplified (e.g., there is no distinction between the different oligomeric species) (Bogush and Zukoski Iv 1991), unless special characterization techniques that distinguish bond coordination (e.g., ^{29}Si-NMR) are used (Shimojima et al. 2005, Wijnen et al. 1989). Some of the most important variables to consider when dealing with sol-gel processes will be sketched here; detailed analysis can be found elsewhere (Brinker and Scherer 1990).

Silica is by far one of the most studied systems due to the commercial availability and the ample library of organoalkoxysilanes. Several parameters allow fine-tuning of the hydrolysis and condensation kinetics:

i) Increasing chain length, bulkiness or inductive effects of the hydrolysable R-groups lowers the hydrolysis rate due to steric factors.

ii) The pH dependence of the hydrolysis reaction in water shows a minimum for the hydrolysis at pH 7, while the condensation shows it at pH 4.5. At low pH, a large number of monomers or small oligomers with reactive Si-OH groups are simultaneously formed and condensation is the rate-determining step. Although strongly alkaline conditions should favour hydrolysis and lower the condensation rate, a solvent change can dramatically minimize the solubility of silicate species. In the case of aerosol synthesis, low pH organoalkoxysilane solutions are usually used.

iii) The solvent has to fulfil multiple requirements: first and foremost, it has to solubilize the starting chemical precursors and the final various silicate and oxo-silicate species; solvent polarity, then, is an important parameter. On the other hand, viscosity and boiling point will be important for atomization conditions.

iv) As the solvent carrier dries during atomization, electrolyte concentration in the colloidal dispersion increases. This has the effect of thinning the electrical double layer around the particles, overcoming repulsion forces and aggregating the particles. Ionic species are present in the precursor solution if acid catalysis is used.

In the case of non-Si based oxides, it must be considered that transitions metals, due to their electropositivity, are more prone to a nucleophilic attack when compared to Si. Moreover, they show higher coordination numbers which result in bigger clusters.

Mesoporous Materials or 'Pore Synthesis'

Whenever one is dealing with industrial and natural chemical processes such as environmental immobilization of aqueous pollutants or the conversion of toxic gases from car exhausts, one would be looking mostly at the solid-gas and/or the solid-liquid interfaces. These fundamental regions of space define chemical environments and properly conditioned can be manipulated to increase the accessible surface area. Alternatively, they can be chemically modified resulting in more adsorption sites or increased catalytic activity. This is highly convenient if looking for increasing the efficiency of a chemical conversion or to immobilize dangerous chemical species in the minimum amount of adsorbent and catalyst. As surface materials technical characterizations exploded from the late 70s (Che and Vedrine 2012), it was recognized that infinitesimal subdivision of particulate matter would directly increase the accessed total surface area. However, this would imply that the material also has to remain dispersed in the continuous phase, just as a simple molecule. This may be suitable if pursuing catalytical efficiency in homogenous processes but, in some cases, heterogeneous processes benefit from the fact that the solid phase may be recoverable, reusable or recyclable. Increasing surface area while keeping solid particles in a size range where settling allows an easy recovery is an added benefit. The only way to increase surface area while keeping the overall dimensions of the object is increasing their porosity. From the perspective of a synthetic chemist, this situation calls for 'making pores' or a 'pore synthesis'. Compared to a natural catalyst, one could find that enzyme pockets hold a cavity where anchored chemical groups carry on a chemical reaction in a concerted way with nanometric precision. As it will be seen below, shaping a pore and their interconnections in a hierarchical way requires a new set of synthetic tools and a change in the paradigmatic view for building macromolecular structures.

When it comes to pore classification the immediate and usual definition employed is that from IUPAC: according to their distribution size pores can be divided in microporous (less than 2 nm diameter), mesoporous (in the 2–50 nm range) and macroporous (larger than 50 nm). Further, pores distributed within materials can be either randomly and disordered or they can have a regular spatial arrangement with a defined symmetry. Historically, natural microporous materials such as zeolites, discovered more than 200 years ago present a highly crystalline structure of a narrow pore system (Xu et al. 2009). Despite being mainly used in the petrochemical industry, the microporous pore dimensions restricts their application for the synthesis/cracking of larger molecules. It was soon recognized that new materials, and consequently a new synthesis approach, must be envisaged. Considered to be the future of novel porous catalysts, the family of mesoporous molecular sieves developed in Mobil Research and Development Corporation (Mobil Composition of Matter, MCM) came from the research on pillared layered-materials and aluminosilicate artificial-zeolite synthesis, holding the promise to bring the solution to the petrochemical industry (Beck et al. 1992, Kresge et al. 2004). However, the initial excitement soon declined as the obtained frameworks had a number of disadvantages: low thermal stability, weak surface acidity, and were easily deactivated, resulting in poor catalysts for oil cracking. Nonetheless, the principles used for designing these materials and the knowledge derived to interpret the synthetic mechanisms soon spilt over various chemical and technological areas which, combined with powerful material characterization techniques, brought new light to the materials science field (Zhao et al. 2012).

Now, how are pores 'synthesized'? If one recalls the enzyme pocket analogy, one should look to chemistries that use a scaffold or molecular interactions that arrange the chemical precursors through self-assembly at a precise distance. In the process of trying to categorize the different strategies for obtaining porous structures, one would find that some of them share similarities; divisions between them become blurred and make the classification harder. The 'pore synthesis' approaches can be divided in three main groups: the use of hard and soft templates, having voids or spatial regions where chemical reactions that end in the formation of solid materials, and the assisted synthesis of using molecular structures and interactions that aid and guide the interconnection of building blocks. Following, some key concepts that use several approaches will illustrate the overall strategies for pore formation.

Hard Templates

The infiltration of chemical precursors within solid porous materials provides a simple starting point for making a porous structure after hard template dissolution. Successful strategies primarily rely on different solubility properties of the filling material and the hard template in various solvents; typical pairs are silica-polymer, semiconductor-silica, metals-polymers where specific solvents are able to selectively etch one of the components. Many of these strategies have been known for a while from the electronic industry for microprocessor fabrication: precise etching rates have been elaborated for silica, oxides, photoresists and metals when patterning integrated transistors (Williams and Muller 1996, Wu et al. 2010). Some researchers coined the term '*nanocasting*' referring to the process (Kyotani et al. 1997) that replicated the inverse structure of zeolites as a hard template for microporous carbon synthesis. One of the reasons for using a hard template is based on its wide accessibility, ease of fabrication and appropriate composition of the framework for later dissolution or removal.

Commercial membrane filters are readily available with pore sizes spanning the tenths of nanometres to microns in diameter and having several microns long (~ 60 µm). Skipping fabrication details, polycarbonate and alumina are among the most used for structure templating where the main differences between them are both their composition and pore density (Al_2O_3: 10^{11} pores/cm^2, polycarbonate: 10^9 pores/cm^2). Arrays of noble metal ultramicrolectrodes (Au, Pt) were prepared by electrochemical deposition within the pores of both types of membrane filters (Foss Jr et al. 1992, Martin 1994, Penner and Martin 1987). Several researchers explored the electrochemical anodization of Al substrates (Anodic Alumina Oxidation, AAO) as high-throughput top-down nanofabrication of 1D structures (nanotubes) or patterning 2D dot arrays (Lee and Park 2014).

One of the essential requirements for choosing a hard template is the possibility of manufacturing a 3D porous structure from simple building blocks that self-assemble spontaneously. In this context, the fabrication of colloidal crystals from sub micrometric spherical particles provides an easy approach for obtaining a highly porous and interconnected structure (Talapin et al. 2010). Multiple approaches have been used to induce the packing of the spheres and the formation of a colloidal crystal: gravity settling, evaporation, electrical fields, spin coating, etc. (Vogel et al. 2015). Regardless of the method used, highly reproducible synthesis of monodisperse silica particles (Stöber et al. 1968) or polystyrene colloids (Bijsterbosch 1978) guarantee the perfect crystallization of highly ordered frameworks with hexagonal or cubic packings. As the particles accommodate, voids are left where metal (Cong and Cao 2004) and metal oxide (Blanford et al. 2001) precursors can react and condensate. Subsequent removal of the starting colloids (e.g., calcination, dissolution) leads to an ordered macroporous material with the oxide material, either crystalline or amorphous, as a pore wall (Subramanian et al. 1999). On the other hand, disordered but highly interconnected porous frameworks are the basis of carbon electrodes for fuel cell applications (Chai et al. 2004, Lee et al. 2006). In this case, SiO_2 colloids are mixed with polymers that are later carbonized; the oxide is finally etched using HF or NaOH (Duraisamy et al. 2019).

Soft Templates

Structured Assisted Templating

During the study of zeolite artificial synthesis, a key concept emerged conceiving that their frameworks were made up from complex building units present in solution (Cundy and Cox 2005). Experimental conditions established that highly porous materials could be obtained after introducing organic molecules (e.g., small quaternary ammonium cations) into the reaction mixtures, where crystallization/condensation reactions occurred around the molecular template (see Fig. 2.2) (Cundy and Cox 2005). Coulombic forces, H-bonding or van der Waals interactions stabilize the inorganic framework/template composite; in addition, template-template interactions can also contribute. After appropriate conditioning of the inorganic backbone, the organic template is removed and pores are obtained. Several techniques tackled the study and analysis of the formation of the nascent building blocks in solution and how cooperative forces operate in the self-assembly with the molecular templates (Burkett and Davis 1994). The preceding molecular scheme was soon extended to bigger templates based on surfactant molecules.

In this context, amphiphilic molecules (surfactants) have a very special role as they display a polar head and a non polar tail; one of these ends tend to aggregate in solvents where one of these domains is insoluble. Depending on the solvent/surfactant ratio highly auto-organized phases appear with a defined geometry and symmetry; typically, with surfactant concentration increase, micellar spheres appear first, followed by cylinders, lamellae, 'inverse' cylinders, and 'inverse' spheres. The different molecular packing is directly related to interfacial curvature energies associated with molecular parameters (a polar charge, chain length, steric hindrance, etc.) (Israelachvili 2015). The variety of self-assembled configurations accessible highlight an exceptional perspective for the development

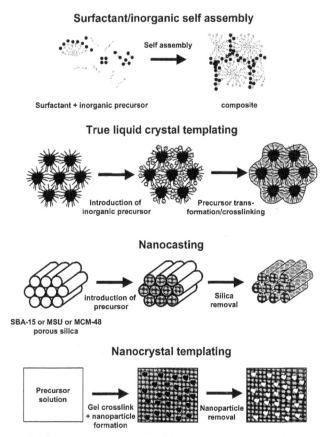

Figure 2.2: Typical routes for obtaining mesoporous materials. Reproduced with permission from Ferdi Schüth 2001. Chemistry of Materials 13(10): 3184–3195. Copyright 2001, American Chemical Society.

of textured materials as there is a wide availability of surfactant molecules in the market, with large tonnes produced each year. Specific interactions appear between the oxide precursor species and the polar head of the surfactant; when ionic surfactants are used as templates, the formation of the mesoporous material is mainly governed by electrostatic interactions. The charges of the surfactant polar head and the mineral precursor are opposite under the pH synthetic conditions, where further condensation of mineral drives the formation of the final material (Soler-Illia et al. 2002). On the other hand, when non-ionic surfactants are used the main interactions between the template and the inorganic species are through H-bonding or dipolar interactions (Soler-Illia et al. 2002). This scheme represented in Fig. 2.2, can be extended to other molecular templates: cyclodextrins (Zheng et al. 2001) or dendrimers (Larsen et al. 2000).

Lyotropic Liquid Crystal Templating

If the solvent fraction is low enough, the surfactant-surfactant interaction drives the formation of a highly auto organized phase and Lyotropic Liquid Crystal (LLC) forms. Interestingly, this defines a new way to drive chemical reactions because the formation of highly organized phases with hydrophobic and hydrophilic regions can be controlled from a self-assembled mixture of a non-ionic surfactant and water with pores in the 2–5 nm diameter range. Then a condensation, precipitation or mineralization that proceeds only in the hydrophilic region of the LLC template, results in an inorganic mesostructured replica of the liquid crystal (Braun et al. 1996, Stupp and Braun 1997). This scheme has been applied to synthesize mesostructured chalcogenide semiconductor particles (Braun et al. 1999), template mesoporous Pt electrodes (Attard et al. 1997) or form mesoporous shells on curved colloidal particles (Wolosiuk et al. 2005, Gough et al. 2009, Son et al. 2009).

Controlled Phase Separations: Spinodal Decomposition

A very interesting approach that requires precise knowledge of the thermo dynamical variables is the spontaneous phase separation of an initially miscible; the phase separation can be controlled in such a way to avoid the coalescence of the individual phases. This strategy is advantageous for synthesizing bicontinuous macrostructures or the nucleation of small regions of the minority phase with nanometre dimensions. Moreover, this concept has been extended to a variety of blends (polymers, solids, alloys, etc.) based on a solid theoretical framework that was developed in the 60s (Cahn 1961, 1962). Commercial Controlled-Pore Glasses (CPG) and VYCOR® glass with pore dimensions in the 4–1000 nm range are a good example of phase separation in a composite alkali-borosilicate material. The structural and textural properties of these porous glasses can be tailor controlled: (i) the starting glass composition, (ii) heat conditioning (temperature, time) and (iii) the dissolution/leaching conditions (Enke et al. 2003). Two different interconnected phases are obtained after heat treatment: the first one is an alkali-rich borate phase soluble in hot mineral acids, water or alcohols, while the second is ~ 96% silica (Inayat et al. 2016).

In the case of hybrid materials, synthesis quenching of miscible systems through temperature drop or solvent removal causes phase separation. In this context, block-, copolymer phase separations are frequently found as the main strategy for the modification of planar surfaces (Walheim et al. 1999), although hierarchical porous materials can also be easily produced starting from this templates (Dorin et al. 2014). The spinodal phase separation of a mesophase acting both as a carbon precursor and as a soft polymer template results in a continuous macro-mesoporous structure after carbonization (Adelhelm et al. 2007). The same principles of phase separation are based in deep eutectic solvents (López-Salas et al. 2016), aggregation of polystyrene colloids with D-fructose as a carbon source forming coral-like structures (Kubo et al. 2013), controlled THF evaporation in block copolymer/oxide precursors/nitric acid for mesoporous Ti-Nb oxide materials for super capacitor applications (Jo et al. 2018). Moreover, small polymers can be used as inducers for phase separation in sol-gel transitions for obtaining highly porous γ-Al_2O_3 frameworks with macro/meso distribution (Passos et al. 2016) or combined for the dewetting of sugar/surfactant samples (Zelcer et al. 2009).

A Special Approach for Templating Colloidal Oxide Materials with High Yields: Aerosol-Based Evaporation-Induced Self-Assembly (EISA)

In 1999, Brinker and colleagues introduced the EISA strategy as a way to produce mesoporous SiO_2 thin films (*Nanoporous materials: science and engineering* 2004). Starting from dilute solutions of a surfactant (soft template), hydrolyzed species or nanometric building blocks (framework), a volatile solvent and a catalyzer (usually an acid or base) aided by the gradual solvent evaporation induced the formation of a liquid crystal mesophase. This mesophase 'assisted' the formation of an inorganic network from hydrolyzed species or building blocks around the soft template. In contrast to the use of a hard template, this is a highly dynamic process, where the kinetics of the inorganic phase mineralization must follow the chemical constraints of the mesostructuring template. This procedure looks promising for avoiding restrained diffusion or pore blocking problems found as chemical precursors infiltrating the liquid crystal phase whenever using hard templates. In addition, the process was adapted to obtain a variety of systems: thin films (Angelomé and Fuertes 2018, Grosso et al. 2004), powders (Kruk et al. 2000), monoliths (Amatani et al. 2005) and for aerosol-based techniques (Checcucci et al. 2018). Besides the sol-gel hybrid structure is finally consolidated after template removal (calcination or solvent extraction) which usually condenses the reticular oxide due to water elimination. In the case of colloidal particles, EISA was soon adapted to an aerosol spray-drying system boosting the opportunities for tailoring the porous material synthesis as sol-gel chemistry is processed under mild conditions (Fig. 2.3). It can be seen that the starting precursors solution can include templating agents that can act as hard templates (e.g., colloidal particles that are later etched) or can self-assemble and assist the condensation of the inorganic matrix during the drying process (i.e., surfactants forming micelles, polymer phase separation).

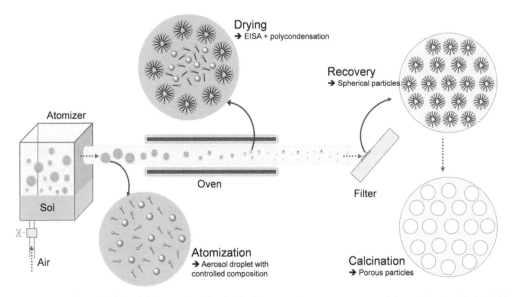

Figure 2.3: Working principle of the aerosol-assisted sol-gel process for mesoporous materials combined with the EISA process. Reproduced with permission from Damien P. Debecker et al. 2018. Chemical Record 18(7): 662–675. Copyright 2017, John Wiley and Sons.

These approach has been renamed as Aerosol-Assisted Sol–Gel processes (AASG) (Debecker et al. 2018).

Spray Drying Process and Devices

Spray Drying Process

In the last few years, spray drying for powder production, based on the evaporation of the solvent from solutions, suspensions or emulsions with chemical precursors, and its use in the synthesis of various advanced materials, such as films, core-shell particles, composite materials, etc. (Arpagaus et al. 2018a, Arpagaus et al. 2018b, Leng et al. 2019, Liu et al. 2015, Pitchumani et al. 2009, Okuyama et al. 2006) has been increasing because it is a simple, reproducible and scalable method.

The spray drying process consists of three relevant steps: (i) formation of small drops of the precursor solution, which are dispersed in the carrier gas (aerosol formation), (ii) the liquid drops come in contact with the drying gas (typically hot air, N_2 or other gases) and (iii) dry product collection. The general scheme of a typical spray dryer can be seen in Fig. 2.4 when each step tailors the characteristics of the final product. Moreover, it is possible to arrange this setup as a closed-loop system, if working with organic solvents and an inert gas without the risk of ignition or in an open-loop system, suitable for working with aqueous solutions (or with low organic content) and with normal air as the drying gas.

The powder obtained can have a different morphology depending on the experimental condition, it can be dense, hollow, porous, non-porous, amorphous or crystalline, etc. The advantage is that once the operating parameters have been identified, spray drying allows powders with defined and constant characteristics to be obtained.

Spray Dryer Components

Atomizers

Atomizers (or nebulizers) are essential parts of any spray drying equipment. Their function is to transform the solution or suspension that enters the equipment into small drops; increased rates of

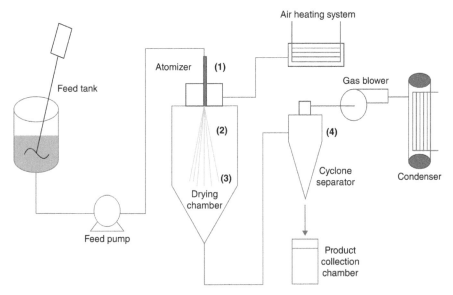

Figure 2.4: Typical spray drying with an open loop. Four main steps can be seen: (1) atomization, (2) contact between droplets and hot gas, (3) solvent evaporation and (4) product separation. Reproduced with permission from Anandharamakrishnan et al. 2015. Spray Drying Techniques for Food Ingredient Encapsulation, Wiley Books. Copyright 2015, John Wiley and Sons.

solvent evaporation are a result of the large increase of external surface, concentrating the chemical precursors within the liquid drop.

The choice of atomizers is large, which one should be used depends on the nature and quantity of the precursor solution and the desired characteristics of the final product:

i. *High-Pressure Nozzle*

In these atomizers the liquid enters a chamber where turbulence is generated at high pressure, after which it passes through a tube that decreases in diameter from the inlet to the outlet. By decreasing the size of the orifice, the liquid pressure decreases, and consequently produces enough kinetic energy that breaks the liquid surface tension producing droplets. Both the size of the chamber and the outlet hole determine the capacity of the atomizer and the size of the droplet. The advantage of these atomizers is that they do not need any gas flow, but the droplet size is usually large (about 10 mm or more).

ii. *Two-Fluid Nozzle*

These types of atomizers are used in pilot-scale spray dryers, requiring an atomizing gas; typically, the most used is compressed air, although N_2 and other gases can also be used. The generation of the drops is achieved when the liquid and the gas interact at a high speed either inside or outside the body of the nozzle. When the atomization occurs outside, a greater velocity of the liquid and a greater pressure of the gas are necessary to achieve it. These types of atomizers achieve a smaller droplet size and narrower distributions when the atomized liquid has a low viscosity (O'Sullivan et al. 2019).

iii. *Rotary Atomizer*

These types of atomizers use centrifugal energy for the production of the drops. The solution or suspension comes into contact with a rotating surface at high speed producing drops in a radial direction. The surface can have various shapes: cup or a flat disc, either with blades or with grooves. The rotational rate alongwith the shape of the rotor control the droplet size. This also depends on the intrinsic properties of the solution or suspension, such as viscosity, density and solvent identity. Under proper operating conditions, a narrow distribution of droplet sizes can be easily obtained.

iv. *Ultrasonic Nebulizer*

These atomizers are composed of a transducer that vibrates at high frequencies, typically between 50 kHz and 2.4 MHz, thus producing the atomization of the liquid that lies on top of the vibrating membrane. The drops obtained can be smaller than 10 mm (O'Sullivan et al. 2019). Essentially there are two approaches for how the ultrasound energy is applied to the liquid phase: (i) the liquid passes through a tube and comes into contact with the transducer and (ii) the transducer is submerged directly into the liquid. The latter is the most widely used in spray drying and allows to obtain droplets with sizes between 2 and 4 mm (Kodas and Hampden-Smith 1999).

Drying Chamber

Drying of the drops begins immediately after atomization in the drying chamber, where the streams of hot gas and atomized liquid come into contact. The evaporation rate of the solvent is a very important parameter that determines the morphological characteristics of the final product. Several factors influence the drops' drying rate such as the temperature of the drying gas, solvent, drop size and mass and heat transfer between the aerosol and the hot gas. Contact between these two phases can take place under three forms: co-current flow, counter-current flow and a combination of both (see Fig. 2.5). Most spray dryers use either co-current flow or mixed flow while counter-current flow is used when the atomizer is a rotary disc.

An essential spray dryer design requirement is to guarantee a uniform contact between the drying air and the sprayed drops. The size of the drying chamber should be sufficient to achieve the successful drying of the drops: smaller drops need less drying time, therefore smaller drying chambers can be used. Nonetheless, the chamber's diameter must be large enough to prevent wet droplets from hitting the chamber's wall. Possible chamber geometries may be studied using Computational Fluid Dynamics (CFD) modelling.

Understanding and modelling drop drying is an important part of the spray drying method. From the seminal works of Ranz and Marshall, droplets drying can be divided into stages (see Fig. 2.5) (Ranz 1952a, 1952b). In the first stage, the liquid in the drop is heated up to its evaporation temperature. When the liquid on the drop's surface begins to evaporate, the liquid inside the drop migrates to the outer surface, producing a contraction of the drop and triggers the precipitation inside the drop. In the case of sol-gel materials, as the reactant/solvent ratio changes, the EISA induces the formation of condensed oxides within the drop. As soon as the liquid in the drop is not enough to maintain the vapour pressure at the surface, nuclei begin to form agglomerates where a solid shell forms on the particle surface halting further size changes. Under certain conditions, liquid in the drop's core may remain because of the higher resistance to evaporation due to the

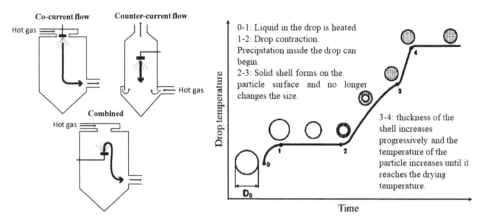

Figure 2.5: Left, Mixing of spray and drying medium (adapted from Buchi Training Papers Spray Drying). Right, stage of droplet drying (adapted from ref (Mezhericher et al. 2010)).

surface solid shell. Finally, the thickness of the shell increases progressively and the temperature of the particle increases until it reaches the drying temperature (Mezhericher et al. 2010). The type of particles obtained (hollow or compact) depends on how all these processes occur; permeability of the shell and the concentration of reagents in the solution to spray are also variables to consider.

If one's attention is drawn to the precipitation of solute within the drop one could find that the initial concentration of chemical precursors is very important. Jayanthi and co-workers (Jayanthi et al. 1993) studied the relationship between the initial concentration of zirconium hydroxy chloride and the morphology of the ZrO_2 particles obtained. When the initial concentration of the feeding precursor solution was low, the concentration necessary to precipitate (critical super saturation) was reached only at the edge of the drop and not in the centre, resulting in a hollow particle. On the other hand, when the concentration was high, the critical super saturation conditions were reached both at the edge as well as inside the droplet, and the solid was distributed throughout the diameter. The nucleation and growth rate of solid within the droplet depends on the relation between solute concentration and the concentration necessary to precipitate (C/Csat) (and other properties of the salt). This determines the morphology of the particles, therefore the precipitation and the type of particle can be limited by the drying rate (time and temperature).

The Peclet number (P_e, Equation 1) is a good estimator for predicting the kind of particle that can be obtained (hollow or solid). This number relates to the drying time (t_d) with the time necessary for the migration of the solid within the drop, calculated from the r_d (drop's radius) and D_{S-E}, the diffusion coefficient obtained from the Stokes-Einstein equation. If $P_e \ll 1$ the drop has enough time for migration of the solid and presents a homogeneous drying. If on the other hand $P_e \gg 1$, the drying is faster than the migration of the solid particles inside the drop, then hollow particles are obtained.

$$P_e = \frac{r_d^2}{t_d D_{S-E}} \tag{1}$$

However, obtaining this number is not easy for a spray drying process under real operating conditions. Complications arise as the droplet size is a distribution, the temperature of the hot drying gas is not homogeneous in the equipment and the heat transfer between the gas and the droplet depends on this temperature. There are numerous works where the parameters that influence the drying of individual drops are studied, where the conditions are extremely controlled (Wulsten and Lee 2008, Kumar et al. 2010, Mondragón et al. 2013). The analysis of single drop evaporation uses a series of dimensionless groups to model this process, Reynolds number (*Re*), Prandtl number (*Pr*), Schmidt number (*Sc*), Nusselt number (*Nu*) and number Sherwood (*Sh*) (Masters 1991) in combination with CFD. Mezhericher and co-workers (Mezhericher et al. 2010) developed a two-dimensional theoretical model of the steady-state spray drying process. They noted a considerable influence of the drying kinetics model used in the expected heat and mass transfer in the drying chamber and concluded that an adequate model of the droplet drying kinetics, as well as a realistic adjustment of the boundary conditions, is crucial to the numerical representation of the actual performance of the spray dryer. In 2011 Mezhericher published a book on the theoretical modelling of spray drying processes that included drying kinetics and 2D and 3D CFD simulations (Mezhericher 2011). Although this technique provides important information about the process, there are several limitations, so CFD is often used in spray drying to obtain information on where problems could appear and help speed up the problem resolution.

Particle Collectors

Once the droplets convert into particles, a separation process is necessary for collecting the dried particles of the desired size. A first separation occurs with the large particles that are generally collected at the bottom of the drying chamber. A second separation step discriminates small particles from the wet gas. The proper collection mechanism depends mainly on the distribution of particle size, particle density and how the carrier gas enters the collector. Below, the most used collectors are presented:

i. *Filtration*

Filters can be designed to be used in almost any situation. It is possible to put them in series, for example, and separate them by different pore size, in order to select the particles by size (Daniel Santos et al. 2018). The airflow containing the dry particles enters the filter under pressure or suction, the dry particles are retained on the surface of the filter while the clean air passes. The particles accumulated on the surface are then collected due to pulses of air injected in counter-current through the filters. The additional step of recovery of the particles increases the production cost, so this type method is usually more expensive than cyclones and therefore is used for particles of 10 microns or less, which cannot be recovered by cyclones (Kodas and Hampden-Smith 1999).

ii. *Cyclones*

Cyclone separators are relatively inexpensive equipments for separating solid particles from a gas carrier. The separation mechanism is based on centrifugal force.

Particles coming from the drying chamber enter the cyclone from the top, then the flow descends and an external vortex is created, as shown in Fig. 2.6. The induced centrifugal force is tangential to the cyclone wall whereby particles are deposited on the wall of the conical cylinder, then they fall by gravity into the collecting container located at the end of the cone. As soon as the gas reaches the bottom, an internal vortex is created at the opposite direction, and the gas exits through the centre of the vortex to the top of the cylinder (Anandharamakrishnan 2015). These devices are efficient for the separation of solid particles of more than 5 microns, although some designs allow the separation of smaller particles, but with less collection efficiency.

Cyclone separators are commercially available and can be custom designed. They are the most widely used devices in powder collection. Quartz cyclones are able to work at relatively low temperatures, generally associated with spray dryer equipment, and metal cyclones permit working at higher temperatures, up to 1000°C (Kodas and Hampden-Smith 1999), associated with synthesis by spray pyrolysis. These devices can be placed in series, in order to separate the larger particles from the smaller diameter particles; as well as before another separation device to decrease the amount of dust that reaches the second.

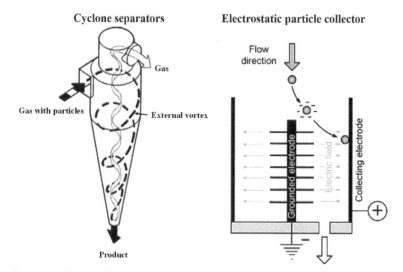

Figure 2.6: Left operating diagram of a cyclone separator. Right, schematic of the electrostatic particle collector that is part of the Büchi Nano Spray Dryer B-90 HP.

iii. *Electrostatic Particle Collector*

In this technique, an electric field is used to separate solid particles from the gas stream. It is necessary to charge the particles so that they are attracted to the collecting wall that has an opposite charge. These collectors are made up of discharge electrodes and collector plates, as represented in Fig. 2.6. High voltage is applied to the discharge electrode to form an electric field between the electrode and the grounded collector plates. When the gas containing the particles passes between the collector plates and the discharge electrodes, the particles in the gas are charged for a short period that is sufficient for them to reach the collector walls. The particles can then either fall into a collection container by gravity or require some additional process to collect them (Ishwarya 2015).

One of the advantages of these collectors is that the efficiency increases with decreasing particle size, since the smaller the size, the greater the mobility (Kodas and Hampden-Smith 1999).

Laboratory Scale Commercial Spray Dryers

One of the most recognized companies in the manufacture of spray dryers is Büchi, which presents two models of dryers, the Mini Spray Dryer B-290 and the Nano Spray Dryer B-90 HP. The B-290 allows to process a greater quantity of sample than the B-90, but with the latter, the particle size obtained is smaller (0.2–5 μm). In the B-290 the maximum inlet temperature is 250°C, while in the B-90 it is 120°C. In the B-290, the atomizer is a two-fluid nozzle 0.7 mm, but they can be exchanged for a two-fluid nozzle with a larger nozzle diameter, a nozzle for two immiscible fluids, and one can also have the option to put in an ultrasonic atomizer. The particle collector is a cyclone collector. The construction material for both the drying chamber and the cyclone is borosilicate glass. And with this model (B-290) the company ensures that it is possible to expand the system to an industrial or pilot scale in a simple way. Model B-90's atomizer is ultrasonic and has three droplet size options (small, medium and large). The particle collector is an electrostatic collector since here, it works with smaller particle sizes. For both spray dryers, there is the option of working in a closed-loop to use solutions with organic solvents safely.

Another company that produces equipment is Shanghai Pilotech Instrument & Equipment, five models are available: Mini Spray Dryer YC-015, YC-500 Laboratory Spray Dryer, YC-1800 Mini Low-Temperature Spray Dryer, YC-2000 lab vacuum spray dryer and Lab YC-018 Spray Dryer. All of them use a 0.7 mm two-fluid nozzle as an atomizer, but it also has nozzles of different diameters available (0.5; 0.75; 1.0; 1.5 & 2.0 mm). In all cases the particle separator is a cyclone separator, in the YC-015 equipment, it is made of borosilicate glass, while in the other types of equipment it is made of stainless steel. In the case of the YC-018 model, the maximum inlet temperature is 350°C. The YC-1800 spray dryer can also be equipped with the fluidized bed drying function.

On the other hand, the MRC laboratory also has lab scale spray drying equipment, Laboratory Scale Spray Dryer SD-15, Low-Temperature Spray Dryer SD-18, Vacuum Spray Dryer SD-20, Mini Spray Dryer SD-8, with characteristics similar to its analogues from Shanghai Pilotech Instrument & Equipment.

It may be noted that in commercial spray dryers, the maximum inlet temperatures are relatively low for what is required in some synthesis as it be seen next. For this reason, as previously mentioned, if this commercial equipment is used, the necessary thermal treatment can be carried out at a later stage outside the equipment. If not, the other option is to design the custom spray dryer in the laboratory, using the aforementioned technologies.

There is also the possibility to make your home-made spray dryer, however, there are several important parameters to consider when designing. The pump that allows the solution or suspension to enter the equipment must be stable and allow variations of the inlet flow. The inlet temperature must be high enough to achieve the drying of the product, but it must be safe for working with the solvents of the solution to be sprayed (keep in mind that the temperature decreases significantly when it comes into contact with the aerosol). The flow of hot air must be proportional to the flow of the solution and they must allow a transfer of mass and energy with the drops in order to achieve drying of the drops throughout the drying chamber. The atomizer must be able to work at different flows of the solution inlet and be suitable for its viscosity. The geometry of the drying chamber depends on several factors, such as the type of atomizer to be used, the interaction of the air hot with spray, etc.; so it is advisable to build on existing drying chambers or use CFD.

Synthesis and Applications

Precursors

In aerosol synthesis, the 'traditional' liquid phase synthesis of mesoporous mixed metal oxides sol-gel materials has been modified so that bond connections of the inorganic oxide and framework condensation occurs as the aerosol droplets dry. Precursor solution composition has to be tuned to avoid undesirable reactions and chemical incompatibilities at the time of mixing. Typically, solutions to be atomized include: (i) an oxide precursor, in the form of an alkoxide or inorganic salt, (ii) a surfactant or colloidal particle that can be etched or calcined later, for pore templating, (iii) a catalyzer, for accelerating the sol-gel hydrolysis or decomposition of the oxide precursors and (iv) a volatile solvent, that keeps all these components homogeneously distributed at the atomization step and is finally evaporated during droplet drying.

Alkoxides

In this context, metal alkoxides emerge as the first option for the synthesis of oxide materials. The templated aerosol production of Si, Ti and Zr-based materials is mainly based on the use of the corresponding Si, Ti or Zr-alkoxide, an inorganic acid (e.g., HCl or acetic acid) that hydrolyzes the precursor and a templating agent as a surfactant (Jin et al. 2012, Jin et al. 2013) (Fig. 2.7A–D). Most of the precursors' solutions used for aerosol production of colloidal particles

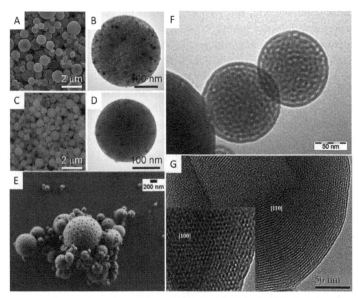

Figure 2.7: A and B, mesoporous TiO$_2$ particles containing Au NP. **C and D,** mesoporous ZrO$_2$ particles containing Au NP. Reproduced with permission from Zhao Jin et al. 2012. Angewandte Chemie – International Edition 51(26): 6406–6410. Copyright 2012, John Wiley and Sons. **E,** organic@silicate core-shell NPs with (Z)-3-(9-ethyl-9H-carbazol-3-yl)-2-(4-nitrophenyl) acrylonitrile as dye. Reproduced with permission from Shenoi-Perdoor et al. 2018. New Journal of Chemistry 42(18): 15353–15360. Copyright 2018, Royal Society of Chemistry. **F,** particles of tin-silicates with a silicon/tin ratio of 74. Reproduced with permission from Godard et al. 2017. ChemCatChem 9(12): 2211–2218. Copyright 2017, John Wiley and Sons. **G,** TEM images of periodic mesoporous organometalsilica spheres with Pd. Reproduced with permission from Zhang et al. 2011. Advanced Functional Materials 21(16): 3189–3197. Copyright 2011, John Wiley and Sons.

share a similar composition as those used for mesoporous thin films synthesis (Angelomé and Fuertes 2018, Sansierra et al. 2019), and this aspect is a good example of the variety of textures that sol-gel chemistry allows exploring. Organoalkoxysilanes mixtures are attractive reagents as they provide a way to embed organic functionalities within the mesoporous framework. For example, thiol groups have been easily incorporated into mesoporous frameworks combining TEOS and 3-MPTMS (3-mercaptopropyltrimethoxysilane) for toxic heavy metal ions adsorption (Suzuki et al. 2010, Suzuki and Yamauchi 2010). Organosilanes mixtures offer the possibility to control the hydrophilic/hydrophobic balance for fluorescent dye encapsulation (Shenoi-Perdoor et al. 2018) (Fig. 2.7E). Synthetic efforts have also been directed to tune the molecular structure of the silane producing organometal-bridged periodic mesoporous with $Pd^{2+}/Rh^+/Ru^{2+}$ centres (Zhang et al. 2011) (Fig. 27G); allyl-modified cyclic silane for chromatography (Ide et al. 2015). The amorphous silica matrix can be doped with diverse cations: for example, stannosilicates were obtained from TEOS hydrolyzed in an acidic aqueous-ethanolic solution and then $SnCl_4 \cdot 5\ H_2O$ and Pluronic P123® were added right before atomization (Godard et al. 2017) (Fig. 2.7F); tungstate sites from phosphotungstic acid hydrate and hydrolyzed TEOS in water and ethanolic/surfactant (Brij 58®), was added to the TEOS solution, mixed and immediately aerosolized (Maksasithorn et al. 2015). Similar principles have also used for introducing Co/Mo sites (Colbeau-Justin et al. 2014).

Ti-based alkoxides for obtaining TiO_2 mostly rely on commercially available titanium (IV) butoxide (Guo et al. 2016, Kim et al. 2019, Pal et al. 2016, Zhang et al. 2010) or Ti-isopropoxide (Oveisi et al. 2010). Easy in-lab preparation of Ti (IV) or Zr (IV) oxide matrices can be achieved from chlorides of Ti and Zr dissolved in ethanol (Araujo et al. 2010a). Complex Ti/Si/Zr oxide frameworks can be prepared by mixing in the advanced alkoxides: Ti-Si from TEOS and Ti-butoxide in tetrapropylammonium hydroxide solutions (Smeets et al. 2019) or TEOS hydrolyzed solutions with Ti-(oiPr)$_2$ (acac)$_2$ (Sachse et al. 2012). This can be extended to more complex frameworks such as SiO_2-Al_2O_3 doped with MoO_3 or WO_3 obtained from TEOS, $AlCl_3$, the corresponding heteropolyacid/chloride and Brij®58 as pore template (Debecker et al. 2012, Debecker et al. 2014). Zr-Si frameworks are also made from mixtures of $ZrCl_4$ and TEOS (Colilla et al. 2010). In other systems, the TEOS hydrolysis in aerosol systems results in preformed zeolite crystals or shortens the time scale required for the fabrication of Sn-bzeolites (Smeets et al. 2020, Meng et al. 2019).

Inorganic Precursors: Nitrates and Chloride Salts

Given that nitrate metal salts are highly soluble in water and have low contaminants contents they are generally preferred as precursors for the preparation of metal/metal oxide-supported catalysts. Moreover, the metal oxides produced under ambient air conditions at temperatures > 400°C can be later converted to the corresponding metal through reduction treatments (Yuvaraj et al. 2003).

Geng's group developed a whole body of work around the use of nitrates decomposition for the obtention of mesoporous oxides from aerosol generated droplets (Kan et al. 2015, Kuai et al. 2014, Kuai et al. 2018, Wang et al. 2015, Wang et al. 2016, Wang et al. 2017). Nitrates are usually dissolved in ethanol along with an amphiphilic surfactant and further decomposed while passing a tube furnace at high temperatures (~ 480°C). As the residence times in the furnace are small, the metal oxides do not have enough time for crystallization resulting then in amorphous products. The simplicity for precursor preparation allows mass-producing mesoporous frameworks spanning a wide range of Fe/Ni ratios for O_2 electrocatalysis or spinel oxides. This has been extended to Au/CeO_2 photocatalysts, starting from $Ce(NO_3)_3$ dissolved in Pluronics P123® along with $AuCl_4^-$ as the noble metal precursor which was further reduced under a stream of H_2 (Jia et al. 2017). Simultaneously, Li et al. showed the possibility to obtain TiO_2-In_2O_3 composites from mixtures of $In(NO_3)_3$ and Ti precursors (Li et al. 2014). While the use of nitrates is an interesting approach, however final temperatures for oxide generation are usually higher than those attainable in commercial spray-dryers, forcing to use in-house designed equipment. Moreover, starting from nitrates of manganese (II) and silver (I), Sun et al. suggested that simple redox chemistry occurs in the droplet where Ag-MnOx hybrid nanostructure forms with the assistance of CTAB as a soft-template (Sun et al. 2019). In this scheme, the authors claim that $AgNO_3$ fulfils a dual role as an oxide generator, oxidizing Mn(II), while Ag nanoparticle electrical properties increase the MnOx capacitance. Recently, the nitrate route has been used for the synthesis of W-doped Al_2O_3 particles: hydrolysis and condensation of Al $(NO_3)_3$ on fumed silica (Aerosil 380, Evonik Industries) on which tungstate sites were introduced starting from ammonium metatungstate $[NH_4]_6H_2W_{12}O_{40}$ (Bukhovko et al. 2020).

On the other hand, the use of chloride salts as a source of metal ions for mesoporous oxide synthesis is limited due to the requirements of higher temperatures for decomposition when compared to nitrate salts (Bhattacharya et al. 1996). Nonetheless, Ir-oxide materials from hydrated iridium(III) chloride (Faustini et al. 2019) and Al_2O_3 from $AlCl_3$ (Maruoka and Kimura 2019, Fiorilli et al. 2016, Kim et al. 2010) have also been obtained.

Miscellaneous Precursors

The widespread use of aerosol technologies in the industry enables the synthesis of various oxides from non-conventional sources: Ni and Zn/Al Layered-Double-Hydroxides (LDH) nanoparticles have been shaped into microspheres (Prevot et al. 2011, Huo et al. 2013); zirconates such as $NaZrO_3$ from sodium and zinc acetate solutions (Bamiduro et al. 2017) and more recently, inexpensive sodium aluminate microspheres, which are insoluble in ethanol, have been prepared from $NaAlO_2$ 25% wt. aqueous solutions as an heterogeneous catalyst for base-catalyzed organic reactions in an alcoholic solvent (Ramesh and Debecker 2017, Ramesh et al. 2018, Rittiron et al. 2019).

Greener Alternatives

The characteristic chemicals employed for EISA-based aerosol synthesis are a volatile (and usually flammable) solvent and a strong inorganic acid, such as HCl, to catalyze the sol-gel and control the condensation reaction. It is evident that with these components there is a two-fold problem: potential corrosion of the equipment and the use of a hazardous and potentially flammable solvent in large quantities at high temperatures. As greener alternatives for spray-drying techniques must comply with industrial and economic demands, there is a quest for finding non toxic solvents and alternative chemical precursors (Avci-Camur et al. 2018, Guo et al. 2015). In this context, Zelcer et al. introduced a mixture of acetylacetonate and acetic acid to Zr- and Ti-isopropoxide aqueous solutions, controlling the alkoxide reactivity (Zelcer et al. 2020). The use of organic acids is an interesting approach as they catalyze the sol-gel hydrolysis, are usually less corrosive than inorganic acids and they also can form complexes with metalloid hydrolyzed species tuning the kinetic parameters.

Templates

Non-ionic surfactants are among the most used templates for the mesoporous oxide synthesis with characteristic pore sizes in the 2–10 nm range. All these molecules have been extensively reviewed and mainly based on non-ionic surfactants of the Brij® (Debecker et al. 2012, Maksasithorn et al. 2015, Lehr et al. 2014, Debecker et al. 2014) or Pluronics® families (Kim et al. 2019, Suzuki et al. 2010, Suzuki and Yamauchi 2010, Godard et al. 2017, Jia et al. 2017, Pal et al. 2016, Kuai et al. 2014, Wang et al. 2017, Kuai et al. 2018, Li et al. 2014, Smeets et al. 2020) and, on the other hand, ionic surfactants as SDS or CTAB (Zhang et al. 2011, Ide et al. 2015, Sun et al. 2019, Guo et al. 2016, Fiorilli et al. 2016). However, it is possible to tailor bigger pores adding a co-template to the formulation to be aerosolized. Recently, Wu et al. demonstrated that a copolymer of ethyl acrylate and methyl methacrylate used as a co-template with a non-ionic tri-block copolymer (Pluronics F127®) increased pore sizes in SBA-15 SiO_2 particles with ordered mesopores of 8.3–10.0 nm and hierarchical large-meso/macropores of 20–100 nm. Moreover, the polymer Eudragit RS 30 D which is used as a pharmaceutical excipient for sustained-release formulations, is readily available as a 30% aqueous dispersion (Evonik Industries AG (Germany)) (Wu et al. 2019). In this case, the authors stated that the role of the RS polymer colloid is three-fold: they modulated the hydrophilic-hydrophobic balance of F127 micelles swelling the pores of the micellar template, and introduced another level of templating as they were distributed throughout the SiO_2 micro particles, and last, as part of a phase separation they introduced 'bumps' and 'pits', that after calcination enable a more open surface structure. Hierarchical macro-meso templating is highly beneficial to enable pore accessibility, a common issue for mesoporous materials (Yuan and Su 2006). As a result of this strategy, they observed faster adsorption rates with no pore blockage and larger adsorption capacity (Wu et al. 2019).

Polymeric beads are very attractive templates for structuring oxide materials as hard templates; several characteristics can be summed up: the availability from high retail manufacturers, easy synthetic schemes and surface functionalization if small scale laboratory provision is desired (Gu et al. 2007) and, most important, orthogonal chemistry for template removal (organic solvent etching or easy calcination) (Grosso et al. 2003). The composition of these sacrificial templates is mostly based on polystyrene (Smeets et al. 2019, Iskandar et al. 2007) or PMMA (Faustini et al. 2019). In this context, the template can be used to manufacture hollow particles (Smeets et al. 2019), as a 'carrier' of metallic nanoparticles (Jin et al. 2013) or simply to generate larger pores in oxide frameworks (Grosso et al. 2003, Faustini et al. 2019).

A typical concern in the manufacture of mesoporous materials is the cost of the template which is ultimately removed. Natural sources as biotemplates are of great interest in an economic context. Chitin, an acetylglucosamine polymer that forms nanorods and a typical waste from shrimp fisheries, has the ability to self-assemble during droplet drying processes and template the condensation of the Si/Ti oligomers (Sachse et al. 2012). Other researchers proposed a simple and good approach based on the etching of cubic NaCl crystals that form during drying and evaporation after NaCl solutions aerosolization (Zhang et al. 2011). Another possibility is to completely avoid the use of a template: structuring and packing of spray-dried Ludox® particles in high melting points or eutectic salt mixtures that allow to obtain macro-meso colloidal particles (Peterson et al. 2010).

Applications

The biggest niche of application for aerosol-produced colloids is in the field of catalysis. The high-volume provisions of these materials for industrial uses impose the most important challenge to the small-bench scientist: the scaling-up of synthetical procedures. Laboratory-scale aerosol technologies help in this task while allowing to analyze the very basics of the chemistry, physics and engineering involved in the process of mesoporous oxide manufacture.

Mesoporous particles enable bigger pores compared to zeolites, an interesting feature if larger molecules are to be chemically transformed catalytically. In this context, the oil and petrol industries fuel this research: titanosilicate molecular sieves and hollow porous TiO_2-SiO_2 as catalysts for cyclohexene epoxidation (Guo et al. 2016, Smeets et al. 2019), supported RuO_2 nanoparticles on TiO_2 for CO_2 methanation (Kim et al. 2019), methanol synthesis from syngas (H_2, CO, CO_2) (Lehr et al. 2014), MoO_3-SiO_2-Al_2O_3 catalysts for high olefin metathesis (Debecker et al. 2012), W-doped Al_2O_3 for the isomerization and metathesis of a 2-butene into propylene (Bukhovko et al. 2020), sulphoxidation of ethyl-phenyl sulphide and dibenzothiophene on Ti-Si (Sachse et al. 2012) or the synthesis of alkyl lactates ('green solvents') from Sn-doped SiO_2 frameworks that oxidize glycerol into dihydroxyacetone (Godard et al. 2017). In the case of Au catalysts embedded in different mesoporous oxide matrices, the $NaBH_4$ reduction of 4-nitrophenol is frequently used as a model reaction (Jin et al. 2012, Kan et al. 2015). Inexpensive $NaAlO_2$ microspheres catalysts have a promising future for the base-catalyzed synthesis of carbonate esters from dimethyl carbonate and a range of alcohols (Ramesh et al. 2018, Ramesh and Debecker 2017, Rittiron et al. 2019) Zhang et al. tested for Ru^{2+}/Pt^{2+}/Rh^+ doped SiO_2 frameworks a series of typical organic reactions: water-medium Barbier reaction, Sonogashira reaction, terminal alkynes acylation, Suzuki reaction, isomerizations, and Miyaura-Michael reaction; reusability and leaching of catalytical sites remain as central issues (Zhang et al. 2011).

High yield aerosol production of photocatalysts mainly benefit from mesoporous particles' increased surface areas. TiO_2 photocatalysis (Oveisi et al. 2010, Araujo et al. 2010b, Iskandar et al. 2007, Li et al. 2014), hybrid $ZnO/ZnAl_2O_4$ microspheres derived from LDH spray-dried precursors (Huo et al. 2013) and Au/CeO_2 photocatalysts made using a household ultrasonic nebulizer (Jia et al. 2017) are a few examples. In the same line, high throughput production of mesoporous oxide materials catalytical materials (MnOx, Fe/Ni oxides, IrO_2, etc.) for electrochemical applications is also a highly active area: oxygen electrocatalysis (Wang et al. 2016, Wang et al. 2017), electrodes for Zn-air batteries (Kuai et al. 2018), proton exchange membrane electrolysers (Faustini et al. 2019) (Fig. 2.8A) and supercapacitors (Sun et al. 2019).

There is a growing interest in generating hybrid heterogeneous catalysts combining enzymes and catalytic particles. Smeets et al. designed a combined system where glucose oxidase immobilized on hollow mesoporous assembled colloids TS-1 zeolite produced *in situ* H_2O_2 from glucose and O_2, that was later utilized by the zeolite for catalyzing the epoxidation of allylic alcohol toward glycidol (Smeets et al. 2020). Immobilization of biomolecules will require bigger pores to avoid crowding, this is where typical template surfactant dimensions have to be increased and are required for the strategies mentioned previously (Wu et al. 2019) (Fig. 2.8B).

Mesoporous particles permit in packing a higher density of functional chemical groups per gram of material, a very attractive feature. Generally, adsorption was aimed towards small ions: recoverable SiO_2 frameworks with Fe_3O_4 nanoparticles modified with thiol groups for heavy metal adsorption (Suzuki et al. 2010), modified SiO_2 for copper adsorption (Lehr et al. 2014) and CO_2 capture (Bamiduro et al. 2017). In the biochemical analysis field, $Fe_3O_4@$ mesoporous TiO_2 were designed for selective phosphopeptide-enrichment activity (Pal et al. 2016). Moreover, they can be engineered as a pH and temperature stable chromatographic packing material (Ide et al. 2015). This calls for further studies that involve physicochemical studies of dissolution and stability of these matrices under operational conditions.

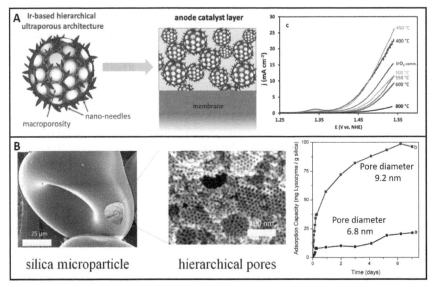

Figure 2.8: A, Macroporous particles with nanoneedles of iridium oxide for proton exchange membrane water electrolyzers, cyclic voltammetry of samples obtained after heat treatment in the air at 400, 450, 500, 550, 600, and 800°C and commercial IrO_2 nanoparticles. Reproduced with permission from Faustini et al. 2019. Adv. Energy Mater. 9(4): 1802136. Copyright 2018, John Wiley and Sons. **B**, hierarchical and ordered mesoporous silica microparticles with enhanced pore accessibility for adsorption of lysozyme. Reproduced with permission from Zhangxiong Wu et al. 2019. Journal of Colloid and Interface Science 556: 529–540. Copyright 2019, Elsevier.

Perspectives

Inspite of the present nanotechnological achievements, one of the challenges in the materials science arena is the upscaling of synthetic techniques for high throughput processes in the obtention of novel materials. Aerosol-based methods have enabled in providing a solution whenever increased yields, in the several grams to kilogram range, are required. It can be envisioned that in the coming years aerosol generated sol-gel materials will certainly help to close the gap between a scientist working on a small-scale laboratory and the industrial demand for the high throughput yield of novel nanomaterials. The mild chemical procedures employed and the gentle operation parameters conditions enable to dream of a myriad of compounds and building blocks that can be integrated into highly functional colloidal objects. While their apparent success and seductive promises, issues such as green chemistry or sustainability connected to this strategy will certainly unfold in the coming years for definite adoption from the industry. Nonetheless, the possibility to mass-produce new and advanced catalyzers and adsorbents taking part in important economical production cycles will outperform the apparent initial drawbacks leading to highly advanced chemical processes.

Acknowledgements

M. V. L., A. V. B. and A. W. are permanent research fellows of CONICET. We thank C.F.K. and N.K. for so many years of continuous moral inspiration.

References

Adelhelm, P., Y. S. Hu, L. Chuenchom, M. Antonietti, B. M. Smarsly and J. Maier. 2007. Generation of hierarchical meso- and macroporous carbon from mesophase pitch by spinodal decomposition using polymer templates. Adv. Mater. 19(22): 4012–4017.

Amatani, T., K. Nakanishi, K. Hirao and T. Kodaira. 2005. Monolithic periodic mesoporous silica with well-defined macropores. Chem. Mater. 17(8): 2114–2119.

Anandharamakrishnan, C. 2014. Drying Techniques for Nanoencapsulation. In Techniques for Nanoencapsulation of Food Ingredients. New York, NY: Springer New York.

Anandharamakrishnan, C. and S. Padma Ishwarya. 2015. Introduction to spray drying. In Spray Drying Techniques for Food Ingredient Encapsulation. John Wiley & Sons, Ltd.

Angelomé, P. C. and G. J. D. A. A. Soler-Illia. 2005. Organically modified transition-metal oxide mesoporous thin films and xerogels. Chem. Mater. 17(2): 322–331.

Angelomé, P. C. and M. C. Fuertes. 2018. Metal nanoparticle-mesoporous oxide nanocomposite thin films. Handbook of Sol-Gel Science and Technology: Processing, Characterization and Applications 2507–2533.

Araujo, P. Z., V. Luca, P. B. Bozzano, H. L. Bianchi, G. J. A. A. Soler Illia and M. A. Blesa. 2010a. Aerosol-assisted production of mesoporous titania microspheres with enhanced photocatalytic activity: The basis of an improved process. ACS Appl. Mater. Inter. 2(6): 1663–1673.

Araujo, P. Z., V. Luca, P. B. Bozzano, H. L. Bianchi, G. J. A. A. Soler Illia and M. A. Blesa. 2010b. Aerosol-assisted production of mesoporous titania microspheres with enhanced photocatalytic activity: the basis of an improved process. ACS Appl. Mater. Inter. 2(6): 1663–73.

Arcos, D. and M. Vallet-Regí. 2010. Sol-gel silica-based biomaterials and bone tissue regeneration. Acta Biomater. 6(8): 2874–2888.

Arpagaus, C., A. Collenberg and D. Rütti. 2018a. Laboratory spray drying of materials for batteries, lasers, and bioceramics. Dry. Technol. 37(4): 426–434.

Arpagaus, C., A. Collenberg, D. Rutti, E. Assadpour and S. M. Jafari. 2018b. Nano spray drying for encapsulation of pharmaceuticals. Int. J. Pharm. 546(1-2): 194–214.

Attard, G. S., P. N. Bartlett, N. R. B. Coleman, J. M. Elliott, J. R. Owen and J. H. Wang. 1997. Mesoporous platinum films from lyotropic liquid crystalline phases. Science 278(5339): 838–840.

Avci-Camur, C., J. Troyano, J. Pérez-Carvajal, A. Legrand, D. Farrusseng, I. Imaz et al. 2018. Aqueous production of spherical Zr-MOF beads: Via continuous-flow spray-drying. Green Chem. 20(4): 873–878.

Bamiduro, F., G. Ji, A. P. Brown, V. A. Dupont, M. Zhao and S. J. Milne. 2017. Spray-dried sodium zirconate: a rapid absorption powder for CO_2 capture with enhanced cyclic stability. ChemSusChem. 10(9): 2059–2067.

Beck, J. S., J. C. Vartuli, W. J. Roth, M. E. Leonowicz, C. T. Kresge, K. D. Schmitt et al. 1992. A new family of mesoporous molecular sieves prepared with liquid crystal templates. J. Am. Chem. Soc. 114(27): 10834–10843.

Bergna, H. E. and W. O. Roberts. 2005. Colloidal Silica: Fundamentals and Applications. CRC Press.

Bhattacharya, A. K., A. Hartridge, K. K. Mallick, C. R. Werrett and J. L. Woodhead. 1996. Low-temperature decomposition of hydrated transition metal chlorides on hydrous gel substrates. J. Mater. Sci. 31(17): 4479–4482.

Bijsterbosch, B. H. 1978. Preparation and characterization of detergent-free monodisperse polystyrene latices of small particle size. Colloid Polym. Sci. 256(4): 343–349.

Blanford, C. F., H. Yan, R. C. Schroden, M. Al-Daous and A. Stein. 2001. Gems of chemistry and physics: Macroporous metal oxides with 3D order. Adv. Mater. 13(6): 401–407.

Bogush, G. H. and C. F. Zukoski Iv. 1991. Studies of the kinetics of the precipitation of uniform silica particles through the hydrolysis and condensation of silicon alkoxides. J. Colloid Interf. Sci. 142(1): 1–18.

Bordoni, A. V., M. V. Lombardo and A. Wolosiuk. 2016. Photochemical radical thiol-ene click-based methodologies for silica and transition metal oxides materials chemical modification: A mini-review. RSC Adv. 6(81): 77410–77426.

Braun, P. V., P. Osenar and S. I. Stupp. 1996. Semiconducting superlattices templated by molecular assemblies. Nature 380(6572): 325–328.

Braun, P. V., P. Osenar, V. Tohver, S. B. Kennedy and S. I. Stupp. 1999. Nanostructure templating in inorganic solids with organic lyotropic liquid crystals. J. Am. Chem. Soc. 121(32): 7302–7309.

Brinker, C. J. and G. W. Scherer. 1990. Chapter 1 – Introduction. *In*: C. J. Brinker and G. W. Scherer [eds.]. Sol-Gel Science. San Diego: Academic Press.

Bukhovko, M. P., B. S. Hanna, T. J. Kucharski and M. L. Ostraat. 2020. An aerosol reactor for the controlled synthesis of heterogeneous catalyst particles. AIChE J. 66(6).

Burkett, S. L. and M. E. Davis. 1994. Mechanism of structure direction in the synthesis of Si-ZSM-5: An investigation by intermolecular 1H-29Si CP MAS NMR. J. Phys. Chem. 98(17): 4647–4653.

Bürki, K., I. Jeon, C. Arpagaus and G. Betz. 2011. New insights into respirable protein powder preparation using a nano spray dryer. Int. J. Pharm. 408(1): 248–256.

Cahn, J. W. 1961. On spinodal decomposition. Acta Metall. Mater. 9(9): 795–801.

Cahn, J. W. 1962. On spinodal decomposition in cubic crystals. Acta Metall. Mater. 10(3): 179–183.

Cattoën, X., A. Noureddine, J. Croissant, N. Moitra, K. Bürglová, J. Hodačová et al. 2014. Click approaches in sol-gel chemistry. J. Sol-Gel Sci. Techn. 70(2): 245–253.

Chai, G. S., I. S. Shin and J. S. Yu. 2004. Synthesis of ordered, uniform, macroporous carbons with mesoporous walls templated by aggregates of polystyrene spheres and silica particles for use as catalyst supports in direct methanol fuel cells. Adv. Mater. 16(22): 2057–2061.

Che, M. and J. C. Vedrine. 2012. Characterization of Solid Materials and Heterogeneous Catalysts: From Structure to Surface Reactivity. Wiley.

Checcucci, S., T. Bottein, J. B. Claude, T. Wood, M. Putero, L. Favre et al. 2018. Titania-based spherical Mie resonators elaborated by high-throughput aerosol spray: single object investigation. Adv. Funct. Mater. 28(31).

Ciriminna, R., A. Fidalgo, V. Pandarus, F. Béland, L. M. Ilharco and M. Pagliaro. 2013. The sol-gel route to advanced silica-based materials and recent applications. Chem. Rev. 113(8): 6592–6620.

Colbeau-Justin, F., C. Boissière, A. Chaumonnot, A. Bonduelle and C. Sanchez. 2014. Aerosol route to highly efficient (Co)Mo/SiO₂ mesoporous catalysts. Adv. Funct. Mater. 24(2): 233–239.

Colilla, M., M. Manzano, I. Izquierdo-Barba, M. Vallet-Reg, C. Boissiére and C. Sanchez. 2010. Advanced drug delivery vectors with tailored surface properties made of mesoporous binary oxides submicronic spheres. Chem. Mater. 22(5): 1821–1830.

Cong, H. and W. Cao. 2004. Macroporous Au materials prepared from colloidal crystals as templates. J. Colloid Interf. Sci. 278(2): 423–427.

Cundy, C. S. and P. A. Cox. 2005. The hydrothermal synthesis of zeolites: Precursors, intermediates and reaction mechanism. Micropor. Mesopor. Mat. 82(1-2): 1–78.

Daniel Santos, A. C. M., Vitor Sencadas, José Domingos Santos, Maria H. Fernandes and Pedro S. Gomes. 2018. Spray drying: An overview. In Biomaterials Physics and Chemistry. New Edition, edited by R. Pignatello. London: IntechOpen.

Debecker, D. P., M. Stoyanova, F. Colbeau-Justin, U. Rodemerck, C. Boissiãre, E. M. Gaigneaux et al. 2012. One-pot aerosol route to MoO₃-SiO₂-Al₂O₃ catalysts with ordered super microporosity and high olefin metathesis activity. Angew. Chem. Int. Edit. 51(9): 2129–2131.

Debecker, D. P., M. Stoyanova, U. Rodemerck, F. Colbeau-Justinc, C. Boissère, A. Chaumonnot et al. 2014. Aerosol route to nanostructured WO₃-SiO₂-Al₂O₃ metathesis catalysts: Toward higher propene yield. Appl. Catal. A—Gen. 470: 458–466.

Debecker, D. P., S. Le Bras, C. Boissière, A. Chaumonnot and C. Sanchez. 2018. Aerosol processing: a wind of innovation in the field of advanced heterogeneous catalysts. Chem. Soc. Rev. 47(11): 4112–4155.

Dorin, R. M., H. Sai and U. Wiesner. 2014. Hierarchically porous materials from block copolymers. Chem. Mater. 26(1): 339–347.

Duraisamy, V., K. Selvakumar, R. Krishnan and S. M. S. Kumar. 2019. Investigation on template etching process of SBA-15 derived ordered mesoporous carbon on electrocatalytic oxygen reduction reaction. Chem. Select 4(8): 2463–2474.

Enke, D., F. Janowski and W. Schwieger. 2003. Porous glasses in the 21st century—a short review. Micropor. Mesopor. Mat. 60(1-3): 19–30.

Faustini, M., M. Giraud, D. Jones, J. Rozière, M. Dupont, T. R. Porter et al. 2019. Hierarchically structured ultraporous iridium-based materials: A novel catalyst architecture for proton exchange membrane water electrolyzers. Adv. Energy Mater. 9(4): 1802136.

Ferdi Schüth. 2001. Non-siliceous mesostructured and mesoporous materials. Chem. Mater. 13(10): 3184–3195.

Fiorilli, S., V. Cauda, C. Pontiroli, C. Vitale-Brovarone and B. Onida. 2016. Aerosol-assisted synthesis of mesoporous aluminosilicate microspheres: The effect of the aluminum precursor. New J. Chem. 40(5): 4420–4427.

Foss Jr, C. A., G. L. Hornyak, J. A. Stockert and C. R. Martin. 1992. Optical properties of composite membranes containing arrays of nanoscopic gold cylinders. J. Phys. Chem. 96(19): 7497–7499.

Gavrilović, T. V., D. J. Jovanović and M. D. Dramićanin. 2018. Chapter 2 – Synthesis of multifunctional inorganic materials: From micrometer to nanometer dimensions. *In*: B. A. Bhanvase, V. B. Pawade, S. J. Dhoble, S. H. Sonawane and M. Ashokkumar [eds.]. Nanomaterials for Green Energy. Elsevier.

Gharsallaoui, A., G. Roudaut, O. Chambin, A. Voilley and R. Saurel. 2007. Applications of spray-drying in microencapsulation of food ingredients: An overview. Food Res. Int. 40(9): 1107–1121.

Godard, N., A. Vivian, L. Fusaro, L. Cannavicci, C. Aprile and D. P. Debecker. 2017. High-yield synthesis of ethyl lactate with mesoporous tin silicate catalysts prepared by an aerosol-assisted sol–gel process. ChemCatChem. 9(12): 2211–2218.

Gómez-Romero, P. and C. Sanchez. 2006. Functional Hybrid Materials. John Wiley & Sons.

Gough, D. V., A. Wolosiuk and P. V. Braun. 2009. Mesoporous ZnS nanorattles: Programmed size selected access to encapsulated enzymes. Nano Lett. 9(5): 1994–1998.

Grosso, D., G. J. D. A. A. Soler Illia, E. L. Crepaldi, B. Charleux and C. Sanchez. 2003. Nanocrystalline transition-metal oxide spheres with controlled multi-scale porosity. Adv. Funct. Mater. 13(1): 37–42.

Grosso, D., F. Cagnol, G. J. d. A. A. Soler-Illia, E. L. Crepaldi, H. Amenitsch, A. Brunet-Bruneau et al. 2004. Fundamentals of mesostructuring through evaporation-induced self-assembly. Adv. Funct. Mater. 14(4): 309–322.

Gu, Z. Z., H. Chen, S. Zhang, L. Sun, Z. Xie and Y. Ge. 2007. Rapid synthesis of monodisperse polymer spheres for self-assembled photonic crystals. Colloid Surf. A Physicochem. Eng. Asp. 302(1-3): 312–319.

Guo, Z., G. Xiong, L. Liu, J. Yin, R. Zhao and S. Yu. 2015. Facile and green aerosol-assisted synthesis of zeolites. RSC Adv. 5(87): 71433–71436.

Guo, Z., G. Xiong, L. Liu, P. Li, L. Hao, Y. Cao et al. 2016. Aerosol-assisted synthesis of hierarchical porous titanosilicate molecular sieve as catalysts for cyclohexene epoxidation. J. Porous Mater. 23(2): 407–413.

Huo, R., Y. Kuang, Z. Zhao, F. Zhang and S. Xu. 2013. Enhanced photocatalytic performances of hierarchical ZnO/ZnAl$_2$O$_4$ microsphere derived from layered double hydroxide precursor spray-dried microsphere. J. Colloid Interf. Sci. 407: 17–21.

Ide, M., E. De Canck, I. Van Driessche, F. Lynen and P. Van Der Voort. 2015. Developing a new and versatile ordered mesoporous organosilica as a pH and temperature stable chromatographic packing material. RSC Adv. 5(8): 5546–5552.

Inayat, A., B. Reinhardt, J. Herwig, C. Küster, H. Uhlig, S. Krenkel et al. 2016. Recent advances in the synthesis of hierarchically porous silica materials on the basis of porous glasses. New J. Chem. 40(5): 4095–4114.

Ishwarya, C. A. S. P. 2015. Spray drying for nanoencapsulation of food components. In Spray Drying Techniques for Food Ingredient Encapsulation. John Wiley & Sons, Ltd.

Iskandar, F., A. B. D. Nandiyanto, K. M. Yun, C. J. Hogan Jr., K. Okuyama and P. Biswas. 2007. Enhanced photocatalytic performance of brookite TiO$_2$ macroporous particles prepared by spray drying with colloidal templating. Adv. Mater. 19(10): 1408–1412.

Israelachvili, J. N. 2015. Intermolecular and Surface Forces. Elsevier Science.

Jafari, S. M., E. Assadpoor, Y. He and B. Bhandari. 2008. Encapsulation efficiency of food flavours and oils during spray drying. Dry. Technol. 26(7): 816–835.

Jayanthi, G. V., S. C. Zhang and G. L. Messing. 1993. Modeling of solid particle formation during solution aerosol thermolysis: The evaporation stage. Aerosol Science and Technology 19(4): 478–490.

Jia, H., X. M. Zhu, R. Jiang and J. Wang. 2017. Aerosol-sprayed gold/ceria photocatalyst with superior plasmonic hot electron-enabled visible-light activity. ACS Appl. Mater. Interf. 9(3): 2560–2571.

Jin, Z., M. Xiao, Z. Bao, P. Wang and J. Wang. 2012. A general approach to mesoporous metal oxide microspheres loaded with noble metal nanoparticles. Angew. Chem. Int. Edit. 51(26): 6406–6410.

Jin, Z., F. Wang, J. Wang, J. C. Yu and J. Wang. 2013. Metal nanocrystal-embedded hollow mesoporous TiO_2 and ZrO_2 microspheres prepared with polystyrene nanospheres as carriers and templates. Adv. Funct. Mater. 23(17): 2137–2144.

Jo, C., J. Hwang, W. G. Lim, J. Lim, K. Hur and J. Lee. 2018. Multiscale phase separations for hierarchically ordered macro/mesostructured metal oxides. Adv. Mater. 30(6).

Kan, E., L. Kuai, W. Wang and B. Geng. 2015. Delivery of highly active noble-metal nanoparticles into microspherical supports by an aerosol-spray method. Chem-Eur J. 21(38): 13291–13296.

Kim, A., C. Sanchez, B. Haye, C. Boissière, C. Sassoye and D. P. Debecker. 2019. Mesoporous TiO_2 support materials for Ru-based CO_2 methanation catalysts. ACS Appl. Nano Mater. 2(5): 3220–3230.

Kim, J. H., K. Y. Jung, K. Y. Park and S. B. Cho. 2010. Characterization of mesoporous alumina particles prepared by spray pyrolysis of $Al(NO_3)_2 \cdot 9H_2O$ precursor: Effect of CTAB and urea. Micropor. Mesopor. Mat. 128(1-3): 85–90.

Kodas, T. T. and M. J. Hampden-Smith. 1999. Aerosol Processing of Materials. Wiley-VCH.

Kresge, C. T., J. C. Vartuli, W. J. Roth and M. E. Leonowicz. 2004. The discovery of ExxonMobil's M41S family of mesoporous molecular sieves. Stud. Surf. Sci. Catal. 148: 53–72.

Kruk, M., M. Jaroniec, C. H. Ko and R. Ryoo. 2000. Characterization of the porous structure of SBA-15. Chem. Mater. 12(7): 1961–1968.

Kuai, L., J. Geng, C. Chen, E. Kan, Y. Liu, Q. Wang et al. 2014. A reliable aerosol-spray-assisted approach to produce and optimize amorphous metal oxide catalysts for electrochemical water splitting. Angew. Chem. Int. Edit. 53(29): 7547–7551.

Kuai, L., E. Kan, W. Cao, M. Huttula, S. Ollikkala, T. Ahopelto et al. 2018. Mesoporous LaMnO3+δ perovskite from spray–pyrolysis with superior performance for oxygen reduction reaction and Zn–air battery. Nano Energy 43: 81–90.

Kubo, S., R. J. White, K. Tauer and M. M. Titirici. 2013. Flexible coral-like carbon nanoarchitectures via a dual block copolymer-latex templating approach. Chem. Mater. 25(23): 4781–4790.

Kumar, R., E. Tijerino, A. Saha and S. Basu. 2010. Structural morphology of acoustically levitated and heated nanosilica droplet. Appl. Phys. Lett. 97(12).

Kyotani, T., T. Nagai, S. Inoue and A. Tomita. 1997. Formation of new type of porous carbon by carbonization in zeolite nanochannels. Chem. Mater. 9(2): 609–615.

Larsen, G., E. Lotero and M. Marquez. 2000. Amine dendrimers as templates for amorphous silicas. J. Phys. Chem. B 104(20): 4840–4843.

Lebeau, B. and P. Innocenzi. 2011. Hybrid materials for optics and photonics. Chem. Soc. Rev. 40(2): 886–906.

Lee, J., J. Kim and T. Hyeon. 2006. Recent progress in the synthesis of porous carbon materials. Adv. Mater. 18(16): 2073–2094.

Lee, W. and S. -J. Park. 2014. Porous anodic aluminum oxide: anodization and templated synthesis of functional nanostructures. Chem. Rev. 114(15): 7487–7556.

Lehr, D., D. Großmann, W. Grünert and S. Polarz. 2014. "Dirty nanostructures": Aerosol-assisted synthesis of temperature stable mesoporous metal oxide semiconductor spheres comprising hierarchically assembled zinc oxide nanocrystals controlled via impurities. Nanoscale 6(3): 1698–1706.

Leng, J., Z. Wang, J. Wang, H. H. Wu, G. Yan, X. Li et al. 2019. Advances in nanostructures fabricated via spray pyrolysis and their applications in energy storage and conversion. Chem. Soc. Rev. 48(11): 3015–3072.

Li, C., T. Ming, J. Wang, J. Wang, J. C. Yu and S. H. Yu. 2014. Ultrasonic aerosol spray-assisted preparation of TiO_2/In_2O_3 composite for visible-light-driven photocatalysis. J. Catal. 310: 84–90.

Li, Z., J. C. Barnes, A. Bosoy, J. F. Stoddart and J. I. Zink. 2012. Mesoporous silica nanoparticles in biomedical applications. Chem. Soc. Rev. 41(7): 2590–2605.

Liu, W., X. D. Chen and C. Selomulya. 2015. On the spray drying of uniform functional microparticles. Particuology 22: 1–12.

Livage, J. 1987. Sol-gel processing of metal oxides. C3 – Chem. Scripta 28(1): 9–13.

Livage, J., M. Henry and C. Sanchez. 1988. Sol-gel chemistry of transition metal oxides. Prog. Solid State Ch. 18(4): 259–341.

López-Salas, N., D. Carriazo, M. C. Gutiérrez, M. L. Ferrer, C. O. Ania, F. Rubio et al. 2016. Tailoring the textural properties of hierarchical porous carbons using deep eutectic solvents. J. Mater. Chem. A 4(23): 9146–9159.

Maksasithorn, S., P. Praserthdam, K. Suriye and D. P. Debecker. 2015. Preparation of super-microporous WO_3-SiO_2 olefin metathesis catalysts by the aerosol-assisted sol-gel process. Micropor. Mesopor. Mat. 213: 125–133.

Martin, C. R. 1994. Nanomaterials: A membrane-based synthetic approach. Science 266(5193): 1961–1966.

Maruoka, H. and T. Kimura. 2019. An effective strategy to obtain highly porous alumina powders having robust and designable extra-large pores. Bull. Chem. Soc. Jpn. 92(11): 1859–1866.

Masters, K. 1991. Spray Drying Handbook. Longman Scientific & Technical.

Meng, Q., J. Liu, G. Xiong, X. Li, L. Liu and H. Guo. 2019. The synthesis of hierarchical Sn-Beta zeolite via aerosol-assisted hydrothermal method combined with a mild base treatment. Micropor. Mesopor. Mat. 287: 85–92.

Mezhericher, M., A. Levy and I. Borde. 2010. Spray drying modelling based on advanced droplet drying kinetics. Chem. Eng. Process. 49(11): 1205–1213.

Mezhericher, M. 2011. Theoretical Modelling of Spray Drying Processes: Volume 1. Drying Kinetics, Two and Three Dimensional CFD Modelling: LAP LAMBERT Academic Publishing.

Mondragón, R., J. E. Juliá, L. Hernández and J. C. Jarque. 2013. Influence of particle size on the drying kinetics of single droplets containing mixtures of nanoparticles and microparticles: modeling and pilot-scale validation. Dry. Technol. 31(7): 759–768.

Murugesan, R. and V. Orsat. 2012. Spray drying for the production of nutraceutical ingredients—A review. Food Bioprocess Tech. 5(1): 3–14.

Nanoporous Materials: Science and Engineering. 2004. Vol. 4. Covent Garden, London, United Kingdom Imperial College Press.

O'Sullivan, J. J., E. -A. Norwood, J. A. O'Mahony and A. L. Kelly. 2019. Atomisation technologies used in spray drying in the dairy industry: A review. J. Food Eng. 243: 57–69.

Okuyama, K., M. Abdullah, I. Wuled Lenggoro and F. Iskandar. 2006. Preparation of functional nanostructured particles by spray drying. Adv. Powder Technol. 17(6): 587–611.

Oveisi, H., N. Suzuki, A. Beitollahi and Y. Yamauchi. 2010. Aerosol-assisted fabrication of mesoporous titania spheres with crystallized anatase structures and investigation of their photocatalitic properties. J. Sol-Gel Sci. Techn. 56(2): 212–218.

Pagliaro, M. and R. S. o. Chemistry. 2009. Silica-based Materials for Advanced Chemical Applications. RSC Pub.

Pal, M., L. Wan, Y. Zhu, Y. Liu, Y. Liu, W. Gao et al. 2016. Scalable synthesis of mesoporous titania microspheres via spray-drying method. J. Colloid Interf. Sci. 479: 150–159.

Passos, A. R., S. H. Pulcinelli, V. Briois and C. V. Santilli. 2016. High surface area hierarchical porous Al_2O_3 prepared by the integration of sol-gel transition and phase separation. RSC Adv. 6(62): 57217–57226.

Penner, R. M. and C. R. Martin. 1987. Preparation and electrochemical characterization of ultramicroelectrode ensembles. Anal. Chem. 59(21): 2625–2630.

Peterson, A. K., D. G. Morgan and S. E. Skrabalak. 2010. Aerosol synthesis of porous particles using simple salts as a pore template. Langmuir 26(11): 8804–8809.

Pitchumani, R., J. J. Heiszwolf, A. Schmidt-Ott and M. O. Coppens. 2009. Continuous synthesis by spray drying of highly stable mesoporous silica and silica–alumina catalysts using industrial raw materials. Micropor. Mesopor. Mat. 120(1-2): 39–46.

Prevot, V., C. Szczepaniak and M. Jaber. 2011. Aerosol-assisted self-assembly of hybrid layered double hydroxide particles into spherical architectures. J. Colloid Interf. Sci. 356(2): 566–572.

Ramesh, S. and D. P. Debecker. 2017. Room temperature synthesis of glycerol carbonate catalyzed by spray dried sodium aluminate microspheres. Catal. Commun. 97: 102–105.

Ramesh, S., K. Indukuri, O. Riant and D. P. Debecker. 2018. Synthesis of carbonate esters by carboxymethylation using $NaAlO_2$ as a highly active heterogeneous catalyst. Org. Process Res. Dev. 22(12): 1846–1851.

Ranz, W. and W. Marshall. 1952a. Evaporation from drops 1. Chem. Eng. Progress 48: 141–146.

Ranz, W. and W. Marshall. 1952b. Evaporation from drops 2. Chem. Eng. Progress 48: 173–180.

Rittiron, P., C. Niamnuy, W. Donphai, M. Chareonpanich and A. Seubsai. 2019. Production of glycerol carbonate from glycerol over templated-sodium-aluminate catalysts prepared using a spray-drying method. ACS Omega 4(5): 9001–9009.

Sachse, A., V. Hulea, K. L. Kostov, N. Marcotte, M. Y. Boltoeva, E. Belamie et al. 2012. Efficient mesoporous silica–titania catalysts from colloidal self-assembly. Chem. Commun. 48(86): 10648–10650.

Sansierra, M. C., J. Morrone, F. Cornacchiulo, M. C. Fuertes and P. C. Angelomé. 2019. Detection of organic vapors using Tamm mode based devices built from mesoporous oxide thin films. ChemNanoMat. 5(10): 1289–1295.

Sheldon, R. A. and S. Van Pelt. 2013. Enzyme immobilisation in biocatalysis: Why, what and how. Chem. Soc. Rev. 42(15): 6223–6235.

Shenoi-Perdoor, S., X. Cattoën, Y. Bretonnière, G. Eucat, C. Andraud, B. Gennaro et al. 2018. Red-emitting fluorescent organic@silicate core-shell nanoparticles for bio-imaging. New J. Chem. 42(18): 15353–15360.

Shimojima, A., Z. Liu, T. Ohsuna, O. Terasaki and K. Kuroda. 2005. Self-assembly of designed oligomeric siloxanes with alkyl chains into silica-based hybrid mesostructures. J. Am. Chem. Soc. 127(40): 14108–14116.

Smeets, V., W. Baaziz, O. Ersen, E. M. Gaigneaux, C. Boissière, C. Sanchez et al. 2020. Hollow zeolite microspheres as a nest for enzymes: A new route to hybrid heterogeneous catalysts. Chem. Sci. 11(4): 954–961.

Smeets, V., C. Boissière, C. Sanchez, E. M. Gaigneaux, E. Peeters, B. F. Sels et al. 2019. Aerosol route to TiO_2-SiO_2 catalysts with tailored pore architecture and high epoxidation activity. Chem. Mater. 31(5): 1610–1619.

Soler-Illia, G. J. D. A. A., C. Sanchez, B. Lebeau and J. Patarin. 2002. Chemical strategies to design textured materials: From microporous and mesoporous oxides to nanonetworks and hierarchical structures. Chem. Rev. 102(11): 4093–4138.

Son, D., A. Wolosiuk and P. V. Braun. 2009. Double direct templated hollow ZnS microspheres formed on chemically modified silica colloids. Chem. Mater. 21(4): 628–634.

Stöber, W., A. Fink and E. Bohn. 1968. Controlled growth of monodisperse silica spheres in the micron size range. J. Colloid Interf. Sci. 26(1): 62–69.

Strutt, J. W. (Lord Rayleigh) 1871. On the scattering of light by small particles. Philos. Mag. 41: 447–454.

Stupp, S. I. and P. V. Braun. 1997. Molecular manipulation of microstructures: Biomaterials, ceramics, and semiconductors. Science 277(5330): 1242–1248.

Subramanian, G., V. N. Manoharan, J. D. Thorne and D. J. Pine. 1999. Ordered macroporous materials by colloidal assembly: A possible route to photonic bandgap materials. Adv. Mater. 11(15): 1261–1265.

Sun, Y. A., L. T. Chen, S. Y. Hsu, C. C. Hu and D. H. Tsai. 2019. Silver nanoparticles-decorating manganese oxide hybrid nanostructures for supercapacitor applications. Langmuir 35(44): 14203–14212.

Suzuki, N., P. Gupta, H. Sukegawa, K. Inomata, S. Inoue and Y. Yamauchi. 2010. Aerosol-assisted synthesis of thiol-functionalized mesoporous silica spheres with Fe_3O_4 nanoparticles. J. Nanosci. Nanotechno. 10(10): 6612–6617.

Suzuki, N. and Y. Yamauchi. 2010. Large-scale aerosol-assisted synthesis of thiol-functionalized mesoporous organosilica. J. Nanosci. Nanotechno. 10(9): 5759–5766.

Talapin, D. V., J. S. Lee, M. V. Kovalenko and E. V. Shevchenko. 2010. Prospects of colloidal nanocrystals for electronic and optoelectronic applications. Chem. Rev. 110(1): 389–458.

Tyndall, J. 1869. On the blue colour of the sky, the polarization of skylight, and on the polarization of light by cloudy matter generally. Proc. R. Soc. London 17: 223–233.

Visca, M. and E. Matijević. 1979. Preparation of uniform colloidal dispersions by chemical reactions in aerosols. I. Spherical particles of titanium dioxide. J. Colloid Interf. Sci. 68(2): 308–319.

Vogel, N., M. Retsch, C. A. Fustin, A. Del Campo and U. Jonas. 2015. Advances in colloidal assembly: the design of structure and hierarchy in two and three dimensions. Chem. Rev. 115(13): 6265–6311.

Waldron, K., W. D. Wu, Z. Wu, W. Liu, C. Selomulya, D. Zhao et al. 2014. Formation of monodisperse mesoporous silica microparticles via spray-drying. J. Colloid Interface Sci. 418: 225–33.

Walheim, S., E. Schäffer, J. Mlynek and U. Steiner. 1999. Nanophase-separated polymer films as high-performance antireflection coatings. Science 283(5401): 520–522.

Wang, J., L. Kuai, J. Wang, T. Ming, C. Fang, Z. Sun et al. 2015. Aerosol-spray diverse mesoporous metal oxides from metal nitrates. Sci. Rep. 5.

Wang, W., J. Geng, L. Kuai, M. Li and B. Geng. 2016. Porous Mn_2O_3: A low-cost electrocatalyst for oxygen reduction reaction in alkaline media with comparable activity to Pt/C. Chem.-Eur. J. 22(29): 9909–9913.

Wang, W., L. Kuai, W. Cao, M. Huttula, S. Ollikkala, T. Ahopelto et al. 2017. Mass-production of mesoporous $MnCo_2O_4$ spinels with manganese(IV)- and cobalt(II)-rich surfaces for superior bifunctional oxygen electrocatalysis. Angew. Chem. Int. Edit. 56(47): 14977–14981.

Wei, H., Y. Wang, J. Guo, N. Z. Shen, D. Jiang, X. Zhang et al. 2015. Advanced micro/nanocapsules for self-healing smart anticorrosion coatings. J. Mater. Chem. A 3(2): 469–480.

Wendel, S. C. and M. Celik. 2005. Spray Drying and Pharmaceutical Applications. In Handbook of Pharmaceutical Granulation Technology. CRC Press.

Wijnen, P. W. J. G., T. P. M. Beelen, J. W. de Haan, C. P. J. Rummens, L. J. M. van de Ven and R. A. van Santen. 1989. Silica gel dissolution in aqueous alkali metal hydroxides studied by 29SiNMR. J. Non-Cryst. Solids 109(1): 85–94.

Williams, K. R. and R. S. Muller. 1996. Etch rates for micromachining processing. J. Microelectromech. S. 5(4): 256–269.

Wolosiuk, A., O. Armagan and P. V. Braun. 2005. Double direct templating of periodically nanostructured ZnS hollow microspheres. J. Am. Chem. Soc. 127(47): 16356–16357.

Wu, B., A. Kumar and S. Pamarthy. 2010. High aspect ratio silicon etch: A review. J. Appl. Phys. 108(5).

Wu, Z., K. Waldron, X. Zhang, Y. Li, L. Wu, W. D. Wu et al. 2019. Spray-drying water-based assembly of hierarchical and ordered mesoporous silica microparticles with enhanced pore accessibility for efficient bio-adsorption. J. Colloid Interf. Sci. 556: 529–540.

Wulsten, E. and G. Lee. 2008. Surface temperature of acoustically levitated water microdroplets measured using infra-red thermography. Chem. Eng. Sci. 63(22): 5420–5424.

Xu, R., W. Pang, J. Yu, Q. Huo and J. Chen. 2009. Chemistry of Zeolites and Related Porous Materials: Synthesis and Structure. Wiley.

Yang, X. Y., L. H. Chen, Y. Li, J. C. Rooke, C. Sanchez and B. L. Su. 2017. Hierarchically porous materials: Synthesis strategies and structure design. Chem. Soc. Rev. 46(2): 481–558.

Yuan, Z. Y. and B. L. Su. 2006. Insights into hierarchically meso-macroporous structured materials. J. Mater. Chem. 16(7): 663–677.

Yuvaraj, S., F. Y. Lin, T. H. Chang and C. T. Yeh. 2003. Thermal decomposition of metal nitrates in air and hydrogen environments. J. Phys. Chem. B 107(4): 1044–1047.

Zelcer, A., E. A. Franceschini, M. V. Lombardo, A. E. Lanterna and G. J. A. A. Soler-Illia. 2020. A general method to produce mesoporous oxide spherical particles through an aerosol method from aqueous solutions. J. Sol-Gel Sci. Techn. 94(1): 195–204.

Zelcer, A., A. Wolosiuk and G. J. A. A. Soler-Illia. 2009. Carbonaceous submicron sized islands: A surface patterning route to hierarchical macro/mesoporous thin films. J. Mater. Chem. 19(24): 4191–4196.

Zhang, F., C. Kang, Y. Wei and H. Li. 2011. Aerosol-spraying synthesis of periodic mesoporous organometalsilica spheres with chamber cavities as active and reusable catalysts in aqueous organic reactions. Adv. Funct. Mater. 21(16): 3189–3197.

Zhang, Y., Y. Shi, Y. H. Liou, A. M. Sawvel, X. Sun, Y. Cai et al. 2010. High performance separation of aerosol sprayed mesoporous TiO_2 sub-microspheres from aggregates via density gradient centrifugation. J. Mater. Chem. 20(20): 4162–4167.

Zhao, D., Y. Wan and W. Zhou. 2012. Ordered Mesoporous Materials. Wiley.

Zheng, J. Y., J. Y. Zheng, J. B. Pang, K. Y. Qiu and Y. Wei. 2001. Synthesis of mesoporous titanium dioxide materials by using a mixture of organic compounds as a non-surfactant template. J. Mater. Chem. 11(12): 3367–3372.

Ziaee, A., A. B. Albadarin, L. Padrela, T. Femmer, E. O'Reilly and G. Walker. 2019. Spray drying of pharmaceuticals and biopharmaceuticals: Critical parameters and experimental process optimization approaches. Eur. J. Pharm. Sci. 127: 300–318.

CHAPTER 3
Nanostructured Semiconducting Oxide Films

Martín Ignacio Broens,[1,2] *Wilkendry Ramos Cervantes,*[1,2]
Andrés Matias Asenjo Collao,[1,2] *Diego Patricio Oyarzun,*[3] *Manuel Lopez Teijelo*[1]
and *Omar Ezequiel Linarez Perez*[1,2,]*

Introduction

In general, metals are unstable in different aqueous and non-aqueous media, which can actively dissolve them in a usually called *corrosion process*. On the other hand, metals may become 'passive' if they resist corrosion under conditions where the clean metal could react significantly. This behaviour is due to inhibition of active dissolution by spontaneous formation of a passive film, usually of limited ionic and electronic conductivity (Schultze and Lohrengel 2000, Vargas et al. 2019). The term 'passive film' is used in connection with films formed in aqueous solutions, but in electronics and other areas, the definition includes all corrosion-protective films, even if they are not generated by anodization.

For several decades now, the electrochemical formation of passive films on metals has been an aspect of intense study due to their diversity of properties and application in areas such as corrosion protection, passivation, materials science, thin films technology, microelectronics, micromechanics and nanotechnology (Schultze and Lohrengel 2000, Macak et al. 2007a, Grimes and Mor 2009, Zoolfakar et al. 2014, Oprea et al. 2018, Vargas et al. 2019). Among passive films, semiconducting oxides are widely studied and employed for many technological applications due to their (photo) conducting properties. The most obvious reasons that allow the electrical conduction through semiconducting passive films are structural and stoichiometric defects as well as their energy band positions. For instance, the point defects can be missing ions/atomic cores in the solid lattice (vacancies), additional ions/atomic cores located between regular positions in a metastable equilibrium state (interstitials), anions at cations sites and vice versa (antisites), and impurity ions instead of compound ions (substitutional impurities) or in interstitial positions (interstitial impurities) (Schultze and Lohrengel 2000, Oprea et al. 2018, Vargas et al. 2019). As these defects are randomly distributed, they generate spatially localized states with energies within the semiconductor energy bandgap, which can give or accept electrons to or from other states.

---

[1] Universidad Nacional de Córdoba. Facultad de Ciencias Químicas. Departamento de Fisicoquímica. Haya de la Torre esq. Medina Allende, X5000HUA. Córdoba, Argentina.
[2] CONICET, INFIQC. Haya de la Torre esq. Medina Allende, X5000HUA. Córdoba, Argentina.
[3] Laboratorio de Nanotecnología, Recursos Naturales y Sistemas Complejos. Facultad de Ciencias Naturales, Departamento de Química y Biología. Universidad de Atacama, Av. Copayapu 485, 1532297, Copiapó, Chile.
* Corresponding author: olinarez@unc.edu.ar

Semiconducting oxide thin films have been extensively studied in the past, and obtaining auto-organized nanostructures of different metal oxides such as Al_2O_3 (Masuda and Fukuda 1995, Chick and Xu 2004, Chen 2008, Patermarakis 2009, Friedman 2011, Sulka et al. 2011, Norek 2016, Salguero Salas et al. 2020), TiO_2 (Mor et al. 2006, Macak et al. 2007a,b, Grimes and Mor 2009, Liang et al. 2010, Brammer et al. 2011, Roguska et al. 2011, Oyarzún et al. 2011, Chen et al. 2013, Pisarek et al. 2013, Lee et al. 2014, Terracciano et al. 2017, Pourandarjani and Nasirpouri 2018, Broens et al. 2020), Nb_2O_5 (Choi et al. 2007, Kang et al. 2015, Liu et al. 2015), CuO/Cu_2O (Musselman et al. 2010, Allam and Grimes 2011, Tahir and Tougaard 2012, Zoolfakar et al. 2014, Gattinoni and Michaelides 2015, Oyarzún et al. 2017, 2018b, Stepniowski and Misiolek 2018) and WO_3 (Gu et al. 2005, Ou et al. 2012, Syrek et al. 2018, Xu et al. 2020) among many others, has acquired increasing interest due to their potential uses for several technological applications. Especially nanostructured semiconductor oxide films and their derivative composite materials show interesting physicochemical properties and improved performance for luminescent modules, piezoelectric transducers, solar cells, chemical sensors and electrocatalysts. In this way, the understanding of the mechanisms involved in the growth of oxide films on metals plays an important role in the control and design of new nanostructures with desired properties for specific applications.

In this chapter, an overview of the anodic film growth, properties, and applications of nanostructured semiconducting metal oxides will be given. A focus on the anodic behaviour and present applications of titanium, copper and antimony oxides as examples of semiconducting materials will also be made.

Anodic Oxide Film Growth

The growth of oxide films on metals involves the transport of matter and charge in at least three phases, the metal, oxide phase and the electrolyte, as well as through the corresponding interfaces, which gives rise to a diversity of kinetic behaviours according to the properties of each region (Dignam 1961, Schultze and Lohrengel 2000, Linarez Pérez et al. 2008, Vargas et al. 2019). Accordingly, the transport of both oxide ions (O^{2-}) from the film/electrolyte and metal cations (M^{n+}) from metal/film interfaces to the bulk of the oxide takes place, allowing for the growth of an oxide film at both metal/film and film/electrolyte interfaces. The composition and structure of the formed oxide are the factors that determine whether the identity of the species that transport the charge inside the film are metal or oxide ions, ion vacancies, interstitials, electrons or holes. Additionally, other important processes that may take place in the metal/film/electrolyte overall system are oxide chemical dissolution, the occurrence of electronic transfer reactions, oxide reduction and capacitive charging, among others (Schultze and Lohrengel 2000, Vargas et al. 2019).

From a general point of view, the electrochemical formation of an oxide film takes place when a metal, M, in contact with an electrolyte is anodized, although the metal may have already a thin layer of oxide on the surface (spontaneously formed oxide). During the anodization process, the metal is oxidized at the metal/oxide interface while the electrons, e^-, circulate through the external electrical circuit, and the injection of metal ions (M^{n+}) into the oxide phase ($MO_{n/2}$) takes place. Meanwhile, the incorporation of oxide ions (O^{2-}) from the water present in the electrolyte occurs at the oxide/electrolyte interface. Thus, the migration of M^{n+} and O^{2-} species within the oxide phase gives rise to the formation and growth of the $MO_{n/2}$ film, according to the following overall reaction (reaction 1 in Fig. 3.1):

$$M_{(s)} + {}^n/_2 H_2O \rightarrow MO_{n/2(s)} + nH^+_{(aq)} + ne^- \tag{1}$$

The so-called 'valve metals' comprise a group of metals that under certain conditions exhibit a relatively simple kinetic behaviour for the formation and growth processes of their oxide films. This group is listed as the passive films of oxides of Hf, Zr, Ta, Nb, Al, Ti, W, Bi and Sb, among others (Dignam 1961, Macagno and Schultze 1984, Hurlen and Gulbrandsen 1994, Schultze and Lohrengel 2000, Pérez and López Teijelo 2004, 2005a,b, Linarez Pérez et al. 2008, 2009, Zaffora et al. 2018,

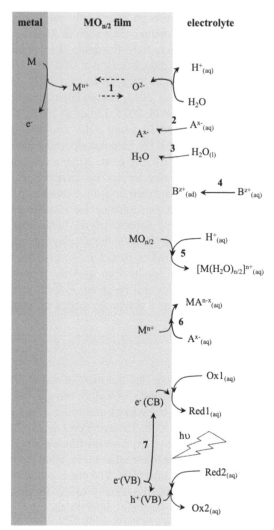

Figure 3.1: Scheme of the processes that may take place during the anodic oxide growth on metals.

Vargas et al. 2019, Almeida Alves et al. 2020), and the film growth kinetics are mainly controlled by the migration of ions within the oxide phase under a high electric field. Controlled valve metals show a characteristic electrochemical response during film growth under the application of potentiostatic, galvanostatic or potentiodynamic perturbations (Dignam 1961, Linarez Pérez et al. 2008). As an example, Fig. 3.2 shows typical potentiodynamic j-E (a), galvanostatic E-t (b) and potentiostatic j-t profiles recorded during the anodization of a W electrode in a 1.0 M Na_2SO_4 + 0.001 M H_2SO_4 solution. Particularly, the potentiodynamic polarization of tungsten electrodes in this electrolytic medium (Fig. 3.2a) shows at low potentials several peaks of anodic currents during the positive scan due to the metal multi-step oxidation until obtaining stable W(VI) species. Later the current reaches a steady-state value in a wide potential range up to ca. 9 V due to the growth of a WO_3 film by ionic migration (see later). At more positive potentials (around 9.5 V), anodic current increases again due to the oxide rupture and the starting of oxygen evolution. When potential scan direction is reversed, the current decreases to approximately zero due to the lowering in the electric field as the applied potential diminishes, assuming that the oxide film thickness remains unchanged. Meanwhile, under galvanostatic conditions (Fig. 3.2b), a linear potential-time evolution is obtained due to the continuous increase of the potential drop inside the film as it thickens, under a high constant electric field. On the

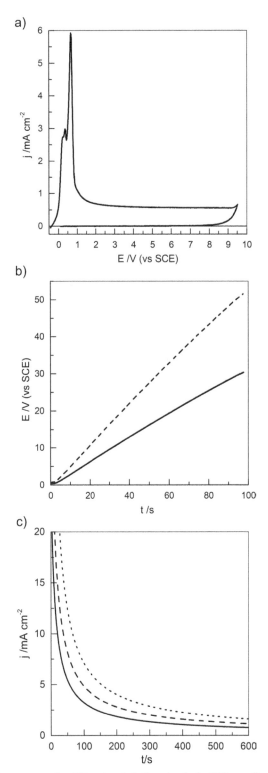

Figure 3.2: Electrochemical response under different perturbation signals for WO$_3$ anodic growth in a 1.0 M Na$_2$SO$_4$ + 0.001 M H$_2$SO$_4$ aqueous solution. (a) Potentiodynamic j-E response at 0.1 V s^{-1}. (b) Galvanostatic E-t evolutionby applying a constant current density of 1 mA cm^{-2} (solid line) and 2 mA cm^{-2} (dashed line). (c) Potentiostatic j-t response during the application of potential steps of 1 V (solid line), 2 V (dashed line), and 3 V (dotted line) V.

other hand, for potentiostatic conditions (Fig. 3.2c), initially the current density is high, and then a continuous and rapid decrease is obtained along the anodization time due to the passive film growth. Meanwhile, the rate of increase in thickness slows down, following an exponential decay.

Usually other processes may take place simultaneously with passive film growth, modifying their morphology and properties and then showing an electrochemical behaviour more complex than that seen in Fig. 3.2. According to the chemical species present in the electrolyte (cations, anions, solvent, or additives), several processes can take place at the film/electrolyte interface, which are also shown schematically in Fig. 3.1, such as the incorporation of anionic species into the film phase (reaction 2), film hydration (reaction 3), adsorption of cationic (or anionic) species from the electrolyte at the film surface (reaction 4), and chemical film dissolution promoted by proton ions (reaction 5) or anionic species (reaction 6).

Other charge transfer reactions may also take place at the film surface, which are promoted by the absorption of light for illuminated semiconductor oxide films (reaction 7). The occurrence of these processes modifies the potential drops in the metal/film/electrolyte system as well as the film growth kinetics and properties (Dignam 1961, Bojinov 1996, 1997a,b, Schultze and Lohrengel 2000, Stancheva and Bojinov 2013, Vargas et al. 2019).

On the other hand, some other metals such as Ag, Cd´and Pb (called 'battery metals'), in aqueous solutions are able of form thick compact anodic films even at low overpotentials through a dissolution-precipitation mechanism. The process depends on the kinetics of the metal dissolution and of the film growth, and frequently the growth is controlled by ions diffusion through the film.

Regarding the acquisition of oxide nanostructures, the direct anodization of metallic surfaces under specific conditions provides a simple and low-cost method to get a high control on the morphology of the nanostructures by modifying the experimental parameters (anodization potential or current, temperature, solvent and electrolyte composition). In this case, it has been reported that the synthesis of nanostructured anodic films on Ti under control of potential (potentiostatic conditions) allows obtaining a better morphology control and higher faradaic efficiencies as compared with the use of galvanostatic conditions (Regonini et al. 2013). Once the anodizing conditions, as well as other experimental parameters required for obtaining self-organization are chosen, either nanoporous or nanotubular structures can be obtained. These anodic films are usually made up of a barrier layer at the metal/oxide interface and an outermost porous layer, where cylindrical pores expand from the barrier layer to the oxide/electrolyte interface. Nanoporous layers may become nanotubes depending mainly on conditions such as solvent, water content and fluoride concentration and anodizing time. In order to obtain oxide nanostructuring of Ti and several other valve-metals, frequently the electrolytes composed of mixtures of an organic solvent as ethylene glycol with water are used, with the addition of ammonium fluoride, which promotes the formation of porous films (see below).

Furthermore, the current evolution with time under potentiostatic conditions usually shows a characteristic shape when nanostructured films are obtained, which is indicative of the different steps that take place during growth. As an example, Fig. 3.3 shows the potentiostatic j/t response of Cu foils anodized at 40 V during 120 seconds in alkaline ethylene glycol + water media containing NaOH and different fluoride ions concentrations (Oyarzún et al. 2017, 2018b). When fluoride ions are absent, the current profile does not present any remarkable differences in comparison to those shown in Fig. 3.2c for tungsten. On the contrary, in the presence of fluoride, the j/t response shows three zones similar to those that have been described to explain the anodic oxidation of titanium and other valve metals, which leads to nanoporous or nanotubular self-organization of the corresponding oxides (Macak et al. 2007a, Grimes and Mor 2009). Initially, the current density increases sharply and then decreases significantly because of passivation. The oxide layer thus formed acts as a barrier layer (reaction 1 in Fig. 3.1), although in the initial stages the active dissolution of the metallic copper substrate ($M(s) \rightarrow M^{n+}(aq) + ne^-$) may take place at some point. Later the dissolution of the oxide layer by the action of fluoride ions present in the electrolyte takes place (reaction 6 in Fig. 3.1). This process promotes the formation of pits and pores in the layer and produces an increase of the

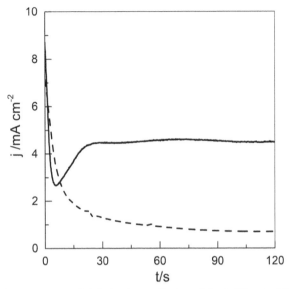

Figure 3.3: Potentiostatic j-t response recorded during the anodization of Cu at 5°C up to 40 V in x% w/V NH_4F + 1 M NaOH in ethylene glycol + 3% V/V H_2O. $x = 0$ (dashed line) and 0.2 (continuous line).

active surface of the substrate, which is observed as an increase of the current density. Subsequently, the current density remains nearly constant with slight oscillations as the nanostructure evolves to form nanoporous, nanotubular or nanofibrillar films (see below) (Oyarzún et al. 2017, 2018b).

Numerous models have been developed to explain the growth of compact or nanostructured oxide films formed on several metals for different conditions. In addition, in some cases information about their physicochemical properties, composition, morphology and band structures have been obtained (Dignam 1961, Chao et al. 1981, Macdonald et al. 1992, 2011, Bojinov et al. 1996, 1997a,b, Schultze and Lohrengel 2000, Macak et al. 2007a, Linarez Pérez et al. 2008, Grimes and Mor 2009, Albu et al. 2012, Ruiquan 2012, Regonini et al. 2013, Stancheva and Bojinov 2013, Chong et al. 2015, Oyarzún et al. 2017, Vargas et al. 2019, Almeida Alves et al. 2020, Broens et al. 2020). However, at present no agreement has been reached on the applicability of a general model that can be used for all systems. Next a summary of the main physical approaches proposed for the understanding of oxide films growth on metallic surfaces by electrochemical anodization is described.

A Brief Historical Review

The *High Field Model* (HFM) developed by Verwey in 1935 (Dignam 1961, Dell'Oca et al. 1971), was the first approach accepted to explain the growth of oxide films on metals. This model assumes that the potential drop in the metal/film/electrolyte system takes place completely within the film, being the generated high electric field and the driving force for the transport of charge carriers. Verwey assumed that the determining step for film growth is the transfer of cations or defects between adjacent sites in the solid lattice within the film. Also, HFM predicts that there is no steady-state for thickness or current under potentiostatic conditions, which has been questioned by many authors.

Later, Mott and Cabrera (Dignam 1961, Dell'Oca et al. 1971, Kirchheim 1987) proposed that the determining step for film growth is the injection of cations from metal to film on the metal/film interface, which are transported through the oxide film to the film/electrolyte interface and react with electrolyte species. In this approach, the potential drop within the film is assumed to be constant and, consequently, independent of the film thickness, meanwhile the migration of cations is assisted by the high and constant electric field obtained inside the film ($E \sim 10^6$ V cm^{-1}). This model has also been criticized as it does not take into account the transport of anions or the dissolution of the passive film, which according to experimental results would play a very important role during oxide growth.

On the other hand, Kirchheim (Kirchheim 1987) described the formation of passive films based on Vetter concepts (Vetter 1971). In the Vetter-Kirchheim model, the kinetics of oxide film formation is described in terms of the processes occurring at both the metal/film and the film/electrolyte interfaces, and the potential gradient within the film. As in the Mott-Cabrera model, it is also assumed that the electric field is high, because its value function of the oxide composition and the current or potential applied during the anodization. In the Vetter-Kirchheim model, the nature of the species transported is not exactly defined, but it was proposed that it takes place by an interstitial mechanism involving either cation or anion vacancies. The most remarkable characteristic of this approach is the coupling of the ionic currents in the bulk of the oxide and those that take place at the interfaces. Also, changes in species concentration within the oxide can be observed as different relaxation time constants when external perturbations are applied.

In the early 80's, Macdonald and collaborators (Chao et al. 1981, Macdonald et al. 1992, Macdonald 2011) developed the *Point Defect Model* (PDM), which is one of the most known and used models for the study of the growth, dissolution as well as predicting several properties of a wide variety of insulating and semiconductor films. The model emphasizes the role of moving charged point defects (oxygen and metal vacancies) through the film and assumes that reactions on metal/film and film/electrolyte interfaces are reversible. Furthermore, the generation of potential drops at the interfaces is instantaneous since no processes of relaxation of charge carriers are taken into account, although subsequently, the authors have recognized the possibility of the existence of such processes (Macdonald 2011). The PDM was initially applied to describe the growth, breakdown and the impedance response of passive films formed on Ni, Fe and alloys (Chao et al. 1981). Later, it was extended to barrier films of different metals in stationary conditions, and its experimental validation for passive films on W, Zr, Ta, Ni and Fe in different media has been demonstrated (Macdonald et al. 1992, Macdonald 2011).

At this point, no model of those detailed above takes into account the possible influence of the spatial or surface charge accumulation, generated by the instantaneous potential drop at the interfaces on the oxide growth kinetics. Accordingly, the basis of PDM has been modified to explain more complex phenomena. An example of this is the *Surface Charge Assisted High Field Model* (SCAHFM) originally proposed by De Wit and extensively developed by Bojinov et al. (De Wit et al. 1996, Bojinov et al. 1996, 1997a,b) to explain the inductive behaviour found in the impedance response for Bi, W, Sb and Mo oxide films, among others. In this model, the nature of the oxide film and the reactions that take place are based on the characteristics introduced by the PDM assuming that the transport of oxygen vacancies is faster than that of metal vacancies. Additionally, the accumulation of defects that can create surface charges close to the interfaces has been considered. As a consequence, the relaxation effects that take place after applying a potential or current perturbation under steady-state conditions are attributed to temporal variations of the surface charges close to one of the interfaces.

On the other hand, it is well known that highly ordered porous structures can be generated under optimized anodization conditions on different metal surfaces such as aluminium and titanium. In the last decade, it has been shown that auto-organized porous structures can be also obtained on several other metallic substrates, including Zr (Lee and Smyrl 2008), Nb (Choi et al. 2007, Liu et al. 2015, Kang et al. 2015), W (Ou et al. 2012, Syrek et al. 2018, Xu et al. 2020), Ta (Wei et al. 2008, Ruckh et al. 2009), Hf (Tsuchiya and Schmuki 2005) and Cu (Musselman et al. 2010, Allam and Grimes 2011, Tahir and Tougaard 2012, Zoolfakar et al. 2014, Gattinoni and Michaelides 2015, Oyarzún et al. 2017, 2018b, Stepniowski and Misiolek 2018). One of the main concerns in the study of nanostructured oxides films preparation has been the understanding of the formation processes leading to the obtainment of nanostructuration phenomenon. Thus, modifications of classical models, as well as new approaches, have been formulated. Within them, Field Assisted

Dissolution (FAD), Plastic Flow (PF) and Oxygen Bubble Mould (OBM) models are the physical approximations usually employed (Garcia-Vergara et al. 2006, Regonini et al. 2013, Zhang et al. 2014, Zhou et al. 2014, Riboni et al. 2016, Tao et al. 2019, Zhu et al. 2019, Wang et al. 2020).

The FAD model was initially proposed to explain the formation and growth of Porous Anodic Alumina (PAA) films and was subsequently adapted to explain the growth of nanotubular TiO_2 films (NT-TiO_2). This model proposes that the formation and subsequent growth of pores is due to the concentration of the electric field in morphological imperfections of the oxide barrier layer generated during the initial stages of anodization. Consequently, the high electric field in these film regions leads to an effective polarization of the M-O bonds at the film/electrolyte interface, which weakens these bonds and facilitates the local dissolution of the oxide film by the action of the anions or proton ions present in the electrolyte (reactions 5–6 in Fig. 3.1) (Regonini et al. 2013, Riboni et al. 2016, Tao et al. 2019, Zhu et al. 2019, Wang et al. 2020). On the other hand, OBM considers that the film growth takes place according to HFM (Diggle et al. 1969, Lohrengel 1993) together with the formation and growth of pores, which occur because of the simultaneous oxygen evolution reaction due to the high voltage applied:

$$2H_2O(l) \rightarrow O_2(g) + 4H^+(aq) + 4e^- \tag{2}$$

In this way, the gas bubbles generated by the electrochemical reaction 2 could act as tube moulds while the Plastic Flow (PF) of the growing oxide phase takes place (Ruiquan et al. 2012, Chong et al. 2015, Zhang et al. 2018). Based on these assumptions, ionic and electronic current contributions are the driving forces for the barrier oxide growth and, as a result, the oxygen evolution plays an important role in the growth of nanopores and nanotubes. Additionally, recent reports on NT-TiO_2 formation employing very different experimental conditions for growth, including multiple anodization steps combining potentiostatic/galvanostatic potential programmes argued against the FAD theory (Yu et al. 2018, Huang et al. 2019, Zhang J. et al. 2019, Zhang K. et al. 2019).

Anodic Behaviour and Nanostructuring of Titanium, Copper, and Antimony

TiO_2

In the last decades, the electrochemical synthesis of auto-organized TiO_2 arrays generated by Ti anodization in different aqueous and non-aqueous electrolytes containing fluoride ions has been widely studied. The versatility of properties of NT-TiO_2 films is due to the high surface/volume relation since their properties strongly depend on size. A wide range of values for preparation conditions such as applied voltage, time of anodization, temperature, nature or solvent, amount of added water, supporting electrolyte, donor fluoride ion species and counterion, pH, viscosity, cathode material, and annealing treatment after anodizing, have been used. Also, a direct correlation between formation conditions and the nanotube geometry and properties has been widely demonstrated. Consequently, NT-TiO_2 films offer potential uses in diverse areas such as the manufacture of gas and biomolecules sensor devices, charge storage devices, biocompatible materials for dental and osseous implants, photocatalysis and photovoltaic systems (Macak et al. 2007a,b, Grimes and Mor 2009, Lee et al. 2014).

One of the most interesting characteristic properties of NT-TiO_2 films is their morphology. The films are composed by ideally ordered hexagonal compact arrays of identical open nanotubes, which are characterized by its length, inner diameter, and wall thickness. Scanning and transmission electron microscopy techniques (SEM and TEM) are employed usually in order to perform the statistical analysis of these geometrical magnitudes. Recently, it was demonstrated that atomic force microscopy (AFM), as well as 3D-Confocal microscopy, are also reliable techniques to perform a detailed morphological analysis of NT-TiO_2 films (Oyarzún et al. 2016).

Figures 3.4a,b show comparatively the Field-mission Scanning lectron Microscopy (FE-SEM) top view images obtained from NT-TiO_2 films anodized up to 40 V during 1.5 and 6 hours in different electrolytic baths. Both images show well-ordered cavities with an inner diameter ranging

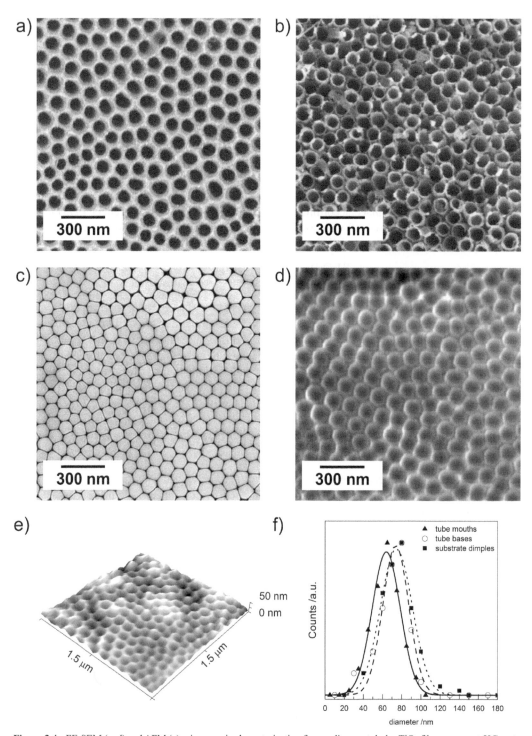

Figure 3.4: FE-SEM (a–d) and AFM (e) microscopic characterization for anodic nanotubular TiO$_2$ films grown at 5°C up to 40 V during 1.5 hours in 0.4% w/V NH$_4$F + 6% V/V water + 94% V/V ethylene glycol media (a) and 6 hours in 0.8% w/V NH$_4$F + 3% V/V water + 20% V/V ethanol + 77% V/V ethylene glycol (b–e) media. (a–b) Film top views (pore/tube mouths). (c) NT-TiO$_2$ film bottom view (tubes base). (d–e) Ti substrate after removal the nanotubular TiO$_2$ film (tube dimples). (f) Statistical analysis of tubes mouth (▲), tubes base (○), and substrate dimples (■) diameter obtained from figures b–e.

from 70 to 80 nm for pores (a) or well-differentiated tubes (b) at the film surface. According to OBM, the application of a given voltage value determines the size of the oxygen bubbles generated during the anodization and consequently, fixes the diameter of the pores/tubes. On the other hand, the differentiation of pores into tubes is the result of the chemical oxide dissolution promoted by fluoride ions taking place during the experiment (reaction 6 in Fig. 3.1), according to:

$$TiO_2(s) + 6F^-(aq) + 4H^+ \rightarrow [TiF_6]^{2-}(aq) + 2H_2O(l) \qquad (3)$$

NT-TiO$_2$ films can be detached from the Ti substrate by mechanical procedures such as adhesive tape exfoliation or ultrasound treatment, allowing to analyze the morphology of the tube bottoms as well as the underneath Ti substrate. Figure 3.4c shows the bottom view of an exfoliated NT-TiO$_2$ film prepared in the same conditions as in Fig. 3.4a, which exhibits the same ordering for the closed tubes film side (tube bottoms). Correspondingly, the exfoliated Ti substrate shows similar dimensions for tube printings, which usually are called 'dimples' (Fig. 3.4d). The topographic analysis from the three-dimensional AFM representation for the Ti exfoliated surface also indicates that the deep of dimples is less than 10 nm (Fig. 3.4e). The statistical diameter distribution obtained for tube mouths, tube bottoms and substrate dimples (Fig. 3.4f) measured from Figs. 3.4b–d show a good correlation between the mean values.

Otherwise, the tube length can be analyzed either by means of lateral FE-SEM images of partially exfoliated NT-TiO$_2$ films (Fig. 3.5a) or through High-Resolution Transmission Electron

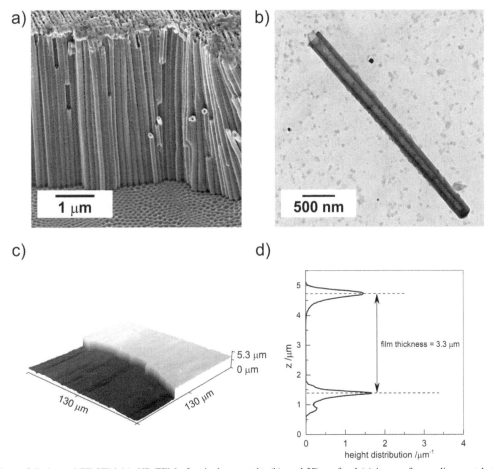

Figure 3.5: Lateral FE-SEM (a), HR-TEM of a single nanotube (b), and 3D-confocal (c) images for anodic nanotubular TiO$_2$ films grown at 5°C up to 40 V during 1.5 hours in 0.4% w/V NH$_4$F + 6% V/V water + 94% V/V ethylene glycol media. (d) Statistical analysis of film thickness from the 3D-confocal image in c.

Microscopy (HR-TEM) images for isolated nanotubes obtained by ultrasonic dispersion of NT-TiO$_2$ films in water (Fig. 3.5b). Alternatively, 3D-confocal microscopy images of partially exfoliated films also allow obtaining the TiO$_2$ film thickness from the statistical height analysis, which depicts the tube lengths (Fig. 3.5c–d) (Oyarzún et al. 2016). From the image analysis performed using the three different methods, a tube length value ranging from 3 to 3.5 mm is obtained for the anodization condition employed in Fig. 3.5.

Chronologically, NT-TiO$_2$ films were first obtained by Ti anodization in aqueous solutions of hydrofluoric acid (Rugoska et al. 2011, Lee et al. 2014, Norek et al. 2016). However, the nanotubes obtained had short lengths (< 500 nm) because of a high acidic oxide film dissolution. Later, HF was replaced by fluoride salts with buffering capability such as NH$_4$F that allows adjusting the pH gradient inside the tubes, and thus, permits obtaining tube lengths up to 5 μm (Macak et al. 2007b, Grimes and Mor 2009, Lee et al. 2014). Later it was found that the use of mixed media based on non-aqueous solvents (ethanol, glycerol, ethylene glycol, dimethyl sulphoxide and propylene carbonate, among others) and small amounts of added water, leads to longer and smoother nanotubes (Mor et al. 2006, Macak et al. 2007b, Grimes and Mor 2009, Oyarzún et al. 2011, 2016, Pisarek et al. 2013, Lee et al. 2014, Broens et al. 2020). Thus, the electrolytic bath composition strongly affects the conductivity and viscosity of the electrolyte that, in turn, directly modifies the fluoride transport in the bulk phase and inside the tubes. Consequently, changes in the rate of the overall chemical dissolution reaction 3 are produced, thus consuming titanium species that are replenished by further metallic titanium oxidation at the metal/oxide interface (HFM growth).

Recently, it was also demonstrated by us that the viscosity in ethylene glycol/water mixtures regulates the 'nanograss' formation during titanium anodization in fluoride-containing media (Broens et al. 2020). The decrease in the viscosity obtained by adding a solvent as ethanol, that has a lower intrinsic viscosity but a similar chemical structure than that of ethylene glycol base solvent promotes both the homogeneous chemical dissolution along the length of the tubes as well as the inhibition of fractures and tubes agglomeration on the film surface, generating structural high-quality NT-TiO$_2$ films.

In brief, a key for self-organized nanotubular TiO$_2$ film growth is the composition of the electrolyte, where both solvent properties and fluoride concentration allow tuning the film morphology and quality. This, in turn generates specific properties for desired uses.

Cu$_2$O/CuO

Among semiconductor materials, Cu$_2$O and CuO films have been extensively studied for their applications in photovoltaic cells as well as for electrocatalysis purposes because of their non-toxic nature, availability, low-cost, antimicrobial properties and narrow band gap (E$_g$ = 1.9–2.2 eV) (Musselman et al. 2010, Tahir and Tougaard 2012, Gattinoni and Michaelides 2015).

Regarding the preparation of nanostructured materials, several synthesis methods for Cu$_2$O, CuO and Cu (OH)$_2$ nanostructures have been reported, e.g., thermal oxidation of metallic copper at high temperatures, template-assisted growth, colloidal methods, electron-beam evaporation and sputtering, and electrochemical anodization (Musselman et al. 2010, Allam and Grimes 2011, Tahir and Tougaard 2012, Zoolfakar et al. 2014, Gattinoni and Michaelides 2015, Oyarzún et al. 2017, 2018b, Stepniowski and Misiolek 2018). High-temperature processes limit the control over the interfacial characteristics of the thin films, which significantly affect the optical and photoelectrochemical properties of the resulting oxides. In addition, the use of templates and colloidal synthesis methods leads to low adherence on conductive substrates, also limiting their use or integration into electronic devices.

Within related bibliography concerning nanostructured anodic copper passive films, Allam et al. (Allam and Grimes 2011) reported the synthesis of various Cu$_2$O nanostructured thin films by anodization of Cu in aqueous and ethylene glycol-based media containing hydroxide, chloride and/or fluoride ions. They found that no stable nanostructured anodic films can be obtained using aqueous alkaline electrolytes. On the one hand, Cu$_2$O crystallites or dendritic structures were obtained in the presence of chloride ions, while in the presence of fluoride ions, nanoporous structures or porous spheroids composed of a mixture of copper hydroxide and copper oxide phases were achieved. On the other hand, for anodization of Cu in ethylene glycol-based electrolytes containing fluoride, leaf-like nanostructures were obtained but no structural film formation was found for the conditions used. More recently (Oyarzún et al. 2017, 2018b), reported that the anodization of copper in alkaline water/ethylene glycol media containing fluoride ions generates nanofibrillar (Fig. 3.6a) or nanograined Cu_2O films as well as highly rough nanofibrillar networks (Fig. 3.6b) or nanoporous (Fig. 3.6c) mixed Cu_2O/CuO films. Raman and X-Ray Photoemission Spectroscopy (XPS) results indicated that in the presence of fluoride, Cu (I) oxide was obtained when anodization takes place applying low voltage values and at a relatively low OH^- concentration. In comparison, the subsequent oxidation to obtain CuO and Cu $(OH)_2$ species was promoted by increasing the OH^- content. The anodization in highly alkaline ethylene glycol media containing fluoride also allows obtaining homogeneous nanoneedles films (Fig. 3.6c). Tt has been briefly proposed that the anodization of copper in alkaline water/ethylene glycol mixtures produces mainly Cu(II) species. Instead, by adding fluoride ions to the electrolytic bath the Cu(I) species are stabilized, thus preventing the direct oxidation from Cu(0) to Cu(II) through adsorbed intermediate Cu(I) species, according to:

$$Cu(sup) + F^-(aq) \rightarrow Cu - F(ad) + e^- \tag{4}$$

$$2Cu - F(ad) + 2OH^-(aq) \rightarrow Cu_2O(s) + H_2O(l) + 2F^- \tag{5}$$

On the other hand, in more alkaline media the subsequent oxidation to Cu(II) may come about, generating oxide/hydroxide mixed films:

$$Cu_2O(s) + 2OH^-(aq) \rightarrow 2CuO(s) + H_2O(l) + 2e^- \tag{6}$$

$$Cu_2O(s)\ 2OH^-(aq) + H_2O(l) \rightarrow 2Cu(OH)_2\ (s) + 2e^- \tag{7}$$

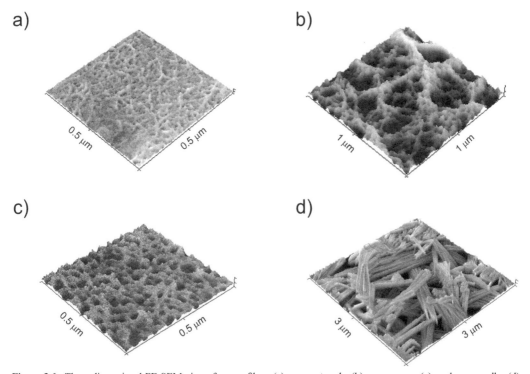

Figure 3.6: Three-dimensional FE-SEM views for nanofibres (a); nanonetworks (b); nanoporous (c); and nanoneedles (d) CuO/Cu_2O mixed anodic films grown potentiostatically at 5°C up to 10 V during different anodization times (t_a) in ethylene glycol media containing 0.1% w/v NH_4F, x% V/V H_2O and yM NaOH. t_a = 3 minutes (a) and 15 minutes (b,c,d). x = 1% (a,b,c); 10% (d). y = 0.1 M (c,d) and 0.5 M (a,b).

Sb₂O₃

Antimony (Sb) has been widely used as an additive as part of lead alloys in the acid battery industry. The absence of Sb in such electrodes has been determined to shorten the battery life, being of increasing interest in understanding the mechanism to control this effect. In addition, the study of the anodic behaviour of metallic Sb was mostly concerned with its use as a pH indicator electrode material, for applications in electrocatalysis, photoconductor surfaces, thin-film capacitors and electrochromic display devices (Metikoš-Huković et al. 1997, 2006). On the other hand, it has been demonstrated that nanostructured antimony trioxide has excellent catalytic performance in photochemistry and superior chemical stability in flame retardance (Pillep et al. 1999, Schubert et al. 2001).

When considering the anodic behaviour of antimony electrodes in aqueous solutions, the corrosion, passivation and immunity regions from the potential/pH relationship in aqueous solutions was determined. It was also proved that Sb is dissolved with the formation of trivalent species, with no evidence of the formation of pentavalent ions (Wikstrom et al. 1975, Laihonen et al. 1990, Pavlov et al. 1991a). Furthermore, Metikoš-Huković et al. determined that antimony oxide films have interesting photoconductor properties, making them an attractive material for several applications but the durability is low due to corrosion (Metikoš-Huković and Lovreček 1980). Additional studies on the processes of corrosion, the formation of the first anodic layers and anodic dissolution of Sb in H_2OS_4 and H_3PO_4 solutions, were also carried out (Laitinen et al. 1991, Pavlov et al. 1991a,b). On the other hand, the impedance response of antimony electrodes in very corrosive acidic aqueous solutions was explained in terms of the SCAHFM considering the establishment of a dual-layer formed by a barrier film and a salt overlayer (Bojinov et al. 1996).

More recently, the kinetics of anodic growth, morphology, chemical composition and stability of antimony oxide films grown in buffered phosphate electrolytes using electrochemical methods, *in situ* ellipsometry, XPS and AFM were studied (Linarez Pérez et al. 2009, 2010). It was determined by us that the kinetic parameters that characterize the dependence between the current growth and the field strength in terms of the HFM. Dissolution current dependence with electrolyte concentration and pH were also obtained (reaction 6 in Fig. 3.1), indicating that antimony oxide dissolution is promoted by phosphate ions and is almost independent of pH.

Figure 3.7 shows the characteristic potentiodynamic and galvanostatic anodic behaviour recorded during the growth of antimony oxide films on Sb electrodes in buffered phosphate and

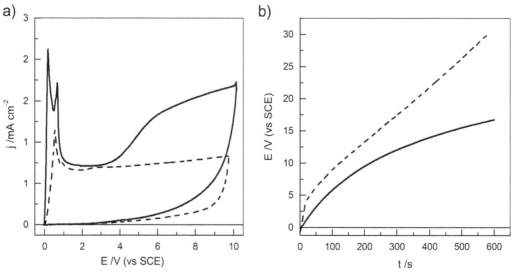

Figure 3.7: Pontentiodynamic j/E profiles at 0.1 V s⁻¹ (a) and galvanostatic E/t transients at 1.4 mA cm⁻² (b) of Sb electrodes in 0.1 M buffered phosphate (pH 7) (continuous lines) and 0.1 M NaCl (dashed lines) solutions.

chloride-containing aqueous solutions. The potentiodynamic j/E behaviour (Fig. 3.7a) is more complex than that observed for tungsten (Fig. 3.2) and also depends on the anionic species present in the electrolyte. After the stepwise electroformation of different antimony soluble species to give Sb_2O_3 observed at low potentials, the steady-state current characteristic for the anodic film growth through an ionic conduction mechanism is obtained. In a phosphate solution, this region extends only up to 4 V. Later the current increases again due to the simultaneous oxygen evolution reaction through a cracked Sb_2O_3 film (Linarez Pérez et al. 2010). Through *in situ* ellipsometry and XPS measurements, it was found that anodic films generated potentiodynamically up to 10 V in phosphate aqueous solutions grow by a characteristic HFM with an electric field strength of 2.25×10^6 V cm^{-1}. The oxide layers are constituted by Sb (III) species, that can dissolve in the electrolysis media, and act as optically anisotropic materials. The complex refractive indices also estimated for the anodic films are lower than those of crystalline antimony oxides due to hydration (reaction 3 in Fig. 3.1), phosphate ions incorporation (reaction 2 in Fig. 3.1) or lack of crystalline structure (Linarez Pérez et al. 2009, 2010).

On the one hand, although the potentiodynamic behaviour obtained for Sb in NaCl solutions seems to be simple (dashed line in Fig. 3.7a), the galvanostatic profiles obtained in both media (Fig. 3.7b) show complex E-t dependences. The lack of a unique E-t slope during the anodization is causally related to the existence of a non-homogeneous oxide dissolution promoted by the type and amount of anions present in the electrolyte, which directly affect the film morphology (Fig. 3.8). It has been found that phosphate ions promote the generation of flat and smoother surfaces (Fig. 3.8a), with grain texture dependent on electrolyte concentration as a consequence of different chemical dissolution rates (Linarez Pérez et al. 2009). On the other hand, the presence of chloride generates porous Sb_2O_3 film morphologies (Fig. 3.8b), as a result caused by local or non-homogeneous film dissolution, which is more important as chloride concentration increases.

Concerning the preparation of nanostructured antimony oxide, there exist a few reports on the synthesis of Sb_2O_3 nanostructures by thermal or sol-gel methodologies in the literature (Deng et al. 2006, Ye et al. 2006, Abdel-Galil et al. 2015, Liu et al. 2019). In addition, to the best of our knowledge, the obtaining of Sb_2O_3 nanostructures by means of anodization methods has not been carried out.

a)   b)

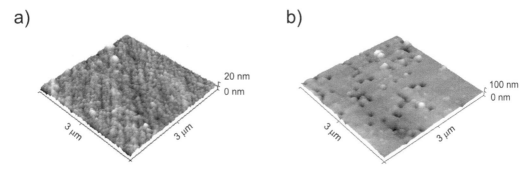

Figure 3.8: 3D AFM images for anodic Sb_2O_3 films grown up to 3 V in 0.002 M phosphate buffer (pH 7) (a) and 0.002 M NaCl (b).

Applications

The versatility of properties of nanostructured semiconductor oxides allows their use as a fundamental component in diverse technological areas. In this regard, it has been reported that according to their morphology, size and shape, they are suitable materials for hybrid solar cells, batteries and supercapacitors as well as in photovoltaic catalysts and sensor devices. Here some literature findings related to these applications will be briefly discussed.

Energy

In recent decades, the global interest in the development of energy conversion or storage devices with high efficiency and power density for its application in new electronic devices with a high technological impact as an alternative to the conventional fossil fuel-driven has increased continuously. One of the most promising aspects of these devices comes from the possibility of producing electricity through renewable alternatives (solar, wind, hydraulic), generating fuels like H_2 in a sustainable manner as well as the possibility of storage in 'custom-made materials'. In this way, photovoltaic or solar cells are capable of efficiently converting solar radiation into electrical energy, while batteries and supercapacitors, which exhibit high power supply, remarkable reversibility, long-term cycling stability and rapid response, stand out among the potential candidates for more promising storage devices (Hotchandani and Kamat 1992, Vogel et al. 1994, Conway 1999, Grätzel 2000, Mor et al. 2006, Kamat 2007, 2013, Macak et al. 2007a, Lee and Smyrl 2008, Gao et al. 2009, Musselman et al. 2010, Oyarzún et al. 2011, Wen et al. 2011, Chen et al. 2013, Come et al. 2014, Hu et al. 2014, Lee et al. 2014, Wang et al. 2014, Chen et al. 2015, Kang et al. 2015, Hong et al. 2015, Torresan et al. 2016, 2017, Vázquez et al. 2016, 2018, Li et al. 2017, Zhou et al. 2017, Tan et al. 2018, Zaffora et al. 2018, Liu et al. 2019, Williamson et al. 2019, Zhu et al. 2019).

These days to increase energy conversion efficiency in solar technology, different materials have been explored. Among them, solar cells based on multiple p-n junctions of semiconductors with different bandgap values allow obtaining the highest efficiencies ($< 46\%$) in comparison to those based on silicon (20–25%), photovoltaic thin films ($< 21\%$), organic polymers ($< 12\%$) or dye-sensitized cells ($< 12\%$) (Vázquez and Iglesias 2018).

Due to low fabrication costs, non-toxic materials used and average efficiencies, Dye-Sensitized Solar Cells (DSSCs) have received much attention over the past 30 years since Grätzel found considerable conversion efficiency (Grätzel 2000). They are based on a transparent n-type semiconducting photoanode, usually composed of TiO_2, SnO_2 or ZnO, covered with a redox dye. At present, the use of binary salt nanocrystals of less than 10 min size of transition metals, so-called *quantum dots* (QDots), of CdS, CdSe, CdTe, PbS, InP, GaAs, GaP, GaN and GaInP, among others (Hotchandani and Kamat 1992, Vogel et al. 1994, Kamat 2007, 2013, Gao et al. 2009, Torresan et al. 2016, 2017, Vázquez et al. 2016, 2018) as optical sensitizers of TiO_2 are of great relevance, leading to the so-called hybrid or third-generation solar cells. The main advantage of QDots compared to redox dyes, is the possibility to tune their optical properties by controlling the size and surface characteristics and then, achieve a better band coupling to the TiO_2 photoanode. Usually, QDots are covered by a capping of stabilizing agents, which may impede the charge transport between the sensitizer and the semiconductor, diminishing the interfacial charge transfer. Recently, it has been shown that optimal conditions of thermal annealing in Ar atmosphere (Gao et al. 2009) or in air (Torresan et al. 2016) of photoanodes composed by TiO_2 nanotubes sensitized with CdSe QDots covered with mercaptopropionic acid and myristic acid/trioctylphosphine, respectively, cause an improvement in the photocurrent generation. In addition, it has also been reported that the amount of CdSe and CdSQDots adsorbed on TiO_2 nanotubes can be tuned by increasing the solvent polarity (Torresan et al. 2017) as well as by performing successive adsorption steps (Vázquez et al. 2016) and then producing an increase in the photocurrent efficiency. Currently, solar cells based on QDots show efficiencies higher than that for DSSCs (up to 14%), indicating that they are promising materials for the development of high-efficiency solar cells (Vázquez and Iglesias 2018).

Regarding metal-ion batteries, various nanostructured semiconductor materials such as TiO_2 and titanates, have been used as anode electrodes due to their adequate redox potential in pressurized liquid media (Wang et al. 2014, Kang et al. 2015, Tan et al. 2018). At present, interest in hybrid capacitors (Li-ion) has increased as they can be built using a battery-type electrode (graphite, titanates) that has a fast response, and a semiconductor oxide film (Nb_2O_5, WO_3) as a supercapacitor electrode that allows to achieve high powers and increase the accumulated energy density (Hu et al. 2014, Wang et al. 2014, Hong et al. 2015, Kang et al. 2015, Tan et al. 2018). Also, post-transition metal oxides, such as tin and antimony oxides (Zoolfakar et al. 2012, Zhou et al. 2017, Williamson et al. 2019) have acquired increasing attention due to their interesting electrochemical behaviour as an anode material. In this way, Sb_2O_3 is one of the more promising materials due to its low cost, simple synthesis procedures and lithium or sodium ions insertion capabilities. Despite this, Sb(III) oxide lacks chemical stability (Linarez Pérez et al. 2009) and shows high volume variation during metal charge-discharge cycling (Liu et al. 2019). It has also been found that the size and shape control of Sb_2O_3 nanostructures, as well as the use of polymeric cross-linked binders, may be the keys to obtain stable electrodes with improved electrochemical performance (Liu et al. 2019, Williamson et al. 2019).

On the other hand, copper oxide films also have interesting (photo)electrochemical properties, which make them promising candidates for developing energy-based devices. In this regard, it was found that CuO films act as a blocking layer that prevents recombination reactions in solar cells based on dye-sensitization (Sze and Ng 2007, Zoolfakar et al. 2012) and $ZnO\text{-}Cu_2O$ heterojunction bilayers have auspicious conversion efficiencies (up 4%) (Musselman et al. 2010, Minami et al. 2011). Nanostructured CuO and Cu_2O also showed high theoretical capacity (> 600 and > 350 mAh g^{-1}, respectively), which makes them interesting materials as anode electrodes in metal-ion batteries (Wang et al. 2010, Xiang et al. 2010, Ji et al. 2011).

Photocatalysts

As it was described earlier, highly ordered nanoporous TiO_2 films can be generated under different anodizing conditions. The construction of surface-modified TiO_2 structures by doping with metals substantially influences the electronic structure allowing high-efficiency photocatalysis under visible light irradiation (Macak et al. 2007a, Grimes and Mor 2009, Oyarzún et al. 2011, Roguska et al. 2011, Pisarek et al. 2014, Lee et al. 2014, Zhu et al. 2019). On the other hand, doping has been one of the routes more used to modify surface structures for photocatalysis, adjusting the conductive properties and the suitable distribution of the induced electron-hole pairs, preventing their recombination without losing the integrity of the crystalline structure (Macak et al. 2007a,b). In this way, in order to increase the versatility of the physicochemical properties for different applications, the development of novel modification strategies of nanostructured TiO_2 films have been investigated (Macak et al. 2007b, Cialla et al. 2012, Tan et al. 2012, Chen et al. 2013, Li et al. 2017).

The surface functionalization of TiO_2 films with inorganic complexes as well as organic compounds has been under great investigation because the formed dye-sensitized cells are chemically stable in extreme pH conditions and high temperatures (Nazeeruddin et al. 2004, Bai et al. 2014, Fattakhova-Rohlfing et al. 2014) as well as due to the low-cost materials (Bourikas et al. 2005, Gázquez et al. 2014). TiO_2 as rutile or anatase-like forms have a skeleton of titanium and oxygen atoms spatially distributed (Alemany et al. 2000, Brumbach et al. 2009, Dahl et al. 2014) establishing a high number of Ti–OH groups on the solid surface (Gawęda et al. 2008, Aryanpour et al. 2009), which allows linking of organometallic compounds and organic chains for the preparation of hybrid materials (Jańczyk et al. 2006, Sun and Xu 2012, Huang et al. 2014, Shakir et al. 2014, Milowska et al. 2015). In this sense, it has been demonstrated by us that tetra(4-carboxyphenyl) porphyrin (TCPP) dye molecules adsorb onto NT-TiO_2 films could be used as an efficient photosensitizer for solar-energy conversion devices (Oyarzún et al. 2011). Photoelectrochemical measurements showed that TCPP facilitates the electron transfer to the conduction band of the semiconductor and presents a low rate of electron-hole recombination, thus contributing to obtain an efficient flow of oxidation current in the external circuit.

On the other hand, numerous studies on electron transfer and transport processes in solar cells based in TiO_2 with ruthenium (II) complexes have demonstrated successful results in energy conversion (Sauve et al. 2000, Pogozhev et al. 2013, Hsu et al. 2014, Chen et al. 2016, Singha et al. 2017). These findings encourage the development of new methods to immobilize ruthenium-based catalysts onto semiconducting or conducting supports with the purpose of obtaining heterogeneous catalysts (Baumann et al. 2010, Pastva et al. 2014, Beloqui Redondo et al. 2015).

These types of catalysts could have important applications in new devices for the conversion of CO_2 to value-added molecules such as CO, which represent an important technological and environmental challenge (Kuninobu and Takai 2011, Valenti et al. 2014, Machan et al. 2015). Since Hawecker et al. reported the efficient obtaining of CO using rhenium (I) tricarbonyl complexes as photocatalysts on vitreous carbon electrodes, a wide range of transition metal complexes have been studied for their photo- and electrocatalytic properties for the reduction of CO_2 (Hawecker et al. 1983, 1984, Gholamkhass et al. 2005, Boston et al. 2014, Ko et al. 2015, Vollmer et al. 2015, Manes and Rose 2015, Wang et al. 2015). A family of the same kind of complexes using dinitrogenated 2,2'-bipyridin ligands with substituents like -COOH, -CH$_3$, t-Bu, -OMe was also prepared and their electrocatalytic properties for the reduction of CO_2 was evaluated. Among these reported compounds, the ruthenium (II) complexes harbouring carbonyl ligands have promising catalytic properties with the choice of suitable bidentate ligands (Huang et al. 2014, Rezayee et al. 2015). In this sense, it was recently shown by us that *cis*-[Ru(bpyC$_4$pyr)(CO)$_2$(CH$_3$CN)$_2$]$^{2+}$, *cis*-[Ru(bpy)$_2$(CO)$_2$]$^{2+}$, *cis*-[Ru(bpyac)(CO)$_2$Cl$_2$] and [Re(bpyac)(CO)$_3$Cl] complexes harbouring carbonyl ligands can be successfully anchored onto NT-TiO$_2$ films (Oyarzún et al. 2018a). The modified NT-TiO$_2$ surfaces were seen to be chemically stable due to complex species are chemically linked to the oxide by esterification of OH$^-$ groups on the TiO$_2$ surface, producing homogeneous layers. The spectroscopic evidence has shown that the anchorage of ruthenium (II) or rhenium (I) complexes that present carboxylate groups as a substituent in the di-nitrogenated ligand are more efficient. Furthermore, the modification of NT-TiO$_2$ films with ruthenium complexes shows interesting photocurrent performance for the oxidation processes, meanwhile, the rhenium complexes are more suitable for the reduction of CO_2 in aqueous solutions. Similarly, in order to generate an electrocatalytic electrode capable to reduce CO_2 or H$^+$, a nanostructured Cu/Cu$_2$O/CuO dendritic electrode presenting high electrocatalytic activity and stability at low overpotentials as the source of electrons for the multi-electron Oxygen Evolution Reaction (OER) in a wide range of conditions, has been proposed elsewhere (Huan et al. 2017). The dense copper oxide layer at the surface of the dendritic supports prevents the corrosion of the basal copper substrate, allowing the electron conductivity of the material.

In the last few decades, efficient acquiring of molecular H$_2$ from water as an energy carrier has led to an increasing interest due to the high global energy demand. As pure water cannot absorb solar radiation, the water-splitting reaction ($H_2O_{(l)} \rightarrow H_{2(g)} + \frac{1}{2} O_{2(g)}$) needs a photoactive material capable of efficiently absorbing solar energy to produce the electrochemical decomposition of water to hydrogen and oxygen. Semiconducting metal oxides, such as KTaO$_3$, SrTiO$_3$, La$_2$(TiO$_3$)$_3$, TiO$_2$, ZnS, CdS and SiC, have accurate band positions and energy bandgap (> 1.23 eV) to produce optimal electron-hole separation and minimum energy loss, which is necessary to promote both reduction and oxidation reactions (Takeuchi et al. 2000, Kitano et al. 2008, Matsuoka et al. 2008, Navarro Yerga et al. 2013, Wang et al. 2018). It has also been demonstrated that composites of TiO$_2$ or ZnO can be achieved with semiconducting QDots, where the electron-hole separation stops the charge recombination and, therefore improves the photocatalytic activity of the semiconductor oxide electrode.

At the same time, it has been stated that noble metal nanoparticles (NPs) constitute a promising new class of catalysts for the use of the full solar energy spectrum because the Localized Resonant Plasmons (LRP) can be tuned and controlled with their size, shape, material and dielectric medium (Homola et al. 1999, Aslam et al. 2018). In addition, the excitation of LRP increases electromagnetic fields on the NPs surface and the free electrons in the conduction band are thermalized and balanced through electron-electron scattering (Brongersma et al. 2015). These 'hot electrons' can induce multiple vibrational transitions in a reacting molecule. Consequently, the vibrational energy stored in a bond is increased and, the activation energy for a determined chemical reaction is then reduced. In this way, the coupling of plasmonic metal NPs and semiconducting materials constitutes promising photocatalytic materials for better efficient conversion of solar energy (Watanabe et al. 2006, Linic et al. 2011, Sun et al. 2012, Spata and Carter 2018, Zhang et al. 2018). It has been found that Ag

NPs with 50 nm diameter embedded into a TiO_2 film (~ 90 nm in thickness) dramatically increase the density of photons in the nearby UV region that overlaps well with the oxide bandgap and then more electron-hole pairs are excited, improving their photolytic activity. As a consequence, the enhanced local plasmonic field increases the rate of the electron-hole pair generation process on the TiO_2 surface, increasing the efficiency of decomposition of methylene blue seven times (Awazu et al. 2008). It was also shown that TiO_2 particles embelished with Au NPs can increase to two orders of magnitude the dissociation rate of H_2 in comparison to pure Au NPs by preventing the recombination of electron-hole pairs in the TiO_2 particles by hot electrons (Mukherjee et al. 2013).

Molecular Sensors

An ideal sensor device must be specific, highly sensitive, respond fast, be portable and have a low fabrication cost. At present, the main interest is obtaining materials that satisfy these characteristics for monitoring global environmental or medical necessities (Capone et al. 2003, Kim et al. 2008). In this way, nanostructured semiconducting oxides have a high surface/volume ratio, tunable crystallographic and surface characteristics and high stability, which may allow the enhancement of molecular selectivity. Several devices based on nanostructured CuO/Cu_2O, ZnO, Sb_2O_3, SnO_2, In_2O_3, WO_3, V_2O_5, TiO_2 have been developed for the detection of H_2 and CO_2 as well as volatile organic compounds such as acetaldehyde, methanol, ethanol, and trimethylamine (Zoolfakar et al. 2013, 2014), among others.

According to the bandgap values, one of the advantages in the use of p-type semiconducting oxides (e.g., CuO or Cu_2O) is that they are suitable for the development of visible optoelectronic sensors against only UV activity of n-types materials. The sensing properties of CuO and Cu_2O can also be improved by decreasing their size to nanoscale dimensions and by adding proper dopants such as Pd, Pt, Ag and Au (Li et al. 2010, Tricoli et al. 2010, Zoolfakar et al. 2013). Copper oxide thin films have been demonstrated to be highly sensitive towards various gas species including ethanol, H_2S, CO and NO_2 (Zhang H. et al. 2007, Zhang Y. et al. 2007, Kim et al. 2008, Barreca et al. 2009, Li et al. 2010, Tricoli et al. 2010, Zhang et al. 2010, Hsueh et al. 2011, Liu et al. 2012, Zoolfakar et al. 2013). Besides, it has been demonstrated that doughnut-like CuO structures possess a high removal capacity for As (III) and can be easily separated and recycled during water treatment processes (Cao et al. 2007).

Furthermore, interest in the detection of Rare Earth Metals (REMs), such as lanthanum (La) and samarium (Sm), has acquired importance of late due to their industrial uses. It has been found that hydrated nanosized antimony trioxide shows monofunctional ion-exchange characteristics that allow recovering REMs from liquid wastes (Abdel-Galil et al. 2015).

Conclusions and Outlook

In this chapter, an overview of the anodic film growth, properties and applications of nanostructured semiconducting metal oxides were presented. The analysis on the relationship existing between the electrochemical formation conditions and the film morphology were also focused. Widespread description of present and promising applications were described.

Due to the diversity of properties, passive films play an important role in recent and future research. Over the last few decades, numerous studies have been conducted involving auto-organized nanostructures of different metal oxides due to their potential technological uses. Among others, nanostructured semiconductor oxide films obtained using optimized electrochemical conditions and derived composite materials, have shown interesting physicochemical properties and improved performance in luminescent modules, piezoelectric transducers, solar cells, chemical sensors and electrocatalysts.

In order to design and tune the desired morphologies and properties of nanostructured semiconducting oxides, several synthesis techniques have been reported. In this way, physical or Chemical Vapour Deposition (PVD, CVD), electrodeposition, hydrothermal/solvothermal, sol-gel or anodization are the methods commonly used. Among them, electrochemical methods generally have a lower cost and allow obtaining better control of the final morphologies.

The anodic behaviour and applications of nanostructured titanium, copper and antimony oxides as examples of semiconducting materials have also been discussed. Within these, TiO_2 is one of the materials most used in energy production and storage devices as well as for photocatalytic applications due to its outstanding semiconducting properties. On the other hand, nanostructured copper oxides, due to their lower bandgap and p-type semiconductor character, are used as materials for water splitting, sensing applications, antimicrobial material and optoelectronics. In the mean time, despite its promising semiconducting properties, nanostructured antimony oxide has been studied less studied. In general, reports on thermal or sol-gel synthesis are found. In this case, the controlled electrochemical anodization is an unexplored methodology for obtaining nanostructured Sb_2O_3 films with improved properties.

Future research on nanostructured semiconducting oxides can be directed towards combining different morphological or chemical surface functionalization methodologies as well as the generation of composites in synergic interaction with other materials to improve their performance in novel, green, feasible and versatile technologies for global needs. In addition, the understanding of the formation mechanisms will allow developing novel bottom-up synthesis strategies to obtain nanostructures for desired applications.

Acknowledgements

We thank the financial support by SECYT-UNC and CONICET. M.I.B., W.R.C. and A.M.A.C. thank CONICET for the fellowships granted. FE-SEM and 3D Confocal Microscopy facilities at LAMARX (FAMAF-UNC), TEM microscopy measurements at CIAP-INTA and AFM topography at LANN (INFIQC-UNC), Sistema Nacional de Microscopía–MINCyT are gratefully acknowledged.

References

Abdel-Galil, E. A., W. M. El-kenany and L. M. S. Hussin. 2015. Preparation of nanostructured hydrated antimony oxide using a sol-gel process. Characterization and applications for sorption of La^{3+} and Sm^{3+} from aqueous solutions. Russ. J. Appl. Chem. 88: 1351–1360.

Albu, S. P., N. Taccardi, I. Paramasivam, K. R. Hebert and P. Schmuki. 2012. Oxide growth efficiencies and self-organization of TiO_2. J. Electrochem. Soc. 159: H697–H703.

Alemany, L. J., M. A. Bañares, E. Pardo, F. Martín-Jiménez and J. M. Blasco. 2000. Morphological and structural characterization of a titanium dioxide system. Mater. Charact. 44: 271–275.

Allam, N. K. and C. A. Grimes. 2011. Electrochemical fabrication of complex copper oxide nanoarchitectures via copper anodization in aqueous and non-aqueous electrolytes. Mater. Lett. 65: 1949–1955.

Almeida Alves, C. F., S. V. Calderon, P. J. Ferreira, L. Marquese and S. Carvalhoaf. 2020. Passivation and dissolution mechanisms in ordered anodic tantalum oxide nanostructures. Appl. Surf. Sci. 513: 145575.

Aryanpour, M., R. Hoffmann and F. J. DiSalvo. 2009. Tungsten-doped titanium dioxide in the rutile structure: theoretical considerations. Chem. Mater. 21: 1627–1635.

Aslam, U., V. G. Rao, S. Chavez and S. Linic. 2018. Catalytic conversion of solar to chemical energy on plasmonic metal nanostructures. Nat. Catal. 1: 656–665.

Awazu, K., M. Fujimaki, C. Rockstuhl, J. Tominaga, H. Murakami, Y. Ohki et al. 2008. A plasmonic photocatalyst consisting of silver nanoparticles embedded in titanium dioxide. J. Am. Chem. Soc. 130: 1676–1680.

Bai, Y., I. Mora-Sero, F. De Angelis, J. Bisquert and P. Wang. 2014. Titanium dioxide nanomaterials for photovoltaic applications. Chem. Rev. 114: 10095–10130.

Barreca, D., E. Comini, A. Gasparotto, C. Maccato, C. Sada, G. Sberveglieri et al. 2009. Chemical vapor deposition of copper oxide films and entangled quasi-1D nanoarchitectures as innovative gas sensors. Sens. Actuators B 141: 270–275.

Baumann, N., P. S. Gamage, T. N. Samarakoon, J. Hodgson, J. Janek and S. H. Bossmann. 2010. A new heterogeneous photocathode based on ruthenium(II)quaterpyridinium complexes at TiO_2 particles. J. Phys. Chem. C 114: 22763–22772.

Beloqui Redondo, A., F. L. Morel, M. Ranocchiari and J. A. van Bokhoven. 2015. Functionalized ruthenium-phosphine metal-organic framework for continuous vapor-phase dehydrogenation of formic acid. ACS Catal. 5: 7099–7103.

Bojinov, M., I. Kanazirski and A. Girginov. 1996. A model for surface charge-assisted Barrier film growth on metals in acidic solutions based on ac impedance measurements. Electrochim. Acta 41: 2695–2705.

Bojinov, M. 1997a. The ability of a surface charge approach to describe barrier film growth on tungsten in acidic solutions. Electrochim. Acta 42: 3489–3498.

Bojinov, M., I. Betova and R. Raicheff. 1997b. Influence of molybdenum on the transpassivity of a Fe + 12% Cr alloy in H_2SO_4 solutions. J. Electroanal. Chem. 430: 169–178.

Boston, D. J., Y. M. Franco Pachon, R. O. Lezna, N. R. de Tacconi and F. M. MacDonnell. 2014. Electrocatalytic and photocatalytic conversion of CO_2 to methanol using ruthenium complexes with internal pyridyl cocatalysts. Inorg. Chem. 53: 6544–6553.

Bourikas, K., M. Stylidi, D. I. Kondarides and X. E. Verykios. 2005. Adsorption of acid orange 7 on the surface of titanium dioxide. Langmuir 21: 9222–9230.

Brammer, K. S., S. Oh, C. J. Frandsen and S. Jin. 2011. Biomaterials and biotechnology schemes utilizing TiO_2 nanotube arrays. pp. 193–210. *In*: R. Pignatello [ed.]. Biomater. Sci. Eng. InTech. London, UK.

Broens, M. I., W. Ramos Cervantes, D. Oyarzún Jerez, M. López Teijelo and O. E. Linarez Pérez. 2020. The keys to avoid undesired structural defects in nanotubular TiO_2 films prepared by electrochemical anodization. Ceramics Int. 46: 13599–13606.

Brongersma, M. L., N. J. Halas and P. Nordlander. 2015. Plasmon-induced hot carrier science and technology. Nat. Nanotechnol. 10: 25–34.

Brumbach, M. T., A. K. Boal and D. R. Wheeler. 2009. Metalloporphyrin assemblies on pyridine-functionalized titanium dioxide. Langmuir 25: 10685–10690.

Cao, A., J. D. Monnell, C. Matranga, J. -M. Wu, L. -L. Cao and D. Gao. 2007. Hierarchical nanostructured copper oxide and its application in arsenic removal. J. Phys. Chem. C 111: 18624–18628.

Capone, S., A. Forleo, L. Francioso, R. Rella, P. Siciliano, J. Spadavecchia et al. 2003. Solid state gas sensors: state of the art and future activities. J. Optoelectron. Adv. Mater. 5: 1335–1348.

Chao, C. Y., L. F. Lin and D. D. Macdonald. 1981. A Point Defect Model for anodic passive films: I. Film growth kinetics. J. Electrochem. Soc. 128: 1187–1194.

Chen, B., J. Hou and K. Lu. 2013. Formation mechanism of TiO_2 nanotubes and their applications in photoelectrochemical water splitting and supercapacitors. Langmuir 29: 5911–5919.

Chen, J., H. Hou, Y. Yang, W. Song, Y. Zhang, X. Yang et al. 2015. An electrochemically anodic study of anatase TiO_2 tuned through carbon-coating for high-performance lithium-ion battery. Electrochim. Acta 164: 330–336.

Chen, W., J. -S. Wu and X. -H. Xia. 2008. Porous anodic alumina with continuously manipulated pore/cell size. ACS Nano 2: 959–965.

Chen, W., F. T. Kong, Z. Q. Li, J. H. Pan, X. P. Liu, F. L. Guo et al. 2016. Superior light-harvesting heterolepticruthenium(II) complexes with electron-donating antennas for high performance dye-sensitized solar cells. ACS Appl. Mater. Interfaces 8: 19410–19417.

Chick, H. and J. M. Xu. 2004. Nanometric superlattices: non-lithographic fabrication, materials and prospects. Mat. Sci. Eng. R. 43: 103–108.

Choi, J., J. H. Lim, J. Lee and K. J. Kim. 2007. Porous niobium oxide films prepared by anodization-annealing-anodization. Nanotechnology 18: 055603.

Chong, B., D. Yu, R. Jin, Y. Wang, D. Li, Y. Song et al. 2015. Theoretical derivation of anodizing current and comparison between fitted curves and measured curves under different conditions. Nanotechnology 26: 145603.

Cialla, D., A. Marz, R. Bohme, F. Theil, K. Weber, M. Schmitt et al. 2012. Surface-enhanced Raman spectroscopy (SERS): progress and trends. Anal. Bioanal. Chem. 403: 27–54.

Come, J., V. Augustyn, J. W. Kim, P. Rozier, P. -L. Taberna, P. Gogotsi et al. 2014. Electrochemical kinetics of nanostructured Nb_2O_5 electrodes. J. Electrochem. Soc. 161: A718–A725.

Conway, B. E. 1999. Electrochemical Supercapacitors: Scientific Fundamentals and Technological Applications. Plenum Press. New York, USA.

Dahl, M., Y. Liu and Y. Yin. 2014. Composite titanium dioxide nanomaterials. Chem. Rev. 114: 9853–9889.

De Wit, J. H. W. and H. J. W. Lenderink. 1996. Electrochemical impedance spectroscopy as a tool to obtain mechanistic information on the passive behaviour of aluminium. Electrochim. Acta 41: 1111–1119.

Dell'Oca, C. J., L. Pulfrey and L. Young. 1971. Anodic oxide films. Phys. Thin films 6: 1–79.

Deng, Z., F. Tang, D. Chen, X. Meng, L. Cao and B. Zou. 2006. A simple solution route to single-crystalline Sb_2O_3 nanowires with rectangular cross sections. J. Phys. Chem. B 110: 18225–18230.

Diggle, J. W., T. C. Downie and C. W. Goulding. 1969. Anodic oxide films on aluminum. Chem. Rev. 69: 365–405.

Dignam, M. J. 1961. The kinetics of the growth of oxides. pp. 247–306. *In*: O'M. Bockris, B. E. Conway, E. Yaeger and R. E. White [eds.]. Comprehensive Treatise of Electrochemistry. New York, USA.

Fattakhova-Rohlfing, D., A. Zaleska and T. Bein. 2014. Three-dimensional titanium dioxide nanomaterials. Chem. Rev. 114: 9487–9558.

Friedman, A. 2011. Porous alumina templates for nanofabrication. pp. 3525–3538. *In*: J. A. Schwarz, S. E. Lyshevski and C. I. Contescu [eds.]. Dekker Encycl. Nanosci. Nanotechnol. CRC Press. Boca Raton, USA.

Gao, X., H. Li, W. Sun, Q. Chen, F. Tang and L. Peng. 2009. CdTe quantum dots-sensitized TiO_2 nanotube array photoelectrodes. J. Phys. Chem. C 113: 7531–7535.

Garcia-Vergara, S. J., P. Skeldon, G. E. Thompson and H. Habazaki. 2006. A Flow Model of porous anodic film growth on aluminium. Electrochim. Acta 52: 681–687.

Gattinoni, C. and A. Michaelides. 2015. Atomistic details of oxide surfaces and surface oxidation: the example of copper and its oxides. Surf. Sci. Rep. 70: 424–447.

Gawęda, S., G. Stochel and K. Szaciłowski. 2008. Photosensitization and photocurrent switching in carminic acid/titanium dioxide hybrid material. J. Phys. Chem. C 112: 19131–19141.

Gázquez, M. J., J. P. Bolívar, R. Garcia-Tenorio and F. Vaca. 2014. A review of the production cycle of titanium dioxide pigment. MSA 5: 441–458.

Gholamkhass, B., H. Mametsuka, K. Koike, T. Tanabe, M. Furue and O. Ishitani. 2005. Architecture of supramolecular metal complexes for photocatalytic CO_2 reduction: ruthenium-rhenium bi- and tetranuclear complexes. Inorg. Chem. 44: 2326–2336.

Grätzel, M. 2000. Perspectives for dye-sensitized nanocrystalline solar cells. Prog. PhotoVoltaics 8: 171–185.

Grimes, C. A. and G. K. Mor. 2009. TiO_2 Nanotube Arrays. Synthesis, Properties and Applications. Springer Sc. Publisher. New York. USA.

Gu, Z., Y. Ma, W. Yang, G. Zhang and J. Yao. 2005. Self-assembly of highly oriented one-dimensional h-WO_3 nanostructures. Chem. Commun. 28: 3597–3599.

Hawecker, J., J. M. Lehn and R. Ziezzel. 1983. Efficient photochemical reduction of CO_2 to CO by visible light irradiation of systems containing $Re(bipy)(CO)_3X$ or $Ru(bipy)_3^{2+}$-Co^{2+} combinations as homogeneous catalysts. J. Chem. Soc., Chemm. Commun. 536–538.

Hawecker, J., J. M. Lehn and R. J. Ziezzel. 1984. Electrocatalytic reduction of carbon dioxide mediated by $Re(bipy)(CO)_3Cl$ (bipy=2,2′-bipyridine). J. Chem. Soc., Chemm. Commun. 328–330.

Homola, J., S. S. Yee and G. Gauglitz. 1999. Surface plasmon resonance sensors: Review. Sens. Actuators B 54: 3–15.

Hong, K. S., D. H. Nam, S. J. Lim, D. Sohn, T. H. Kim and H. Kwon. 2015. Electrochemically synthesized Sb/Sb_2O_3 composites as high-capacity anode materials utilizing a reversible conversion reaction for Na-ion batteries. ACS Appl. Mater. Interfaces 7: 17264–17271.

Hotchandani, S. and P. V. Kamat. 1992. Charge-transfer processes in coupled semiconductor systems. Photochemistry and photoelectrochemistry of the colloidal cadmium sulfide-zinc oxide system. J. Phys. Chem. 96: 6834–6839.

Hsu, H. Y., C. W. Cheng, W. K. Huang, Y. P. Lee and E. W. -G. Diau. 2014. Femtosecond infrared transient absorption dynamics of benzimidazole-based ruthenium complexes on TiO_2 films for dye-sensitized solar cells. J. Phys. Chem. C 118: 16904–16911.

Hsueh, H. T., S. J. Chang, F. Y. Hung, W. Y. Weng, C. L. Hsu, T. J. Hsueh et al. 2011. Ethanol gas sensor of crabwise CuO nanowires prepared on glass substrate. J. Electrochem. Soc. 158: J106–J109.

Hu, M. J., Y. Z. Jiang, W. P. Sun, H. T. Wang, C. H. Jin and M. Yan. 2014. Reversible conversion-alloying of Sb_2O_3 as a high-capacity, high-rate, and durable anode for sodium ion batteries. ACS Appl. Mater. Interfaces 6: 19449–19455.

Huan, T. N., G. Rousse, S. Zanna, I. T. Lucas, X. Xu, N. Menguy et al. 2017. A dendritic nanostructured copper oxide electrocatalyst for the oxygen evolution reaction. Angew. Chem. Int. Ed. 56: 1–6.

Huang, J., J. Chen, H. Gao and L. Chen. 2014. Kinetic aspects for the reduction of CO_2 and CS_2 with mixed-ligand ruthenium(II) hydride complexes containing phosphine and bipyridine. Inorg. Chem. 53: 9570–9580.

Huang, L., K. E. Gubbins, L. Li and X. Lu. 2014. Water on titanium dioxide surface: A revisiting by reactive molecular dynamics simulations. Langmuir 30: 14832–14840.

Huang, W. Q., H. Q. Xu, Z. R. Ying, Y. Dan, Q. Zhou, J. Zhang et al. 2019. Split TiO_2 nanotubes – Evidence of oxygen evolution during Ti anodization. Electrochem. Commun. 106: 106532.

Hurlen, T. and E. Gulbrandsen. 1994. Growth of anodic films on valve metals. Electrochim. Acta 39: 2169–2172.

Jańczyk, A., E. Krakowska, G. Stochel and W. Macyk. 2006. Singlet oxygen photogeneration at surface modified titanium dioxide. J. Am. Chem. Soc. 128: 15574–15575.

Ji, L., Z. Lin, M. Alcoutlabi and X. Zhang. 2011. Recent developments in nanostructured anode materials for rechargeable lithium-ion batteries. Energy Environ. Sci. 4: 2682–2699.

Kamat, P. 2007. Meeting the clean energy demand: Nanostructure architectures for solar energy conversion. J. Phys. Chem. C 111: 2834–2860.

Kamat, P. 2013. Quantum dot solar cells. The next big thing in photovoltaics. J. Phys. Chem. Lett. 4: 908–918.

Kang, S. H., C. -M. Park, J. Lee and J. -H. Kim. 2015. Electrochemical lithium storage kinetics of self-organized nanochannel niobium oxide electrodes. J. Electroanal. Chem. 746: 45–50.

Kim, P., J. Albarella, J. Carey, M. Placek, A. Sen, A. Wittrig et al. 2008. Towards the development of a portable device for the monitoring of gaseous toxic industrial chemicals based on a chemical sensor array. Sens. Actuators B: Chem. 134: 307–312.

Kim, Y., I. -S. Hwang, S. -J. Kim, C. -Y. Lee and J. -H. Lee. 2008. CuO nanowire gas sensors for air quality control in automotive cabin. Sens. Actuators B 135: 298–303.

Kirchheim, R. 1987. Growth kinetics of passive films. Electrochim. Acta 32: 1619–1629.

Kitano, M., K. Iyatani, K. Tsujimaru, M. Matsuoa, M. Takeuchi, M. Ueshima et al. 2008. Recent advances in visible-light-responsive photocatalysts for hydrogen production and solar energy conversion – from semiconducting TiO_2 to MOF/PCP photocatalysts. M. Top. Catal. 49: 24–31.

Ko, C. -C., A. W. -Y. Cheung and S. M. Yiu. 2015. Synthesis, photophysical and electrochemical study of diisocyano-bridged homodinuclear rhenium(I) diimine complexes. Polyhedron 86: 17–23.

Kuninobu, Y. and K. Takai. 2011. Organic reactions catalyzed by rhenium carbonyl complexes. Chem. Rev. 111: 1938–1953.

Laihonen, S., T. Laitinen, G. Sundholm and A. Yli-Penti. 1990. The anodic behaviour of Sb and Pb-Sb eutectic in sulphuric acid solutions. Electrochim. Acta 35: 229–238.

Laitinen, T., G. Sundholm, J. K. Vilhunen, D. Pavlov and M. Bojinov. 1991. Electrochemical behaviour of the antimony electrode in sulphuric acid solutions-III. Identification of corrosion products after long-term polarization. Electrochim. Acta 36: 2093–2102.

Lee, K., A. Mazare and P. Schmuki. 2014. One-dimensional titanium dioxide nanomaterials: Nanotubes. Chem. Rev. 114: 9385–9454.

Lee, W. -J. and W. H. Smyrl. 2008. Oxide nanotube arrays fabricated by anodizing processes for advanced material application. Curr. App. Phys. 8: 818–821.

Li, D., J. Hu, R. Q. Wu and J. G. Lu. 2010. Conductometric chemical sensor based on individual CuO nanowires. Nanotechnology 21: 485502.

Li, Y., S. Wang, Y. B. He, L. Tang, Y. V. Kaneti, W. Lv et al. 2017. Li-ion and Na-ion transportation and storage properties in various sized TiO_2 spheres with hierarchical pores and high tap density. J. Mater. Chem. A 5: 4359–4367.

Liang, H. C., X. Z. Li and J. Nowotny. 2010. Photocatalytical properties of TiO_2 nanotubes. Solid State Phenom. 162: 295–328.

Linarez Pérez, O. E., V. C. Fuertes, M. A. Pérez and M. López Teijelo. 2008. Characterization of the anodic growth and dissolution of oxide films on valve metals. Electrochem. Commun. 10: 433–437.

Linarez Pérez, O. E., M. A. Pérez and M. López Teijelo. 2009. Characterization of the anodic growth and dissolution of antimony oxide films. J. Electroanal. Chem. 632: 64–71.

Linarez Pérez, O. E., M. D. Sánchez and M. López Teijelo. 2010. Characterization of growth of anodic antimony oxide films by ellipsometry and XPS. J. Electroanal. Chem. 645: 143–148.

Linic, S., P. Christopher and D. B. Ingram. 2011. Plasmonic-metal nanostructures for efficient conversion of solar to chemical energy. Nat. Mater. 10: 911–921.

Liu, M., C. Yan and Y. Zhang. 2015. Fabrication of Nb_2O_5 nanosheets for high-rate lithium ion storage applications. Sci. Rep. 5: 8326.

Liu, X. -W., F. -Y. Wang, F. Zhen and J. -R. Huang. 2012. *In situ* growth of Au nanoparticles on the surfaces of Cu_2O nanocubes for chemical sensors with enhanced performance. RSC Adv. 2: 7647–7651.

Liu, Y., H. Wang, K. Yang, Y. Yang, J. Ma, K. Pan et al. 2019. Enhanced electrochemical performance of Sb_2O_3 as an anode for lithium-ion batteries by a stable cross-linked binder. Appl. Sci. 9: 2677.

Lohrengel, M. M. 1993. Thin anodic oxide layers on aluminium and other calve metals: High field regime. Mater. Sci. Eng. R 11: 243–294.

Macagno, V. and J. W. Schultze. 1984. The growth and properties of thin oxide layers on tantalum electrodes. J. Electroanal. Chem. 180: 150–157.

Macak, J. M., H. Tsuchiya, A. Ghicov, K. Yasuda, R. Hahn, S. Bauer et al. 2007a. TiO_2 nanotubes: self-organized electrochemical formation, properties and applications. Curr. Op. Solid State Mater. Sci. 11: 3–18.

Macak, J. M., M. Zlamal, J. Krysas and P. Schmuki. 2007b. Self-organized TiO_2 nanotube layers as highly efficient photocatalysts. Small 3: 300–304.

Macdonald, D. D., S. R. Biaggio and H. Song. 1992. Steady-state passive films: Interfacial kinetic effects and diagnostic criteria. J. Electrochem. Soc. 139: 170–177.

Macdonald, D. D. 2011. The history of the Point Defect Model for the passive state: a brief review of film growth aspects. Electrochim. Acta 56: 1761–1772.

Machan, C. W., S. A. Chabolla and C. P. Kubiak. 2015. Reductive disproportionation of carbon dioxide by an alkyl-functionalized pyridine monoimine Re(I) *fac*-tricarbonyl electrocatalyst. Organometallics 34: 4678–4683.

Manes, T. A. and M. J. Rose. 2015. Redox properties of a bis-pyridine rhenium carbonyl derived from an anthracene scaffold. Inorg. Chem. Commun. 61: 221–224.

Masuda, H. and K. Fukuda. 1995. Ordered metal nanohole arrays made by a two-step replication of honeycomb structures of anodic alumina. Science 268: 1466–1468.

Matsuoka, M., M. Kitano, S. Fukumoto, K. Iyatani, M. Takeuchi and M. Anpo. 2008. The effect of the hydrothermal treatment with aqueous NaOH solution on the photocatalytic and photoelectrochemical properties of visible light-responsive TiO_2 thin films. Catal. Today 132: 159–164.

Metikoš-Huković, M. and B. Lovreček. 1980. Electrochemical behaviour of the oxide covered antimony. Electrochim. Acta 25: 717–723.

Metikoš-Huković, M., R. Babić and S. Brinić. 1997. Influence of antimony on the properties of the anodic oxide layer formed on Pb-Sb alloys. J. Power Sources 64: 13–19.

Metikoš-Huković, M., R. Babić and S. Brinić. 2006. EIS-*in situ* characterization of anodic films on antimony and lead-antimony alloys. J. Power Sources 157: 563–570.

Milowska, K., A. Rybczynska, J. Mosiolek, J. Durdyn, E. M. Szewczyk, N. Katir et al. 2015. Biological activity of mesoporous dendrimer-coated titanium dioxide: insight on the role of the surface-interface composition and the framework crystallinity. ACS Appl. Mater. Interfaces 7: 19994–20003.

Minami, T., Y. Nishi, T. Miyata and J. Nomoto. 2011. High-efficiency oxide solar cells with ZnO/Cu_2O heterojunction fabricated on thermally oxidized Cu_2O sheets. Appl. Phys. Express 4: 62301.

Mor, G. K., O. K. Varghese, M. Paulose, K. Shankar and C. A. Grimes. 2006. A review on highly ordered, vertically oriented TiO_2 nanotube arrays: fabrication, material properties, and solar energy applications. Sol. Energy Mat. Sol. Cells 90: 2011–2075.

Mukherjee, S., F. Libisch, N. Large, O. Neumann, L. V. Brown, J. Cheng et al. 2013. Hot electrons do the impossible: Plasmon-induced dissociation of H_2 on Au. Nano Lett. 13: 240–247.

Musselman, K. P., A. Wisnet, D. C. Iza, H. C. Hesse, C. Scheu, J. L. MacManus-Driscoll et al. 2010. Strong efficiency improvements in ultra-low-cost inorganic nanowire solar cells. Adv. Mater. 22: E254–E258.

Navarro Yerga, R. M., M. C. Alvarez-Galván, F. Vaquero, J. Arenales and J. L. García Fierro. 2013. Hydrogen production from water splitting using photo-semiconductor catalysts. pp. 43–61. *In*: L. M. Gandía, G. Arzamendi and P. M. Diéguez [eds.]. Renewable Hydrogen Technologies. Elsevier B.V. Amsterdam, Netherlands.

Nazeeruddin, Md. K., R. Humphry-Baker, D. L. Officer, W. M. Campbell, A. K. Burrell and M. Grätzel. 2004. Application of metalloporphyrins in nanocrystalline dye-sensitized solar cells for conversion of sunlight into electricity. Langmuir 20: 6514–6517.

Norek, M., W. J. Stępniowski and D. Siemiaszko. 2016. Effect of ethylene glycol on morphology of anodic alumina prepared in hard anodization. J. Electroanal. Chem. 762: 20–28.

Oprea, A., D. Degler, N. Barsan, A. Hemeryck and J. Rebholz. 2018. Basics of semiconducting metal oxide-based gas sensors. pp. 61–165. *In*: N. Barsan and K. Schierbaum [eds.]. Gas Sensors Based on Conducting Metal Oxides. Elsevier Inc. Amsterdam, Netherlands.

Ou, J. Z., S. Balendhran, M. R. Field, D. G. McCulloch, A. S. Zoolfakar, R. A. Rani et al. 2012. The anodized crystalline WO_3 nanoporous network with enhanced electrochromic properties. Nanoscale 4: 5980–5988.

Oyarzún, D., R. Córdova, O. Linarez Pérez, E. Muñoz, R. Henríquez, M. López Teijelo et al. 2011. Morphological, electrochemical and photoelectrochemical characterization of nanotubular TiO_2 synthetized electrochemically from different electrolytes. J. Solid State Electrochem. 15: 2265–2275.

Oyarzún, D. P., O. E. Linarez Pérez, M. López Teijelo, C. Zúñiga, E. Jeraldo, D. A. Geraldo et al. 2016. Atomic force microscopy (AFM) and 3D confocal microscopy as alternative techniques for the morphological characterization of anodic TiO_2 nanoporous layers. Mater. Lett. 65: 67–70.

Oyarzún, D. P., M. López Teijelo, W. Ramos Cervantes, O. E. Linarez Pérez, J. Sánchez, G. del C. Pizarro et al. 2017. Nanostructuring of anodic copper oxides in fluoride-containing ethylene glycol media. J. Electroanal. Chem. 807: 181–186.

Oyarzún, D. P., S. Chardon-Noblat, O. E. Linarez Pérez, M. López Teijelo, C. Zúñiga, X. Zarate et al. 2018a. Comparative study of the anchorage and the catalytic properties of nanoporous TiO_2 films modified with ruthenium(II) and rhenium(I) carbonyl complexes. Chem. Phys. Lett. 694: 40–47.

Oyarzún, D. P., M. I. Broens, O. E. Linarez Pérez, M. López Teijelo, R. Islas and R. Arratia-Perez. 2018b. Simple and rapid one-step electrochemical synthesis of nanogranular Cu_2O films. Chemistry Select 3: 8610–8614.

Pastva, J., K. Skowerski, S. J. Czarnocki, N. Zilkova, J. Cejka, Z. Bastl et al. 2014. Ru-based complexes with quaternary ammonium tags immobilized on mesoporous silica as olefin metathesis catalysts. ACS Catal. 4: 3227–3236.

Patermarakis, G. 2009. The origin of nucleation and development of porous nanostructure of anodic alumina films. J. Electroanal. Chem. 635: 39–50.

Pavlov, D., M. Bojinov, T. Laitinen and G. Sundholm. 1991a. Electrochemical behaviour of the antimony electrode in sulphuric acid solutions-I. Corrosion processes and anodic dissolution of antimony. Electrochim. Acta 36: 2081–2086.

Pavlov, D., M. Bojinov, T. Laitinen and G. Sundholm. 1991b. Electrochemical behaviour of the antimony electrode in sulphuric acid solutions-II. Formation and properties of the primary anodic layer. Electrochim. Acta 36: 2087–2092.

Pérez, M. A. and M. López Teijelo. 2004. Ellipsometric study of WO_3 films dissolution in aqueous solutions. Thin Solid Films 449: 138–146.

Pérez, M. A. and M. López Teijelo. 2005a. Cathodic behaviour of bismuth. I. Ellipsometric study of the electroreduction of thin Bi_2O_3 films. J. Electroanal. Chem. 583: 212–220.

Pérez, M. A. and M. López Teijelo. 2005b. Ellipsometric study of dissolution of anodic WO_3 films in aqueous solutions. 2. Reaction mechanism. J. Phys. Chem. B 109: 19369–19376.

Pillep, B., P. Behrens, U. -A. Schubert, J. Spengler and H. Közinger. 1999. Mechanical and thermal spreading of antimony oxides on the TiO_2 surface: dispersion and properties of surface antimony oxide species. J. Phys. Chem. B 103: 9595–9603.

Pisarek, M., A. Roguska, A. Kudelski, M. Andrzejczuk, M. Janik-Czachor and K. J. Kurzydłowski. 2013. The role of Ag particles deposited on TiO_2 or Al_2O_3 self-organized nanoporous layers in their behaviour as SERS-active and biomedical substrates. Mat. Chem. Phys. 139: 55–65.

Pogozhev, D. V., M. J. Bezdek, P. A. Schauer and C. P. Berlinguette. 2013. Ruthenium(II) complexes bearing a naphthalimide fragment: a modular dye platform for the dye sensitized solar cell. Inorg. Chem. 52: 3001–3006.

Pourandarjani, A. and F. Nasirpouri. 2018. Tuning substrate roughness to improve uniform growth and photocurrent response in anodic TiO_2 nanotube arrays. Ceram. Int. 44: 22671–22679.

Regonini, D., C. R. Bowen, A. Jaroenworaluck and R. Stevens. 2013. A review of growth mechanism, structure and crystallinity of anodized TiO_2 nanotubes. Mater. Sci. Eng. R Rep. 74: 377–406.

Rezayee, N. M., C. A. Huff and M. S. Sanford. 2015. Tandem amine and ruthenium-catalyzed hydrogenation of CO_2 to methanol. J. Am. Chem. Soc. 137: 1028–1031.

Riboni, F., N. T. Nguyen, S. So and P. Schmuki. 2016. Aligned metal oxide nanotube arrays: Key-aspects of anodic TiO_2 nanotube formation and properties. Nanoscale Horiz. 1: 445–466.

Roguska, A., M. Pisarek, M. Andrzejczuk, M. Dolata, M. Lewandowska and M. Janik-Czachor. 2011. Characterization of a calcium phosphate-TiO_2 nanotube composite layer for biomedical applications. Mater. Sci. Eng. C 31: 906–914.

Ruckh, T., J. R. Porter, N. K. Allam, X. Feng, C. A. Grimes and K. C. Popat. 2009. Nanostructured tantala as a template for enhanced osseointegration. Nanotechnology 20: 45102.

Ruiquan, Y., J. Longfei, Z. Xufei, S. Ye, Y. Dongliang and H. Aijun. 2012. Theoretical derivation of ionic current and electronic current and comparison between fitting curves and measured curves. RSC Adv. 2: 12474–12481.

Salguero Salas, M. A., J. M. De Paoli, O. E. Linarez Pérez, N. Bajales and V. C. Fuertes. 2020. Synthesis and characterization of alumina-embedded $SrCo_{0.95}V_{0.05}O_3$ nanostructured perovskite: an attractive material for supercapacitor devices. Micropor. Mesopor. Mat. 293: 109797.

Sauve, G., M. E. Cass, G. Coia, S. J. Doig, I. Lauermann, K. E. Pomykal et al. 2000. Dye sensitization of nanocrystalline titanium dioxide with osmium and ruthenium polypyridyl complexes. J. Phys. Chem. B 104: 6821–6836.

Schubert, U. -A., F. Anderle, J. Spengler, J. Zühlke, H. -J. Eberle, R. K. Grasselli et al. 2001. Possible effects of site isolation in antimony oxide-modified vanadia/titania catalysts for selective oxidation of oxylene. Top. Catal. 15: 195–200.

Schultze, J. W. and M. M. Lohrengel. 2000. Stability, reactivity and breakdown of passive films. Problems of recent and future research. Electrochim. Acta 45: 2499–2513.

Shakir, M., N. Iram, M. Shoeb Khan, S. I. Al-Resayes, A. Ali Khan and U. Baig. 2014. Electrical conductivity, isothermal stability, and ammonia-sensing performance of newly synthesized and characterized organic-inorganic polycarbazole-titanium dioxide nanocomposite. Ind. Eng. Chem. Res. 53: 8035–8044.

Singha, K., P. Laha, F. Chandra, N. Dehury, A. L. Koner and S. Patra. 2017. Long-lived polypyridyl based mononuclear ruthenium complexes: synthesis, structure, and azo dye decomposition. Inorg. Chem. 56: 6489–6498.

Spata, V. A. and E. A. Carter. 2018. Mechanistic insights into photocatalyzed hydrogen desorption from palladium surfaces assisted by localized surface plasmon resonances. ACS Nano 12: 3512–3522.

Stancheva, M. and M. Bojinov. 2013. Interfacial and bulk processes during oxide growth on titanium in ethylene glycol-based electrolytes. J. Solid State Electrochem. 17: 1271–1283.

Stepniowski, W. J. and W. Z. Misiolek. 2018. Review of fabrication methods, physical properties, and applications of nanostructured copper oxides formed via electrochemical oxidation. Nanomaterials 8: 379.

Sulka, G., L. Zaraska and W. J. Stępniowski. 2011. Anodic porous alumina as a template for nanofabrication. pp. 261–349. *In*: H. S. Nalwa [ed.]. Encyc. Nanosci. Nanotechnol. American Science Pub. USA.

Sun, M. and H. Xu. 2012. A novel application of plasmonics: plasmon-driven surface-catalyzed reactions. Small 8: 2777–2786.

Sun, Z., L. Xu, W. Guo, B. Xu, S. Liu and F. Li. 2014. Enhanced photoelectrochemical performance of nanocomposite film fabricated by self-assembly of titanium dioxide and polyoxometalates. J. Phys. Chem. C 114: 5211–5216.

Syrek, K., L. Zaraska, M. Zych and G. D. Sulka. 2018. The effect of anodization conditions on the morphology of porous tungsten oxide layers formed in aqueous solution. J. Electroanal. Chem. 829: 106–115.

Sze, S. M. and K. K. Ng. 2007. Physics of Semiconductor Devices. John Wiley and Sons. New Jersey, USA.

Tahir, D. and S. Tougaard. 2012. Electronic and optical properties of Cu, CuO and Cu_2O studied by electron spectroscopy. J. Phys. Condens. Matter. 24: 175002–175010.

Takeuchi, M., H. Yamashita, M. Matsuoka, M. Anpo, T. Hirao and N. Iwamoto. 2000. Photocatalytic decomposition of NO under visible light irradiation on the Cr-Ion-implanted TiO_2 thin film photocatalyst. Catal. Lett. 66: 185–187.

Tan, E. -Z., P. G. Yin, T. T. You, H. Wang and L. Guo. 2012. Three dimensional design of large-scale TiO_2 nanorods scaffold decorated by silver nanoparticles as SERS sensor for ultrasensitive malachite green detection. ACS Appl. Mater. Interfaces 4: 3432–3437.

Tan, Y., L. Chen, H. Chen, Q. Hou and X. Chen. 2018. Synthesis of a symmetric bundle-shaped Sb_2O_3 and its application for anode materials in lithium ion batteries. Mater. Lett. 212: 103–106.

Tao, B., Y. Deng, L. Song, W. Ma, Y. Qian, C. Lin et al. 2019. BMP2-loaded titania nanotubes coating with pH-responsive multilayers for bacterial infections inhibition and osteogenic activity improvement. Colloids Surf. B 177: 242–252.

Terracciano, M., V. Galstyan, I. Rea, M. Casalino, L. De Stefano and G. Sbervegleri. 2017. Chemical modification of TiO_2 nanotube arrays for label-free optical biosensing applications. Appl. Surf. Sci. 419: 235–240.

Torresan, M. F., A. M. Baruzzi and R. A. Iglesias. 2016. Thermal annealing of photoanodes based on CdSeQdots sensitized TiO_2. Sol. Energy Mater. Sol. Cells 155: 202–208.

Torresan, M. F., A. M. Baruzzi and R. A. Iglesias. 2017. Enhancing the adsorption of CdSe quantum dots on TiO_2 nanotubes by tuning the solvent polarity. Sol. Energy Mater. Sol. Cells 164: 107–113.

Tricoli, A., M. Righettoni and A. Teleki. 2010. Semiconductor gas sensors: Dry synthesis and application. Angew. Chem., Int. Ed. 49: 7632–7659.

Tsuchiya, H. and P. Schmuki. 2005. Self-organized high aspect ratio porous hafnium oxide prepared by electrochemical anodization. Electrochem. Commun. 7: 49–52.

Valenti, G., M. Panigati, A. Boni, G. D'Alfonso, F. Paolucci and L. Prodi. 2014. Diazine bridged dinuclear rhenium complex: new molecular material for the CO_2 conversion. Inorg. Chim. Acta 417: 270–273.

Vargas, R., D. Carvajal, B. Galavis, A. Maimone, L. Madriz and B. R. Scharifker. 2019. High-field growth of semiconducting anodic oxide films on metal surfaces for photocatalytic application. Int J. Photoenergy 2571906.

Vázquez, C. I., A. M. Baruzzi and R. A. Iglesias. 2016. Charge extraction from TiO_2 nanotubes sensitized with CdS quantum dots by SILAR method. IEEE J. Photovolt. 6: 1515–1521.

Vázquez, C. I. and R. A. Iglesias. 2018. Engineered nanomaterials in energy production industry. *In*: C. M. Hussain [ed.]. Handbook of Nanomaterials for Industrial Applications. Elsevier Inc. Amsterdam, Netherlands.

Vetter, K. J. 1971. General kinetics of passive layers on metals. Electrochim. Acta 16: 1923–1937.

Vogel, R., P. Hoyer and H. Weller. 1994. Quantum-sized PbS, CdS, Ag_2S, Sb_2S_3, and Bi_2S_3 particles as sensitizers for various nanoporous wide-bandgap semiconductors. J. Phys. Chem. 98: 3183–3188.

Vollmer, M. V., C. W. Machan, M. L. Clark, W. E. Antholine, J. Agarwal, H. F. Schaefer III et al. 2015. Synthesis, spectroscopy, and electrochemistry of (α-diimine)M(CO)$_3$Br, M = Mn, Re, complexes: ligands isoelectronic to bipyridyl show differences in CO_2 reduction. Organometallics 34: 3–12.

Wang, B., X. -L. Wu, C. -Y. Shu, Y. -G. Guo and C. -R. Wang. 2010. Synthesis of CuO/graphene nanocomposite as a high-performance anode material for lithium-ion batteries. J. Mater. Chem. 20: 10661–10664.

Wang, B., M. Anpo and X. Wang. 2018. Visible light-responsive photocatalysts from TiO_2 to carbon nitrides and boron carbon nitride. Adv. Inorg. Chem. 72: 49–92.

Wang, J., W. Li, F. Wang, Y. Y. Xia, A. M. Asiri and D. Y. Zhao. 2014. Controllable synthesis of SnO_2@C yolk-shell nanospheres as a high-performance anode material for lithium ion batteries. Nanoscale 6: 3217–3222.

Wang, W., Y. Himeda, J. T. Muckerman, G. F. Manbeck and E. Fujita. 2015. CO_2 hydrogenation to formate and methanol as an alternative to photo- and electrochemical CO_2 reduction. Chem. Rev. 115: 12936–12973.

Wang, X., M. Sun, M. Muruganantham, Y. Zhang and L. Zhang. 2020. Electrochemically self-doped WO_3/TiO_2 nanotubes for photocatalytic degradation of volatile organic compounds. Appl. Catal. B Environ. 260: 118205.

Watanabe, K., D. Menzel, N. Nilius and H. -J. Freund. 2006. Photochemistry on metal nanoparticles. Chem. Rev. 106: 4301–4320.

Wei, W., J. M. Macak and P. Schmuki. 2008. High aspect ratio ordered nanoporous Ta_2O_5 films by anodization of Ta. Electrochem. Commun. 10: 428–432.

Wen, H., Z. Liu, J. Wang, Q. Yang, Y. Li and J. Yu. 2011. Facile synthesis of Nb_2O_5 nanorod array films and their electrochemical properties. App. Surf. Sci. 257: 10084–10088.

Wikstrom, L. L., N. T. Thomas and K. Nobe. 1975. Electrode kinetics of antimony in acidic chloride solutions. J. Electrochem. Soc. 122: 1201–1206.

Williamson, G. A., V. W. Hu, T. B. Yoo, M. Affandy, C. Opie, E. K. Paradis et al. 2019. Temperature-dependent electrochemical characteristics of antimony nanocrystal alloying electrodes for Na-ion batteries. ACS Appl. Energy Mater. 2: 6741–6750.

Xiang, J. Y., J. P. Tu, L. Zhang, Y. Zhou, X. L. Wang and S. J. Shi. 2010. Self-assembled synthesis of hierarchical nanostructured CuO with various morphologies and their application as anodes for lithium ion batteries. J. Power Sources 195: 313–319.

Xu, Q., Y. Yin, T. Gao, G. Cao, Q. Chen, C. Lan et al. 2020. Sputter deposition of Ag-induced WO_3 nanoisland films with enhanced electrochromic properties. J. Alloys Compd. 829: 154431.

Ye, C., G. Wang, M. Kong and L. Zhang. 2006. Controlled synthesis of Sb_2O_3 nanoparticles, nanowires, and nanoribbons. J. Nanomater. 95670.

Yu, M. S., Y. Chen, C. Li, S. Yan, H. M. Cui, X. F. Zhu et al. 2018. Studies of oxide growth location on anodization of Al and Ti provide evidence against the field-assisted dissolution and field-assisted ejection theories. Electrochem. Commun. 87: 76–80.

Zaffora, A., R. Macaluso, H. Habazaki, I. Valov and M. Santamaria. 2018. Electrochemically prepared oxides for resistive switching devices. Electrochim. Acta 274: 103–111.

Zhang, F., A. W. Zhu, Y. P. Luo, Y. Tian, J. H. Yang and Y. Qin. 2010. CuO nanosheets for sensitive and selective determination of H_2S with high recovery ability. J. Phys. Chem. C 114: 19214–19219.

Zhang, H., Q. S. Zhu, Y. Zhang, Y. Wang, L. Zhao and B. Yu. 2007. One-pot synthesis and hierarchical assembly of hollow Cu_2O microspheres with nanocrystals-composed porous multishell and their gas-sensing properties. Adv. Funct. Mater. 17: 2766–2771.

Zhang, J., W. Q. Huang, K. Zhang, D. Z. Li, H. Q. Xu and X. F. Zhu. 2019. Bamboo shoot nanotubes with diameters increasing from top to bottom: evidence against the field-assisted dissolution equilibrium theory. Electrochem. Commun. 100: 48–51.

Zhang, K., S. K. Cao, S. Li, J. Qi, L. F. Jiang, J. Zhang et al. 2019. Rapid growth of TiO_2 nanotubes under the compact oxide layer: evidence against the digging manner of dissolution reaction. Electrochem. Commun. 103: 88–93.

Zhang, S., D. L. Yu, D. D. Li, Y. Song, J. F. Che, S. Y. You et al. 2014. Forming process of anodic TiO_2 nanotubes under a preformed compact surface layer. J. Electrochem. Soc. 161: E135–E141.

Zhang, S., S. Xu, D. Hu, C. Zhang, J. Che and Y. Song. 2018. Formation of TiO_2 nanoribbons by anodization under high current density. Mater. Res. Bull. 103: 205–210.

Zhang, Y., X. He, J. Li, H. Zhang and X. Gao. 2007. Gas-sensing properties of hollow and hierarchical copper oxide microspheres. Sens. Actuators B 128: 293–298.

Zhang, Y., S. He, W. Guo, Y. Hu, J. Huang, J. R. Mulcahy et al. 2018. Surface-plasmon-driven hot electron photochemistry. Chem. Rev. 118: 2927–2954.

Zhou, X., N. T. Nguyen, S. Özkan and P. Schmuki. 2014. Anodic TiO_2 nanotube layers: why does self-organized growth occur—A mini review. Electrochem. Commun. 46: 157–162.

Zhou, X. Z., Z. F. Zhang, J. W. Wang, Q. T. Wang, G. F. Ma and Z. Q. Lei. 2017. Sb_2O_4/reduced graphene oxide composite as high-performance anode material for lithium ion batteries. J. Alloys Compd. 699: 611–618.

Zhu, W., Y. Liu, A. Yi, M. Zhu, W. Li and N. Fu. 2019. Facile fabrication of open-ended TiO_2 nanotube arrays with large area for efficient dye-sensitized solar cells. Electrochim. Acta 299: 339–345.

Zoolfakar, A. S., R. A. Rani, A. J. Morfa, S. Balendhran, A. P. O'Mullane, S. Zhuiykov et al. 2012. Enhancing the current density of electrodeposited ZnO-Cu_2O solar cells by engineering their heterointerfaces. J. Mater. Chem. 22: 21767–21775.

Zoolfakar, A. S., M. Z. Ahmad, R. A. Rani, J. Z. Ou, S. Balendhran, S. Zhuiykov et al. 2013. Nanostructured copper oxides as ethanol vapour sensors. Sensor Actuat. B-Chem. 185: 620–627.

Zoolfakar, A. S., R. A. Rani, A. J. Morfa, A. P. O'Mullaned and K. Kalantar-zadeh. 2014. Nanostructured copper oxide semiconductors: a perspective on materials, synthesis methods and applications. J. Mater. Chem. C 2: 5247–5270.

CHAPTER 4

Implications of Structure in Properties of Micro and Mesoporous Carbons

Francisco J. García-Soriano,[1] *Manuel Otero,*[1] *M. Laura Para,*[1] *Alexis Paz,*[2]
M. Belén Suárez Ramazin[1] and *C. Andrea Calderón*[1,*]

Introduction

Porous solid materials are very interesting both from a scientific and technological point of view due to their ability to interact with different types of atoms and ions not only through their high surface area but also through the bulk material. Advances in synthesis techniques now allow the design of these porous materials at a nanometric scale. Therefore, these materials have great versatility, a wide range of pore sizes and morphology, a variety of porous networks, controllable length scales for well-defined morphologies and chemical functionality. In this manner, they are able to meet a wide range of requirements and properties for different applications (Zhao et al. 2006). In a porous material, not only are the structures generated by the void of the pores important but also the solid structure of the atoms that constitute the bulk, which would be responsible for the specific properties of the material itself. Although the porous structure can be found in various types of materials such as zeolites, silica, etc., in this chapter the focus will be particularly on carbonaceous materials.

Carbon is a widely used element in material science as it is an abundant element in the biosphere, which makes it more sustainable than other materials. It can be prepared even from organic waste making it cheap, there are simple synthetic methods which can be used to scale the production and it has good chemical stability. These are some characteristics that make this material used on a wide spectrum of technological applications (Titirici et al. 2015). Nanostructured porous carbons are materials with structure and texture controlled at the nanometer scale, with very interesting properties such as large Specific Surface Areas (SSA) and pore volumes, excellent thermal, chemical and mechanical stabilities. Therefore, this type of material, according to its nanoarchitectures in terms of pore sizes and surface chemistry, has demonstrated exceptional performance in a variety of applications related to energy storage, environmental cleaning and sense issues, like water purification, gas adsorption and separation, catalyst supports, energy conversion and storage, electrocatalysis, among others (Tian et al. 2020).

In this chapter, different types of porous carbons are presented with their classification according to structures and characteristics. In addition, a review of the different synthesis methods is carried out, with the advantages and disadvantages of each, as well as a discussion about which

[1] IFEG, Facultad de Matemática Astronomía y Física, Universidad Nacional de Córdoba-CONICET. 5000 Córdoba, Argentina.
[2] INFIQC, Facultad de Ciencias Químicas, Universidad Nacional de Córdoba-CONICET. 5000 Córdoba, Argentina.
* Corresponding author: acalderon@famaf.edu.ar

methods are more appropriate to obtain porous carbons of specific characteristics, such as pore size, hierarchical structure, chemical functionalization, etc. Different templating methods are discussed as well as some synthetic routes using biomass as a precursor as a physical or chemical activation steps. For nanostructured materials in technological applications, a deep understanding of the different reaction mechanisms that occur in the device is necessary to design the optimal material for this purpose. Thus, theoretical simulations become crucial in nanostructured materials studies. Therefore, a summary of how the different theoretical models are built as well as their applications for the study of different properties such as structures, adsorption, mass diffusion, conductivity, etc., can be made. Finally, different applications of these porous carbon materials are discussed, particularly in the field of renewable energy. An analysis of the use of nanostructured porous carbon on several energy storage devices as Lithium-Ion Batteries (LIBs), Lithium Sulfur Batteries (LSBs), sodium-ion batteries (NIBs) and supercapacitors is presented. For each device, the energy storage mechanisms as well as how the different characteristics of the porous carbons influence the material behavior will be discussed. Finally, some strategies used to obtain greater storage capacities, better electrochemical behavior and better stability are summarized.

Structure and Synthesis Methods of Porous Carbons

Structure and Classification

Porous carbons have high SSA produced by nanopore networks inside the material. According to IUPAC, nanopores are cavities with apertures up to around 100 nm which can be categorized as microporous (pore size $\leq$ 2 nm), mesoporous (2 nm $\leq$ pore size $\leq$ 50 nm) and macroporous (pore size > 50 nm) depending on their pore size (Rouquerol et al. 1994). Additionally, microporous can be sub-divided into super (> 0.7 nm) and ultra-microporous (< 0.7 nm). When different types of pores are present following a hierarchical pattern, the carbon is called hierarchically porous, with the micropores branching off from the mesopores, and those from the macropores, which in turn open out to the external surface of the particles. Compared to conventional porous materials consisting of uniform pore dimensions, these interconnected structures have a synergistic effect for different applications: they can exhibit minimized diffusive resistance to mass transport from macropores and high active surface area from micro- and/or mesopores dispersion (Dutta et al. 2014). Furthermore, the crystalline structure of the carbon can be ordered or disordered, emerging as an extra classification of graphitic/amorphous materials (Tian et al. 2020).

Another important issue is the elemental composition and surface chemistry of the nanoporous carbon. Although it is well known that carbon atoms are the major element present in the porous skeleton, the elemental composition of these materials also includes hydrogen, oxygen, nitrogen, sulfur or phosphorus-containing groups, depending on the precursor, preparation route and post-synthesis functionalization. The elemental composition and type of surface groups of a nanoporous carbon influence its performance either in gaseous or liquid phase processes, due to specific interactions with the adsorptive or solvent. This ability of porous carbon materials could be advantageous or not, depending on the desired application (Mestre and Carvalho 2018).

Synthetic Methods

There have been many efforts responsibly for the analysis and the synthesis on demand regarding pore sizes, surface chemistry and structure of porous carbon materials. Materials with different characteristics and properties can be produced with different synthetic approaches. It is important to understand the effect that different synthesis conditions have on the properties to design the appropriate materials for different applications (Tian et al. 2020). Here strategies for synthesizing, controlling and functionalizing porous carbons are summarized.

Templating Methods

The templating method is a key synthesis strategy to introduce controlled and ordered porosity on a material. The templating processes are divided according to the template material. When hard and stiff materials are used as a template the process is a hard-templating method. Nanocasting is a type of hard templating method where void spaces on a rigid mold to be filled with the material are in nanoscale and is one of the most effective approaches to obtain order nanoporous materials (Lu and Schüth 2006). This synthetic route consists of different steps: first, synthesize the hard templates; second, fill the hard templates voids with a carbon precursor; third chemical or thermal treatments to transform the precursors into solid carbon; and fourth remove the sacrificial templates by a chemical etching (Knossalla et al. 2017). The advantage of the hard-templating method is that it can produce negative replicas of many different templates which allows precise control on pore size and morphology. Nevertheless, is a time-consuming method and needs hazardous chemicals to remove the template. As an example, CMK3 is a popular nanoporous carbon material prepared by a hard-template method, where template material is mesoporous silica SBA-15 which is impregnated with a carbon source (usually a carbohydrate), carbonized with heat treatment in an inert atmosphere and finally washed with NaOH or HF solution to dissolve the silica (Jun et al. 2000). The CMK3 obtained is a highly ordered mesoporous material with high SSA and structured pores (Ryoo et al. 2001).

Salt Melt Synthesis (SMS) use molten inorganic salts as a medium to produce nanostructured carbons. Different types of salts (LiCl, KCl, NaCl, etc.), can be used as hard templates, however, the drawback is that the operating temperature should be high enough to work with the melted salt. The advantage of this method is that salts are water-soluble and many of them are not toxic which facilitate the product separation. They can also be collected after the synthesis for reuse (Liu et al. 2013).

There are synthesis protocols that propose some modifications to these two general methods, for example, *in situ* templates and soft templates (Petkovich and Stein 2013). The hard templated can be prepared *in situ* from a solvent or solute and used in combination with soft templates. The advantages of this method compared to pre-synthesized hard templates is the elimination of infiltration and coating steps. It is important to have precise control of template synthesis to prepare high-quality porous carbon materials.

To prepare mesoporous carbon with a soft-templating method pore-forming and carbon-yielding components are mixed in a solvent. Both components need to have the ability to self-assemble into nanostructures, and a pore-forming component needs to be stable during carbon-yielding component but decompose at a carbonization temperature so it can be readily removed (Liang et al. 2008). The surfactant used as the pore-forming component can be divided into anionic, cationic and non-ionic surfactants (e.g., Triton X-100, Pluronics) (Petkovich and Stein 2013). In general, the smallest pores are obtained with ionic surfactants (micropores or small mesopores), while non-ionic surfactants generate mesopores from several to over 10 nm (Soler-Illia et al. 2002). Attempts to obtain mesoporous carbons by the soft template method using amphiphilic molecules as surfactant has not been successful as yet because it is difficult to obtain a stable structure after carbonization (Liang et al. 2008). Highly ordered porous carbon films were prepared by a soft-templating method using a self-assembled block polymer system prepared with a PS-P4VP/resorcinol–formaldehyde composes (Liang et al. 2004).

Multiple templating methods are conformed by protocols that combine hard and soft templating. It allows the production of hierarchically porous, although they are usually more expensive and time-consuming because of the multiple synthesis steps (Tian et al. 2020).

Different types of porous carbons can be obtained with templating methods by correctly choosing the template, carbon source, carbonization temperature, etc. (Zhao et al. 2006). Microporous carbon with graphitic walls is a very interesting material since it combines the good conduction of graphite with a porous structure, but its preparation is not simple. This type of carbon was obtained by nanocasting using zeolites as a template by Kyotani et al. (Kyotani et al. 1997). The synthesis of this material involves techniques as CVD allowing the formation of hexagonal atomic C rings, and temperature treatments to allow the carbon inside the zeolite pores and its graphitization.

Disordered microporous carbon structures can be transferred by hard templating, followed by a chemical activation (by adding $FeCl_3$, alkali hydroxide or salts with alkali ions). It is important to note that the carbonization step has a crucial impact on the properties of the samples. A microporous carbon can be acquired, although in general it has both meso and macropores. Micropores are highly favored in different applications such as carbon capture and also enhancing the specific capacities in electrochemical energy storage devices (Tian et al. 2020).

Synthesis from Biomass

Templating methods allow designing the properties of the carbon porous material as texture, structure, morphology and surface chemistry by choosing synthetic conditions. Nevertheless, the harmful nature of the carbon precursors traditionally used in these syntheses has motivated research on more sustainable alternative approaches, based on renewable raw materials. In this manner, biomass or biowaste obtained from agricultural residues, industrial or domestic wastes, emerge as an economical and renewable material precursor of nanoporous carbons (Figueiredo 2018). A usual synthetic route to prepare nanoporous carbon from biomass involves pyrolysis, activation and carbonization under oxygen-deficient conditions steps (Chen et al. 2020). Materials obtained through this synthetic route have been extensively studied because of their naturally diverse and high porosity, conductivity, versatility for modification, as well as the abundance and renewable nature of the starting materials, and feasibility to be produced on a large scale. Lignocellulosic precursors are widely used as biomass for carbon production because they are abundant and renewable, are of low cost and diverse. Furthermore, due to the high reactivity of the biomass, the synthesis is relatively simple and it is possible to obtain a high carbon yield and porous product (Yu et al. 2019, González-García 2018). During the pyrolysis step, biomass is heated at a temperature between 300–1000°C in the absence of oxygen, discomposing in gases, liquid bio-oil, and solid bio-charcoal. Pyrolysis is a complex physicochemical process and its behavior depends on biomass characteristics, used parameters, reactor condition, etc. (Chen et al. 2020). High-temperature direct carbonization involves the pyrolysis of biomass in an inert gas atmosphere at high temperature as a unique synthesis step. It is a facile and green method to convert biomass into value-added carbon materials without using any expensive chemical reagents and complex installations, including high-temperature direct carbonization, physical and chemical activations and hydrothermal carbonization. During this particular pyrolysis C, H and O bonds are broken, and then recombines into H_2O, CO_2, and other gases, producing porous carbon materials with a large number of different pore sizes. However, this method has certain requirements for the composition and structure of biomass starting materials and it is difficult to prepare carbon materials with high porosity and large SSA, thus the products generally require further activation treatment in order to change the internal structure of carbon materials and improve their properties (Liu et al. 2019).

To increase the carbon SSA and to improve the porous and graphitic structures, activation methods are used. These methods can be divided into physical and chemical activation, depending on the activator nature. Physical activation mainly includes a high-temperature activation step. During the process, steam, CO_2 and other gases are introduced to react with the material to produce the porous structure on the surface of the material (Liu et al. 2019, Li et al. 2020). Physical activation is a simple and clean method because carbon products do not need to be cleaned and can be used directly. However, the reactivity of the gas activator selected is relatively weak, and it is difficult to form a developed pore structure (Mestre and Carvalho 2018).

Chemical activation has advantages over the physical process related to the use of lower temperatures, usually higher yields, shorter activation time (hours) and higher SSA and pore volume. In the chemical activation method, the carbon precursors or biochar is mixed with the activating agent and the mix is further heat-treated in an inert atmosphere at temperatures ranging from 400 to 900°C, depending on the selected activating agent. The solid product obtained needs to be washed with water to remove the chemicals and dried before storage. The most common activators used are $ZnCl_2$, H_3PO_4, KOH, NaOH, K_2CO_3. It is worth noting that the mechanism of pore formation is dependent on the chemical agent, and consequently, the microstructure obtained will depend on it (Molina-Sabio and Rodríguez-Reinoso 2004). Normally H_3PO_4 and $ZnCl_2$ are used to activate lignocellulosic materials that have not previously been carbonized, however, the reaction mechanism is different: $ZnCl_2$ promotes the removal of water molecules from the lignocellulosic structures of the raw material while H_3PO_4 chemically combines y with them (Nakagawa et al. 2007). H_3PO_4 also in comparison with $ZnCl_2$ has fewer restrictions on environmental and toxicological contamination and requires a lower activation temperature (Heidarinejad et al. 2020). On the other hand, KOH and other metal compounds are used to activate the precursors of charcoal. In this case, the mechanism consists of the charcoal etching by the K-containing species (KOH, K_2CO_3) through redox reaction, resulting in a great amount of micro/mesopores. Intermediate metallic potassium which can intercalate into the carbon framework and expand the carbon lattices, leading to a highly porous structure. Water vapor also produced during the activation process contributes to the gasification of carbon, which further helps to develop the porosity (Wang and Kaskel 2012).

In porous carbons preparation from biomass, activator agents, the biomass type, as well as the ratio between them and the activation conditions will have a great impact on carbon porosity and surface chemical structure (Li et al. 2020). For instance, Wu et al. 2017 synthesized four kinds of biomass carbon materials obtained from mushrooms using H_3PO_4, K_2CO_3, KOH and $ZnCl_2$ as activators, showing that the activator can effectively tailor pore size, pore volume and surface area of biomass carbon (Wu et al. 2017). On the other hand, Apaydin-Varol and Erülken studied the structural and morphological changes occurring during pyrolysis and activation steps of an arid-land plant Euphorbia Rigida. The solid product obtained was impregnated with different chemical activates (HCl, KOH, K_2CO_3, H_2SO_4 H_3PO_4, NaOH and $ZnCl_2$) and a second thermal treatment was applied under different activation atmospheres. It was observed that the impregnation material, impregnation ratio and activation techniques have a strong influence on the yield and porous texture of the resulting carbons, with K_2CO_3 being the most effective chemical agent producing a carbon with 1079 $m^2 g^{-1}$ of SSA and 0.443 $cm^3 g^{-1}$ of micropore volume (Apaydin-Varol and Erülken 2015).

The process temperature is another important factor which will determine the microstructure of porous carbon. Yuan et al. studied the temperature effect on the porous structure of carbon materials got from peanut dregs through simple carbonization and chemical activation with different temperatures, using KOH as an activator agent. They obtained a highly-disordered carbon material with an ultra high specific superficial area with narrow pore size distribution and low graphitization degree at a low temperature, while a mesoporous carbon with graphene-like structure and high graphitization degree, could be procured at high temperature (Yuan et al. 2019).

A hydrothermal method is an alternative way to obtain porous carbon from a biomass source. This synthesis process is carried out by mixing a carbon source and an aqueous solution in a closed system with an inert atmosphere which is heated and pressurized. The critical conditions increase the activity of ions and molecules and accelerate the carbonization process converting the biomass into carbon materials with various oxygen-containing functional groups and a good morphology structure. However, products from hydrolysis and degradation of biomass cannot get away from the system and the porosity and SSA of the porous carbon materials are generally low (Liu 2019).

Virtual Porous Carbons (VPC)

The focus here will be on molecular models of carbon materials and in particular on Virtual Porous Carbons (VPCs). There are some considerations that should be kept in mind. First, this not an exhaustive review on VPCs. Many other interesting works in the bibliography can serve for such proposes (Biggs and Buts 2006, Bandosz et al. 2003, Bahamon et al. 2019, Gonciaruk and Siperstein 2015, Terzyk et al. 2012). These works also constitute the main references used in this text. Here an attempt has been made to give a brief introduction that is considered of broad interest to the general reader together with a selection of different VPCs applications. Second, the different theories and analytical models traditional used to describe porosity and physical adsorption are not addressed.

These theories are frequently used to determine the surface area and pore size of materials by the interpretation of the experimental adsorption isotherms. The interested reader can refer to many other books and reviews that cover these theories to (Condon 2019, Terzyk et al. 2012, Nicholson 1996) and references therein.

Building VPC Models

The most simple molecular model used to treat nanoporous carbons is the slit pore model (Emmett 1947, Nicholson 1996, Lastoskie et al. 1993, Seaton et al. 1989, Seaton et al. 1997). In its basic form, this model considers each porous as the space between two infinite blocks of graphite. Different pores are isolated (unconnected) and have a width distribution given by the experimental Pore Size Distribution (PSD) of the material. This model has been extensively used because of its simple interpretation and the low computational cost. However, the model does not consider the extent, thickness, deformation, and edges of the pore walls nor the pore system topology, making it frequently inaccurate for the consideration of many adsorptions, diffusion, isotherm hysteresis or functional groups effects. These shortcomings have led to many different extensions of the original slit model to include many characteristics like pore wall thickness distributions, chemical heterogeneity, etched surfaces, finite sizes, pore networks, and/or corrugation (Jorge et al. 2002, Vishnyakov et al. 1998, Seaton et al. 1997, Bhatia 2002, Jagiello et al. 2011, Jagiello and Olivier 2013).

The term 'Virtual Porous Carbon' (VPC) was introduced by Biggs et al. (Biggs and Buts 2006, Bandosz et al. 2003) to refer to the computer-based models of nanoporous carbons that are susceptible of being n used in molecular simulations. These authors emphasize that VPCs goes beyond the ubiquitous slit pore model and distinguish two different ways for building VPC models: the mimetic and reconstruction approaches. The first one attempts to mimic the process used to manufacture the solid. The last one is used to build computational models that are required to match certain experimental structural data. The structural and chemical complexity of the different carbon materials make the mimetic approach a difficult task, which also requires to designing an ad-hoc simulation scheme for each one. On the other hand, mimetic approaches provide a deeper understanding of the synthesis process which could result in fine tuning of material properties or additional protocol improvements.

Following Biggs and Butts (Biggs and Buts 2006), reconstruction approaches can be further divided into two groups according to the kind of experimental properties aimed to reproduce in the target solid. The named 'top-down' reconstructions are used to create VPCs that match characteristics such as the density or the porosity of the material. On the other hand, if the VPCs are built from considering atomic characteristics such as the pair distribution function, the methods are considered 'bottom-up' reconstructions. In principle, these approaches could generate the real atomistic structure of the material, although they require larger computational resources or otherwise be restricted to smaller VPCs volumes.

VPC models developed by Biggs and coworkers are constructed using a set of predefined Basic Building Elements (BBEs) that are assembled into VPC using different methodologies (Biggs and Buts 2006). The first of these models is the result of a 'top-down' reconstruction methodology where a set of graphite crystallites were randomly distributed in a cubic lattice to give a solid with a particular porosity (Biggs and Agarwal 1992). Since then, many BBEs like graphene layers, polyaromatic molecules or the addition of functional groups has been used (Biggs et al. 2004a,b, Cai et al. 2008). Different algorithms to assembly the BBEs have also been proposed (Faulon et al. 1993, Biggs et al. 2004b, Biggs and Buts 2006, Thomson and Gubbins 2000). In the case of 'bottom-down' reconstruction methods, these algorithms aim to match or minimize the difference between the experimental and modelled atomic Pair Distribution Functions (PDF) (Petkov et al. 1999).

Thomson and Gubbins (Thomson and Gubbins 2000) built a VPCs combining two reconstruction steps. First, they followed a top-down method to obtain a solid of a specific density, using as BBEs polyaromatic plates of variable shapes and sizes. They then used a method named reverse Monte

Carlo that creates new aromatic rings from the plate borders. Plates are also rotated and translated and deleted or created to compensate the density changes. This process is iterated until the target experimental PDF is obtained, as in a 'bottom-down' reconstruction. For further details on VPC models see references (Biggs and Buts 2006, Furmaniak 2013, Terzyk et al. 2012, Bahamon et al. 2019).

Once the VPC is created, it can be used in molecular simulations to study equilibrium thermodynamics and kinetics associated with these systems. Adsorption and diffusion, separation of different adsorbent materials, the behavior of confined fluids, phase equilibria, and structural properties are some of the processes that can be studied (Bahamon et al. 2019, Gonciaruk and Siperstein 2015, Do and Do 2006, Wongkoblap and Do 2007, Cai et al. 2008, Herrera and Do 2009). Molecular simulation can shed light on the relevance of the many different variables of these processes, variables that experiments are not able to control or analyze in a direct and systematic study. Molecular Dynamics (MD) and Monte Carlo (MC) simulations, the two main groups of simulation methods, have many applications in porous carbons. Although the length scale of these materials is hierarchical by nature, the progress in high-performance computing combined with the constant develops in advanced simulation techniques like multiscale simulations, allows to describe the micro and mesoporous process, improving one's understanding of these systems, the experimental interpretations, and the prediction and design of new materials.

VPC Applications

Virtual Porous Carbon (VPC) models have a wide variety of applications. The constant development of new models and the increase in computing power allows addressing larger, more complex and realistic systems. During the last few decades progress has developed the application of VPC models to the characterization and industrial design of new materials. The different uses of these models have been classified into four categories (Biggs and Buts 2006): structural elucidation, fundamental study, assessment of simpler models and design.

Earlier works had focused on the use of VPC models as a complementary tool to analyze experimental results with the attempt to elucidate structural characteristics like porosity, density, average inter-layer spacing, among others. The work of Peterson et al. in 2004 (Petersen et al. 2004) correlated experimentally obtained pair correlation functions and a Hybrid Reverse Monte Carlo (HRMC) model to describe an industrial disordered carbonaceous material in the nanoscale. The modeled structure included disordered graphitic sheets and a small percentage of interstitial carbon atoms in a diamond-like arrange. Previously Foley and coworkers (Acharya et al. 1999, Petkov et al. 1999) had used the PDF to model and describe the influence of the pyrolysis temperature in the structure of nanoporous carbon obtained from polyfurfuryl alcohol precursors. Nevertheless, they pointed out that this analysis had shortcomings in the case of highly disordered carbons and more experimental techniques should be taken into account. Later the work of Mi et al. (Mi and Shi 2013) concluded that the Pore Size Distribution (PSD) obtained from experimental isotherms were not discernible influenced by topological imperfections as pentagon rings, turning them ineffective to elucidate the defects in the structure of nanoporous carbons (NPC). They proposed the combination of VPC models with structure factor S(q) measurements from X-ray experiments as an effective method to study these defects. Using different virtual synthesis procedures, they generated a large dataset of realistic NPC models that could be used to elucidate the structure of a particular material from experimental results. In parallel to HRMC methods (Palmer and Gubbins 2012), described how the Quench Molecular Dynamics (QMD) method allows modeling defective NPC. This method is based on a pseudo-mimetic approach: carbon is thermally quenched from a starting liquid of vapor phase into a metastable disordered NPC. Since the simulation quench rate plays an analogous role in the experimental synthesis temperature, QMD methods allow a detailed study of the temperature effect in the final metastable configurations obtained.

Several researchers used VPC models to perform fundamental studies such as adsorption, capillary condensation, mass transport, and even studying the influence of structure in chemical reactions. Thomson et al. (Thomson and Gubbins 2000) presented a realistic carbon model using rigid carbon basal plates as building blocks and used it to obtain simulated N_2 isotherms. The model evidenced the existence of localized capillary condensation in pores of around 14.5Å, which cannot be observed in the simpler and more used slit pore models. Subsequently, this phenomenon was explained by further VPC simulations considering the pore space convexity concept (Biggs et al. 2004b). Regarding more complex systems (Brennan et al. 2002), studied the adsorption of water in activated and non-activated carbon. They showed that lactose groups adsorbing even small amounts of water may block the pores and reduce the accessible porosity, which is a characteristic feature observed in experimental results. Furmaniak et al. (Furmaniak et al. 2009) extended the analysis to study the effect of oxidation on Pore Size Distribution (PSD) curves determined from Ar, N_2 and CO_2 isotherms. The early work of (Biggs and Agarwal 1992); considered atomic and diatomic gases mass transport in microscopic carbon pores and found a sub-diffusive behavior as the porosity approximates the percolation threshold. Later (Pikunic and Gubbins 2003) studied the diffusion of fluids using equilibrium molecular dynamics and observed how self-diffusion increases with temperature and presents a maximum with respect to loading. Recently (Furmaniak et al. 2009, 2017), presented an extensive analysis of the effect of confinement on the reactions that take place inside activated porous carbon. They systematically change the porosity of the modeled samples and studied its influence in isomerization (A↔B), dimerization (A+A↔B) and synthesis (A+B↔C) reactions. The results allowed to determine the general qualitative relationships between porous geometry and chemical equilibria.

In the last decade, the increase in computing power enabled the development of more complex and detailed models to emulate real carbons with more accuracy. Additionally, these models allowed a further assessment of simpler models, which are still widely used since they are much faster and require lower expertise from the user. Biggs and coworkers employed VPC models to evaluate the adsorption-based porous solid characterization methods (Biggs et al. 2004b) and the Intersection Capillaries Model (ICM) (Cai et al. 2007). They pointed out the strengths and weaknesses of each method, helping future users to evaluate the precision and implications of the obtained results. For example, they concluded that ICM gives a correct prediction of pores with a width below 10 Å, but there are significant discrepancies for larger pores including some omission. Furmaniak et al. (Furmaniak et al. 2013) used molecular simulations to show that the empirical parameters of the DA (Dubinin-Astakhov) adsorption isotherms equation are related to the average micropore diameter. Gonciaruk and Siperstein (Gonciaruk and Siperstein 2015) evaluated the simple slit pore model application for the prediction of gas adsorption properties in nanoporous carbons (NPC). They highlighted that the simple model makes an accurate description when the contribution of pores formed by edges of graphite platelets is small. However, the results of the slit pore model give a poor characterization if the edge effects are significant. Recently (Loi et al. 2020) proposed an improvement of the simple slit pore model by the inclusion of wedge pores, based on experimental X-ray observation. They concluded that wedge geometry gives a better description of NPC as it takes into account linear connectivity between pores spaces of different sizes.

The industry is constantly developing and synthesizing new material to increase capabilities and expand possible applications. Some of the required qualities are higher storage capacities, separation of gases and selective adsorption. The experimental design and quantification of nanoscale effects, which have shown to have a great influence on these properties, faces a huge practical complexity: controlling and systematically changing nanoscale parameters. Theoretical and computational studies have recently shifted its focus to emulate real application conditions. Since these techniques allow to describe and model materials from an atomic level, they currently assist and even guide experimental design. The particular use of molecular simulations to optimize CO_2 capture and separation by adsorption on carbonaceous material, attracted researchers' attention in the last few years (Bahamon et al. 2019). Nitrogen doping of carbon is experimentally proposed as an effective design strategy to improve CO_2 adsorption. The studies by (Kumar et al. 2015) on VPC shows that N-doping has a small effect on CO_2 uptake, but improves CO_2/N_2 selectivity. The same year, the work of (Liu et al. 2015) demonstrated that N-doped nanopores allow separating gases with a flux sequence of $H_2 > CO_2 > N_2 > Ar > CH_4$, generally following the kinetic diameter trend. Fatemi et al. 2017 reported the recent investigation of carbon nanotubes (CNTs) and graphene as membranes for CO_2 capture. Pore size control shows to be an effective way to achieve high CO_2 selectivity, due to the small kinetic diameter of this molecule, compared to most gases. It is important to highlight that this process requires the synthesis of graphene with ultramicropores. Although such technologies are not yet available in the industry, they may be possible in the future. The separation of CO_2 from CO_2/CH_4 mixtures is also important for practical applications since it improves natural gas caloric content. Tenney and Lastoskie 2006 showed that an effective strategy to increase CO_2 adsorption is to increase the density of oxidation groups in NPC. Moreover (Liu and Wilcox 2013), studied the oxygen groups effect and concluded that the interaction between the surface with induced polarity and polar/nonpolar molecules increase CO_2 selectivity over CH_4 and N_2. Di Biase and Sarkisov (Di Biase and Sarkisov 2013) supported these conclusions showing that the type of functional group does not affect CO_2 adsorption, indicating that the phenomenon is mainly due to the increase in the polarity of the material. Subsequently, they evaluated the influence of water impurities (which is also a polar molecule) in CO_2 separation (Di Biase and Sarkisov 2015). They demonstrated that at low concentration water has no significant influence since both species do not compete for the same porous space sub-region. A molecular simulation study made by (Müller 2013) showed that water adsorption largely depends on the location and type of surface groups. Moreover, CO_2 selectivity presents a small increase at low pressures because water-binding offers additional adsorption sites for CO_2, although at higher pressures selectivity falls due to competitive adsorption between both polar molecules. Beyond CO_2 adsorption, the study of water in VPC has great importance for purification and desalination. Computational simulations allowed understanding that water behavior, unlike N_2 and CO_2, is determined by three factors: strong self-interaction (water-water), weak water-carbon interaction, and the formation of hydrogen bonds with carbon oxygenated groups (Brennan et al. 2001). Gauden et al. [(Gauden et al. 2014)] simulate the adsorption in VPC of organic compounds (benzene, phenol and paracetamol) in aqueous solutions. They confirmed the experimental observation that the incorporation of oxygen functionalities decreases organics adsorption and showed that this is due to water accumulation and blocking of pore entrances. They also demonstrated that this effect decreases as the pore diameter increases and vanishes for values larger than 0.7 nm. In the last few years Bahamon et al. 2017 and Pereira et al. 2019 employed the GCMC method to show how activated carbons can be used to effectively remove pharmaceuticals impurities from water. Interestingly, nanoporous graphene oxides has recently shown to outperform water desalination of currently used membranes by two or three orders of magnitude (Cohen-Tanugi and Grossman 2012, Konatham et al. 2013, Cohen-Tanugi et al. 2016).

Applications on Renewable Energy

The development of energy storage devices of higher energy density and higher power is one of the greatest technological challenges of recent years due to the imminent energy transition that the world will have to face to alleviate climate change and pollution caused by the use and abuse of fossil fuels energy sources. Energy storage needs to be improved not only to be able to switch the world energy matrix to renewable sources but also because they are critical for several electronic portable devices, including electric vehicles (Dai et al. 2019). The possible applications of energy storage devices are varied, so there are different types of technologies at different stages of development and application: LIBs, LSBs, NIBs, potassium ion batteries (KIBs), supercapacitors, etc. Functionalized porous carbons are widely used in energy storage devices because of their properties: excellent chemical stability, good electrical conductivity and easily engineered chemical/physical properties. Crystalline structure, design over porosity, surface chemistry, electronic structure and morphology

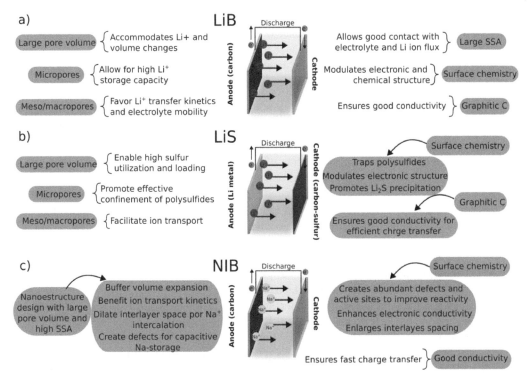

Figure 4.1: Structure dependent properties of porous carbons as electrode materials in (a) LIBs, (b) LSBs, (c) NIBs.

of porous carbons are critical for a good electrochemical performance to be used on energy storage technologies (Tian et al. 2020). Here the use of porous carbons on several energy storage devices: LIBs, LSBs, NIBs and supercapacitors will be summarized.

Lithium-Ion Batteries (LIBs)

Since their commercialization started in 1991 by Sony, LIBs have been the state of art on storage energy devices for portable electronics, electric vehicles, power tools, storage of sustainable energy, etc. (Tarascon and Armand 2001). Carbonaceous materials are major material for LIBs anodes since both good theoretical capacity and more negative redox potential than most of the other anodic materials are presented. Graphite is the most widely used because it has a flat working potential, good conductivity, low cost, and a long cycle life. However, it has a theoretical capacity of 372 mAhg^{-1} and poor behavior in high-density currents working conditions, which prevent its use in high power and high energy density batteries. Porous carbons with different porous sizes are promising anode materials for LIBs due to their open structures and high SSA, with prominently increased capacities in comparison with traditional graphitic carbons. These types of materials present several interesting characteristics for LIBs: (a) percolation networks from interconnected nanopores which allow effective diffusion pathways for Li$^+$, (b) a large number of active sites for Li$^+$ storage because of their high SSA, (c) good electrical conductivity because of well-interconnected carbon walls, (d) high electrode/electrolyte interface which improves charge transfer, (e) available free space to accommodate volume change during lithium storage and (f) a 3D stable structure which remains stable during battery cycling (Ji et al. 2011). As meso/macropores favor electrolyte mobility and Li$^+$ transfer kinetic and micropores promote reversible intercalation/de-intercalation of Li$^+$, a hierarchy of pores is necessary to have rapid ion transport with improved rate capability (Roberts et al. 2014). Figure 4.1a outlines which characteristics of LIBs batteries are dependent on the structural properties of the carbons.

Several porous carbon materials have been applied as anodes for Li-ion batteries with different results. CMK3 prepared from SBA-15 templated as anode material for Li-ion battery reported high specific capacity and good cycling performance for the first time in 2003 (Zhou et al. 2003). CMK3 material prepared with uniform pore sizes of 4 nm, small amounts of crystalline graphite phases and a high SSA. A cell with CMK3 as anodic material cycled about 20 times with a capacity around 850–900 mAh g^{-1} and very good reversibility. The 3D ordered porous structure seems to allow this good electrochemical performance, although the mechanism was still not clear.

Lithium storage mechanism on porous carbons is still debatable due to their complicated structures. Several mechanisms have been proposed based on cavities filling, adsorption of Li$^+$ on active sites onto carbon surface, intercalation of Li$^+$ on porous carbon walls or a combination of these processes (Kasuh et al. 1997, Wu et al. 1999, Sato et al. 1994, Dahn et al. 1995). Yang et al. studied Li$^+$ storage in a porous carbon prepared from sucrose pyrolysis. They observed that Li$^+$ adsorption is a dominant process through a potential range used during cycling. Insertion reactions can likely occur through a wide range so there is no defined potential limit between adsorption or insertion reactions. They also observed structural degradation of the carbon due to Li$^+$ intercalation in hydrogen-containing graphene layers, which provokes a potential hysteresis and capacity loss at low potentials. Optimization of potential window has been proposed to avoid anode degradation and to improve cycling performance (Yang et al. 2020).

Today, even if the mechanism is still not completely clear, the increase of specific capacity for nanoporous carbon compared to graphite is attributed to the dual effect of interspersed nanoporous filling and Li$^+$ adsorption onto the carbon (Li et al. 2013). The specific capacity of porous carbons is improved by doping with heteroatoms B, N, O, S, etc. But the mechanism for lithium storage depends on heteroatom on the surface. On boron-doped carbon Li$^+$ energy adsorption decreases because, as B is electron deficient, it can accept Li$^+$ electron according to first principles studies (Zhou et al. 2004). For nitrogen-doped carbons, the first principles studies showed that Li$^+$ energy adsorption is higher in respect to graphite. However higher electronegativity and hybridization of it lone pair electron with p electron from carbon, are beneficial to Li$^+$ interaction with carbon surface. Furthermore, when superficial N atoms have a pyridinic structure Li$^+$ adsorption is favored with respect to graphitic structure (Ma et al. 2012).

Composites preparation is another strategy to improve the electrochemical performance of material for LIBs anodes. Composites can be prepared with several materials as metal (Wang et al. 2009), metal oxides (Zhou et al. 2010), silicon (Chou et al. 2010), etc. Lener et al. prepared a composite based on a hard carbon structure to improve electronic and ionic conductivity with a highly ordered SiO$_2$ from CMK3 and SBA-15 and tested as an anode for LIBs (Lener et al. 2018). They observed a synergistic effect between CMK3 and SBA-15 with a combination of good conductivity from the carbon resulting in a high specific capacity from the silica. Structured porous carbon also contains the volume expansion of silica during cycling, which improves the mechanical properties of the active material.

Lithium-Sulfur Batteries (LSBs)

Lithium-Sulfur (Li-S) batteries have been extensively studied in the past few years for a next-generation electrochemical energy storage system due to their high theoretical specific capacity (1672 mAh g^{-1}) and specific energy ($\approx$ 2600 Wh kg^{-1}) (Lim et al. 2015). Furthermore, sulfur is naturally abundant, environmentally benign and relatively cheap; these attributes could allow its application in electric vehicles, artificial intelligence and grid-scale stationary storage system, among others (Li et al. 2019, Zubair et al. 2019). However, there are several drawbacks to be solved before a practical application of Li-S batteries. Sulfur is naturally insulated, in the lithiation process suffers from a volumetric expansion of 80%, and has a rapid capacity decay. The loss in capacity is mostly due to the formation of polysulfides. During discharge sulfur reacts with lithium forming long-chain polysulfides, then short-chain polysulfides and solid Li_2S_2/Li_2S at the end of the cycle, showing a reaction profile with two voltage plateaus. Polysulfides are soluble in the liquid electrolyte so they can diffuse to the anode surface and be further reduced there in a process known as shuttle effect (Ma et al. 2014, Sun et al. 2017). As the active material is changing from solid to liquid and to solid again, the battery works through a solid-liquid-solid mechanism (Wang et al. 2013).

Several strategies have been studied to generate a matrix with electric and ionic conductivity to host sulfur while allowing the volume changes that sulfur undergoes in the charge/discharge process (Guler et al. 2018, Zhang et al. 2015, Han and Manthiram 2017). Carbon porous materials are considered as the most promising hosts for sulfur loading because of their lightweight, structural diversity and electrical conductivity, which can be changed by surface modification. In addition, carbon is abundant on the Earth's crust and can be prepared by several methods with designed characteristics like structure, porosity, functionalization, etc. (Lener et al. 2018, Zubair et al. 2017).

The structure of porous carbon materials is usually engineered according to the following principles: (a) high SSA to improve electronic transfer and sulfur utilization; (b) proper pore size to retain intermediate products; (c) enough space to accommodate sulfur and its volumetric expansion. Figure 4.1b outlines which characteristics of LISs batteries are dependent on the structural properties of the porous carbons. Although these considerations have led to promising results with high capacities and relatively good cyclability, there is still capacity fading in these systems because the carbon matrix by itself is not effective enough to avoid the shuttle effect. Some others strategies as separator coatings, hierarchically porous carbon and ultra-micropores, etc., are used to retain polysulfides on the cathodic side of the battery (Zheng et al. 2016, Balach et al. 2015). Furthermore, the surface of carbon may be modified with different functional groups which can interact with polysulfides trapping them, avoiding the shuttle effect and improving cycling performance (Qu et al. 2016).

Zheng et al. synthesized a hierarchical porous carbon (HPCR) using MgO microrods as a template and chemical vapor deposition method to growth graphene-like nanosheets (Zheng et al. 2016). The synthesis method allowed the carbon to maintain the porous structure of the template and therefore obtain a hierarchical porous structure with vertically oriented porous graphene-like nanosheets. HPCR showed mesoporous solids characteristics with 2226 m^2g^{-1} of SSA and 4.9 cm^3g^{-1} total pore volume. This material provides a large surface area for electronic transfer and a very large pore volume for high sulfur loading. The pore size distribution according to the BJH method showed an average pore size of 9 nm, which improves the physical confinement of polysulfides. The electrochemical characterization showed an initial specific capacity of 1430 mAh g^{-1}, meaning 86% of the theoretical specific capacity of sulfur and a high sulfur utilization. After 300 charge/discharge cycles, the capacity was 750 mAh g^{-1} indicating a small capacity fading of 0.02% per cycle.

Most of the strategies to avoid shuttle effect aim to limit or suppress polysulfides dissolution on the electrolyte. Ultra-microporous carbon is very interesting due to the confinement of sulfur in very narrow pores that can completely change the reaction mechanism. Theoretical studies show that only small sulfur molecules (S_{2-4}) have the appropriate size to accommodate in these smaller pores, so long-chain polysulfides cannot be formed if the sulfur is confined in ultra-micropores (Xin et al. 2012). Zhou et al. synthesized a bio-carbon using peanut shells as a natural precursor activated at 800°C by K_2FeO_4 which works not only as an activating agent but also as a catalyst to graphitize the carbon sample (Zhou et al. 2018). The resulting carbon showed ultra-microporous structure with a median pore width smaller than 0.4 nm and sulfur is located inside the pores, so only small molecules of sulfur are present and polysulfides are not formed. Use of carbonate electrolytes, which are far more environmentally friendly and cheaper, produces a Solid Electrolyte Interface (SEI) by reaction between carbon and carbonates during the first discharge cycle. The formed SEI is a surface film which separates sulfur hosted inside of the pores from the liquid electrolyte so reaction changes to a quasi-solid mechanism which can be evidenced on the potential profile with only one plateau (Markevich et al. 2015). The use of these types of carbon materials and electrochemical treatments to improve SEI formation can effectively suppress the shuttle effect improving the cycling performance of LSBs notably.

Porous carbons are also used on an interlayer between a sulfur cathode and lithium anode to avoid polysulfides diffusion as a strategy to alleviate the shuttle effect. The purpose of this film is to adsorb polysulfides dissolved in the electrolyte, leaving them on the cathodic side of the cell and thus reducing their diffusion and the loss of active material. Balach et al. synthesized a mesoporous carbon by a polymerization-carbonization process using resorcinol resin and commercial colloidal silica as carbon precursor and template, respectively (Balach et al. 2015). The obtained carbon showed uniform spherical particles with a high degree of pore size uniformity with a pore size distribution centred at 12 nm according to the QSDFT method (Hanzawa et al. 2002, Apaydin-Varol and Erülken 2015). A free-standing interlayer was prepared and placed between the sulfur cathode and the separator. The Li-S battery had an initial specific capacity of 1388 mAh g^{-1} and displayed a capacity fading of 0,03% per cycle, which is a substantial improvement compared with the cell without the interlayer which showed an initial capacity of 1038 mAh g^{-1} and a capacity fading of 0,06%.

Sodium-Ion Batteries (NIBs)

Interest in the development of NIBs is growing, even if they have a lower energy density than LIBs because sodium is abundant, of low cost and its sources are widely dispersed around the world (Balogun et al. 2016). Carbon has emerged as one of the most promising materials for anodes in NIBs since is chemically inert, a good conductor, abundant and easy to prepare with a wide variety of structures, among other characteristics. Even if there are some similarities between LIBs and NIBs, the different nature of lithium and sodium causes the same materials to have different behaviors on these devices. Graphite is the most used anodic material for LIBs, lithium storage occurs through intercalation of Li$^+$ between the graphene sheets. But graphite cannot be used as an anode for NIBs because of sodium's higher ionic radio (0.102 nm Na$^+$ vs. 0.076 nm Li$^+$) prevents intercalation of Na$^+$ on graphite (Zhang et al. 2016). Therefore other forms of carbon have been studied as anode material for NIBs like graphene (Wang et al. 2013) or hard porous carbons (Stevens and Dahn 2000). Correlations between structure characteristics and electrochemical properties of porous carbons as electrode materials for NIBs are presented in Fig. 4.1c.

The mechanism by which sodium can be stored in carbons include: (a) ionic adsorption on surface and defects, (b) nanopores filling and (c) intercalation between parallel or nearly parallel graphene layers with enough d-spacing (for Na$^+$ intercalation d-spacing needs to be at least 0.37 nm, d-spacing for graphite is 0.34 nm) (Zhang et al. 2016). The sodium storage onto hard porous carbons has been studied but the storage mechanism is still not clear (Stevens and Dahn 2000, 2001). A potential response during a typical charge/discharge process for a porous carbon anode in a NIB generally consists on a potential slope between 1.0 and 0.2 V and a flat potential plateau around 0.1 V vs. Na/Na$^+$ (Zhang et al. 2016). The sloping voltage region is associated with Na$^+$ insertion between graphene layers (Stevens and Dahn 2001) and adsorption onto the surface vacancy (Cao et al. 2012). A potential variation in this storage process is because there is a wide distribution on the insertion potential for the different sites as well an electrostatic repulsion with previously-stored Na$^+$ (Bommier and Ji 2015). During the potential plateau at 0.1 V, storage process occurs by nanopore filling (in a process analogue to adsorption) and insertion between graphene expanded layers (Stevens and Dahn 2001, Zhang et al. 2016, Cao et al. 2012). The plateau is the highest contribution to capacity, the potential is 0.1 V which is near the sodium plating potential. In addition, sodium chemical potential in nanopores is close to sodium chemical potential in the metal itself, so sodium metallic deposition could occur on the pores (Stevens and Dahn 2001). This provokes safety issues especially for fast charge and discharge (Yan et al. 2014).

Supercapacitors

Supercapacitors, electrochemical capacitors or ultracapacitors have attracted significant attention as a next-generation energy-storage device for their application in miniaturized portable electric systems. Additionally, they have some advantages compared to batteries such as their high power density, fast charge/discharge rate, long cycle life, safety and environmental friendliness (Jiang et al. 2016, Long et al. 2015, Sankar et al. 2019). For an objective comparison, it should be noted that the volumetric energy density of commercially available supercapacitors is 5 to 8 Whl^{-1}, 10 times smaller compared with batteries (> 50 Whl^{-1}) (Yang et al. 2013). Among supercapacitors, Electric Double-Layer Capacitors (EDLCs) are advantageous because of their superior cycle stability and higher power density. The overall electrochemical performance of supercapacitors depends on the electrode material because the charge accumulation at the electrode-electrolyte interface is related with its porosity, surface area and electric conductivity (Lee et al. 2006, Huang et al. 2017).

Recently, Sankar et al. synthesized graphitic carbon nanoflakes by using green-tea wastes with initial carbonization and a KOH activation process combined with water treatment (Sankar et al. 2019). The FE-SEM images of the carbon displayed an interconnected structure of ultrathin nanoflakes which the authors considered promising due to the higher specific surface area that this two-dimensional morphology possesses. The latter was confirmed by N_2 adsorption-desorption isotherm measurements, the BET SSA was 1058 $m^2\ g^{-1}$, a total pore volume of 0.4713 $cm^3\ g^{-1}$ and average pore diameter of 1.66 nm. The supercapacitor devised with the carbon nanoflakes showed 162 $F\ g^{-1}$ of specific capacity and capacity retention of 121% after 5000 cycles.

Conclusion

Porous carbonaceous materials are very versatile materials, friendly to the environment, very stable and whose properties can be designed very precisely by carefully choosing the synthesis pathway. Their properties are closely related not only to the characteristics of the massive material that composes them, but also to the characteristics of the empty space that the pore structures form. To better understand both their properties and the reactions that occur when they are used in various devices, it is necessary to complement experimental studies with theoretical studies, which use VPC models. Within the wide spectrum of applications in which these materials can be used, renewable energy storage systems stand out since the large number of reaction sites available given their high surface area, allow them to interact with different ionic species very effectively. Due to their particular properties porous carbon materials are used in Li-ion, Li-S, Na-ion batteries, supercapacitors, among other devices on renewable energy fields.

Acknowledgements

The authors thank to CONICET Argentina, Programa BID-Foncyt and YPF-Tecnología (YTEC) for financial support.

References

Acharya, M., M. S. Strano, J. P. Mathews, S. J. L. Billinge, V. Petkov, S. Subramoney et al. 1999. Simulation of nanoporous carbons: A chemically constrained structure. Philos. Mag. B.79: 1499–1518.

Apaydin-Varol, E. and Y. Erülken. 2015. A study on the porosity development for biomass based carbonaceous materials. J. Taiwan Inst. Chem. E. 54: 37–44.

Bahamon, D., L. Carro, S. Guri and L. F. Vega. 2017. Computational study of ibuprofen removal from water by adsorption in realistic activated carbons. J. Colloid Interface Sci. 498: 323–334.

Bahamon, D., M. R. M. Abu-Zahra and L. F. Vega. 2019. Molecular simulations of carbon-based materials for selected CO_2 separation and water treatment processes. Fluid Phase Equilib. 492: 10–25.

Balach, J., T. Jaumann, M. Klose, S. Oswald, J. Eckert and L. Giebeler. 2015. Mesoporous carbon interlayers with tailored pore volume as polysulfide reservoir for high-energy lithium-sulfur batteries. J. Phys. Chem. C. 119: 4580–4587.

Balogun, M. S., Y. Luo, W. Qiu, P. Liu and Y. Tong. 2016. A review of carbon materials and their composites with alloy metals for sodium ion battery anodes. Carbon 98: 162–178.

Bandosz, T. J., M. J. Biggs, K. E. Gubbins, Y. Hattori, T. Iiyama, K. Kaneko et al. 2003. Molecular models of porous carbons. Chem. Phys. Carbon. 28: 41–228.

Bhatia, S. K. 2002. Density functional theory analysis of the influence of pore wall heterogeneity on adsorption in carbons. Langmuir 18: 6845–6856.

Biggs, M. and P. Agarwal. 1992. Mass diffusion of atomic fluids in random micropore spaces using equilibrium molecular dynamics. Phys. Rev. A. 46: 3312–3318.

Biggs, M. and P. Agarwal. 1994. Mass diffusion of diatomic fluids in random micropore spaces using equilibrium molecular dynamics. Phys. Rev. E. 49: 531–540.

Biggs, M. J., A. Buts and D. Williamson. 2004a. Absolute assessment of adsorption-based porous solid characterization methods: comparison methods. Langmuir 20: 7123–7138.

Biggs, M. J., A. Buts and D. Williamson. 2004b. Molecular simulation evidence for solidlike adsorbate in complex carbonaceous micropore structures. Langmuir 20: 5786–5800.

Biggs, M. J. and A. Buts. 2006. Virtual porous carbons: what they are and what they can be used for. Mol. Simulat. 32: 579–593.

Bommier, C. and X. Ji. 2015. Recent development on anodes for Na-Ion batteries. Isr. J. Chem. 55: 486–507.

Brennan, J. K., T. J. Bandosz, K. T. Thomson and K. E. Gubbins. 2001. Water in porous carbons. Colloids and surfaces. Colloids Surf. A 187: 187–188.

Brennan, J. K., K. T. Thomson and K. E. Gubbins. 2002. Adsorption of water in activated carbons: Effects of pore blocking and connectivity. Langmuir 18: 5438–5447.

Cai, Q., A. Buts, M. J. Biggs and N. A. Seaton. 2007. Evaluation of methods for determining the pore size distribution and pore-network connectivity of porous carbons. Langmuir 23: 8430–8440.

Cai, Q., A. Buts, N. A. Seaton and M. J. Biggs. 2008. A pore network model for diffusion in nanoporous carbons: validation by molecular dynamics simulation. Chem. Eng. Sci. 63: 3319–3327.

Cao, Y., L. Xiao, M. L. Sushko, W. Wang, B. Schwenzer, J. Xiao et al. 2012. Sodium ion insertion in hollow carbon nanowires for battery applications. Nano Lett. 12: 3783–3787.

Chen, Q., X. Tan, Y. Liu, S. Liu, M. Li, Y. Gu et al. 2020. Biomass-derived porous graphitic carbon materials for energy and environmental applications. J. Mater. Chem. A. 8: 5773–5811.

Chou, S. L., J. Zhao Wang, M. Choucair, H. K. Liu, J. A. Stride and S. X. Dou. 2010. Enhanced reversible lithium storage in a nanosize silicon/graphene composite. Electrochem. Commun. 12: 303–306.

Cohen-Tanugi, D. and J. C. Grossman. 2012. Water desalination across nanoporous graphene. Nano Lett. 12: 3602–3608.

Cohen-Tanugi, D., L. C. Lin and J. C. Grossman. 2016. Multilayer nanoporous graphene membranes for water desalination. Nano Lett. 16: 1027–1033.

Condon, J. 2019. Surface Area and Porosity Determinations by Physisorption: Measurement, Classical Theories and Quantum Theory. Second. Elsevier Science.

Dahn, J. R., T. Zheng, Y. Liu and J. S. Xue. 1995. Mechanisms for lithium insertion in carbonaceous materials. Science 270: 590–593.

Dai, Q., J. C. Kelly, L. Gaines and M. Wang. 2019. Life cycle analysis of lithium-ion batteries for automotive applications. Batteries. 5: 48–63.

Di Biase, E. and L. Sarkisov. 2013. Systematic development of predictive molecular models of high surface area activated carbons for adsorption applications. Carbon 64: 262–280.

Di Biase, E. and L. Sarkisov. 2015. Molecular simulation of multi-component adsorption processes related to carbon capture in a high surface area, disordered activated carbon. Carbon 94: 27–40.

Do, D. D. and H. D. Do. 2006. Modeling of adsorption on nongraphitized carbon surface: GCMC simulation studies and comparison with experimental data. J. Phys. Chem. B 110: 17531–17538.

Dutta, S., A. Bhaumik and K. C. W. Wu. 2014. Hierarchically porous carbon derived from polymers and biomass: effect of interconnected pores on energy applications. Energy Environ. Sci. 7: 3574–3592.

Emmett, P. H. 1947. Adsorption and pore-size measurements on charcoals and whetlerites. Chem. Rev. 43: 69–148.

Fatemi, S. Mahmood, A. Baniasadi and M. Moradi. 2017. Recent progress in molecular simulation of nanoporous graphene membranes for gas separation. J. Korean Phys. Soc. 71: 54–62.

Faulon, J. L., G. A. Carlson and P. G. Hatcher. 1993. Statistical models for bituminous coal: A three-dimensional evaluation of structural and physical properties based on computer-generated structures. Energ. Fuel. 7: 1062–1072.

Figueiredo, J. L. 2018. Nanostructured porous carbons for electrochemical energy conversion and storage. Surf. Coat. Tech. 350: 307–12.

Furmaniak, S., A. P. Terzyk, P. A. Gauden, P. J. F. Harris and P. Kowalczyk. 2009. Can carbon surface oxidation shift the pore size distribution curve calculated from Ar, N_2 and CO_2 adsorption isotherms? Simulation results for a realistic carbon model. J. Phys. Condens. Matter. 21: 315005–315015.

Furmaniak, S. 2013. New virtual porous carbons based on carbon EDIP potential and monte carlo simulations. CMST 19: 47–57.

Furmaniak, S., A. P. Terzyk, P. A. Gauden, P. Kowalczyk, P. J. F. Harris and S. Koter. 2013. Applicability of molecular simulations for modelling the adsorption of the greenhouse gas CF4 on carbons. J. Phys. Condens. Matter. 25: 15004–15013.

Furmaniak, S., P. A. Gauden, P. Kowalczyk and A. Patrykiejew. 2017. Monte carlo study of chemical reaction equilibria in pores of activated carbons. RSC Adv. 7: 53667–53679.

Gauden, P. A., A. P. Terzyk, S. Furmaniak, J. Włoch, P. Kowalczyk and W. Zieliński. 2014. MD simulation of organics adsorption from aqueous solution in carbon slit-like pores. Foundations of the pore blocking effect. J. Phys. Condens. Matter. 26: 55008–55022.

Gonciaruk, A. and F. R. Siperstein. 2015. *In silico* designed microporous carbons. Carbon 88: 185–195.

González-García, P. 2018. Activated carbon from lignocellulosics precursors: A review of the synthesis methods, characterization techniques and applications. Renew. Sust. Energ. Rev. 82: 1393–1414.

Guler, A., S. O. Duman, D. Nalci, M. Guzeler, E. Bulut, M. Oguz Guler et al. 2018. Graphene assisted template based $LiMn_2O_4$ flexible cathode electrodes. Int. J. Energ. Res. Feb: 1–12.

Han, P. and A. Manthiram. 2017. Boron- and nitrogen-doped reduced graphene oxide coated separators for high-performance Li-S batteries. J. Power Sources 369: 87–94.

Hanzawa, Y., H. Hatori, N. Yoshizawa and Y. Yamada. 2002. Structural changes in carbon aerogels with high temperature treatment. Carbon 40: 575–581.

Heidarinejad, Z., M. H. Dehghani, M. Heidari, G. Javedan, I. Ali and M. Sillanpää. 2020. Methods for preparation and activation of activated carbon: A review. Environ. Chem. Lett. 18: 393–415.

Herrera, L. F. and D. D. Do. 2009. Effects of surface structure on the molecular projection area. Adsorption of Argon and Nitrogen onto Defective Surfaces: AA Contribution on the Occasion of 60th Birthday of Professor Mietek Jaroniec. Adsorption 15: 240–246.

Huang, Y., J. He, Y. Luan, Y. Jiang, S. Guo, X. Zhang et al. 2017. Promising biomass-derived hierarchical porous carbon material for high performance supercapacitor. RSC Adv. 7: 10385–10390.

Jagiello, J., J. Kenvin, J. P. Olivier, A. R. Lupini and C. I. Contescu. 2011. Using a new finite slit pore model for NLDFT analysis of carbon pore structure. Adsorpt. Sci. Technol. 29: 769–780.

Jagiello, J. and J. P. Olivier. 2013. 2D-NLDFT adsorption models for carbon slit-shaped pores with surface energetical heterogeneity and geometrical corrugation. Carbon 55: 70–80.

Jeewoo, L., J. Pyun and K. Char. 2015. Recent approaches for the direct use of elemental sulfur in the synthesis and processing of advanced materials. Angew. Chem. Int. Ed. 54: 3249–3258.

Ji, Liwen, Z. Lin, M. Alcoutlabi and X. Zhang. 2011. Recent developments in nanostructured anode materials for rechargeable lithium-ion batteries. Energy Environ. Sci. 4: 2682–26889.

Jiang, Y., J. Yan, X. Wu, D. Shan, Q. Zhou, L. Jiang et al. 2016. Facile synthesis of carbon nanofibers-bridged porous carbon nanosheets for high-performance supercapacitors. J. Power Sources 307: 190–98.

Jorge, M., C. Schumacher and N. A. Seaton. 2002. Simulation study of the effect of the chemical heterogeneity of activated carbon on water adsorption. Langmuir 18: 9296–9306.

Jun, S., S. H. Joo, R. Ryoo, M. Kruk, M. Jaroniec, Z. Liu et al. 2000. Synthesis of new, nanoporous carbon with hexagonally ordered mesostructure. J. Am. Chem. Soc. 11: 10712–10713.

Kasuh, T., A. Mabuchi, K. Tokumitsu and H. Fujimoto. 1997. Recent trends in carbon negative electrode materials. J. Power Sources 68: 99–101.

Knossalla, J., D. Jalalpoor and F. Schüth. 2017. Hands-on guide to the synthesis of mesoporous hollow graphitic spheres and core-shell materials. Chem. Mater. 29: 7062–7072.

Konatham, D., J. Yu, T. A. Ho and A. Striolo. 2013. Simulation insights for graphene-based water desalination membranes. Langmuir 29: 11884–11897.

Kumar, K. Vasanth, K. Preuss, L. Lu, Z. X. Guo and M. M. Titirici. 2015. Effect of nitrogen doping on the CO_2 adsorption behavior in nanoporous carbon structures: A molecular simulation study. J. Phys. Chem. C 119: 22310–22321.

Kyotani, T., T. Nagai, S. Inoue and A. Tomita. 1997. Formation of new type of porous carbon by carbonization in zeolite nanochannels. Chem. Mater. 9: 609–615.

Lastoskie, C., K. E. Gubbins and N. Quirke. 1993. Pore size heterogeneity and the carbon slit pore: A density functional theory model. Langmuir 9: 2693–2702.

Lastoskie, C. M. and K. E. Gubbins. 2001. Characterization of porous materials using molecular theory and simulation. Adv. Chem. Engineer. Sci. 28: 203–250.

Lee, J., J. Kim and T. Hyeon. 2006. Recent progress in the synthesis of porous carbon materials. Adv. Mater. 18: 2073–2094.

Lener, G., A. A. Garcia-Blanco, O. Furlong, M. Nazzarro, K. Sapag, D. E. Barraco et al. 2018. A silica/carbon composite as anode for lithium-ion batteries with a large rate capability: Experiment and theoretical considerations. Electrochim. Acta 279: 289–300.

Li, R., Y. Zhou, W. Li, J. Zhu and W. Huang. 2020. Structure engineering in biomass-derived carbon materials for electrochemical energy storage. Research 2020: 8685436.

Li, Zhi, Z. Xu, X. Tan, H. Wang, C. M. B. Holt, T. Stephenson et al. 2013. Mesoporous nitrogen-rich carbons derived from protein for ultra-high capacity battery anodes and supercapacitors. Energy Environ. Sci. 6: 871–878.

Li, Z., L. Ma, Z. Li and W. Ni. 2019. Multi-energy cooperative utilization business models: A case study of the solar-heat pump water heater. Renew. Sust. Energ. Rev. 108: 392–397.

Liang, C., K. Hong, G. A. Guiochon, J. W. Mays and S. Dai. 2004. Synthesis of a large-scale highly ordered porous carbon film by self-assembly of block copolymers. Angew. Chem. Int. Ed. 43: 5785–5789.

Liang, C., Z. Li and S. Dai. 2008. Mesoporous carbon materials: synthesis and modification. Angew. Chem. Int. Ed. 47: 3696–3717.

Lim, Z. and C. Liang. 2015. Lithium–sulfur batteries: from liquid to solid cells. J. Mater. Chem. A 3: 936–958.

Liu, H., Z. Chen, S. Dai and D. E. Jiang. 2015. Selectivity trend of gas separation through nanoporous graphene. J. Solid State Chem. 224: 2–6.

Liu, P., Y. Wang and J. Liu. 2019. Biomass-derived porous carbon materials for advanced lithium sulfur batteries. J. Energ. Chem. 34: 171–185.

Liu, X., N. Fechler and M. Antonietti. 2013. Salt melt synthesis of ceramics, semiconductors and carbon nanostructures. Chem. Soc. Rev. 42: 8237–65.

Liu, Y. and J. Wilcox. 2013. Molecular simulation studies of CO_2 adsorption by carbon model compounds for carbon capture and sequestration applications. Environ. Sci. Technol. 47: 95–101.

Loi, Q. K., L. Prasetyo, J. S. Tan, D. D. Do and D. Nicholson. 2020. Wedge pore modelling of gas adsorption in activated carbon: Consistent pore size distributions. Carbon 166: 414–426.

Long, C., L. Jiang, X. Wu, Y. Jiang, D. Yang, C. Wang et al. 2015. Facile synthesis of functionalized porous carbon with three-dimensional interconnected pore structure for high volumetric performance supercapacitors. Carbon 93: 412–420.

Lu, B. A. and F. Schüth. 2006. Nanocasting: A versatile strategy for creating nanostructured porous materials. Adv. Mater. 18: 1793–1805.

Ma, C., X. Shao and D. Cao. 2012. Nitrogen-doped graphene nanosheets as anode materials for lithium ion batteries: A first-principles study. J. Mater. Chem. 22: 8911–8915.

Ma, G., Z. Wen, Q. Wang, C. Shen, J. Jin and X. Wu. 2014. Enhanced cycle performance of a Li–S battery based on a protected lithium anode. J. Mater. Chem. A. 2: 19355–19359.

Markevich, E., G. Salitra, A. Rosenman, Y. Talyosef, F. Chesneau and D. Aurbach. 2015. The effect of a solid electrolyte interphase on the mechanism of operation of lithium–sulfur batteries. J. Mater. Chem. A. 3: 19873–19883.

Mestre, A. S. and A. P. Carvalho. 2018. Nanoporous carbon synthesis: An old story with exciting new chapters. pp. 37–68. *In*: Taher Ghrib [ed.]. Porosity—Process, Technologies and Applications. InTechopen. London.

Mi, X. and Y. Shi. 2013. Topological defects in nanoporous carbon. Carbon 60: 202–214.

Molina-Sabio, M. and F. Rodríguez-Reinoso. 2004. Role of chemical activation in the development of carbon porosity. Colloid. Surface. 241: 15–25.

Müller, E. A. 2013. Purification of water through nanoporous carbon membranes: A molecular simulation viewpoint. Curr. Opin. Chem. Eng. 2: 223–228.

Nakagawa, Y., M. Molina-Sabio and F. Rodríguez-Reinoso. 2007. Modification of the porous structure along the preparation of activated carbon monoliths with H_3PO_4 and $ZnCl_2$. Micropor. Mesopor. Mat. 103: 29–34.

Nicholson, D. 1996. Using computer simulation to study the properties of molecules in micropores. J. Chem. Soc. 92: 1–9.

Palmer, J. C. and K. E. Gubbins. 2012. Atomistic models for disordered nanoporous carbons using reactive force fields. Micropor. Mesopor. Mat. 154: 24–37.

Pereira, J. M., V. Calisto and S. M. Santos. 2019. Computational optimization of bioadsorbents for the removal of pharmaceuticals from water. J. Mol. Liq. 279: 669–676.

Petersen, T., I. Yarovsky, I. Snook, D. G. McCulloch and G. Opletal. 2004. Microstructure of an industrial char by diffraction techniques and reverse monte carlo modelling. Carbon 42: 2457–2469.

Petkov, V., R. G. Difrancesco, S. J. L. Billinge, M. Acharya and H. C. Foley. 1999. Local structure of nanoporous carbons. Philos. Mag. B. 79: 1519–1530.

Petkovich, N. D. and A. Stein. 2013. Controlling macro- and mesostructures with hierarchical porosity through combined hard and soft templating. Chem. Soc. Rev. 42: 3721–3739.

Pikunic, J. and K. E. Gubbins. 2003. Molecular dynamics simulations of simple fluids confined in realistic models of nanoporous carbons. Eur. Phys. J. E. 12: 35–40.

Qu, J., S. Lv, X. Peng, S. Tian, J. Wang and F. Gao. 2016. Nitrogen-doped porous 'green carbon' derived from shrimp shell: Combined effects of pore sizes and nitrogen doping on the performance of lithium sulfur battery. J. Alloy. Compd. 671: 17–23.

Roberts, A. D., X. Li and H. Zhang. 2014. Porous carbon spheres and monoliths: morphology control, pore size tuning and their applications as Li-Ion battery anode materials. Chem. Soc. Rev. 43: 4341–56.

Rouquerol, J., D. Avnir, D. H. Everett, C. Fairbridge, M. Haynes, N. Pernicone et al. 1994. Guidelines for the characterization of porous solids. Stud. Surf. Sci. Catal. 87: 1–9.

Ryoo, R., S. H. Joo, M. Kruk and M. Jaroniec. 2001. Ordered mesoporous carbons. Adv. Mater. 13: 677–681.

Sankar, S., A. T. Aqueel Ahmed, A. I. Inamdar, H. Im, Y. B. Im, Y. Lee et al. 2019. Biomass-derived ultrathin mesoporous graphitic carbon nanoflakes as stable electrode material for high-performance supercapacitors. Mater. Desig. 169: 107688–107697.

Sato, K., M. Noguchi, A. Demachi, N. Oki and M. Endo. 1994. A mechanism of lithium storage in disordered carbons. Science 264: 556–558.

Seaton, N. A., J. P. R. B. Walton and N. quirke. 1989. A new analysis method for the determination of the pore size distribution of porous carbons from nitrogen adsorption measurements. Carbon 27: 853–861.

Seaton, N. A., S. P. Friedman, J. M. D. MacElroy and B. J. Murphy. 1997. The molecular sieving mechanism in carbon molecular sieves: A molecular dynamics and critical path analysis. Langmuir 13: 1199–1204.

Soler-Illia, G., C. Sanchez, B. Lebeau and J. Patarin. 2002. Chemical strategies to design textured materials: from microporous and mesoporous oxides to nanonetworks and hierarchical structures. Chem. Rev. 102: 4093–4138.

Stevens, D. A. and J. R. Dahn. 2000. High capacity anode materials for rechargeable sodium-ion batteries. J. Electrochem. Soc. 147: 1271–1273.

Stevens, D. A. and J. R. Dahn. 2001. The mechanisms of lithium and sodium insertion in carbon materials. J. Electrochem. Soc. 148: A803–A811.

Sun, Z., J. Zhang, L. Yin, G. Hu, R. Fang, H. Cheng et al. 2017. Conductive porous vanadium nitride/graphene composite as chemical anchor of polysulfides for lithium-sulfur batteries. Nature Comm. 8: 1–8.

Tarascon, J. -M. and M. Armand. 2001. Issues and challenges facing rechargeable lithium batteries. Nature. 414: 359–367.

Tenney, C. M. and C. M. Lastoskie. 2006. Molecular simulation of carbon dioxide adsorption in chemically and structurally heterogeneous porous carbons. Environ. Prog. Sustani. 25: 343–54.

Terzyk, A. P., S. Furmaniak, P. A. Gauden, P. J. F. Harris and P. Kowalczyk. 2012. Virtual porous carbons. pp. 61–104. *In*: J. M. D. Tascon [ed.]. Novel Carbon Adsorbents. Elsevier Ltd. Amsterdam.

Thomson, Kendall T. and Keith E. Gubbins. 2000. Modeling structural morphology of microporous carbons by reverse Monte Carlo. Langmuir 16: 5761–5773.

Tian, W., H. Zhang, X. Duan, H. Sun, G. Shao and S. Wang. 2020. Porous carbons: Structure-oriented design and versatile applications. Adv. Funct. Mater. 1909265: 1909265–1909265.

Titirici, M. M., R. J. White, N. Brun, V. L. Budarin, D. Sheng Su, F. Del Monte et al. 2015. Sustainable carbon materials. Chem. Soc. Rev. 44: 250–290.

Vishnyakov, A., E. M. Piotrovskaya and E. N. Brodskaya. 1998. Capillary condensation and melting/freezing transitions for methane in slit coal pores. Adsorption 4: 207–224.

Wang, D. W., Q. Zeng, G. Zhou, L. Yin, F. Li, H. Ming Cheng et al. 2013. Carbon-sulfur composites for Li-S batteries: Status and prospects. J. Mater. Chem. A 1: 9382–9394.

Wang, G., B. Wang, X. Wang, J. Park, S. Dou, H. Ahn et al. 2009. Sn/graphene nanocomposite with 3D architecture for enhanced reversible lithium storage in lithium ion batteries. J. Mater. Chem. 19: 8378–8384.

Wang, J. and S. Kaskel. 2012. KOH activation of carbon-based materials for energy storage. J. Mater. Chem. 22: 23710–23725.

Wang, Y. X., S. Lei Chou, H. Kun Liu and S. Xue Dou. 2013. Reduced graphene oxide with superior cycling stability and rate capability for sodium storage. Carbon 57: 202–8.

Wongkoblap, A. and D. D. Do. 2007. Characterization of cabot non-graphitized carbon blacks with a defective surface model: Adsorption of argon and nitrogen. Carbon 45: 1527–1534.

Wu, H., Y. Deng, J. Mou, Q. Zheng, F. Xie, E. Long et al. 2017. Activator-induced tuning of micromorphology and electrochemical properties in biomass carbonaceous materials derived from mushroom for lithium-sulfur batteries. Electrochim. Acta 242: 146–158.

Wu, Y. P., C. Rong Wan, C. Yin Jiang, S. Bi Fang and Y. Yan Jiang. 1999. Mechanism of lithium storage in low temperature carbon. Carbon 37: 1901–1908.

Xin, S., L. Gu, N. H. Zhao, Y. Xia Yin, L. Jie Zhou, Y. Guo Guo et al. 2012. Smaller sulfur molecules promise better lithium' sulfur batteries. J. Am. Chem. Soc. 134: 18510–185113.

Yan, Y., Y. X. Yin, Y. Guo Guo and L. Jun Wan. 2014. A sandwich-like hierarchically porous carbon/graphene composite as a high-performance anode material for sodium-ion batteries. Adv. Energy Mater. 4: 2–6.

Yang, G., X. Li, Z. Guan, Y. Tong, B. Xu, X. Wang et al. 2020. Insights into lithium and sodium storage in porous carbon. Nano Letters 20: 3836–3843.

Yang, X., C. Cheng, Y. Wang, L. Qiu and D. Li. 2013. Liquid-mediated dense integration of graphene materials for compact capacitive energy storage. Science 341: 534–537.

Yu, F., S. Li, W. Chen, T. Wu and C. Peng. 2019. Biomass-derived materials for electrochemical energy storage and conversion: Overview and perspectives. Energy Environ. Mater. 2: 55–67.

Yuan, G., H. Li, H. Hu, Y. Xie, Y. Xiao, H. Dong et al. 2019. Microstructure engineering towards porous carbon materials derived from one biowaste precursor for multiple energy storage applications. Electrochim. Acta 326: 134974–134983.

Zhang, B., C. Matei Ghimbeu, C. Laberty, C. Vix-Guterl and J. M. Tarascon. 2016. Correlation between microstructure and Na storage behavior in hard carbon. Adv. Energy Mater. 6: 1–9.

Zhang, Y., L. Chen, Y. Meng, J. Xie, Y. Guo and D. Xiao. 2016. Lithium and sodium storage in highly ordered mesoporous nitrogen-doped carbons derived from honey. J. Power Sources 335: 20–30.

Zhang, Z., G. Wang, Y. Lai and J. Li. 2015. A freestanding hollow carbon nanofiber/reduced graphene oxide interlayer for high-performance lithium-sulfur batteries. J. Alloy. Compd. 663: 501–503.

Zhao, X. S., F. Su, Q. Yan, W. Guo, X. Ying Bao, L. Lv et al. 2006. Templating methods for preparation of porous structures. J. Mater. Chem. 16: 637–648.

Zheng, Z., H. Guo, F. Pei, X. Zhang, X. Chen, X. Fang et al. 2016. High sulfur loading in hierarchical porous carbon rods constructed by vertically oriented porous graphene-like nanosheets for Li-S batteries. Adv. Funct. Mater. 26: 8952–8959.

Zhou, G., D. Wei Wang, F. Li, L. Zhang, N. Li, Z. Shuai Wu et al. 2010. Graphene-wrapped Fe_3O_4 anode material with improved reversible capacity and cyclic stability for lithium ion batteries. Chem. Mater. 22: 5306–5313.

Zhou, H., S. Zhu, M. Hibino, I. Honma and M. Ichihara. 2003. Lithium storage in ordered mesoporous carbon (CMK-3) with high reversible specific energy capacity and good cycling performance. Adv. Mater. 15: 2107–2111.

Zhou, J., Y. Guo, C. Liang, J. Yang, J. Wang and Y. Nuli. 2018. Confining small sulfur molecules in peanut shell-derived microporous graphitic carbon for advanced lithium sulfur battery. Electrochim. Acta 273: 127–135.

Zhou, Z., X. Gao, J. Yan, D. Song and M. Morinaga. 2004. Enhanced lithium absorption in single-walled carbon nanotubes by boron doping. J. Phys. Chem. B 108: 9023–9026.

Zubair, U., A. Anceschi, F. Caldera, M. Alidoost, J. Amici, C. Francia et al. 2017. Dual confinement of sulphur with RGO-wrapped microporous carbon from β-cyclodextrin nanosponges as a cathode material for Li–S batteries. J. Solid State Electr. 21: 3411–3420.

Zubair, U., J. Amici, S. Martinez, C. Francia and S. Bodoardo. 2019. Rational design of porous carbon matrices to enable efficient lithiated silicon sulfur full cell. Carbon 145: 100–111.

CHAPTER 5

Metal-Organic Frameworks (MOFs): Multi-Functionality within Order

Material's Definitions, Descriptions. Current Trends and Applications

Alejandro M. Fracaroli

Introduction

By definition Coordination Polymers (CPs) are extended structures where coordination entities are repeated in 1, 2 or 3 dimensions (Batten et al. 2013). They have been known for centuries, being one of the earliest examples the Hofmann clathrates (Hofmann and Küspert 1897). These clathrates were found to be 2D extended layers of octahedral and square planar nickel (II) cations linked by cyanide anions (Fig. 5.1a). Additional ammonia ligands pointing towards parallel layers define cavities filled by benzene molecules encapsulated during the material's synthesis. 3D coordination networks were reported several years later by Saito et al. 1959 (Kinoshita et al. 1959). In this case, the prepared material comprised of infinite three-dimensional networks of tetrahedrally coordinated Cu^+ ions with 2-connected adiponitrile (ADI) and nitrate ions ($[Cu(ADI)_2]NO_3$, Fig. 5.1b). This connectivity leads to the formation of 3D frameworks encompassing open spaces filled with non-bonded and intertwined identical 3D frameworks (interpenetration (Ockwig et al. 2005)).

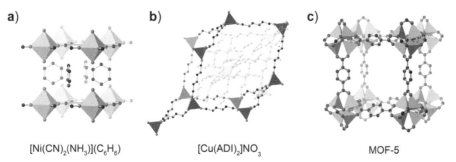

a) $[Ni(CN)_2(NH_3)](C_6H_6)$ b) $[Cu(ADI)_2]NO_3$ c) MOF-5

Figure 5.1: Schematic representation of a Hofmann clathrate (a), a 3D coordination network (b) and the structure of MOF-5 (c). While cavities of clathrates and void spaces of 3D networks are filled with benzene and 6-fold interpenetrated structures respectively, MOF's pores can be made available by a procedure called activation.

Instituto de Investigaciones en Físicoquímica de Córdoba, INFIQC-CONICET, Facultad de Ciencias Químicas, Departamento de Química Orgánica, Universidad Nacional de Córdoba, Ciudad Universitaria, X5000HUA Córdoba, Argentina.
Email: a.fracaroli@unc.edu.ar

Originated from the CPs field, but rather a newly developed subset of materials are Metal-Organic Frameworks (MOFs). MOFs are crystalline extended structures constructed by linking organic struts with inorganic units (Yaghi et al. 2019). Often the term coordination polymer by itself is used to refer to MOFs. In addition, a disturbing number of abbreviations and terminology are used to define new MOF's structures that make the literature in the field very confusing in terms of definitions. For these reasons, it is important to mention at this point that in 2009 the International Union Pure and Applied Chemistry (IUPAC) organized a task group named *Coordination Polymers and Metal-Organic Frameworks: Terminology and Nomenclature*. The objective was to discuss this topic and agree on the correct definitions and the usage of the term in the field (Batten et al. 2013). The final report from this IUPAC task group recommends that the term MOF should be specifically used to refer to coordination networks where organic ligands are involved and which contain potential void spaces. Therefore, and in comparison with the above presented Hofmann clathrates and $[Cu(ADI)_2]NO_3$, MOFs are a subset of coordination networks incorporating additional structural features such as rigidity and porosity.

The term MOF was first used by Yaghi and coworkers when reporting the solvothermal synthesis of $[Cu(BIPY)_{1.5}]NO_3$ (Yaghi and Li 1995). However, it was not until the beginning of 2000, that the concept attracted the attention of researchers mainly from inorganic chemistry and porous materials fields. Until then, scientists saw MOFs simply as another example of the already known CPs. In a report from the same group in 1999, charged chelating organic struts and polynuclear inorganic clusters (now known as secondary building units or SBUs), were introduced, thus achieving a structure with enhanced chemical and thermal stability, called MOF-5 (Fig. 5.1c) (Li et al. 1999). The nowadays iconic MOF-5 was stable enough to empty its pores after the synthesis demonstrating the so-called permanent porosity (*vide infra*). The procedure, including the washing of the prepared crystals by solvent exchange and evaporation by applying heat and vacuum, allowed obtaining fully desolvated crystals with a Langmuir surface area of 2900 $m^2.g^{-1}$ and pore volumes surpassing all the already reported zeolites.

An additional consideration on the used terminology is that coordination polymers do not need to be crystalline, while the majority of the reported MOF structures are. In fact, MOF crystallinity results in an important feature of these materials as it allowed to explore previously unknown conceptual areas such mechanistic aspects of heterogeneous catalysis (Reinares-Fisac et al. 2016, Trickett et al. 2019), X-ray structural elucidation non-crystalline molecules (Inokuma et al. 2013, Lee et al. 2016) and the design and installation of enzyme-inspired catalytic sites within the pores of the structures by the material's functionalization with crystallographic precision (Ji et al. 2020, Fracaroli et al. 2016).

MOFs: Nanostructured Materials

Reticular Chemistry

Except for nano-MOFs (or simply nMOFs), where the MOF's crystal size is reduced to the nanometer range (Choi et al. 2015), most of the reported MOFs crystallize within micrometer to millimeter size. Hence and strictly speaking, MOFs are not really 'nanomaterials' but rather 'nanostructured materials', since many of their structural features are defined in the nano-regime (e.g., pore geometry and apertures, metrics between organic struts or inorganic clusters, among others).

In the last two decades, MOFs have attracted enormous scientific interest due to their high crystallinity, exceptional porosity, structural versatility, and ease of functionalization achieved both pre- and post-synthetically (Hong-Cai and Kitagawa 2014, Furukawa et al. 2013). The MOF's constructional flexibility, based on the nearly endless combination of building units (organic struts and inorganic clusters), allows accessing a large variety of structures. By the end of 2016, more than 75,000 different MOFs were included in the Cambridge Structural Database (version 5.38 of the CSD) (Moghadam et al. 2017). To completely understand the reason behind this number, it is necessary to consider the geometric design principles for this subset of coordination networks.

As mentioned in the introduction of this chapter, the uniqueness of MOFs is the structural rigidity coming from the rigidity and well-defined-geometry and connectivity present in both organic struts and inorganic clusters. These geometrically well-defined building units can be connected in a predictable topology resulting in a new level of synthetic control, previously unknown in other coordination networks.

Topological terms have been widely used to describe extended structures. Specifically, net topologies have been applied to MOFs in order to understand and explain the connectivity of the building units. The nomenclature of the nets consists of three-letter codes compiled in the Reticular Chemistry Structure Resource (RCSR) database. Although the codes are assigned arbitrarily on discovery, often they are related to naturally occurring names of minerals (http://rcsr.net/nets). The topological analysis of a given MOF can be performed by deconstructing its structure into vertices (or nodes) and edges (or links), both of which can have different connections with other building units (points of extension). As an example of this analysis, the connectivity found in MOF-5 is described in terms of its topology and coordination net (O'Keeffe et al. 2000, Yaghi et al. 2003, O'Keeffe and Yaghi 2012). The Secondary Building Units (SBU) of MOF-5 are cationic Zn_4O clusters hexa-coordinated by carboxylate anionic ligands. As each Zn is tetrahedrally coordinated and the cluster has one oxygen bridging all four of them, the carboxylate coordination is established at the vertices of an octahedron (Fig. 5.2). Therefore, the SBUs in the structure (or vertex in topological terms), can be represented by an octahedron with six points of extension. On the other hand, the organic struts in this MOF are terephthalates (benzenedicarboxylates or BDC), which are links with two points of extension (or ditopic links). As they are also rigid and linear molecules, they can be simply represented by sticks. The connectivity between vertex and links develops the MOF-5 cubic structure that is deconstructed in a primitive cubic net or **pcu** (Fig. 5.2).

The topological analysis shown above is useful to perform a comprehensive analysis and clearly describe the MOFs structures. However, as proposed earlier by O'Keeffe and Yaghi (O'Keeffe et al. 2000, Yaghi et al. 2003, O'Keeffe and Yaghi 2012), and more recently by Eddaudi and coauthors (Chen et al. 2020), the knowledge gained on MOF net-topologies is also a powerful design tool for new MOFs. The strong bonding between molecular building units with defined geometries and points of extension, allow to synthetically achieve predetermined structures with repetitive units throughout the 1, 2 o 3D space. This directed synthesis of crystalline extended solids is known as *reticular chemistry* and has been extensively used to accelerate the discovery of coordination networks, in particular, MOFs. To illustrate this concept, a recently reported example of a material that undergoes a single crystal-to-single crystal transformation to achieve the desired properties could be considered (Abtab et al. 2018). In this case, achromium (III)-based MOF is targeted. However, the desired topologies could not be directly prepared by the coordination of Cr (III) clusters and

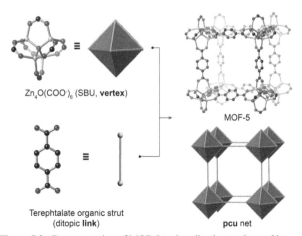

$Zn_4O(COO^-)_6$ (SBU, **vertex**)

MOF-5

Terephtalate organic strut
(ditopic **link**)

pcu net

Figure 5.2: Deconstruction of MOF-5 to describe the topology of its net.

the corresponding carboxylate organic struts for synthetic reasons. Thus, the deconstruction of its net topology allows to design a stepwise synthetic strategy where an *isoreticular* (having the same topology, the concept is further explained in the section 'Isoreticular expansions' below), iron-based MOF was prepared and named Fe-soc-MOF-1, as a precursor of the Cr-MOF. In this synthesis, a square planar tetratopic organic strut and hexacoordinated trigonal prismatic $[Fe_3(\mu_3-O)(COO^-)_6]$ clusters were used. As the geometry and points of extension for the Fe(III) cluster are equivalent to the targeted $[Cr_3(\mu_3-O)(COO^-)_6]$ clusters, both SBUs should have the same **soc** net topology. After the preparation of Fe-soc-MOF-1, the SBUs were replaced post-synthetically by transmetallation of Fe(III) ions by Cr(III) ions with 98% efficiency. This single crystal-to-single crystal transformation was only possible because of the predictability in the connectivity of the building units, which allowed to replace one SBU by another with the same geometry and points of extension.

Permanent Porosity

The possibility of accessing the pores of the material and customizing them in terms of their size and affinities is a key feature of MOFs and has broadened their application in different fields. Due to their porosity, since the first reports, MOFs have been explored as promising candidates for gas adsorption and storage applications. Different studies on the capacity and selectivity towards capture and storage of hydrogen (Rosi et al. 2003, Gómez-Gualdrón et al. 2017, García-Holley et al. 2018, Allendorf et al. 2018), methane (Bin et al. 2016, Mason et al. 2014, 2015, Gándara et al. 2014), carbon dioxide (Millward and Yaghi 2005, Fracaroli et al. 2014, McDonald et al. 2015, Ding et al. 2019, Kim et al. 2020), and water (Hanikel et al. 2020, Kim et al. 2017, Furukawa et al. 2014, Cavinet et al. 2014, Logan et al. 2020), was greatedly impacted in the field. This effect attracted the attention of different chemical companies worldwide (e.g., BASF, Mosaic, Water Harvesting Inc., etc.), who began and are now exploring the material's potential in these industrial applications. Nowadays, several patents have been already transferred to companies and different MOFs are being produced in large scales for their pilot testing.

Although molecular compounds connected by supramolecular bonds are able to show porosity (Hasell and Cooper 2016, Rabone et al. 2010, Lim et al. 2008), often on uptake-release cycling of users such as nitrogen, they undergo phase transitions to less porous or even dense non-porous forms. This is likely explained by considering the weak non-covalent interactions that hold the structure together, which usually cannot withstand the capillary forces produced in the cycling processes or the evacuation of the pores. Hence, in literature, the term *permanent porosity* is used to refer to porous materials in which pore's environments can be made accessible to different user molecules. Additionally, materials with permanent porosity typically should show reversible gas adsorption isotherms (adsorption-desorption), at low pressures and temperatures using rather inert molecules such as liquid nitrogen or argon. In this regard, IUPAC defines the *specific surface area* as the accessible area of solid surface per unit mass of material, and list gas adsorption methods as a widely used technique for their characterization (Rouquerol et al. 1994). Some of the important and related definitions included in the same reference are worth mentioning here:

Adsorption isotherm: the amount of substance (*adsorbate*) adsorbed in the solid (adsorbent), at equilibrium to the pressure in the gas phase.

Physisorption: adsorption through intermolecular forces which do not involve a significant change in the participating specie's electronic orbitals.

Chemisorption: adsorption in which the forces involved are valence forces of the same kind as those operating in the formation of chemical bonds.

Monolayer adsorption: all the adsorbed molecules are in contact with the surface layer of the adsorbent.

Multilayer adsorption: the adsorption space accommodates more than one layer of molecules and not all adsorbed molecules are in contact with the adsorbent surface.

Hysteresis: refers to a mismatch in the adsorption curve with respect to the desorption, where the latter process does not proceed exactly in the reverse adsorption pathway.

Capillary condensation: a process that takes place when multilayer adsorption from vapor proceeds to the point at which the pores are filled with liquid separated by the gas by menisci.

Texture: geometry of the void space in the particles.

Pore size (generally pore width): according to the distance between opposite walls, *micropores* (not exceeding 2 nm), *mesopores* (between 2 and 50 nm) and *macropores* (exceeding 50 nm).

Specific surface area: accessible area of solid surface per unit of mass of material.

Pore size distribution: distribution of pore volume with respect to pore size (represented by the derivatives dV_p/dr_p or dA_p/dr_p, where V_p and A_p are volume and wall area of the pores, respectively).

IUPAC has also differentiated six different isotherm types based on their shapes. Among these, Type I isotherm is typically found in microporous materials, Types IV and V are associated with mesoporosity and Types II, III and IV are associated with non-porous materials or macroporosity. Usually, MOFs present Types I and IV isotherms that can include hysteresis loops related to material's texture (topology, pore geometries, etc.). An important consideration at this point is that the observed gas adsorption properties in MOFs are strongly dependent on the material's activation. A part of the sample preparation process to determine the surface area by gas adsorption consists of emptying the pores or void spaces in the structure. At the end of the MOF synthesis, these pores are filled with unreacted materials such as solvents, organic struts of inorganic species. These 'byproducts' have to be removed in order to make this space available for other guest molecules such as the N_2 or Ar used as adsorbates when measuring the adsorption-desorption isotherms to determine the material's surface area. The process of vacating the pores is called *activation*, and involves several steps and techniques planned while considering the material's stability. Thus, some MOFs can be activated by simply applying vacuum and heat, while others may require solvent exchange procedures before applying vacuum, or supercritical CO_2 activation. The activation process is an important step not only for the surface area determination but also for the MOF characterization and is typically an iterative process where the crystallinity of the sample has to be revised throughout the process. Other techniques can provide some hints on the best activation strategies. For instance, thermogravimetric analysis (TGA) can provide partial but relevant information on the thermal stability of the sample and the amount of solvent remaining at a given temperature.

Once the adsorption-desorption isotherm is obtained for a particular MOF there are essentially two models that can help to calculate its specific surface area (SA): Langmuir and Brunauer–Emmett–Teller (BET) models (Lowell et al. 2004). While the first model approaches the calculation of the SA by considering the formation of an adsorbed gas monolayer on the adsorbent surface, the second is not restricted to the formation of monolayers, therefore considering multilayers of adsorbate molecules on the adsorbent surface. The Langmuir model was found to better describe microporous materials in which the specific guest adsorption usually involves higher energies. However, it tends to overestimate the SA in materials in which the adsorption process is driven by weak interactions, such as van der Waals forces. Considering this observation, the BET model was more widely applied to the MOF's surface areas calculation, particularly to better account for the multilayer adsorption taking place in their mesopores. Bearing in mind the existence of these two models, comparison of SA values for different materials is sometimes challenging as reported values are indistinctively calculated according to different models. Thus reporting the model used to calculate the informed SA values and the measured isotherms is critical to avoid confusion, when comparing MOFs or other porous materials. An example of this, is MOF-200 which was reported to have a Langmuir SA of 10400 $m^2.g^{-1}$, but according to the BET model its SA is equivalent to 4530 $m^2.g^{-1}$ (Furukawa et al. 2010). As many of the MOFs applications are related to their high surface area, this consideration becomes relevant.

After the synthesis of MOF-5, several new MOFs structures were reported breaking previous records for the material's SA, and even beyond MOFs, surpassing all the reported SA of previously known porous materials. Later, and given the achieved SA values, the concept of *ultra-high porosity* (500 < BET SA > 10,000 m^2.g^{-1}) was coined (Furukawa et al. 2010 cited above). A summary of some of the best performers with regards to and their reported SA and dates of the original reports, is presented in Table 5.1.

It is interesting to note here that while MOFs are a relatively new class of materials they are in constant development. Therefore, the reported BET SA values for a given material can be later updated or rectified in a following reports (sometimes years later). This usually happens on optimization of its synthesis, crystallinity and/or activation procedures. For instance, this was the case of MOF-5, the first MOF entry in the table above, which almost eight after its discovery was reported to have a four-times higher BET SA. The same phenomenon can be observed in other MOFs like MOF-177, among others.

Table 5.1: Representative examples of high and ultrahigh surface area MOFs in the context of the average SA found in other porous materials.

Material/year	BET surface area/m^2.g^{-1}	Reference
Zeolites*	~ 900	Choi et al. 2006, Kresge et al. 1992, Na and Somorjai 2015
Silicas*	~ 1000	Goto et al. 2019
Carbons*	~ 1500	Inagaki 2000
MOF-5 (2003/2007)	950/3800	Huang et al. 2003, Kaye et al. 2007
MIL-101 (2005)	3750	Férey et al. 2005
MOF-177 (2004/2007)	4500/4630	Chae et al. 2004, Furukawa et al. 2007
UMCM-2 (2009)	5200	Koh et al. 2009
PCN-68 (2010)	5110	Yuan et al. 2010
MOF-210 (2010)	6240	Furukawa et al. 2010
NU-110E (2012)	7140	Farha et al. 2012

* The average BET SA of representative materials in each of these classes is listed as a reference.

Building Units and Pore Geometries

Organic Struts

As described earlier geometrically well-defined building units and their connection in a topologically predictable manner are used for the construction of MOFs. Hence it is worth summarizing here the variability of shapes and connectivities explored so far in organic struts used for preparing these coordination networks.

The organic struts (also called linkers) employed in the construction of MOFs are polyfunctional organic molecules, generally featuring a certain rigidity (shape persistent molecules) and high symmetry. Different functional groups have been used so far in these organic struts for binding the SBUs and holding the structure together. Among them, N-heterocycles were extensively used during the early development of MOFs. For instance, bipyridyl organic struts with different length were employed for the synthesis of a variety of high-dimensionality coordination structures including MOFs. Recently, bipyridyl struts have been mostly used in the construction of 3D pillared-layer structures, where 2D coordination layers are extended to 3D frameworks by the introduction of these N-heterocycle 'bridges' (Seo et al. 2009). Although these pyridyl struts can add interesting features to MOFs such as structure flexibility and a dynamic adsorption behavior, compared to other bridging coordination modes, they possess a relatively weak donor-ability, which translates to instability in the MOFs structures (Lu et al. 2014). Due to this reason, other N-heterocycles such as azolate-derivatives were more attractive for researchers in the construction of MOF. Imidazolate rings were used in the preparation of metal–imidazolate frameworks called Zeolitic Imidazolate Frameworks

(ZIFs) due to their similarity to zeolite topologies. In ZIFs, imidazole derivatives used as organic struts (called HIM) deprotonate to form IM⁻ anions (or simply IM) that bind tetrahedral metal centres with M-IM-M bond angles close to 145°, which resemble Si-O-Si linkage in zeolites (Park et al. 2006, Hayashi et al. 2007, Eddaoudi et al. 2015). The strong bonds between IM and the divalent metal cations (e.g., Zn^{2+}, Fe^{2+}, Co^{2+}), along with the hydrophobicity of the formed cavities, provided ZIFs with excellent thermal and chemical stability, as well as gas separation properties (Nguyen et al. 2014). A few years ago the scope of this chemistry reported that ZIFs with cavities up to 4.6 nm and BET surface areas above 1400 $m^2.g^{-1}$ (ZIF-414, $Zn(nbIM)_{0.91}(mIM)_{0.62}(IM)_{0.4}$). In the case of the reported ZIF-414, the steric effects provided by the chosen combination imidazolate-derivatives (nbIM: 6-nitrobenzimidazolate, mIM: 2-methylimidazolate and IM: imidazolate) led to the crystallization of structures with extra-large pore openings (Yang et al. 2017).

Other azolate functionalities incorporated to MOFs organic struts are pyrazolates, $1H$-1,2,3-triazolates and tetrazolates (Masciocchi et al. 2010, Demessence and Long 2010, Dincă et al. 2006, respectively). The high basicity of these azolate functionalities led to MOF structures with enhanced strength in the metal–nitrogen bonds and, concomitantly, high thermal and chemical stability even when boiled in water or the presence of strongly acidic media. The chemistry of the azolate ligands in relation to the MOF construction has been recently reviewed (Zhang et al. 2012), highlighting the advantages of incorporating building blocks which are already widely used in medicine, agriculture and industry, in general. In addition, their sp^2 N act as donors providing coordination modes analogous to pyridines.

Despite the variety of functional groups that can be used for the coordination of metals and metal clusters in MOF's SBUs (phosphonates, sulfonates and others were used but are not discussed here), poly-carboxylate anions are by far the most frequently employed. The large number of polycarboxylate molecules used as organic struts in the synthesis of MOFs can be better described and classified in terms of their geometry and points of extension or *topicity* (*vide supra*). For instance, terephthalic acid, the organic strut for the construction of MOF-5, can be described as a linear ditopic (two points of extension) linker. Although the organic strut's topicity defines the number of SBUs they can be linked to, their spatial distribution (molecular geometry) determine the connectivity. As such, the topology found in a MOF incorporating square planar tetratopic struts might not be achieved by tetratopic tetrahedral organic linker (Fig. 5.3). The first and most commonly used organic struts in the construction of MOFs have been the linear ditopic molecules like the terephthalic acid mentioned above. Elongated versions of these ditopic struts, such as *p*-terphenyl-4,4″-dicarboxylic acid shown in Fig. 5.3, also fall in this category. It is worth noting, that although there may be free rotation or additional functionalization in their central phenylene units, this does not affect the strut shape or connectivity and therefore neither the MOF underlying topology. This fact becomes relevant when discussing multivariate functionalities in MOFs.

In general, the carboxylate organic struts are prepared by performing C-C coupling reactions (Heck reaction or Suzuki or Sonogashira couplings) on a functionalized core. In most cases, the carboxylic acids present in the intermediates have to be protected using their ester forms in order to avoid undesired reactions (Fracaroli et al. 2014). While intensive organic synthesis to prepare elaborated organic struts may seem impractical from the industrial application perspective, unprecedented properties and drastic differences in pore affinities were achieved thanks to this structural flexibility and the extensive collaboration between scientists from different areas, and taking advantage of the accumulated knowledge in the field of organic synthesis.

Secondary Building Units (SBUs)

The polynuclear inorganic clusters of MOFs that comprise the Secondary Building Units or SBUs are typically formed *in situ*, from the inorganic precursors used in the synthesis. Under the reaction's conditions, these molecular building units are gradually and reversibly connected to the organic struts to form the MOF infinite coordination networks. Since the SBUs can be constructed from a large variety of periodic table elements they are largely responsible for the nearly endless variety of

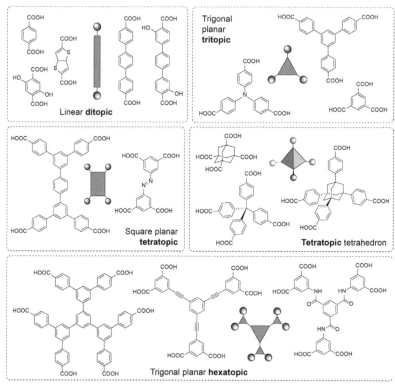

Figure 5.3: Representative examples of organic struts of different topicity that were employed to build MOFs (for a comprehensive list of organic struts), please see Chapter 3 in (Yaghi et al. 2019).

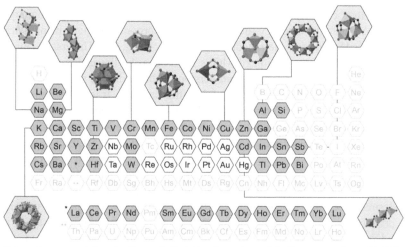

Figure 5.4: Schematic representation for the periodic table elements used in the construction of MOF's SBU (gray-background highlighted). Those elements with a white background and black symbols were incorporated into the MOF structures through organic struts metalation. This figure was adapted from Yaghi Laboratory's research website (http://yaghi.berkeley.edu/research.html).

plausible frameworks (Fig. 5.4). SBUs, as well as the organic struts, are geometrically well-defined making the synthesis of MOFs by designed topologies feasible.

As a large number of MOF structures are neutral and organic struts are in general negatively charged, the SBUs are typically polycationic. As such, a material like MOF-5 can be understood as a salt formed by a conjugate base of a weak acid (terephthalate)$^{2-}$ and a positively charged

$(Zn_4O)^{6+}$ cluster. In this case, the SBU formula can be described as $Zn_4O(COO)_6$ where each of the carboxylates appears from a different terephthalate strut.

Considering that this chapter's aim is not to fully describe the reported SBU's structures, but is only a brief and general description for the variety of existing SBUs. For a comprehensive and detailed list of the already reported clusters used in MOFs, please refer to references by O'Keeffe and Yaghi (Tranchemontagne et al. 2009, Li et al. 2014, Schoedel et al. 2016).

As in the case of the organic struts, SBUs can also be defined by their points of extension, which correspond to the maximum number of possible connections between them and the struts. For multinuclear SBUs, the minimum number of point of extension is 3 (Cheng et al. 2004), but there are known molecular clusters linked to carboxylates that contain up to 66 points of extension (Tasiopoulos et al. 2004). The most commonly used in MOF structures have between 3 and 12.

As an example of the extraordinary variability of shapes and connectivities found in the MOF's SBUs, hexa-zirconium clusters can be highlighted. Metal carboxylate clusters of zirconium with the general formula $Zr_6(OH)_4O_4(COO)_{12}$ are well-known as molecular inorganic compounds. In order to reticulate this molecular clusters into extended structures the strong metal-carboxylate bonds formed between them and the polycarboxylate organic struts are formed *in situ* from an appropriate metal source such as $ZrCl_4$. The notable oxophylicity of zirconium allowed to report some of the most thermally and chemically stable structures within this field. However, it is worth mentioning here that the stability found for these structures is not only due to the Zr-(COO) bond energy but also because of the formation of these polynuclear SBUs involving several of these bonds in the structure. To exemplify this, it is possible to picture the energy necessary to remove the one metal ion (M) from in the cluster mentioned above. To achieve this it would be necessary to break 12 M-L bonds, as opposed to removing a metal ion from a coordination polymer where often the coordination is established with a single atom.

It is necessary to mention here that although each SBU has a maximum number of possible connections, lesser points of extension can be found in the structure. For instance, the 12 connected (or simply 12c) $Zr_6(OH)_4O_4(L)_{12}$ SBU mentioned above, was also reported to form MOFs in which it has 4, 6 and 8 points of extension (4c, 6c and 8c, respectively). While 12c clusters are fully coordinated by the carboxylate organic struts, in the 4-8c the vacant coordination sites in the cluster are occupied by other ligands that are not able to reticulate the structures such as mono-carboxylate species, hydroxide or even water. The case of MOF-808 (Furukawa et al. 2014) is typical. The molecular formula for the MOF-808 is $Zr_6O_4(OH)_4(BTC)_2(HCOO)_6$ (BTC = benzene-1,3,5-tricarboxylate) showing that its SBUs are only 6c (BTC has 3 carboxylates each), and the remaining coordination sites are occupied by formic acid used as a modulator during the synthesis (Fig. 5.5).

Interestingly, the bond energy difference between the BTC and HCOO⁻ coordination can also be used to functionalize the pores of these ordered structures. In the case of MOF-808 crystals, their treatment with sulfuric acid solutions allows in almost completely replacing the formiates in the

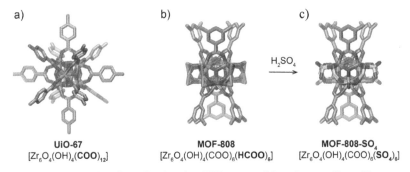

Figure 5.5: Schematic representation of: (a) 12 czirconium SBUs extracted from the crystallographic structure of UiO-67, (b) 6c Zr-SBU from the crystal structure of MOF-808, and (c) MOF-808 Zr-SBUs after formiate anions were replaced by sulfates, generating a super acidic solid.

structure, by sulfate. This transformation occurs without losing the MOF crystallinity or porosity in the MOF (Jiang et al. 2014). The resultant material called MOF-808-2.5SO$_4$, possess the same color and crystal morphology (by SEM analysis) than its precursor, but it behaves as a superacid solid (Hammett acidity $H_0 \leq -14.5$).

MOFs' Pore Geometries and Affinities by Design

Isoreticular Expansions

As mentioned above, one of the features that allow MOFs to be categorized as nanostructured materials is their pore shape and metrics within the nano-regime. Permanent porosity has been perhaps the main motivation for companies to bring these materials to the industry, but it was the variability of design tools and nanostructures which inspired many researchers to work with these new materials in their laboratories. The uniqueness of MOF is the crystallographic precision with which the pore geometry and environment can be varied. Due to the intrinsic composition of the material the void spaces in the structures can be adjusted to target particular properties. One of the construction principles that has made this possible is the *isoreticular expansion* (enlarging the pore dimensions without changing the underlying topology), first reported in 2002 (Eddaoudi et al. 2002). Through this principle, the pore size of a given MOF can be expanded using the same SBU, and by simply extending the length of the organic strut while maintaining its shape, points of extension and connectivity. This pore expansion is performed without changing the net-topology of the MOF. Nonetheless, and as can be observed in the work reported by Yaghi et al. 2002 and later works by (Schaate et al. 2011) among others, this strut elongation often results in interpenetrated frameworks (mechanically interlocked), thus decreasing the pore size rather than increasing their dimensions (see IRMOF-16 and PIZOF structures in Fig. 5.6). So far, three main strategies have

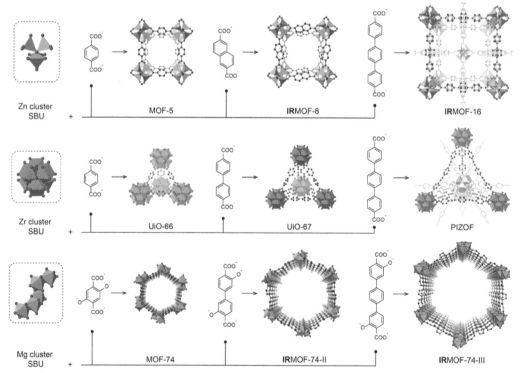

Figure 5.6: Different examples of isoreticular expansions in MOFs. While in the first two cases the elongation of the organic struts leads to interpenetrated structures (MOF-5 and UiO-66), in the third case (MOF-74), it allows achieving one of the lowest density materials ever known.

been applied to overcome interpenetration in the MOF structures: (i) the use of templates (Shekhah et al. 2009), (ii) geometric considerations (Deng et al. 2012, Li et al. 2018), and (iii) lattice expansion by post-synthetic linker-exchange (Feng et al. 2019, Li et al. 2013). The developers of the first strategy (i) state that avoiding interpenetration from the topological viewpoint is difficult because different networks bear translational symmetry and are, therefore, equivalent within the bulk. Thus, they propose a strategy based on the soft-templating of frameworks involving Self-Assembled Monolayer (SAM) of 4,4-pyridyl-benzenemethanethiol (PBMT) on a gold substrate. This template mimics one of the organic strut's rim (4,4'-bipyridine or 4,4'-bipy) used in the synthesis leading to the epitaxial growth of the MOF-508[Zn(BDC)-(4,4'-bipy)$_{0.5}$]. The second approach (ii) was demonstrated by selecting MOFs in which the distance between the organic struts attached to the SBUs was short enough to avoid multiple intergrowths within the same structure. This is the case of MOF-74 (Fig. 5.6), constructed by infinite rod-shaped metal oxide SBUs and linear ditopic organic struts such as 2,5-dihydroxyterephthalic acid (DOT), in an **etb** topology.

In this work, the original organic strut (DOT, 7 Å long) is elongated to incorporate up to 11 phenylene units in a ditopic linear strut reaching a length of c.a. 50 Å. The use of these extended struts in the MOF structures allows to expand the unidimensional hexagonal pores from dimensions of 10 × 14 Å in MOF-74, to 85 × 98 Å in IRMOF-74-XI. It is interesting to note here that expanding the pore dimensions in this case, also means enlarging the pore apertures in the material, due to its topology. As such, the material with the largest pore dimensions reported in this work, IRMOF-74-XI, possess pore apertures close to 10 nm and it is composed of 282 atoms. IRMOF-74-XI in its activated form also corresponds to one of the lowest density crystals ever reported ($\delta = 0.195$ g.cm^{-3}). Materials featuring these large pore dimensions are suitable for the incorporation of complex and large molecules in their pores. This is the case of Green Fluorescent Protein (GFP) passing through the hexagonal pores of IRMOF-74-IV and above in the series, and lactate dehydrogenase (LDH) encapsulated in an isoreticularly expanded NU-100x (Li et al. 2018). In the latter case, in order to immobilize LDH, the tetratopic organic struts were extended by maintaining their geometry to achieve materials with **csq** topology featuring hexagonal pores with 67 Å in diameter (NU-1007). It is important to highlight here that larger pore dimensions in MOFs do not always mean higher surface areas, as the amount of adsorbed guest molecules is proportional to the adsorption sites and not to the pore volumes. As demonstrated earlier, in MOFs the strongest adsorption sites are located close to the SBUs and organic struts (Rowsell et al. 2005) and therefore larger pores represent larger 'dead volumes' where the adsorbates do not interact with the framework.

It was demonstrated that isoreticular expansion can also be achieved even when the use of templates or geometry principles do not apply. In this case, larger pore dimensions with the same underlying topology can be obtained post-synthetically by a linker-exchange strategy (strategy iii, *vide supra*). When first reported, this methodology was thought to be applicable only to materials with low chemical stability, as it is necessary to break the bonds that hold the structure together in order to be able to install longer organic struts. Accordingly, it was first successfully demonstrated for low–valent metal-based MOFs with relatively labile M(II)–carboxylate coordination bonds. However, recently, a sequential linker reinstallation approach allowed to manipulate post-synthetically more stable frameworks featuring strong M(III/IV)–carboxylate bonds (Feng et al. 2019). As mentioned above, the one–pot synthesis of isoreticularly expanded UiO-66 lead to interpenetrated structures called PIZOF (see Fig. 5.6). Nevertheless, this strategy allows achieving this expanded framework with no interpenetration using imine-based labile intermediate linkers during the synthesis. Thus by using organic struts with a similar length but less stable bonding, one can synthesize MOFs that can be post-synthetically exchanged by stronger bond-forming phenylene dicarboxylate analogues. This sequential replacement is demonstrated to occur for structures where the length difference between the organic struts is less than 2.5 Å, which is also believed to be the maximum structure flexibility.

Pore Environments and Multivariate Functionalities (MTV-MOFs)

In addition to the structural diversity that comes from the nearly endless combination of building units, MOF's ability to crystallize allowed scientists to resolve their single crystal structures, in order to know these materials down to the atomic level. As other crystalline materials, MOFs are composed of highly ordered and repetitive units, extended in all dimensions. However, a unique feature of MOFs with respect to other porous materials, is that functionality can be installed within their highly ordered structures, with crystallographic precision. This multi-functionality within an order can be achieved by using different strategies (Fig. 5.7).

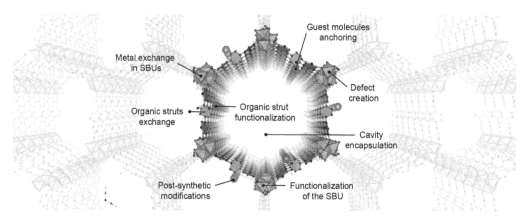

Figure 5.7: Different strategies for achieving multi-functionality within order in MOF structures.

Pre-synthetic Functionalization of Organic Struts. Multivariate Functionalities in MOFs (MTV-MOFs)

Functionalizing porous solids has been a long-standing goal because installing different functionalities in the material's surface has the potential to change the pore's identity and their affinity for different guest molecules. However, introducing functional groups post-synthetically in materials like zeolites or mesoporous carbons is challenging because the whole material's surface is chemically identical. Accordingly, reactive molecules aiming to bind to the material's surface react randomly throughout the surface often clogging the pores or generating a heterogeneous distribution of functional groups. An example of this is observed in the functionalization of mesoporous ZSM-5 type zeolite with tetraethylenepentamine (TEPA) (Wang et al. 2017). While the pristine ZSM-5 has a relatively high BET surface area (400 $m^2.g^{-1}$), after the TEPA impregnation procedure the porosity is significantly reduced (20 $m^2.g^{-1}$).

On the other hand, taking into consideration that the MOF's underlying topology solely depends on the geometry and points of extension of its building units, the frameworks can be functionalized by simply adding functional groups to the organic struts before making the MOF. This was demonstrated earlier by Yaghi and coworkers using MOFs derived from zinc and magnesium (Eddaudi et al. 2002, Deng et al. 2010). In these reports, the organic struts are pre-synthetically functionalized without altering the geometry and points of extension in the struts. The functionalized organic struts are later used to construct MOFs with the same topology than the non-functionalized MOF. The latter reported examples of several functional groups that are added to the MOF-5 pores by the means of this strategy achieving multivariate functionalities of MTV-MOF-5. Up to four different organic struts were mixed in the synthesis of the MTV-MOF, obtaining phase pure crystals (according to the X-ray analysis) with the same underlying topology in the prepared framework. Moreover, the prepared MTV-MOF-5 was proven to retain a relatively large BET surface area (2860 $m^2.g^{-1}$ for the so called MTV−MOF-5-ABCD, where A = terephthalate, B = 2-amino-terephthalate, C = 2-bromo-terephthalate and D = 2,5-dichloro-terephthalate). Although X-ray analysis cannot differentiate if the functionalities are mixed well or forming macroscopic domains, it was later found by NMR and simulations that the functionalities are rather well dispersed throughout the crystals (Kong et al. 2013). This level of control over the functionalization of MOF-5 allowed finding materials featuring interesting synergistic properties of those functionalities in the pore. For instance, MTV−MOF-5−EHI, where E = 2-nitro-terephthalate, H = 2,5-bis(allyloxy) terephthalate and I = 2,5-bis(benzyloxy) terephthalate, shows 400% enhancement in carbon dioxide selectivity over carbon monoxide, when compared to pristine MOF-5 structure.

The MTV concept is shown in Fig. 5.8 below, by using the MOF-5 cubic structure.

Furthermore, this MTV approach can be used to dial-in or to 'dilute' the functionalities in the framework which results in attractive applications such as heterogeneous catalysis. In some MOF structures to achieve this 'dilution' it is only necessary to change the organic strut's molar ratio during the synthesis thus causing the desired incorporation of different amounts of functionalities within the structure (see Fracaroli et al. 2016). This experimental observation is extremely useful when installing reactive groups in the MOF pores to later perform post-synthetic transformations. Post-synthetic reactions that take place within the pore environments require the rather smooth diffusion of both reagents and products to and from the pores. A fully functionalized framework can cause pore blocking and/or slow reagents diffusion. However, the 'dilution' of the reactive sites improves the reaction yields. Similarly, when installing 'synthetically expensive' or complex catalytic centers in the pores, dialling-in the necessary amount of active sites result particularly attractive (Yuan et al. 2016).

A limitation of this approach is that sometimes the functionalities present in the organic struts affect the MOF crystallization process. This is the case of secondary amines when incorporated to the organic struts prevent the MOF structure to form (Lun et al. 2011, Deshpande et al. 2011). Possibly these reactive groups are chemically transformed under the MOF synthetic conditions (e.g., protonation or metal ion coordination) affecting the crystallization process. Nevertheless, this obstacle can be overcome by using photo- or thermo-labile protecting groups that can be removed post-synthetically. This strategy has been efficiently demonstrated by Telfer and coworkers (Lun et al. 2011).

Given the possibility to combine multiple organic struts in a single MOF structure and in different ratios, a high throughput computational screening can be applied for finding the best performer arrangements of struts for a particular application such as CO_2 capture. Recently, grand

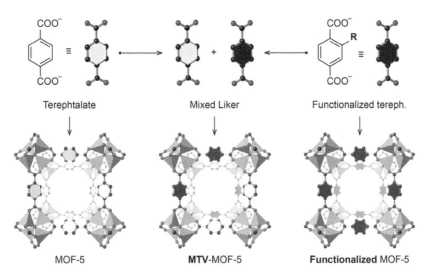

Figure 5.8: Schematic representation of the MTV–MOF concept. MTV–MOF-5 is prepared by using a mixture of organic struts with the same geometry and points of extension.

canonical Monte Carlo simulations were performed by Randall Snurr, constructing a database of ~ 10,000 hypothetical structures. In this study, the authors found that not only the introduction of a certain combination of functionalized struts increase the CO_2 uptake selectivity with respect to the parent non-functionalized structures, but also their effect is sharper when pores are smaller (Li et al. 2017).

Post-Synthetic Modifications (PSM)

As described above, reactive functional groups cannot always can be introduced pre-synthetically by modifying the organic struts. Sometimes, these functionalities prevent the MOFs to form either because of their reactivity under the MOF synthetic conditions (Li et al. 2017) or simply because of electronic or steric factors. For example, the synthesis of MOF-177 was observed to be unsuccessful when using the nitro-substituted organic struts [1,3,5-tris(3-nitro-4-carboxyphenyl) benzene] (Zhang et al. 2015). A possible explanation for this experimental observation is that the nitro groups are probably too close to the carboxylates coordinating the Zn clusters (steric effects). Additionally, the presence of these electron-withdrawing groups affects the organic strut electron density and concomitantly, its pKa. In general, the preparation of highly functionalized MOFs is limited by the solvothermal synthetic methods under which ligands featuring reactive or labile functionalities or simply poor solubility, cannot be incorporated by this MTV method. In these cases, Post-Synthetic Modifications (PSM) prove to be particularly useful. PSM can not only overcome the limitations of installing reactive functionalities in the MOF pores, but also open the possibility to explore chemical reactions in the confined spaces of the framework's pores. The component of an organic building unit present in the MOF structure provides the possibility of performing a vast range of organic transformations to achieve complexity both in the inner and outer material's surface, in contrast to other known inorganic solids, such as Quantum Dots (QDs), metal nanoparticles (MNPs).

Different types of PSM have been reported so far and can be classified depending on the type of bond involved in the transformation. Hence, a modification performed on an SBU's coordination site is called *coordinative or dative* PSM, as generally a metal-ligand bond is formed or broken. Alternatively, if the modification involves the formation or breaking of covalent bonds, the transformation is called *covalent* PSM (Cohen 2012, 2017, Tanabe and Cohen 2011), where this last type of PSM is generally targeting the organic struts in the structure.

A good example of coordinative PSM was demonstrated when suspending an expanded MOF-74 structure [M_2(dobpdc) with M = Zn^{2+} or Mg^{2+} and dobpdc = 4,4'-dioxide-3,3'-biphenyldicarboxylate] in a solution of N,N'-dimethylethylenediamine (DMEDA). One of the secondary amine functionalities present in DMEDA binds the metal center through the nitrogen lone pair, occupying the previously vacant coordinatively site in the SBU, while the remaining amine group is available for further reactions. The prepared material, denoted as Mg_2(dobpdc)(mmen)$_{1.6}$(H_2O)$_{0.4}$, displayed an extremely high affinity for CO_2 at low partial pressures, relevant to the CO_2 capture application. In fact, this material is able to adsorb 15 times more CO_2 than the same MOF before the coordinative PSM (McDonald et al. 2012).

The post-synthetic exchange of organic struts is another useful coordinative PSM strategy. Often active sites for catalysis cannot be incorporated to the MOF structures by direct synthesis because they decompose under synthetic conditions. Therefore, post-synthetic linker exchange allows installing these active sites once the MOF is formed. In an example by Pullen et al. 2013, thermally unstable diiron carbonyl complex with a bridging thiocatecholato ligand, were installed in the UiO-66 structure. After UiO-66 preparation, its crystals were suspended in a solution of a modified thiocatecholato ligand. The coordinative PSM resulted in a material with 14% of the catalyst incorporated, via ligand exchange with the native terephthalate struts (Fig. 5.9a) (Pullen et al. 2013).

Another example of coordinative PSM is the exchange of terminal OH^- and H_2O ligands on 8c Zr_6O_8 SBUs (*vide supra*) by a protocol called SALE or Solvent-Assisted Ligand Exchange, which consists of soaking a MOF in a concentrated solution of a ligand at elevated temperatures

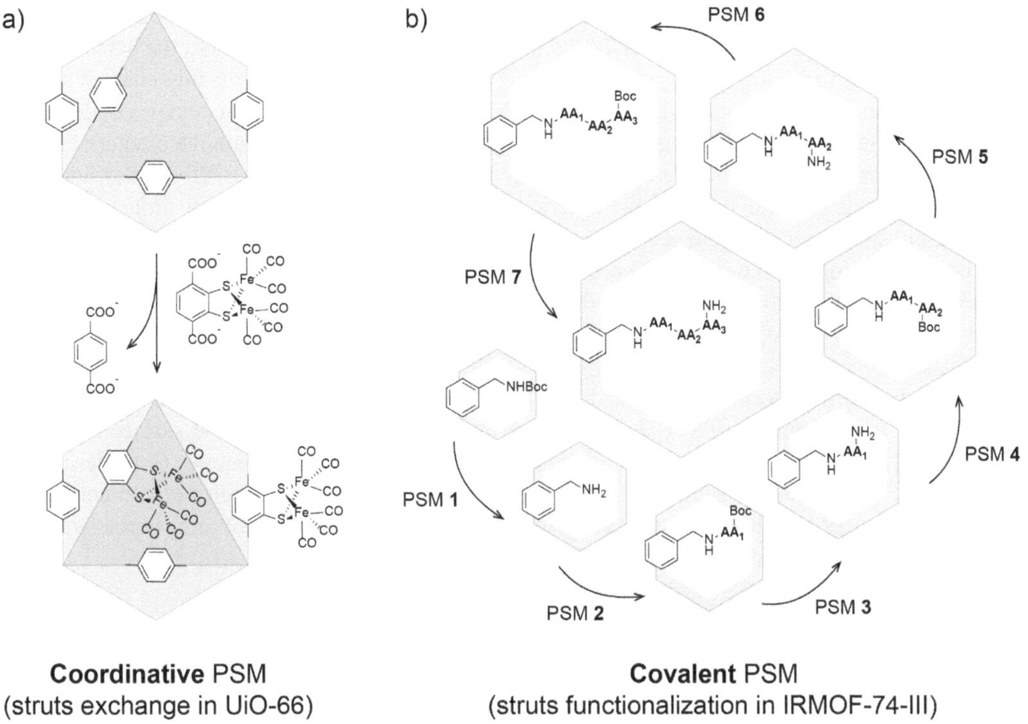

Figure 5.9: Examples of PSM: (a) organic strut exchange to incorporate a labile diiron carbonyl catalyst to UiO-66, and (b) 7 covalent PSM in *tandem* to install tripeptides that resemble the TEV enzyme active sites.

for a prolonged period. Through this strategy optimized by the group of Hupp and Farha, OH⁻ and H_2O are exchanged by perfluoroalkyl carboxylic acids in NU-1000 achieving a material with enhanced CO_2 capture capacities attributed to the presence of strong C–F dipole moments in the pores (Karagiaridi et al. 2014).

Early examples of covalent PSM were performed using amine or aldehyde functionalized organic struts. The term 'post-synthetic modification' itself was coined by Cohen and coworkers, while describing the post-synthetic reaction of IRMOF-3 (amino-functionalized MOF-5) with acetic anhydride to obtain the acetylated derivative (Wang and Cohen 2007). This seminal work demonstrated the scope of this strategy which has the potential to drastically change the pore environments without compromising the crystallinity and porosity of the material. Later, the same group extended the scope of this strategy by performing up to 4 covalent PSM in tandem to install urea derivatives (Garibay et al. 2009). Recently using this strategy, it was possible to install an enzyme inspired organocatalyst in isoreticularly expanded IRMOF-74-III. In this work, MTV-IRMOF-74-III functionalized with *tert*-butyloxycarbonyl protected primary amines underwent 7 covalent PSM in tandem with no loss of crystallinity or porosity to install the functional complexity previously observed only in enzyme pockets. In a manner resembling solid-phase peptide synthesis, tripeptides with the sequence H_2N-Cys-His-Asp-COOH were covalently attach to the organic struts present in the one-dimensional hexagonal pores of this MOF. After the reactions, the resultant material featuring c.a. 2.5 nm diameter pores and a complexity resembling *Tobacco Etch Virus* protease (TEV) enzyme pocket, was able to catalyze highly selective peptide bond cleavages (Fig. 5.9b) (Fracaroli et al. 2016).

Pore Heterogeneity within Order

Thanks to the contributions of researchers from many different fields, the MOF chemistry has matured to the point where scientists can accurately predict the framework's structures and their desirable properties targeting a particular application. Tt is natural to speculate about the perspectives in the field and strengths of the material (Allendorf and Stavila 2015). In this case, the heterogeneity that can be achieved within the crystalline frameworks plays a key role. In this chapter several unique features for these types of materials were described. Some of them uncovered levels of design and synthetic control previously unseen for other porous or inorganic nanomaterials, such as crystallographic precision for the functionality incorporation to the pores. If one considers the complexity of nature and the specific functionality achieved by combining a finite number of building units in a specific sequence, one can envision the coming challenges in the MOF field that will be able to program the material's properties by incorporating encoded heterogeneity either by multivariate functionalities or defects in the structure. MTV and PSM strategies described earlier, allow to break the homogeneity given by the crystallinity of the frameworks, however the control over the sequence of these functional groups has not yet been achieved. Mixed linker materials where there is only a little information about the different struts apportionment within a crystal (Kong et al. 2013), and incomplete PSM leaving behind 'byproducts' anchored to the surface (Fracaroli et al. 2016), are examples of chaotic agents that generate this heterogeneity. Considering the definition of *heterogeneity* as "the quality or state of being diverse in character or content", other strategies proposed to achieve this diversity were listed by Yaghi and coworkers (Furukawa et al. 2015), and are highlighted below and in Fig. 5.10:

a) mixing metal ions in the MOF Secondary Building Units (SBUs),

b) mixing of both the SBUs and organic linkers within the same MOF backbone,

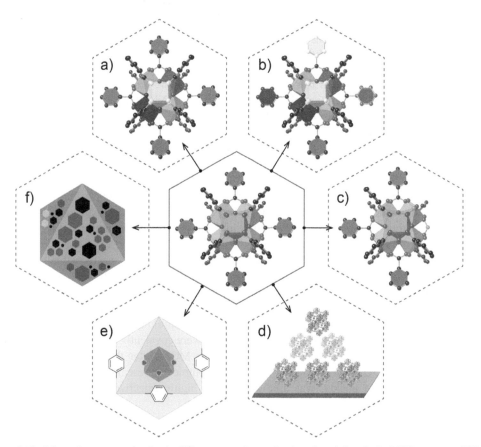

Figure 5.10: Schematic representation for the different strategies used to introduce defects in the MOF structures (UiO-66 SBU as an example). The MTV-approach is not included here, as it was represented in Fig. 5.9.

c) the generation of random and ordered defects,

d) integrating MOFs to functional surfaces,

e) combining inorganic nanocrystals and MOFs,

f) MOFs with heterogeneous pores.

Multi-metallic MOFs (MM-MOFs, Fig. 5.10a)

In an analogous approach to mix organic struts within the same MOF and without altering the underlying topology (MTV-MOFs), different metal cations were mixed within the same MOFs (MM-MOFs). For instance, up to 10 different metals were mixed within MOF-74 structure (Wang et al. 2014). PXRD and energy-dispersive X-ray spectroscopy (EDS) mapping on an M10M-MOF-74 micrometer-sized crystal showed the incorporation of the 10 metals used in the synthesis within the same crystal phase. Although, ICP-OES measurements allowed in finding M10M-MOF-74 empirical formula equal to $Mg_{0.269}Ca_{0.022}Sr_{0.030}Ba_{0.075}Mn_{0.234}Fe_{0.422}Co_{0.272}Ni_{0.282}Zn_{0.199}Cd_{0.196}$ (DOT) $(H_2O)_{7.8}$, where DOT stands for 2,5-dioxidoterephthalate, the control over the metal distribution in the MOFs structures has only recently been achieved (Castillo-Blas et al. 2019). It is interesting to highlight that in most reported cases of MM-MOFs, the materials featuring multiple metal cations in their SBUs exhibit properties that differ from their single-metal counterparts or the physical mixtures of the parent single-metal MOFs such as enhanced catalytic activity (Liu et al. 2016).

As it could be expected, the different strategies mentioned above are not exclusive as one can combine organic struts mixing (MTV-MOF) and metal cations mixing (MM-MOF) at the same time and within the same structure creating a new level of complexity. Zheng et al. 2011, prepared a cage-within-cage MOF named CPM-24 (CPM = crystalline porous material) (Zheng et al. 2011). $[(CH_3)_2NH_2]_{10}[Co_2(IN)_4(Ac)_2]–[Co_2(BTC)_2(H_2O)]_4[Co_2(OH)(IN)_3]_4$ $4H_2O$ (BTC = 1,3,5-benzenetricarboxylate, IN = isonicotinic acid anion) or simply CPM-24, incorporates two types of cobalt dimers [Co_2 paddlewheel and V-shaped $Co_2(OH)$] and two types of organic struts (BTC and IN) representing a unique structural complexity in a framework-in-framework porous material.

Defects in MOFs Structure (Fig. 5.10c)

Although one is used to imagining the flawless periodic structures represented in MOF figures from literature, in reality, MOF crystals contain an important number of defects. According to recent definitions by Fischer and coworkers, defects are "sites that locally break the regular periodic arrangement of atoms or ions of the crystalline parent framework because of missing or dislocated atoms or ions" (Dissegna et al. 2018). For instance, in MOFs, a typical source of defects is a missing organic strut in the coordination of the SBU. The amount and nature of these defects is of impact as it can drastically modify the material's properties and therefore its application. The application of MOFs as heterogeneous catalysts is typical. The large surface area and tunability of pore environments are appealing points for MOFs to be employed in catalysis. Often, the catalytic activity of these materials is attributed to the presence of Lewis acid sites in the SBUs (coordination vacancies in the metal clusters) or the Brønsted acidity generated by the coordination of water of hydroxide binding these vacant sites. These vacant sites can be generated by a missing organic strut or defect as mentioned above. Recently, the correlation between these vacant coordination sites or 'defects' and the catalytic activity of a series of Zr-MOFs were studied (Liu et al. 2016). It was found that 6c Zr-clusters, $Zr_6(OH)_4O_4(L)_6$, where L = carboxylate organic strut, is 50% more active as heterogeneous catalysts than the 12c$Zr_6(OH)_4O_4(L)_{12}$ in the epoxide ring-opening reaction. These defects present in Zr-SBUs were also characterized by single-crystal X-ray diffraction (Trickett et al. 2015). The crystallographic studies on UiO-66 identified defective sites in the SBUs where missing linkers are replaced by water molecules and hydroxide anions present in the pores balance the framework charge.

Furthermore, the importance of these defects in the MOFs structures in relation to MOF properties, motivated researchers to actually design and synthetically generate these defects in a controlled manner. Different approaches can be used to introduce defects in the frameworks: (i) use of terminal organic struts (monocarboxylate molecules) in the synthesis (Choi et al. 2011), (ii) post-synthetic exchange using modified terminal organic struts (Karagiaridi et al. 2012), (iii) post-synthetic etching using reactive solvent or molecules that can hydrolyze carboxylate-metal bonds (Lee et al. 2016).

Concluding Remarks

In the last two decades, MOF chemistry has grown to the point where the structure, functionality and porosity can be designed for a specific application. The crystallographic precision with which functionality can be installed in the frameworks has led to new levels of control over materials synthesis. In this chapter, fundamental concepts, properties and strategies have been summarized to give a brief look of this rich chemistry. Recently reported examples show that MOF research is pushing the frontiers of material science as this new level of control in the synthesis of porous structures call for the development of new technologies for their analysis (Ji et al. 2020). An example of this is that it remains as a challenge to characterize the functional groups apportionment within an MTV-MOF crystal down to the atomic level where the generated periodic disorder make the X-ray analysis difficult to perform.

In the chapter, an attempt was made to highlight the MOF uniqueness among other classes of materials and, particularly, among other coordination polymers and porous materials. Their rational design calls to creatively explore different geometries and connectivities provided by the nearly endless combination of organic struts and SBUs. It is anticipated that the reticular chemistry approach and the new levels of control over the synthesis, functionalization, defects design among others will play an important role in finding efficient materials for diverse applications such as gas capture and storage and heterogeneous catalysis. In the last part of the chapter, and by showing that multifunctional heterogeneity can be incorporated within the ordered structures of MOF, it is envisioned that further complexity and synergistic properties will be uncovered in next generations of these materials, perhaps anticipating inspiration from nature.

Acknowledgements

We acknowledge the Office of Naval Research Global (ONRG, award N62909-20-1-2025), the Agencia Nacional de Ciencia y Tecnología (FONCyT, PICT-2017-2487), and the Consejo Nacional de Investigaciones Científicas y Técnicas for funding our group research.

The author also would like to thank Prof. Omar Yaghi and Prof. Rita H. de Rossi for their constant mentoring and support, and Dr. H. Furukawa and K. Cordova for the helpful and enriching discussions.

References

Abtab, S. M. T., D. Alezi, P. M. Bhatt, A. Shkurenko, Y. Belmabkhout, H. Aggarwal et al. 2018. Reticular chemistry in action: a hydrolytically stable MOF capturing twice its weight in adsorbed water. Chem. 4: 94–105.

Allendorf, M. D. and V. Stavila. 2015. Crystal engineering, structure–function relationships, and the future of metal–organic frameworks. CrystEngComm. 17: 229–246.

Allendorf, M. D., Z. Hulvey, T. Gennett, A. Ahmed, T. Autrey, J. Camp et al. 2018. An assessment of strategies for the development of solid-state adsorbents for vehicular hydrogen storage. Energy Environ. Sci. 11: 2784–2812.

Batten, S. R., N. R. Champness, X. -M. Chen, J. Garcia-Martinez, S. Kitagawa, L. Öhrström et al. 2013. Terminology of metal–organic frameworks and coordination polymers (IUPAC Recommendations 2013). Pure Appl. Chem. 85: 1715–1724.

Bin, L., H. -M. Wen, W. Zhou, J. Q. Xu and B. Chen. 2016. Porous metal-organic frameworks: promising materials for methane storage. Chem. 1: 557–580.

Canivet, J., A. Fateeva, Y. Guo, B. Coasne and D. Farrusseng. 2014. Water adsorption in MOFs: fundamentals and applications. Chem. Soc. Rev. 43: 5594–5617.

Castillo-Blas, C., N. López-Salas, M. C. Gutiérrez, I. Puente-Orench, E. Gutiérrez-Puebla, M. L. Ferrer et al. 2019. Encoding metal–cation arrangements in metal–organic frameworks for programming the composition of electrocatalytically active multimetal oxides. J. Am. Chem. Soc. 141: 1766–1774.

Chae, H. K., D. Y. Siberio-Pérez, J. Kim, Y. Go, M. Eddaoudi, A. J. Matzger et al. 2004. A route to high surface area, porosity and inclusion of large molecules in crystals. Nature 427: 523–527.

Chen, Z., H. Jiang, M. Li, M. O'Keeffe and M. Eddaoudi. 2020. Reticular chemistry 3.2: typical minimal edge-transitive derived and related nets for the design and synthesis of metal–organic frameworks. Chem. Rev. (in press) DOI: https://dx.doi.org/10.1021/acs.chemrev.9b00648.

Cheng, D., M. A. Khan and R. P. Houser. 2004. Structural variability of cobalt(II) coordination polymers: three polymorphs of $Co_3(TMA)_2$ [TMA = trimesate, $C_6H_3(COO)_3^{3-}$]. Cryst. Growth Des. 4: 599–604.

Choi, K. M., H. J. Jeon, J. K. Kang and O. M. Yaghi. 2011. Heterogeneity within order in crystals of a porous metal organic framework. J. Am. Chem. Soc. 133: 11920–11923.

Choi, K. M., K. Na, G. A. Somorjai and O. M. Yaghi. 2015. Chemical environment control and enhanced catalytic performance of platinum nanoparticles embedded in nanocrystalline metal-organic frameworks. J. Am. Chem. Soc. 137: 7810–7816.

Choi, M., H. S. Cho, R. Srivastava, C. Venkatesan, D. -H. Choi and R. Ryoo. 2006. Amphiphilic organosilane-directed synthesis of crystalline zeolite with tunable mesoporosity. Nature Mater. 5: 718–723.

Cohen, S. M. 2012. Postsynthetic methods for the functionalization of metal-organic frameworks. Chem. Rev. 112: 970–1000.

Cohen, S. M. 2017. The postsynthetic renaissance in porous solids. J. Am. Chem. Soc. 139: 2855–2863.

Demessence, A. and J. R. Long. 2010. Selective gas adsorption in the flexible metal–organic frameworks Cu(BDTri) L (L = DMF, DEF). Chem. Eur. J. 16: 5902–5908.

Deng, H., C. J. Doonan, H. Furukawa, R. B. Ferreira, J. Towne, C. B. Knobler et al. 2010. Multiple functional groups of varying ratios in metal-organic frameworks. Science 327: 846–850.

Deng, H., S. Grunder, K. E. Cordova, C. Valente, H. Furukawa, M. Hmadeh et al. 2012. Large-pore apertures in a series of metal-organic frameworks. Science 336: 1018–1023.

Deshpande, R. K., G. I. N. Waterhouse, G. B. Jameson and S. G. Telfer. 2011. Photolabile protecting groups in metal–organic frameworks: preventing interpenetration and masking functional groups. Chem. Commun. 48: 1574–1576.

Dincă, M., A. F. Yu and J. R. Long. 2006. Microporous metal–organic frameworks incorporating 1,4-benzeneditetrazolate: syntheses, structures, and hydrogen storage properties. J. Am. Chem. Soc. 128: 8904–8913.

Ding, M., R. W. Flaig, H. Jiang and O. M. Yaghi. 2019. Carbon capture and conversion using metal–organic frameworks and MOF-based materials. Chem. Soc. Rev. 48: 2783–2828.

Dissegna, S., K. Epp, W. R. Heinz, G. Kieslich and Roland A. Fischer. 2018. Defective metal-organic frameworks. Adv. Mater. 30: 1704501–1704524.

Eddaoudi, M., J. Kim, N. L. Rosi, D. T. Vodak, J. Wachter, M. O'Keeffe et al. 2002. Systematic design of pore size and functionality in isoreticular metal-organic frameworks and application in methane storage. Science 295: 469–472.

Eddaoudi, M., D. F. Sava, J. F. Eubank, K. Adila and V. Guillerm. 2015. Zeolite-like metal–organic frameworks (ZMOFs): design, synthesis, and properties. Chem. Soc. Rev. 44: 228–249.

Farha, O. K., I. Eryazici, N. C. Jeong, B. G. Hauser, C. E. Wilmer, A. A. Sarjeant et al. 2012. Metal–organic framework materials with ultrahigh surface areas: is the sky the limit? J. Am. Chem. Soc. 134: 15016–15021.

Feng, L., S. Yuan, J. -S. Qin, Y. Wang, A. Kirchon, D. Qiu et al. 2019. Lattice expansion and contraction in metal-organic frameworks by sequential linker reinstallation. Matter. 1: 156–167.

Férey, G., C. Mellot-Draznieks, C. Serre, F. Millange, J. Dutour, S. Surblé et al. 2005. A Chromium terephthalate-based solid with unusually large pore volumes and surface area. Science 309: 2040–2042.

Fracaroli, A. M., H. Furukawa, M. Suzuki, M. Dodd, S. Okajima, F. Gándara et al. 2014. Metal-organic frameworks with precisely designed interior for carbon dioxide capture in the presence of water. J. Am. Chem. Soc. 136: 8863–8866.

Fracaroli, A., P. Siman, D. Nagib, M. Suzuki, H. Furukawa, F. D. Toste et al. 2016. Seven post-synthetic covalent reactions in tandem leading to enzyme-like complexity within metal-organic framework crystals. J. Am. Chem. Soc. 138: 8352–8355.

Furukawa, H., M. A. Miller and O. M. Yaghi. 2007. Independent verification of the saturation hydrogen uptake in MOF-177 and establishment of a benchmark for hydrogen adsorption in metal–organic frameworks. J. Mater. Chem. 17: 3197–3204.

Furukawa, H., N. Ko, Y. B. Go, N. Aratani, S. B. Choi, E. Choi et al. 2010. Ultra-high porosity in metal-organic frameworks. Science 239: 424–428.

Furukawa, H., K. E. Cordova, M. O'Keeffe and O. M. Yaghi. 2013. The chemistry and applications of metal-organic frameworks. Science 341: 1230444–12.

Furukawa, H., F. Gándara, Y. -B. Zhang, J. Jiang, W. L. Queen, M. R. Hudson et al. 2014. Water adsorption in porous metal-organic frameworks and related materials. J. Am. Chem. Soc. 136: 4369–4381.

Furukawa, H., U. Müller and O. M. Yaghi. 2015. "Heterogeneity within order" in metal–organic frameworks. Angew. Chem. Int. Ed. 54: 3417–3430.

Gándara, F., H. Furukawa, S. Lee and O. M. Yaghi. 2014. High methane storage capacity in aluminum metal-organic frameworks. J. Am. Chem. Soc. 136: 5271–5274.

García-Holley, P., B. Schweitzer, T. Islamoglu, Y. Liu, L. Lin, S. Rodriguez et al. 2018. Benchmark study of hydrogen storage in metal–organic frameworks under temperature and pressure swing conditions. ACS Energy Lett. 3: 748–754.

Garibay, S. J., Z. Wang, K. K. Tanabe and S. M. Cohen. 2009. Postsynthetic modification: a versatile approach toward multifunctional metal-organic frameworks. Inorg. Chem. 48: 7341–7349.

Gómez-Gualdrón, D. A., T. C. Wang, P. García-Holley, R. M. Sawelewa, E. Argueta, R. Q. Snurr et al. 2017. Understanding volumetric and gravimetric hydrogen adsorption trade-off in metal–organic frameworks. ACS Appl. Mater. Interfaces 9: 33419–33428.

Goto, Y., N. Mizoshita, M. Waki, M. i. Ikai, Y. Maegawa and Shinji Inagaki. 2019. Synthesis and applications of periodic mesoporous organosilicas. pp. 1–25. *In*: A. Douhal and M. Anpo [eds.]. Chemistry of Silica and Zeolite-based Materials Synthesis, Characterization and Applications. ScienceDirect Elsevier.

Hanikel, N., M. S. Prévot and O. M. Yaghi. 2020. MOF water harvesters. Nat. Nanotechnol. 15: 348–355.

Hasell, T. and A. Cooper. 2016. Porous organic cages: soluble, modular and molecular pores. Nat. Rev. Mater. 1: 16053.

Hayashi, H., A. P. Côté, H. Furukawa, M. O'Keeffe and O. M. Yaghi. 2007. Zeolite a imidazolate frameworks. Nature Mater. 6: 501–506.

Hofmann, K. and F. Küspert. 1897. Verbindungen von kohlenwasserstoffen mit metallsalzen. Zeitschrift für Anorganische Chemie 15: 204–207.

Hong-Cai, J. Z. and S. Kitagawa. 2014. Metal–organic frameworks (MOFs). Chem. Soc. Rev. 43: 5415–5418.

https://www.ccdc.cam.ac.uk/support-and-resources/support/case/?caseid=9833bd2c-27f9-4ff7-8186-71a9b415f012, accessed July 29th, 2020.

http://rcsr.net/nets, accessed July 29th, 2020.

http://yaghi.berkeley.edu/research.html, accessed July, 29th, 2020.

Huang, L., H. Wang, J. Chen, Z. Wang, J. Sun, D. Zhao et al. 2003. Synthesis, morphology control, and properties of porous metal–organic coordination polymers. Microporous Mesoporous Mater. 58: 105–114.

Inagaki, M. 2000. New Carbons—Control of Structure and Functions. ScienceDirect Elsevier.

Inokuma, Y., S. Yoshioka, J. Ariyoshi, T. Arai, Y. Hitora, K. Takada et al. 2013. X-ray analysis on the nanogram to microgram scale using porous complexes. Nature 495: 461–466.

Ji, Z., T. Li and O. M. Yaghi. 2020. Sequencing of metals in multivariate metal-organic frameworks. Science 369: 674–680.

Ji, Z., H. Wang, S. Canossa, S. Wuttke and O. M. Yaghi. 2020. Pore chemistry of metal-organic frameworks. Adv. Funct. Mater. (in press) DOI: 10.1002/adfm.202000238.

Jiang, J., F. Gándara, Y. -B. Zhang, K. Na, O. M. Yaghi and W. G. Klemperer. 2014. Superacidity in sulfated metal-organic framework-808. J. Am. Chem. Soc. 136: 12844–12847.

Kalmutzki, M. J., N. Hanikel and O. M. Yaghi. 2018. Secondary building units as the turning point in the development of the reticular chemistry of MOFs. Sci. Adv. 4: eaat9180.

Karagiaridi, O., M. B. Lalonde, W. Bury, A. A. Sarjeant, O. K. Farha and J. T. Hupp. 2012. Opening ZIF-8: a catalytically active zeolitic imidazolate framework of sodalite topology with unsubstituted linkers. J. Am. Chem. Soc. 134: 18790–18796.

Karagiaridi, O., W. Bury, J. E. Mondloch, J. T. Hupp and O. K. Farha. 2014. Solvent-assisted linker exchange: an alternative to the *de novo* synthesis of unattainable metal–organic frameworks. Angew. Chem. Int. Ed. 53: 4530–4540.

Kaye, S. S., A. Dailly, O. M. Yaghi and J. R. Long. 2007. Impact of preparation and handling on the hydrogen storage properties of $Zn_4O(1,4$-benzenedicarboxylate$)_3$ (MOF-5). J. Am. Chem. Soc. 129: 14176–14177.

Kim, H., S. Yang, S. R. Rao, S. Narayanan, E. A. Kapustin, H. Furukawa et al. 2017. Water harvesting from air with metal-organic frameworks powered by natural sunlight. Science 356: 430–434.

Kim, E. J., R. L. Siegelman, H. Z. H. Jiang, A. C. Forse, J. -H. Lee, J. D. Martell et al. 2020. Cooperative carbon capture and steam regeneration with tetraamine-appended metal–organic frameworks. Science 369: 392–396.

Kinoshita, Y., I. Matsubara, T. Higuchi and Y. Saito. 1959. The crystal structure of bis(adiponitrilo)copper(I) nitrate. Bull. Chem. Soc. Japan 32: 1221–1226.

Koh, K., A. G. Wong-Foy and A. J. Matzger. 2009. A porous coordination copolymer with over 5000 m^2/g BET surface area. J. Am. Chem. Soc. 131: 4184–4185.

Kong, X., H. Deng, F. Yan, J. Kim, J. A. Swisher, B. Smit et al. 2013. Mapping of functional groups in metal-organic frameworks. Science 341: 882–885.

Kresge, C. T., M. E. Leonowicz, W. J. Roth, J. C. Vartuli and J. S. Beck. 1992. Ordered mesoporous molecular sieves synthesized by a liquid-crystal template mechanism. Nature 359: 710–712.

Lee, S. J., C. Doussot, A. Baux, L. Liu, G. B. Jameson, C. Richardson et al. 2016. Multicomponent metal−organic frameworks as defect-tolerant materials. Chem. Mater. 28: 368–375.

Lee, S., E. Kapustin and O. M. Yaghi. 2016. Coordinative alignment of molecules in chiral metal-organic frameworks. Science 353: 808–811.

Li, H., M. Eddaoudi, M. O'Keeffe and O. M. Yaghi. 1999. Design and synthesis of an exceptionally stable and highly porous metal-organic framework. Nature 402: 276–279.

Li, T., M. T. Kozlowski, E. A. Doud, M. N. Blakely and N. L. Rosi. 2013. Stepwise ligand exchange for the preparation of a family of mesoporous MOFs. J. Am. Chem. Soc. 135: 11688–11691.

Li, M., D. Li, M. O'Keeffe and O. M. Yaghi. 2014. Topological analysis of metal-organic frameworks with polytopic linkers and/or multiple building units and the minimal transitivity principle. Chem. Rev. 114: 1343–1370.

Li, S., Y. G. Chung, C. M. Simon and R. Q. Snurr. 2017. High-throughput computational screening of multivariate metal−organic frameworks (MTV-MOFs) for CO_2 capture. Phys. Chem. Lett. 8: 6135–6141.

Li, P., Q. Chen, T. C. Wang, N. A. Vermeulen, B. L. Mehdi, A. Dohnalkova et al. 2018. Hierarchically engineered mesoporous metal-organic frameworks toward cell-free immobilized enzyme systems. Chem. 4: 1022–1034.

Lim, S., H. Kim, N. Selvapalam, K. -J. Kim, S. J. Cho, G. Seo et al. 2008. Cucurbit[6]uril: organic molecular porous material with permanent porosity, exceptional stability, and acetylene sorption properties. Angew. Chem. Int. Ed. 47: 3352–3355.

Liu, Q., H. Cong and H. Deng. 2016. Deciphering the spatial arrangement of metals and correlation to reactivity in multivariate metal−organic frameworks. J. Am. Chem. Soc. 138: 13822–13825.

Liu, Y., R. C. Klet, J. T. Hupp and O. Farha. 2016. Probing the correlations between the defects in metal-organic frameworks and their catalytic activity by an epoxide ring-opening reaction. Chem. Commun. 52: 7806–7809.

Logan, M. W., S. Langevin and Z. Xia. 2020. Reversible atmospheric water harvesting using metal–organic frameworks. Sci. Report 10: 1492–1503.

Lowell, S., J. E. Shields, M. A. Thomas and M. Thommes. 2004. Characterization of porous solids and powders: surface area, pore size and density. Springer Science.

Lu, W., Z. Wei, Z. -Y. Gu, T. -F. Liu, J. Park, J. Park et al. 2014. Tuning the structure and function of metal–organic frameworks via linker design. Chem. Soc. Rev. 43: 5561–5593.

Lun, D. J., G. I. N. Waterhouse and S. G. Telfer. 2011. A general thermolabile protecting group strategy for organocatalytic metal-organic frameworks. J. Am. Chem. Soc. 133: 5806–5809.

Masciocchi, N., S. Galli, V. Colombo, A. Maspero, G. Palmisano, B. Seyyedi et al. 2010. Cubic octanuclear Ni(II) clusters in highly porous polypyrazolyl-based materials. J. Am. Chem. Soc. 132: 7902–7904.

Mason, J. A., M. Veenstra and J. R. Long. 2014. Evaluating metal–organic frameworks for natural gas storage. Chem. Sci. 5: 32–51.

Mason, J. A., J. Oktawiec, M. K. Taylor, M. R. Hudson, J. Rodriguez, J. E. Bachman et al. 2015. Methane storage in flexible metal–organic frameworks with intrinsic thermal management. Nature 527: 357–361.

McDonald, T. M., W. R. Lee, J. A. Mason, B. M. Wiers, C. S. Hong and J. R. Long. 2012. Capture of carbon dioxide from air and flue gas in the alkylamine-appended metal–organic framework mmen-Mg_2(dobpdc). J. Am. Chem. Soc. 134: 7056–7065.

McDonald, T. M., J. A. Mason, X. Kong, E. D. Bloch, D. Gygi, A. Dani et al. 2015. Cooperative insertion of CO_2 in diamine-appended metal-organic frameworks. Nature 519: 303–308.

Moghadam, P. Z., A. Li, S. B. Wiggin, A. Tao, A. G. P. Maloney, P. A. Wood et al. 2017. Development of a Cambridge structural database subset: a collection of metal–organic frameworks for past, present, and future. Chem. Mater. 29: 2618–2625.

Millward, A. R. and O. M. Yaghi. 2005. Metal-organic frameworks with exceptionally high capacity for storage of carbon dioxide at room temperature. J. Am. Chem. Soc. 127: 17998–17999.

Na, K. and G. A. Somorjai. 2015. Hierarchically nanoporous zeolites and their heterogeneous catalysis: current status and future perspectives. Catal. Lett. 145: 193–213.

Nguyen, N. T. T., H. Furukawa, F. Gándara, H. T. Nguyen, K. E. Cordova and O. M. Yaghi. 2014. Selective capture of carbon dioxide under humid conditions by hydrophobic chabazite-type zeolitic imidazolate frameworks. Angew. Chem. Int. Ed. 53: 10645–10648.

Ockwig, N. W., O. Delgado-Friedrichs, M. O'Keeffe and O. M. Yaghi. 2005. Reticular chemistry: occurrence and taxonomy of nets and grammar for the design of frameworks. Acc. Chem. Res. 38: 176–182.

O'Keeffe, M., M. Eddaoudi, T. Reineke, H. Li and O. M. Yaghi. 2000. Frameworks for extended solids: geometrical design principles. J. Solid State Chem. 152: 3–20.

O'Keeffe, M. and O. M. Yaghi. 2012. Deconstructing the crystal structures of metal-organic frameworks and related materials into their underlying nets. Chem. Rev. 112: 675–702.

Park, K. S., Z. Ni, A. P. Côté, J. Y. Choi, R. Huang, F. J. Uribe-Romo et al. 2006. Exceptional chemical and thermal stability of zeolitic imidazolate frameworks. P. Natl. Acad. Sci. USA 103: 10186–10191.

Pullen, S., H. Fei, A. Orthaber, S. M. Cohen and S. Ott. 2013. Enhanced photochemical hydrogen production by a molecular diiron catalyst incorporated into a metal–organic framework. J. Am. Chem. Soc. 135: 16997–17003.

Rabone, J., Y. -F. Yue, S. Y. Chong, K. C. Stylianou, J. Bacsa, D. Bradshaw et al. 2010. An adaptable peptide-based porous material. Science 329: 1053–1057.

Reinares-Fisac, D., L. M. Aguirre-Díaz, M. Iglesias, N. Snejko, E. Gutiérrez-Puebla, M. A. Monge et al. 2016. A mesoporous indium metal–organic framework: remarkable advances in catalytic activity for Strecker reaction of ketones. J. Am. Chem. Soc. 138: 9089.

Rosi, N. L., M. Eddaoudi, D. T. Vodak, J. Eckert, M. O'Keeffe and O. M. Yaghi. 2003. Hydrogen storage in microporous metal-organic frameworks. Science 300: 1127–1129.

Rouquerol, J., D. Avnir, C. W. Fairbridge, D. H. Everett, J. M. Haynes, N. Pernicone et al. 1994. Recommendations for the characterization of porous solids (Technical Report). Pure & Appl. Chem. 66: 1739–1758.

Rowsell, J. L. C., E. C. Spencer, J. Eckert, J. A. K. Howard and O. M. Yaghi. 2005. Gas adsorption sites in a large-pore metal-organic framework. Science 309: 1350–1354.

Schaate, A., P. Roy, T. Preuße, S. J. Lohmeier, A. Godt and P. Behrens. 2011. Porous interpenetrated zirconium-organic frameworks (PIZOFs): a chemically versatile family of metalorganic frameworks. Chem. Eur. J. 17: 9320–9325.

Schoedel, A., M. Li, D. Li, M. O'Keeffe and O. M. Yaghi. 2016. Structures of metal–organic frameworks with rod secondary building units. Chem. Rev. 116: 12466–12535.

Seo, J., R. Matsuda, H. Sakamoto, C. Bonneau and S. Kitagawa. 2009. A pillared-layer coordination polymer with a rotatable pillar acting as a molecular gate for guest molecules. J. Am. Chem. Soc. 131: 12792–12800.

Shekhah, O., H. Wang, M. Paradinas, C. Ocal, B. Schüpbach, A. Terfort et al. 2009. Controlling interpenetration in metal-organic frameworks by liquid-phase epitaxy. Nature Mater. 8: 481–484.

Tanabe, K. K. and S. M. Cohen. 2011. Postsynthetic modification of metal–organic frameworks—a progress report. Chem. Soc. Rev. 40: 498–519.

Tasiopoulos, A. J., A. Vinslava, W. Wernsdorfer, K. A. Abboud and G. Christou. 2004. Giant single-molecule magnets: a {Mn84} torus and its supramolecular nanotubes. Angew. Chem. Int. Ed. 116: 2169–2173.

Tranchemontagne, D. J., J. L. Mendoza-Cortes, M. O'Keeffe and O. M. Yaghi. 2009. Secondary building units, nets and bonding in the chemistry of metal-organic frameworks. Chem. Soc. Rev. 38: 1257–1283.

Trickett, C. A., K. J. Gagnon, S. Lee, F. Gándara, H. -B. Bürgi and O. M. Yaghi. 2015. Definitive molecular level characterization of defects in UiO-66 crystals. Angew. Chem. Int. Ed. 54: 11162–11167.

Trickett, C. A., T. M. Osborn Popp, J. Su, C. Yan, J. Weisberg, A. Huq et al. 2019. Identification of the strong Brønsted acid site in a metal–organic framework solid acid catalyst. Nature Chem. 11: 170–176.

Wang, L. J., H. Deng, H. Furukawa, F. Gándara, K. E. Cordova, D. Peri et al. 2014. Synthesis and characterization of metal-organic framework-74 containing 2, 4, 6, 8, and 10 different metals. Inorg. Chem. 53: 5881–5883.

Wang, Y., T. Du, Y. Song, S. Che, X. Fang and L. Zhou. 2017. Amine-functionalized mesoporous ZSM-5 zeolite adsorbents for carbon dioxide capture. Solid State Sci. 73: 27–35.

Wang, Z. and Seth M. Cohen. 2007. Postsynthetic covalent modification of a neutral metal-organic framework. J. Am. Chem. Soc. 129: 12368–12369.

Yaghi, O. M. and H. Li. 1995. Hydrothermal synthesis of a metal-organic framework containing large rectangular channels. J. Am. Chem. Soc. 117: 10401–10402.

Yaghi, O. M., M. O'Keeffe, N. W. Ockwig, H. K. Chae, M. Eddaoudi and J. Kim. 2003. Reticular synthesis and the design of new materials. Nature 423: 705–714.

Yaghi, O. M., M. J. Kalmutzki and C. S. Diercks. 2019. Introduction to Reticular Chemistry: Metal-Organic Frameworks and Covalent Organic Frameworks. Wiley-VCH, Weinheim.

Yang, J., Y. Zhang, Q. Liu, C. A. Trickett, E. Gutierrez-Puebla, M. Á. Monge et al. 2017. Principles of designing extra-large pore openings and cages in zeolitic imidazolate frameworks. J. Am. Chem. Soc. 139: 6448–6455.

Yuan, D., D. Zhao, D. Sun and H. -C. Zhou. 2010. An isoreticular series of metal-organic frameworks with dendritic hexacarboxylate ligands and exceptionally high gas-uptake capacity. Angew. Chem. Int. Ed. 49: 5357–5361.

Yuan, J., A. M. Fracaroli and W. G. Klemperer. 2016. Convergent synthesis of a metal−organic framework supported olefin metathesis catalyst. Organometallics 35: 2149–2155.

Zhang, J. -P., Y. -B. Zhang, J. -B. Lin and X. -M. Chen. 2012. Metal azolate frameworks: from crystal engineering to functional materials. Chem. Rev. 112: 1001–1033.

Zhang, Y. -B., H. Furukawa, N. Ko, W. Nie, H. J. Park, S. Okajima et al. 2015. Enhancement of gas adsorption in multivariate metal–organic framework-177. J. Am. Chem. Soc. 137: 2641–2650.

Zheng, S. -T., T. Wu, B. Irfanoglu, F. Zuo, P. Feng and X. Bu. 2011. Multicomponent self-assembly of a nested $Co_{24}@Co_{48}$ metal–organic polyhedral framework. Angew. Chem. Int. Ed. 50: 8034–8037.

CHAPTER 6

Supported Metal Nanoparticles in Catalysis

Anabel Estela Lanterna

Introduction

The first use of colloidal metal particles for catalysis was published at the beginning of the 20th century when Rampino and Nord (Rampino and Nord 1941) reported on the preparation of Pd and Pt colloidal solutions and studied hydrogenation kinetics. Thus, scientists have recognized the importance of the size of materials in their applications long before the concept of 'nanotechnology' appeared in the literature. Since the development of electronic microscopy techniques early in the 80s, a deeper understanding on the relationship between structure and activity has triggered applications of nanotechnology in different fields, from optics and communication to medicine and catalysis (Butet et al. 2015, Hartshorn et al. 2018, Liu and Corma 2018, Peng et al. 2020). Why are nanomaterials so unique? As described in the introduction of this book, the nanoscale size of these particles confers them properties that are frequently absent in bulk. Particularly, the percentage of surface atoms increases drastically with the decrease of the particle size (see Fig. 1.1 in Chapter 1). This increases not only the number of exposed active sites but also their reactivity. That is, the atoms on the surface of a nanoparticle (NP) have fewer direct neighbours than the atoms in the bulk do. As a consequence, smaller particles have a low mean coordination number (number of nearest neighbours – CN) (Roduner 2006) and they are more reactive (Wang et al. 2018) as shown in Fig. 6.1 – the resemblance with Fig. 1.1 in Chapter 1 should also be noted. In general, the catalytic performance will be discussed in terms of turnover number (TON = number of moles of product/number of moles of an active catalyst), turnover frequency (TOF = TON/time of reaction) or activation energy (Ea); where high TON/TOF values or low Ea are the desirable outcomes.

Fields such as optics, nanomedicine and nanotechnology define nanomaterials as those materials with sizes from 1 to 100 nm. However, applications in catalysis require materials in a narrower range of sizes, typically below 20 nm. In a landmark publication, Haruta and coworkers (Haruta et al. 1987) reported supported AuNP of 3–5 nm diameter exhibiting surprisingly high catalytic efficiency towards oxidation of CO. Nowadays, one knows that sub-nanoparticles (diameter < 2 nm) of late transition metals are also catalytically active. Additionally, decreasing the particle size down to a single atom has contributed to developing atom-economic processes, which have large advantages particularly when using precious metals as catalysts (Li et al. 2020). Therefore, when studying metal NP as catalysts, it is important to recognize all plausible catalytic species that can drive catalytic processes such as (a) metal Isolated Single-Atom Sites (ISAS), (b) atomically

School of Chemistry, University of Nottingham, University Park, Nottingham, NG7 2RD, England.
Email: anabel.lanterna@nottingham.ac.uk

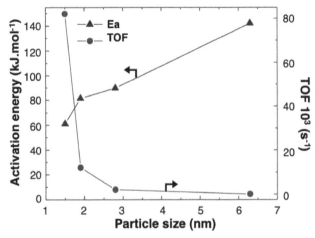

Figure 6.1: Activation energy (Ea) and turnover frequency (TOF) for the ethane oxidation as a function of the Pt/C particle size. Adapted with permission from (Isaifan et al. 2013) Copyright 2013 Elsevier.

precise metal nanoclusters (NC) or (c) metal NPs. Although this chapter is focused in the latter case, it is important to distinguish key differences between these three species (Liu and Corma 2018). Recognizing the true catalytic active species and their mechanism of action are fundamental aspects of the design and development of novel and efficient catalytic materials.

Metal ISAS were discovered due to the investigation of unusual behaviour of certain metal-supported catalysts. For instance, Fu et al. discovered that catalytic activity of Au and Pt supported on ceria was preserved after removal of metal NP (Fu et al. 2003). Further, Zhang and collaborators showed an unexpected increase of catalytic efficiency when decreasing the metal loading (Zhang et al. 2005) on Au supported on ZrO_2. Additionally, it was demonstrated that metal ISAS and metal NPs may coexist in supported catalysts (Kuai et al. 2020). Advanced characterization techniques with atomic resolution have facilitated the direct observation of the metal ISAS and triggered the development of a new family of catalysts, characterized by maximized atom utilization and defined active centres (Li et al. 2020).

Moving from a single atom to a group of atoms, the term cluster—or nanocluster—refers to molecularly precise polymetallic molecules with a defined crystalline structure, frequently stabilized with organic ligands (Astruc 2020). NC are constructed with a few hundred atoms and are typically smaller than 2 nm. When metal particles reach sizes about 2 nm or smaller, the metal density of states becomes discrete and the particle loses its metallic properties. This threshold can be calculated using the free-electron theory (Qian et al. 2012) as follows: the spacing between energy levels can be approximately expressed as $\delta = 4/3\ \varepsilon_F/N$, where ε_F is the Fermi level energy and N is the number of atoms in the particle. As N decreases with particle size, the spacing δ becomes larger and different physicochemical properties are expected as the particles get smaller. Using the thermal energy ($k_B T$) at room temperature (~ 298 K) as a criterion, if δ is larger than $k_B T$ the particle may lose its metallic properties. For instance, if one replaces the Fermi energy value of Au in the equation ($\varepsilon_F = 5.5$ eV) one finds that when the particle has less than 300 Au atoms—equivalent to ~ 2.1 nm diameter—the electronic levels are discrete. Therefore, below this size the electronic energy quantization becomes important and the collective plasmon mode (*vide infra*) is inactive. Overall, NC present ultra-small size, unique crystalline structures, discrete energy levels and abundant unsaturated active sites that set their fundamentally different physicochemical properties compared with their larger NP counterparts (Du et al. 2020). NC are extremely sensitive to their composition and therefore they are usually denoted by the formula $M_n L_m$, where *n* is the number of Metal (M) atoms and *m* is the number of organic ligands.

Finally, the term 'nanoparticle' is used for mixtures of somewhat polydisperse large nanoclusters (> 2 nm). Metal NP are frequently used for catalysis due to their simple preparation and high catalytic activity. Usually, capping agents used for NP stabilization are weakly bonded to the metal surface—as opposed to NC and therefore, substrates can easily displace them from the catalytic active surface. Nevertheless, the use of surface capping ligands may passivate the catalytic surface and decrease the catalytic activity. To overcome this problem, most of the catalytic research is done after deposition of the particles in different kinds of supports, where post-synthetic protocols can be applied to remove the organic ligands used for the colloidal synthesis. The use of NP anchored onto different supports will be discussed next for both thermal- and photocatalytic processes.

Nanoparticles in Heterogeneous Catalysis

Nanoparticles are at the verge between homogeneous and heterogeneous processes. Although colloidal metal NPs can catalyze organic transformations, they are not stable and are prone to aggregate during the catalytic process, losing their catalytic activities. Therefore, most metal NP catalysts are supported on a solid support, usually metal oxides, carbon materials and porous structures. Traditional heterogeneous catalysis has merged with nanomaterials to supply a platform to support the nanoparticles in place improving their stability and facilitating their separation and potential recyclability. Materials such as zeolites, micro-structured silica, alumina and other oxides can stabilize not only NP but also subnanoclusters and ISAS (Li et al. 2020). Additionally, 1D, 2D and 3D materials such as nanotubes, graphene derivatives (Gerber and Serp 2020), polymers and Metal-Organic Frameworks (MOFs (Wang and Astruc 2020)) are being used as catalysts supports. These NP-support ensembles are frequently known as 'nanocomposites'. Although the main role of the support is to stabilize the NP and hold them in place, many of these materials are frequently non-passive 'spectators' and can play an active role in the catalytic activity shown by the nanocomposite. The supports physical and chemical properties can synergistically assist metal NP catalysis adding functionalities to the nanocomposite. Before exploring the catalytic activity of nanocomposites, the use of noble metal NP (i.e., gold, palladium, platinum) in catalysis and the importance of some critical characteristics of the metal NP that make them suitable for uses in catalysis will be discussed first.

Role of Nanoparticle Morphology in Catalysis

Gold was for a long time considered an inert metal due to its inability to chemisorb and dissociate H_2 and O_2, a reaction that can be readily performed on palladium (Pd) or platinum (Pt) surfaces (Hammer and Norskov 1995). However, pioneer independent work from (Hutchings 1985) and (Haruta et al. 1987) demonstrated early in the 80's that cationic as well as nanometric gold can be used as catalysts. This has had a tremendous impact in industry. For instance, the industrial use of highly toxic mercury chloride catalysts for the hydrochlorination of acetylene to vinyl chloride has been replaced by less toxic carbon-supported gold catalysts. Since Haruta et al. reported the crucial role of particle size for the catalytic activity of Au nanoparticles (Au NP), there has been a large number of studies on the use of Au NP as catalysts for different types of reactions such as oxidation and hydrogenation in both gas (gaseous reagents) or liquid (reagent in solution) phase. Detailed mechanistic studies on the oxidation of CO suggest that AuNP supported on reducible metal oxides (e.g., TiO_2, ZnO, ZrO_2 (Ishida et al. 2020)) exhibited high catalytic activity for CO oxidation that is very sensitive to NP size and shape. Particularly, hemispherical AuNP (on TiO_2) show greater activity compared to spherical NPs. The authors believe that the more catalytic active sites are located at the perimeter of the NP, i.e., at the Au-support interface (Tsubota et al. 1998). Overall, every catalytic process can be affected by the metal NP shape and size as discussed below.

Nanoparticle Size Effect

The role of the NP size in different catalytic processes is frequently attributed to both geometrical and electronic effects (Li et al. 2020). For instance, when considering a truncated Au octahedron, the fraction of edge atoms increases when NP size decreases from 10 to 4 nm, whereas the number of corner atoms increases considerably when the NP is smaller than 4 nm (Fig. 6.2A). As mentioned

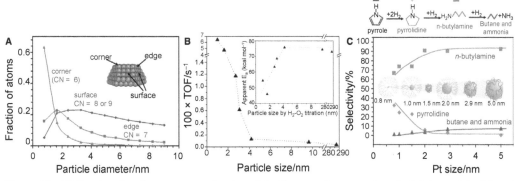

Figure 6.2: (A) Fraction of surface atoms as a function of the diameter of a truncated octahedral Au particle. Adapted with permission from (Hvolbæk et al. 2007) Copyright 2007 Elsevier. (B) Dependence of TOF and apparent Ea for ethane hydrogenolysis on the size of Pt particles. Adapted with permission from (Song et al. 2006) Copyright 2006 American Chemical Society. (C) Product selectivity of pyrrole hydrogenation as a function of the Pt particle size. Adapted with permission from (Kuhn et al. 2008) Copyright 2008 American Chemical Society.

earlier, atoms with lower coordination numbers are more reactive and, in this case (Hvolbæk et al. 2007), their fraction was seen to scale approximately with their catalytic activity. Additionally, edge and corner atoms provide different coordination geometries for the adsorbed molecules compared with the atoms on a terrace, which can also affect the strength of the bond between the reaction intermediates and the catalyst site (Chen et al. 2014). Finally, as discussed above, size changes lead to different local electronic structures; that is, the metal electronic Density Of States (DOS) becomes into discrete energy levels as particle size decreases, showing the transition from the metal into their non-metal state (*vide supra*) (Boronat et al. 2014).

The size effects can affect all types of reactions in either a gas or liquid phase. Important industrial processes have been studied for the particle size effect. Among them, CO oxidation plays an important role in many industrial processes, from the removal of CO from the automobile exhaust gas to the synthesis of pure gases, ethanol or other fuel production (Chen and Goodman 2004, Toyoshima et al. 2012); the Fischer-Tropsch synthesis can convert synthesis gas into long-chain hydrocarbons (Torres Galvis et al. 2012, Bezemer et al. 2006); the Water-Gas Shift (WGS) reaction ($CO + H_2O \rightarrow H_2 + CO_2$) is a key step for the production of high purity hydrogen (Shekhar et al. 2012, Williams et al. 2010); and hydrogenation reactions which are involved in important transformations in the chemical industry of food, additives, pharmaceuticals, etc. These transformations have been extensively explored using precious metals (e.g., Pt, Pd) as the best catalytic performers. For instance, ethane hydrogenolysis over Pt NP show higher TOF numbers when particle sizes are below 4 nm, whereas the apparent Ea increases linearly from 1.7 to 4 nm and remains constant for PtNP larger than 4 nm (Fig. 6.2B) (Song et al. 2006). Pyrrole hydrogenation to form pyrrolidine shows no relation to the catalyst structure, however, ring-opening to *n*-butylamine was structure-sensitive (Fig. 6.2C) (Kuhn et al. 2008). The authors suggest that more electron-rich N in *n*-butylamine compared to pyrrole and pyrrolidine can more strongly bond to the metal surface and inhibit the turnover. Therefore, more reactive surface atoms in small NP are more susceptible to this chemical poisoning, leading to their selectivity towards cyclic products.

Nanoparticle Shape Effect

According to the shape of the NP, the surface area can expose different crystalline planes, which can deliver different catalytic performances. Since the atomic packings on different crystalline planes are essentially different, they can lead to different geometrical (different distance of adjacent atom and coordination number of surface atoms) and electronic (different d-band centres) properties (Li et al. 2020). For instance, for face-centred cubic (fcc) metals the atomic packing follows the order of (111) > (100) > (110); whereas for bulk-centred cubic (bcc) metals the atomic-packing order is

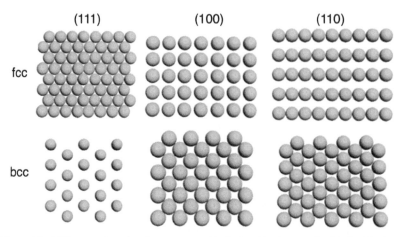

Figure 6.3: Different surfaces in a face-centred cubic (fcc) and body-centred cubic (bcc) lattices.

(110) > (100) > (111) (Fig. 6.3). Bratlie et al. have shown that the hydrogenation of benzene over cubic Pt (100) facets produces only cyclohexane, while the same reaction run over cubooctahedral Pt (111) and (100) facets leads to a mixture of cyclohexane and cyclohexene (Bratlie et al. 2007).

Supported Nanoparticles in Catalysis

When designing new nanocomposites, the interaction between the metal NP and the support is important to determine not only the stability of the material but also the overall catalytic performance. For this, some considerations regarding the nature of the material support and the plausible interactions with the metal NP have to be taken into account. There are several different types of supports reported in the literature, from the use of carbon-based supports (carbon nanotubes, carbon nitrides, graphite, graphene, etc.), to transition metal oxides (reducing and non-reducing metal oxides) (Ishida et al. 2020). This chapter will focus on the use of transition metal oxides and how their physicochemical properties can infer different catalytic properties to the nanocomposites. For instance, for metal oxide supports, the Fermi level and the isoelectric point (IEP) of the oxide are critical to determine the interaction with the metal NP. If a support has a high Fermi level, the electronic interactions with the metal NP can inhibit particle agglomeration leading to small particle sizes. Small NP can also be favoured by using supports with low IEP, which leads to better adsorption of metal ions on the surface by electrostatic interactions generating more metal nuclei (Han et al. 2018). Electrostatic effects can be tuned externally, taking advantage of the ease of polarization of metal oxide surfaces when they are in suspension. For instance, SiO_2 surface can be positively charged by suspending the particles in a solution of pH below their IEP, while at pH above IEP the deprotonation of hydroxyl groups will negatively charge the surface. Now, selecting the right metal precursor, one can induce their strong electrostatic adsorption onto the SiO_2 promoting the formation of ultrasmall particles (Zanella et al. 2006). Besides their role as NP stabilizers and help in maintaining the NP high dispersion under harsh reaction conditions, supports can actively modify the catalytic properties of metal NP by metal-support interactions and charge transfer mechanisms.

Nanoparticle Stabilization

NP are considered to be stable over a given support when the diffusion barriers are high and sintering rates are slow. The presence of surface defects (oxygen vacancies, impurities, surface steps, etc.), on transition metal oxides are considered as anchoring sites for NP, decreasing the sintering processes during the reaction. Many efforts are focused on tuning the surface structures of supports to improve NP stability. For instance, oxygen vacancies can be easily introduced in reducible metal oxides by addition of dopants (e.g., replacement of metal cation by another metal cation with lower valence generates O vacancies to retain neutrality) or by chemical reduction (Sarkar and Khan 2019).

Strong Metal-Support Interaction (SMSI)

It was seen that strong interaction with the supports can frequently alter the chemisorption of the substrates onto the metal NP, in many cases suppressing them at high temperatures (> 500°C). Scanning Tunnelling Microscopy (STM) and Transmission Electron Microscopy (TEM) studies suggest that reduced metal oxides can migrate on top of the metal surface during high-temperature reduction, covering the NP surface and blocking the active sites. An interesting *in situ* study by Bennett et al. proved that by alternative oxidation and reduction treatments at high temperatures (Bennett et al. 1999) the encapsulation process can be reversed.

Electronic Metal-Support Interaction (EMSI)

Once the metal NP comes in contact with the support surface, the metal/oxide interface undergoes a charge distribution that can flow from the metal surface to the oxide support, or vice versa, which results in electron-rich or electron-deficient active sites. This plays a critical role in determining the adsorption/desorption behaviours of reacting species on the active sites. Charge transfer processes are also critical in light-induced processes and are discussed in depth later in this chapter.

Synergistic Metal-Support Interaction

Supports can also be involved in the mechanism of the catalytic reaction, working in cooperation with metal NP. This synergistic effect can enhance reactivity as well as selectivity of the catalytic reaction. For instance, hydrogenation reactions can occur through a hydrogen spillover process, where H species can migrate from metal NP to the metal oxide surface, facilitating the selective reduction of carbonyl groups on α,β-unsaturated aldehydes (Taniya et al. 2012). In an interesting study, Karim et al. demonstrated that H spillover is much slower and limited to short distances on non-reducible metal oxides (e.g., Al_2O_3) than it is on reducible metal oxides (e.g., TiO_2) (Karim et al. 2017). It is worth noting that not only hydrogen but other atomic or molecular species can also move across from one surface to another (e.g., oxygen) (Ruiz Puigdollers et al. 2017).

Metal-Support Interfacial Sites

Mechanistic studies suggest that the active sites for many catalytic reactions are located at the interface between the metal NP and the support. Many studies tried to demonstrate this hypothesis by changing the size of AuNP, however, this approach makes it difficult to separate these observations from the effects due to different particle sizes. An excellent study by Yan et al. prepared an inverse Fe_2O_3/Au (111) catalyst to study CO oxidation. By increasing the areas covered by iron oxide on an initially inert Au (111) surface the authors observed CO oxidation rate increased with the Fe_2O_3 coverage (for coverage < 0.5 monolayers) and then decreased (Yan et al. 2012). Further, Zhou et al. designed Au/CeO_2 nanotowers consisting of alternating gold and ceria layers and exposing only the interface sides. They found a linear relationship between the CO oxidation rate and the number of Au/CeO_2 interfaces (Zhou et al. 2008).

Surface Dynamics

The shape and size of the NP can change dynamically during reaction conditions; thus, the static structure model of solid surfaces is imperfect when describing them under reactive environments. For instance, Kuwauchi et al. demonstrated that AuNP can move back and forth about 0.09 nm when deposited on ceria support (Kuwauchi et al. 2013). This indicates that AuNP are loosely bound to the support. Furthermore, mobile small NP can be fused onto larger ones, a process known as particle sintering. This effect is common when catalysts operate at high temperatures or under plasmon-induced heating conditions (Goodman et al. 2017, Lanterna et al. 2015). Strategies to prevent particle sintering range from the design of core-shell structures (Lee et al. 2016) to the use of redispersion processes under controlled thermal and atmosphere conditions (Lopez et al. 2004). For instance, Hirata et al. demonstrated that Pt NP can sinter after ageing at 750°C under vacuum or reducing atmospheres, whereas redispersion into small particles is produced under oxygen (Hirata et al. 2011). In this case, the authors suggest Pt NP are oxidized into Pt oxides, which can easily migrate and get trapped by ceria support.

Furthermore, for liquid-phase catalysis, the dynamic changes on the NP surface can frequently be responsible for the observation of induction periods before reactions accelerate. A plausible explanation is that NP may undergo structural changes at the beginning of the reaction preparing active catalytic sites to react. For instance, Corma et al. demonstrated that copper NP act as a reservoir for Cu(I)NC; and thus, they observed an induction period on C–X coupling (X: C, N, O, S) reactions when using CuNP. As time is necessary for the NP to redisperse and form small Cu(I) NC, the reaction speeds up when the catalytic species are formed (Oliver-Messeguer et al. 2015). Examples can also be found for other cross-coupling reactions catalyzed by PdNP. Although some evidence supports that the reaction occurs on the solid surface, a few studies suggest that the PdNP are reservoirs for soluble Pd species which are responsible for the catalytic activity (Phan et al. 2006). For instance, working with $PdAu_2$ alloys Niu et al. showed that after reaction a mixture of $PdAu_4$ and Pd NP was found suggesting Pd leaching process during the catalytic cycle of a Suzuki-Miyaura coupling reaction (Niu et al. 2012). As the changes occurred in the presence of aryl halides it was hypothesized that oxidative addition of aryl halides to Pd triggers the metal leaching.

Nanoparticles in Heterogeneous Photocatalysis

Early in the 20th century, Giacomo Ciamician explored the use of light to drive organic reactions opening a new field in chemistry: 'photochemistry'. His vision of a future that replaces fossil fuels by the use of sunlight energy is yet to be realized, with the widespread implementation of photochemistry in the industry as one of this century's challenges. Industrial heterogeneous catalytic reactions usually run at relatively high temperatures compromising the energy efficiency of the processes and the long-term stability of the catalysts. In contrast, the use of sunlight can enable mild reaction conditions and activation conditions that allow for more direct reactions improving atom economy. Therefore, the potential to convert light into valuable chemicals is attracting the chemical industry's attention, especially after the development of inexpensive and highly efficient light sources (i.e., LEDs). In this context, effective use of heterogeneous photocatalysis is the potential solution for developing economically-, atom- and energy-efficient catalytic processes (Scaiano and Lanterna 2019).

Heterogeneous photocatalysis has been applied in different fields such as water remediation, H_2 generation, CO_2 reduction and fine chemicals synthesis (Lanterna and Scaiano 2017, Ali-Elhage et al. 2019, Gisbertz and Pieber 2020, Li et al. 2016). The main challenge in the field is to develop robust and long-lasting photocatalysts that show good sunlight absorption and charge separation to photocatalyze redox reactions. Semiconductor (SC) materials are good candidates for these purposes because of their high light absorption, photostability and heterogeneous nature. Modifications to these SC materials are frequently focused on suppressing charge recombination and improving the charge utilization of the SC to optimize solar energy conversion. Although there are multiple approaches to improve the photocatalytic activity of SC materials (Tan et al. 2019), this chapter focuses only on the decoration of SC surface with metal NP.

In a typical photocatalytic process, the excited SC catalyst can undergo a series of redox steps involving electron-acceptor and electron-donor molecules as shown in Fig. 6.4 left. Briefly, light excitation (1) of energy greater (or equal) to the bandgap energy (E_g) can promote an electron (2) from the Valence Band (VB) into the Conduction Band (CB) forming an exciton ($e^-_{CB} + h^+_{VB}$). This exciton can recombine to take the SC to its ground state (3) or trigger redox reactions on the SC surface. Excited SC have shown highly oxidizing properties (Hainer et al. 2019) (4) being able to oxidize organic substrates (D) adsorbed on the surface or produce reactive oxygen species,

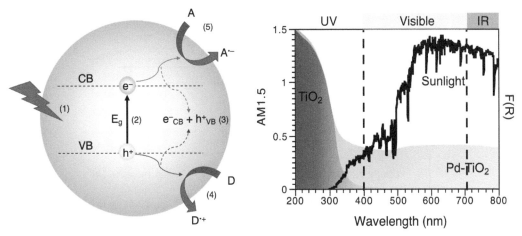

Figure 6.4: Left: representative photocatalytic process in a semiconductor material. Right: representative absorption profile of bare TiO$_2$ versus Pd NP supported on TiO$_2$. Solar emission spectrum under AM1.5 (black line) is shown for reference. UV, visible and near IR regions are separated by dash lines.

ROS, forming ˙OH radicals through water oxidation (Tan et al. 2019). Additionally, the excited electron can reduce organic molecules with electron-accepting properties (A) that are adsorbed on the surface of the photocatalyst (5). This same pathway can promote the formation of superoxide radicals (O$_2$˙$^-$) by reduction of O$_2$ under air or the formation of H$_2$ gas by reduction of H$^+$ under an inert atmosphere.

Titanium dioxide (TiO$_2$) is by far one of the most studied SC in photocatalysis. Despite being an excellent light absorber, TiO$_2$ can only absorb light in the UV region of the solar spectrum (< 400 nm) due to its large band gap energy value (E$_g$ ~ 3.2 eV). As discussed later on this chapter, metal NP decoration of SC materials can not only extend their absorption profile into the visible and near IR regions (as shown in Fig. 6.4 right) but also slow down the electron-hole recombination rates (step 3); overall, favouring the photocatalytic activity of the SC material (Wang et al. 2016, Marina et al. 2018, Elhage et al. 2017).

Mechanisms of Plasmon-Enhanced Photocatalysis

The use of metal NP in photocatalysis has emerged in the last two decades, particularly taking advantage of their broad absorption spectrum that can be excited with a large fraction of the solar spectrum (Han et al. 2018). That is, when light interacts with a metal NP, the incoming electromagnetic (EM) field generates the collective oscillation of free electrons in the metal NP, polarizing the NP with charges accumulating at opposite ends of its structure. A depolarization field creates a restoring force (induced EM) in the electron cloud that reinstates the charges to the original state. If the incident EM field oscillates at the same frequency of the resonant mode, both EM fields can sync and enhance the amplitude of the total EM field near the regions where charges are accumulated (i.e., NP surface). This phenomenon is known as Localized Surface Plasmon Resonance (LSPR), Fig. 6.5 top. As a result, the photon energy is confined to the surface of the NP for longer than photons would normally spend in the same volume travelling at the speed of light and NP can enhance the intensity of the incident light to various orders of magnitude (Kale et al. 2014).

For coinage metals (Au, Ag, and Cu), the LSPR absorption band has its maximum in the visible region of the solar spectrum (Peiris et al. 2016). The lifetime of the excited electrons (plasmon) is about ~ 5–100 fs. The relaxation of these excited plasmons can occur by two competitive dephasing processes: (A) radiative emission of photons and (B) non-radiative relaxation of electrons—i.e., Landau damping and Chemical Interface Damping (CID) (Kale et al. 2014, Ma et al. 2016, Gellé et al. 2020). These processes (Fig. 6.5 bottom) strongly depend on the NP geometry, composition and

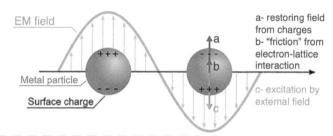

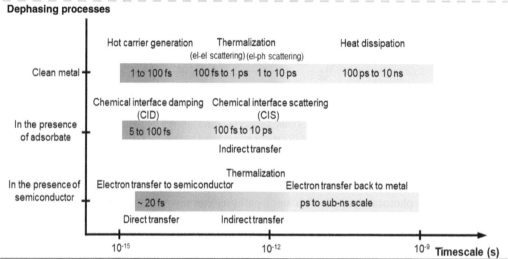

Figure 6.5: Top: Diagram showing the generation of the Localized Surface Plasmon Resonance (LSPR). An incoming electric field excites the free electrons in the metal NP (plasmons) which follows the external electric field orientation. A restoring electric field is generated by the generated surface charges and the ionic network causes damping ('friction' from the electron-lattice interaction). Adapted with permission from (Gellé et al. 2020) Copyright 2020 American Chemical Society. Bottom: Time scale for the non-radiative dephasing processes of the LSPR in the absence and the presence of an adsorbate, i.e., molecule or semiconductor (via Schottky contact, vide infra). Adapted with permission from (Zhang et al. 2018) Copyright 2018 American Chemical Society.

local environment; and therefore, these parameters need to be considered for mechanistic studies. Throughout these relaxation pathways, molecules in the proximity to the NP surface (adsorbates) can be excited through energy or electron transfer processes, giving rise to the photocatalytic properties of plasmonic NP.

During the radiative plasmon decay process, adsorbates can undergo energy transfer processes by absorption of photons from intense, reradiated photon fluxes from the plasmonic NP. For this, the optical properties of the NP and the adsorbate need to be compatible. Specifically, the molecule absorption spectrum should overlap with the plasmon band of the NP, i.e., molecules with allowable electronic transitions with the energy of similar magnitude to the energy of reemitted photons (Kale et al. 2014). This process is analogous to the well-known Surface-Enhanced Raman Scattering (SERS) and Metal Enhanced Fluorescence (MEF), however, it has been hardly explored in the field of photocatalysis. This is likely due to the need of UV photons to excite the adsorbate intramolecular electronic transitions; only a few examples are found using the visible regime, particularly when dye-like molecules are used (Mori et al. 2010).

Non-radiative relaxation pathways (Landau damping and CID) are frequently dominant in small NP and are caused by transferring the plasmon energy into the NP itself or to the immediate environment. Landau damping results in the conversion of the photon energy into an energetic electron and hole (exciton) and occurs *ca.* 10 fs after the initial plasmon excitation. The temporal evolution of this exciton results in different phenomena. During the first hundred fs, the excited primary electrons interact with other electrons through Coulombic inelastic scattering (electron-electron scattering) transferring the energy across the electrons in the metal NP above the Fermi level energy (E_F). Further, within a timescale of *ca.* 1 ps, low energy electrons can couple to phonon modes, transferring the electronic kinetic energy to the metal lattice heating the NP and the surroundings (electron-phonon energy relaxation). Additionally, an adsorbate (i.e., a molecule or a support) present on the metal NP surface can induce the ultrafast (*ca.* 5 fs) dephasing pathway CID through the direct transfer of energetic charge carriers to unpopulated adsorbate electronic states (Kale et al. 2014, Ma et al. 2016, Gellé et al. 2020).

When considering electronic transitions in metal NP it is important to distinguish intraband from interband transitions. Plasmonic NP (e.g., Au, Ag, Cu) are characterized by fully-filled d-bands placed well below E_F, and therefore, their plasmon decay occurs by 'intraband' transitions of electrons within the sp band of the metal (Fig. 6.6A). These energetic electrons may occupy a continuum of states in the conduction band distributed within $E_F < E_e < (E_F + h\nu)$ and are called 'hot' electrons. This transition is responsible for their bright colours. Furthermore, plasmonic NP have strong absorption on the high-energy side of their plasmon resonances (Fig. 6.6B) due to intrinsic 'interband' transitions; i.e., electrons from the d-band promoted to the sp-band. These electrons have a much lower kinetic energy than those generated by plasmon excitation ($E_e \sim E_F$), however, are generated concomitantly with 'hot' holes in the d band with low absolute energy values, and therefore, with great oxidation potential. Although interband transitions are not a plasmonic effect, they can occur at the same time and can also trigger catalytic processes. Furthermore, non-plasmonic NP (e.g., Pd, Pt) can also be excited under UV and visible irradiation, however, their unfilled d-orbital induce interband transitions showing no plasmonic resonance. Nevertheless, they can also generate hot charge carriers and participate in catalytic processes.

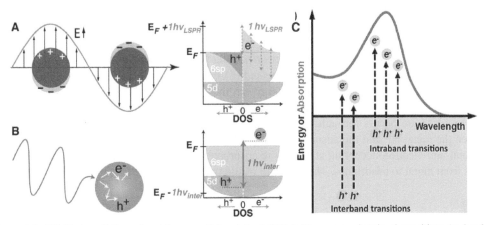

Figure 6.6: (A) Photoexcitation of metal nanoparticles at the LSRP (left) promotes intraband transitions (sp-band) to generate hot electrons (right) at various energies ($E_e > E_F$). (B) Photoexcitation of metal nanoparticles with a higher energy photon (left) promotes the interband transitions and directly generates an e⁻-h⁺ pair (right). Adapted with permission from (Zhao et al. 2017) Copyright 2017 American Chemical Society (https://pubs.acs.org/doi/10.1021/acscentsci.7b00122 further permission related to the material excerpted should be directed to the ACS). (C) The absorption spectrum of a plasmonic metal nanoparticle overlapping a schematic representation of the interband (high energy) and intraband (low energy) transitions. Adapted with permission from (Zhang et al. 2018) Copyright 2018 American Chemical Society.

LSPR Thermal Effects

Initially, it was assumed that the main photocatalytic activity of plasmonic NP was due to heating (Meng et al. 2014). That is, a 'photothermal' process triggered by plasmon excitation is expected to show an exponential relationship between the reaction rate and the illumination intensity. Although the temperature increase due to light absorption can be approximated to be proportional to the intensity of the light, the reaction rate constant k follows the Arrhenius temperature dependence,

$k = Ae^{-Ea/RT}$, where R is the gas constant, T is the temperature, Ea is the molar activation energy and A is the pre-exponential constant factor. As a consequence, the reaction rate for the photothermal processes follows an exponential dependence on the illumination power (Baffou et al. 2020). In contrast, a process driven by hot electron transfer would show a linear relationship. This has been used as a rational method to determine which of the LSPR dephasing pathways is driving photocatalysis (Gellé et al. 2020). However, recent studies suggest that when a metal NP of 5 nm diameter is irradiated with low light intensities (i.e., five times of mean irradiance of sunlight) and assuming that all absorbed photons are converted into heat, the temperature increase on the NP surface is only 1 K. Considering that typical photocatalytic processes are performed under simulated sunlight irradiation (1000 Wm²), these results suggest that photothermal effect plays a minimal role in photocatalysis for small particle sizes. An outstanding summary of the different experimental approaches that can be followed to determine the nature of the photocatalytic processes has been recently published by Baffou et al. (Baffou et al. 2020).

LSPR Field Enhancement Effects

Plasmonic NP can interact with the immediate surrounding by near-field interaction. This phenomenon has been dubbed as 'nanoantennae', analogous to the way a radio antenna works. Although this has been extensively used for sensing (SERS, MEF, etc.), examples in photocatalysis are rare (Mori et al. 2010, Zhang et al. 2014).

Charge Transfer Mechanisms

The use of plasmon NP for charge transfer processes has gained great interest in the last two decades, however, due to the concomitance of different charge generation and transfer processes in the NP surface, mechanistic studies are usually challenging. Different mechanisms considering: (1) indirect and (2) direct (CID) hot charge carrier transfer from NP to the adsorbate will be explored next. Note that for both charge transfer processes, metals and adsorbate orbitals must overlap (Zhang et al. 2018).

As shown in Fig. 6.7A, the indirect electron transfer is a two-step process where (i) hot electron-hole pairs are generated on the metal surface and (ii) thermalized hot carriers adopt a Fermi-Dirac distribution. Hot electrons with enough energy can then be transferred to the Lowest Unoccupied Molecular Orbital (LUMO) of the adsorbate. On the other hand, direct electron transfer from metal to adsorbate occurs through CID. This requires the formation of hybridized surface states between the metal NP and the adsorbate (Fig. 6.7B). If the electron-donating states are located on the metal and electron-accepting states at the adsorbate, it is assumed the direct electron transfer occurs from metal to adsorbates. The opposite would transfer an electron from the adsorbate to the

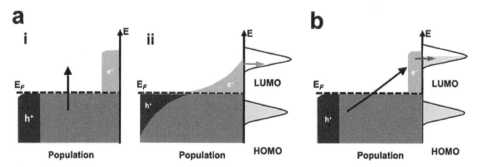

Figure 6.7: Hot electron transfer from metals to adsorbates. (a) Indirect electron transfer: (i) hot electron-hole pairs are generated on the metal surface and (ii) thermalized hot carriers adopt a Fermi-Dirac distribution. Energetic electrons can be transferred to the LUMO of adsorbed molecules. (b) Direct electron transfer: substrates are coupled to the metal surface, can form adsorbate-induced states close to the EF, inducing shift and broadening of the LSRP band. Thus, plasmon-generated hot electrons can transfer directly from metal to adsorbate. Adapted with permission from (Zhang et al. 2018) Copyright 2018 American Chemical Society.

metal NP, although few examples of hot hole-mediated reactions have been reported. As mentioned earlier, CID is one of the LSPR dephasing processes and occurs within the first 5 fs after light absorption. In contrast, indirect electron transfer competes with energy loss processes (electron-electron scattering) and is expected to have lower electron-transfer efficiency comparing with CID. In general, thermalization and back-transfer processes (vide infra) can compete with the electron transfer processes because they occur at comparable time scales. These are the processes that can contribute to decreasing plasmonic photochemistry efficiency (Fig. 6.5 bottom).

Supported Nanoparticles in Photocatalysis

As mentioned earlier, different supports can assist the chemical functionality going beyond their role as substrates. In addition to the physical and chemical properties described above, photophysical properties of the supports are critical to understanding their role in photocatalytic processes. Note that semiconductor supports can participate during the photon absorption and charge carrier transfer processes and therefore, it is important to understand how they interact with metal NP. When the metal and the semiconductor surfaces are in contact, their Fermi levels (E_F) align to provide electronic continuity in the system (Fig. 6.8 top). Between 1938 and 1939 Schottky and Mott described this phenomenon as a 'band bending effect' which is induced by the metal/semiconductor contact. Once the contact is made, free electrons can be transferred between the metal and semiconductor according to the difference of their work functions. As shown in Fig. 6.8, when the metal work function (ϕ_m) is higher than that of the semiconductor (ϕ_s) the electrons flow from the metal to the semiconductor until the equilibrium is reached (E_F levels are aligned). In other words, when the Fermi level of the metal (E_F^m) is higher than the Fermi level of the semiconductor (E_F^s) charges flow from the metal into the semiconductor decreasing the E_F^M (charge carriers flow in the opposite direction is expected when $E_F^m < E_F^s$). Under equilibrium, the Helmholtz double layer is established at the metal/semiconductor interface, where the negatively charged carriers are accumulated at the metal surface whereas the positively charged are collected at the semiconductor. This shows a depletion of the free charge carriers' concentration at the semiconductor surface compared with the bulk, generating a 'space charge region'. When considering n-type semiconductors, if $\phi_m > \phi_s$ the space charge region is called 'depletion layer'; whereas when $\phi_m < \phi_s$, electrons are accumulated in the semiconductor surface, the region is called 'accumulation layer'. This provokes a bending in the semiconductor energy band edges that reflects the electrostatic energy experienced by an electron as it moves through the interface: (i) bending up toward the interface when $\phi_m > \phi_s$, (ii) bending down toward the interface when $\phi_m < \phi_s$. Band bending (V_{BB}) can be calculated as the difference between the metal and the semiconductor work functions ($V_{BB} = \phi_m - \phi_s$). Additionally, when considering n-type semiconductors, if $\phi_m > \phi_s$ an extra barrier is created at the metal/surface interface called Schottky barrier (ϕ_{SB}) and can be calculated as the difference between the metal work function and the semiconductor electron affinity (χ_S). No barrier is formed when $\phi_m < \phi_s$ and consequently, the metal/semiconductor contact is ohmic.

The possibility to transport charge carriers through the metal/semiconductor interface can help to increase the spatial separation of the electron-hole pair photogenerated, extending the effective lifetime of the charge carriers. Electron injection from metal into the semiconductor can occur only if the carriers have enough energy to surpass the ϕ_{SB}, which sets the threshold of electrons that can be transferred (Fig. 6.8 bottom) (Clavero 2014). That is, low Schottky barrier (ϕ_{SB}) height will allow injection of a larger absolute number of carriers for a given plasmonic excitation wavelength. Note that the conduction band curvature at the depletion layer has to be sufficiently high to impede back electron transfer. This will provide spatial separation for the electrons and holes and extend their lifetime giving time for chemical transformations to take place.

The electron transfer between metal/semiconductor contacts can undergo a direct or indirect injection process. Frequently, indirect electron transfer shows low efficiency due to the broad energy distribution of plasmon-generated charge carriers which do not all have an energy sufficient to

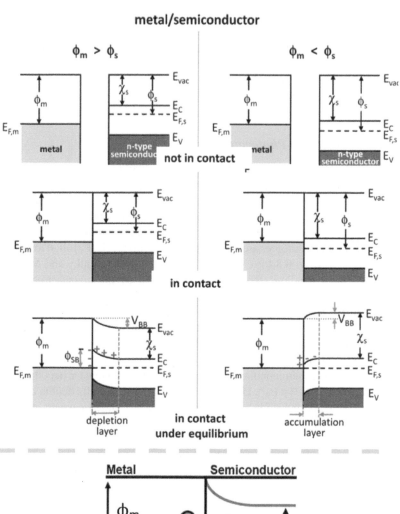

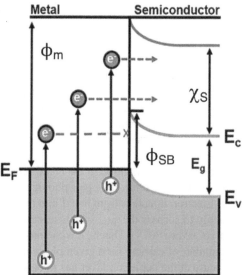

Figure 6.8: Top: Energy band diagrams for metal and n-type semiconductor contacts. E_{vac}, vacuum energy; E_c, energy of conduction band minimum; E_v, energy of valence band maximum; ϕ_m, metalwork-function; ϕ_s, semiconductor work function; χ_s, electron affinity of the semiconductor; ϕ_{SB}, Schottky barrier. Adapted with permission from (Zhang and Yates 2012) Copyright 2012 American Chemical Society. Bottom: Photoexcitation of metal nanoparticles promotes the generation of energetic charge carriers (electrons and holes). Only electrons with enough energy to surpass the ϕ_{SB} will be injected into the semiconductor. Adapted with permission from (Marchuk and Willets 2014) Copyright 2014 Elsevier.

overcome the Schottky barrier (White and Catchpole 2012). In contrast, during direct injection electrons spend less time in the metal lattice and are directly ejected into the SC, leaving holes in the plasmonic metal (Fig. 6.8 bottom). Similar to CID, direct promotion to the semiconductor CB largely depends on the metal–semiconductor state hybridization at the interface, which is more favourable when their wave function extends into both material systems.

The electron injection from metal NP into the semiconductors allows the use of excitation wavelengths in the visible and near IR regions for photocatalysis, even when using semiconductors that only absorb in the UV (e.g., TiO_2). Due to the broad absorption spectrum of the nanocomposites, one can expect the use of different irradiation sources to trigger various electron transfer processes that are directly correlated with the photon energy of the excitation wavelength. Semiconductor supports can play an active role in the mechanism depending on the excitation wavelength used (Elhage et al. 2019, Marina et al. 2018, Elhage et al. 2017, Elhage et al. 2018). Figure 6.9 summarizes plausible scenarios that can be expected when using metal NP-decorated semiconductors as photocatalysts. Figure 6.9a shows that UV excitation can promote an electron from the semiconductor VB to the CB, where it can be hampered by the metal nanoparticles. This induces charge separation, delaying the electron-hole recombination process, and favours electron transfer (eT) processes with electron-Donor (D) and electron-Acceptor (A) molecules (Fig. 6.4 left). For example, PdNP supported on TiO_2 (Elhage et al. 2017) were used to yield hydrogenation of alkenes under UV irradiation using methanol solvent as the only H source (no H_2 gas pressures needed). The mechanism of the reaction was rationalized with methanol working as a Sacrificial Electron Donor (SED) that can trap the hole in the semiconductor VB, whereas releasing H^+ that can be reduced at the Pd NP surface. This environmentally-friendly approach circumvents the use of hazardous pressurized H_2 gas and elaborates experimental setups and provides accessible, inexpensive, and easy to handle hydrogen sources (Elhage et al. 2017, Wang et al. 2019). Interestingly, when the same reaction is subjected to visible light irradiation, only alkene isomerization products are obtained, suggesting a different mechanism is taking place. Further mechanistic analyses suggest the reaction under visible irradiation is triggered by the local heat generated on the NP surface (Fig. 6.9c). Additionally, visible light photoexcitation of metal nanoparticles at the LSRP could transfer hot electrons to the semiconductor CB, where the SC can act as an electron shuttle facilitating eT reactions (Fig. 6.9b). For instance, it was demonstrated that Au NP supported on TiO_2 can generate hydrogen gas under both UV and visible light excitation. The mechanism under the different irradiation wavelengths was rationalized as electron transfer from the semiconductor into the metal NP under UV excitation and the injection of hot electrons from the metal NP into the semiconductor CB under visible irradiation (Gomes Silva et al. 2011). Finally, visible light photoexcitation of metal nanoparticles at the LSRP can generate local heating and hot electrons on the surface of the metal NP which are responsible

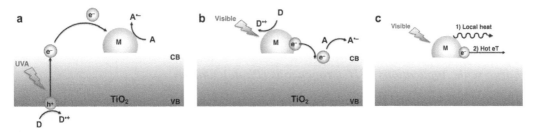

Figure 6.9: Plausible photocatalytic mechanisms using different excitation sources. UV excitation can promote an electron from the semiconductor VB to the CB, where it can be hampered by the metal nanoparticles (a). This induces charge separation, delaying the electron-hole recombination process, and favours electron transfer (eT) processes with electron-Donor (D) and electron-Acceptor (A) molecules (Fig. 6.4 left). Visible light photoexcitation of metal nanoparticles at the LSRP can generate local heating and hot electrons on the surface of the NP. Hot electrons can be transferred to the semiconductor CB, where the SC can act as an electron shuttle facilitating eT reactions (b). When the SC is not involved in the reaction mechanism, heat or hot electrons at the NP surface are responsible for the photocatalytic properties of the material and SC is only used as the catalyst support (c).

for the photocatalytic activity. In this case, the SC is not involved in the reaction mechanism and is, therefore, only used as heterogeneous support for metal NP (Fig. 6.9c).

This ability to switch the electron flow by simply changing the irradiation conditions has been used to improve selectivity in other organic transformations. Furthermore, it was also shown that visible light can change product selectivity when working with C-C coupling reactions such as Ullmann cross-coupling (Marina et al. 2018). What is more, the versatility of this material allows several uses and recyclability strategies approaching circular economy principles (Elhage et al. 2019).

One way to follow the reaction dependence with the excitation wavelength, and thus, gain insights into the mechanism, is to construct an action spectrum. That is, to determine the wavelength dependence of the Apparent Quantum Yield (AQY) of the reaction. The AQY is the number of product molecules formed by each photon used to excite the photocatalyst and can be calculated as AQY (%) = 100% (Y_{light} – Y_{dark})/(incident number of photons), where Y_{light} and Y_{dark} are the yields obtained with and without light irradiation, respectively. By overlapping the AQY at different excitation wavelengths with the photocatalyst absorption spectrum, one can gain information regarding how the energy is transferred to the molecular species (Sarina et al. 2017). For instance, Fig. 6.10 shows how Sonogashira reaction catalyzed by Pd NP supported on TiO_2 can work under visible light excitation (> 400 nm), suggesting direct excitation of Pd NP is involved in the reaction mechanism. However, direct excitation of the semiconductor (UV light irradiation) totally shield Pd-NP from light absorption and results in a sudden drop of the photonic efficiency of the reaction in the UV region (Elhage et al. 2018).

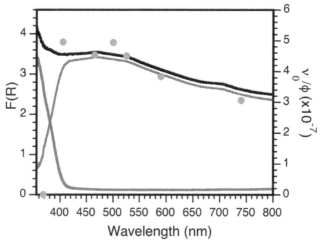

Figure 6.10: Action spectrum (dots) of Sonogashira reaction catalyzed by Pd NP supported on TiO_2 overlapped with the optical absorption profile of TiO_2 (a), Pd NP supported on TiO_2 (b), and their difference (c). Adapted with permission from (Elhage et al. 2018). Copyright 2018, American Chemical Society.

Large-scale Applications of Heterogeneous (Photo)catalysis

The use of heterogeneous (photo)catalysts on a large scale relies largely on their ability to become resilient materials under different reaction conditions. That is long catalysts lifetimes that enable multiple reusability cycles. Additionally, the vast majority of the heterogeneous (photo)catalysts explored to date are based on the use of precious metals that are not only expensive but also scarce. For large-scale applications, less expensive, earth-abundant materials such as Co, Ni or Mo are more appealing as they show high cost-effectiveness. Particularly for heterogeneous photocatalysis, the scale-up of photochemical methods presents technical challenges, as light penetration cannot be scaled up in the way thermal reactions are. According to the Beer-Lambert law, light attenuation (A) increases when the light path (*l*) increases and hence as batch reactors become bigger irradiation efficiencies are lower. One way to overcome this problem is by using continuous flow processes, a promising venue for exploitation in photocatalyzed processes. Flow chemistry has been used for decades in the production of commodities, such as petroleum-derived products (e.g., processes such as fluid catalytic cracking) and over the last 20 years as an important tool for liquid-phase reactions and to scale up organic reactions at the bench level. However, as mentioned in literature "the use of heterogeneous photo-catalysts in continuous-flow is not trivial" (Cambie et al. 2016). Although flow systems are relatively easy to scale up, in photochemical systems any scale-up will require major surface incrementation, as stacking photoreactors may pose a challenge for the irradiation efficiency. In general, continuous flow chemistry has marked advantages compared to analogous batch reactor protocols such as reduce reaction times and increase process productivity. In terms of light penetration, flow reactors are characterized by small reactor channel dimensions which reduce the path length of light necessary to irradiate the photocatalyst, hence increasing light exposure. Three main reactor categories have been described by (Thomson et al. 2020) as the most common approaches to continuous flow chemistry for heterogeneous photocatalysis: (i) fixed bed reactors, (ii) coated reactors, and (iii) suspension reactors. Fixed bed reactors are typically transparent cylinders with frits stoppers that allow liquids to flow through the reactor whereas the catalyst is kept in place. Since the catalyst is retained within the reactor, it can be continuously reused. One main disadvantage of this setup is that irradiation is not uniform. In contrast, coated reactors are made of heterogeneous photocatalysts immobilized as a thin film on the surface of the microfluidic device, providing greater excitation cross-section and hence, more efficient irradiation. Finally, free-flowing suspension reactors can use flow reactors designed for homogeneous systems as long as the particle size is low (< 100 μm) and a syringe or peristaltic pump is used in replacement of HPLC piston pumps, which are not compatible with solids.

Conclusions

The use of metal nanoparticles in the field of heterogeneous catalysis has introduced a vast landscape of properties that involve multiple physical/chemical processes over different length scales. A fundamental understanding of these properties is necessary to be considered carefully in catalyst design. In particular, this chapter has introduced the effect of size, shape and metal-support interactions in heterogenous catalysis, discussed the mechanism of plasmon induced photocatalysis, and described how supported nanoparticles can interact with light, introducing basic concepts about catalytic wavelength dependence. Although this chapter was not a comprehensive summary of the current literature, it provides the reader with the essential readings to acquire basic knowledge in the field of nanocatalysis.

Acknowledgements

The author is grateful to the University of Nottingham for its valuable support and to former colleagues at the University of Ottawa, especially to Prof. J. C. Scaiano, for their experimental and intellectual contributions to the various publications cited in this chapter. These publications are the result of work supported for many years by the Natural Sciences and Engineering Research Council of Canada, by the Canada Research Chairs program and by the Canada Foundation for Innovation.

References

Astruc, D. 2020. Introduction: Nanoparticles in catalysis. Chem. Rev. 120: 461–463.
Baffou, G., I. Bordacchini, A. Baldi and R. Quidant. 2020. Simple experimental procedures to distinguish photothermal from hot-carrier processes in plasmonics. Light: Sci. Appl. 9: 108.
Bennett, R. A., P. Stone and M. Bowker. 1999. Pd nanoparticle enhanced re-oxidation of non-stoichiometric TiO_2: STM imaging of spillover and a new form of SMSI. Catal. Lett. 59: 99–105.
Bezemer, G. L., J. H. Bitter, H. P. C. E. Kuipers, H. Oosterbeek, J. E. Holewijn, X. Xu et al. 2006. Cobalt particle size effects in the Fischer–Tropsch reaction studied with carbon nanofiber supported catalysts. J. Am. Chem. Soc. 128: 3956–3964.

Boronat, M., A. Leyva-Pérez and A. Corma. 2014. Theoretical and experimental insights into the origin of the catalytic activity of subnanometric gold clusters: Attempts to predict reactivity with clusters and nanoparticles of gold. Acc. Chem. Res. 47: 834–844.

Bratlie, K. M., H. Lee, K. Komvopoulos, P. Yang and G. A. Somorjai. 2007. Platinum nanoparticle shape effects on benzene hydrogenation selectivity. Nano Lett. 7: 3097–3101.

Butet, J., P. -F. Brevet and O. J. F. Martin. 2015. Optical second harmonic generation in plasmonic nanostructures: From fundamental principles to advanced applications. ACS Nano 9: 10545–10562.

Cambie, D., C. Bottecchia, N. J. W. Straathof, V. Hessel and T. Noel. 2016. Applications of continuous-flow photochemistry in organic synthesis, material science, and water treatment. Chem. Rev. 116: 10276–10341.

Chen, M. S. and D. W. Goodman. 2004. The structure of catalytically active gold on titania. Science 306: 252.

Chen, C., Y. Kang, Z. Huo, Z. Zhu, W. Huang, H. L. Xin et al. 2014. Highly crystalline multimetallic nanoframes with three-dimensional electrocatalytic surfaces. Science 343: 1339.

Clavero, C. 2014. Plasmon-induced hot-electron generation at nanoparticle/metal-oxide interfaces for photovoltaic and photocatalytic devices. Nat. Photonics 8: 95–103.

Du, Y., H. Sheng, D. Astruc and M. Zhu. 2020, Atomically precise noble metal nanoclusters as efficient catalysts: A bridge between structure and properties. Chem. Rev. 120: 526–622.

Elhage, A., A. E. Lanterna and J. C. Scaiano. 2017. Tunable photocatalytic activity of palladium-decorated TiO$_2$: Non hydrogen-mediated hydrogenation or isomerization of benzyl substituted alkenes. ACS Catal. 7: 250–255.

Elhage, A., A. E. Lanterna and J. C. Scaiano. 2018. Light-induced Sonogashira C–C coupling under mild conditions using supported palladium nanoparticles. ACS Sustainable Chem. Eng. 6: 1717–1722.

Elhage, A., A. E. Lanterna and J. C. Scaiano. 2019. Catalytic farming: Reaction rotation extends catalyst performance. Chem. Sci. 10: 1419–1425.

Fu, Q., H. Saltsburg and M. Flytzani-Stephanopoulos. 2003. Active nonmetallic Au and Pt species on ceria-based water-gas shift catalysts. Science 301: 935.

Gellé, A., T. Jin, L. de la Garza, G. D. Price, L. V. Besteiro and A. Moores. 2020. Applications of plasmon-enhanced nanocatalysis to organic transformations. Chem. Rev. 120: 986–1041.

Gerber, I. C. and P. Serp. 2020. A theory/experience description of support effects in carbon-supported catalysts. Chem. Rev. 120: 1250–1349.

Gisbertz, S. and B. Pieber. 2020. Hetergeneous photocatalysis in organic synthesis. ChemPhotoChem. 4: 456–475.

Goodman, E. D., J. A. Schwalbe and M. Cargnello. 2017. Mechanistic understanding and the rational design of sinter-resistant heterogeneous catalysts. ACS Catal. 7: 7156–7173.

Gomes Silva, C., R. Juárez, T. Marino, R. Molinari and H. García. 2011. Influence of excitation wavelength (UV or visible light) on the photocatalytic activity of titania containing gold nanoparticles for the generation of hydrogen or oxygen from water. J. Am. Chem. Soc. 133: 595–602.

Hainer, A., N. Marina, S. Rincon, P. Costa, A. E. Lanterna and J. C. Scaiano. 2019. Highly electrophilic titania hole as a versatile and efficient photochemical free radical source. J. Am. Chem. Soc. 141: 4531–4535.

Hammer, B. and J. K. Norskov. 1995. Why gold is the noblest of all the metals. Nature 376: 238–240.

Han, P., W. Martens, E. R. Waclawik, S. Sarina and H. Zhu. 2018. Metal nanoparticle photocatalysts: Synthesis, characterization, and application. Part. Part. Syst. Charact. 35: 1700489.

Hartshorn, C. M., M. S. Bradbury, G. M. Lanza, A. E. Nel, J. Rao, A. Z. Wang et al. 2018. Nanotechnology strategies to advance outcomes in clinical cancer care. ACS Nano 12: 24–43.

Haruta, M., T. Kobayashi, H. Sano and N. Yamada. 1987. Novel gold catalysts for the oxidation of carbon monoxide at a temperature far below 0°C. Chem. Lett. 16: 405–408.

Hirata, H., K. Kishita, Y. Nagai, K. Dohmae, H. Shinjoh and S. i. Matsumoto. 2011. Characterization and dynamic behavior of precious metals in automotive exhaust gas purification catalysts. Catal. Today 164: 467–473.

Hutchings, G. J. 1985. Vapor phase hydrochlorination of acetylene: Correlation of catalytic activity of supported metal chloride catalysts. J. Catal. 96: 292–295.

Hvolbæk, B., T. V. W. Janssens, B. S. Clausen, H. Falsig, C. H. Christensen and J. K. Nørskov. 2007. Catalytic activity of Au nanoparticles. Nano Today 2: 14–18.

Isaifan, R. J., S. Ntais and E. A. Baranova. 2013. Particle size effect on catalytic activity of carbon-supported Pt nanoparticles for complete ethylene oxidation. Appl. Catal., A 464-465: 87–94.

Ishida, T., T. Murayama, A. Taketoshi and M. Haruta. 2020. Importance of size and contact structure of gold nanoparticles for the genesis of unique catalytic processes. Chem. Rev. 120: 464–525.

Kale, M. J., T. Avanesian and P. Christopher. 2014. Direct photocatalysis by plasmonic nanostructures. ACS Catal. 4: 116–128.

Karim, W., C. Spreafico, A. Kleibert, J. Gobrecht, J. VandeVondele, Y. Ekinci et al. 2017. Catalyst support effects on hydrogen spillover. Nature 541: 68–71.

Kuai, L., Z. Chen, S. Liu, E. Kan, N. Yu, Y. Ren et al. 2020. Titania supported synergistic palladium single atoms and nanoparticles for room temperature ketone and aldehydes hydrogenation. Nat. Commun. 11: 48.

Kuhn, J. N., W. Huang, C. -K. Tsung, Y. Zhang and G. A. Somorjai. 2008. Structure sensitivity of carbon–nitrogen ring opening: Impact of platinum particle size from below 1 to 5 nm upon pyrrole hydrogenation product selectivity over monodisperse platinum nanoparticles loaded onto mesoporous silica. J. Am. Chem. Soc. 130: 14026–14027.

Kuwauchi, Y., S. Takeda, H. Yoshida, K. Sun, M. Haruta and H. Kohno. 2013. Stepwise displacement of catalytically active gold nanoparticles on cerium oxide. Nano Lett. 13: 3073–3077.

Lanterna, A. E., A. Elhage and J. C. Scaiano. 2015. Heterogeneous photocatalytic C-C coupling: Mechanism of plasmon-mediated reductive dimerization of benzyl bromides by supported gold nanoparticles. Catal. Sci. Technol. 5: 4336–4340.

Lanterna, A. E. and J. C. Scaiano. 2017. Photoinduced hydrogen fuel production and water decontamination technologies. Orthogonal strategies with a parallel future? ACS Energy Lett. 2: 1909–1910.

Lee, S., J. Seo and W. Jung. 2016. Sintering-resistant $Pt@CeO_2$ nanoparticles for high-temperature oxidation catalysis. Nanoscale 8: 10219–10228.

Li, K., B. Peng and T. Peng. 2016. Recent advances in heterogeneous photocatalytic CO_2 conversion to solar fuels. ACS Catal. 6: 7485–7527.

Li, Z., S. Ji, Y. Liu, X. Cao, S. Tian, Y. Chen et al. 2020. Well-defined materials for heterogeneous catalysis: From nanoparticles to isolated single-atom sites. Chem. Rev. 120: 623–682.

Liu, L. and A. Corma. 2018. Metal catalysts for heterogeneous catalysis: From single atoms to nanoclusters and nanoparticles. Chem. Rev. 118: 4981–5079.

Lopez, N., J. K. Nørskov, T. V. W. Janssens, A. Carlsson, A. Puig-Molina, B. S. Clausen et al. 2004. The adhesion and shape of nanosized Au particles in a Au/TiO_2 catalyst. J. Catal. 225: 86–94.

Ma, X. C., Y. Dai, L. Yu and B. B. Huang. 2016. Energy transfer in plasmonic photocatalytic composites. Light: Sci. Appl. 5.

Marchuk, K. and K. A. Willets. 2014. Localized surface plasmons and hot electrons. Chem. Phys. 445: 95–104.

Marina, N., A. E. Lanterna and J. C. Scaiano. 2018. Expanding the color space in the two-color heterogeneous photocatalysis of Ullmann C–C coupling reactions. ACS Catal. 8: 7593–7597.

Meng, X. G., T. Wang, L. Q. Liu, S. X. Ouyang, P. Li, H. L. Hu et al. 2014. Photothermal conversion of CO_2 into CH_4 with H_2 group VIII nanocatalysts: An alternative approach for solar fuel production. Angew. Chem. Int. Edit. 53: 11478–11482.

Mori, K., M. Kawashima, M. Che and H. Yamashita. 2010. Enhancement of the photoinduced oxidation activity of a ruthenium(II) complex anchored on silica-coated silver nanoparticles by localized surface plasmon resonance. Angew. Chem. Int. Ed. 49: 8598–8601.

Mott, N. F. 1938. Note on the contact between a metal and an insulator or semiconductor. Math. Proc. Cambridge Philos. Soc. 34: 568–572.

Mott, N. F. 1939. The theory of crystal rectifiers. Proc. R. Soc. Lond. A 171: 27–38.

Niu, Z., Q. Peng, Z. Zhuang, W. He and Y. Li. 2012. Evidence of an oxidative-addition-promoted Pd-leaching mechanism in the suzuki reaction by using a Pd-nanostructure design. Chem. Eur. J. 18: 9813–9817.

Oliver-Messeguer, J., L. Liu, S. García-García, C. Canós-Giménez, I. Domínguez, R. Gavara et al. 2015. Stabilized naked sub-nanometric cu clusters within a polymeric film catalyze C–N, C–C, C–O, C–S, and C–P bond-forming reactions. J. Am. Chem. Soc. 137: 3894–3900.

Peiris, S., J. McMurtrie and H. -Y. Zhu. 2016. Metal nanoparticle photocatalysts: Emerging processes for green organic synthesis. Catal. Sci. Technol. 6: 320–338.

Peng, T., X. Li, K. Li, Z. Nie and W. Tan. 2020. DNA-modulated plasmon resonance: Methods and optical applications. ACS Appl. Mater. Interfaces 12: 14741–14760.

Phan, N. T. S., M. Van Der Sluys and C. W. Jones. 2006. On the nature of the active species in palladium catalyzed Mizoroki–Heck and Suzuki–Miyaura couplings—homogeneous or heterogeneous catalysis, a critical review. Adv. Synth. Catal. 348: 609–679.

Qian, H., M. Zhu, Z. Wu and R. Jin. 2012. Quantum sized gold nanoclusters with atomic precision. Acc. Chem. Res. 45: 1470–1479.

Rampino, L. D. and F. F. Nord. 1941. Preparation of palladium and platinum synthetic high polymer catalysts and the relationship between particle size and rate of hydrogenation. J. Am. Chem. Soc. 63: 2745–2749.

Roduner, E. 2006. Size matters: Why nanomaterials are different. Chem. Soc. Rev. 35: 583–592.

Ruiz Puigdollers, A., P. Schlexer, S. Tosoni and G. Pacchioni. 2017. Increasing oxide reducibility: The role of metal/oxide interfaces in the formation of oxygen vacancies. ACS Catal. 7: 6493–6513.

Sarkar, A. and G. G. Khan. 2019. The formation and detection techniques of oxygen vacancies in titanium oxide-based nanostructures. Nanoscale 11: 3414–3444.

Sarina, S., E. Jaatinen, Q. Xiao, Y. M. Huang, P. Christopher, J. C. Zhao et al. 2017. Photon energy threshold in direct photocatalysis with metal nanoparticles: Key evidence from the action spectrum of the reaction. J. Phys. Chem. Lett. 8: 2526–2534.

Scaiano, J. C. and A. E. Lanterna. 2019. A green road map for heterogeneous photocatalysis. Pure Appl. Chem. 92: 63–73.

Schottky, W. 1938. Halbleitertheorie der Sperrschicht. Naturwissenschaften 26: 843.

Schottky, W. 1939. Zur Halbleitertheorie der Sperrschicht- und Spitzengleichrichter. Z. Physik 113: 367–414.

Shekhar, M., J. Wang, W. -S. Lee, W. D. Williams, S. M. Kim, E. A. Stach et al. 2012. Size and support effects for the water–gas shift catalysis over gold nanoparticles supported on model Al_2O_3 and TiO_2. J. Am. Chem. Soc. 134: 4700–4708.

Song, H., R. M. Rioux, J. D. Hoefelmeyer, R. Komor, K. Niesz, M. Grass et al. 2006. Hydrothermal growth of mesoporous SBA-15 silica in the presence of pvp-stabilized Pt nanoparticles: Synthesis, characterization, and catalytic properties. J. Am. Chem. Soc. 128: 3027–3037.

Tan, H. L., F. F. Abdi and Y. H. Ng. 2019. Heterogeneous photocatalysts: An overview of classic and modern approaches for optical, electronic, and charge dynamics evaluation. Chem. Soc. Rev. 48: 1255–1271.

Taniya, K., H. Jinno, M. Kishida, Y. Ichihashi and S. Nishiyama. 2012. Preparation of sn-modified silica-coated Pt catalysts: A new ptsn bimetallic model catalyst for selective hydrogenation of crotonaldehyde. J. Catal. 288: 84–91.

Thomson, C. G., A. -L. Lee and F. Vilela. 2020. Heterogeneous photocatalysis in flow chemical reactors. Beilstein J. Org. Chem. 16: 1495–1549.

Torres Galvis, H. M., J. H. Bitter, T. Davidian, M. Ruitenbeek, A. I. Dugulan and K. P. de Jong. 2012. Iron particle size effects for direct production of lower olefins from synthesis gas. J. Am. Chem. Soc. 134: 16207–16215.

Toyoshima, R., M. Yoshida, Y. Monya, Y. Kousa, K. Suzuki, H. Abe et al. 2012. *In situ* ambient pressure xps study of CO oxidation reaction on Pd(111) surfaces. J. Phys. Chem. C 116: 18691–18697.

Tsubota, S., T. Nakamura, K. Tanaka and M. Haruta. 1998. Effect of calcination temperature on the catalytic activity of Au colloids mechanically mixed with TiO_2 powder for CO oxidation. Catal. Lett. 56: 131–135.

Wang, Q. and D. Astruc. 2020. State of the art and prospects in metal–organic framework (MOF)-based and MOF-derived nanocatalysis. Chem. Rev. 120: 1438–1511.

Wang, B., J. Durantini, J. Nie, A. E. Lanterna and J. C. Scaiano. 2016. Heterogeneous photocatalytic click chemistry. J. Am. Chem. Soc. 138: 13127–13130.

Wang, S., N. Omidvar, E. Marx and H. Xin. 2018. Coordination numbers for unraveling intrinsic size effects in gold-catalyzed CO oxidation. Phys. Chem. Chem. Phys. 20: 6055–6059.

Wang, B., K. Duke, J. C. Scaiano and A. E. Lanterna. 2019, Cobalt-molybdenum co-catalyst for heterogeneous photocatalytic H-mediated transformations. J. Catal. 379: 33–38.

White, T. P. and K. R. Catchpole. 2012. Plasmon-enhanced internal photoemission for photovoltaics: Theoretical efficiency limits. Appl. Phys. Lett. 101: 073905.

Williams, W. D., M. Shekhar, W. -S. Lee, V. Kispersky, W. N. Delgass, F. H. Ribeiro et al. 2010. Metallic corner atoms in gold clusters supported on rutile are the dominant active site during water–gas shift catalysis. J. Am. Chem. Soc. 132: 14018–14020.

Yan, T., D. W. Redman, W. -Y. Yu, D. W. Flaherty, J. A. Rodriguez and C. B. Mullins. 2012. CO oxidation on inverse Fe_2O_3/Au(111) model catalysts. J. Catal. 294: 216–222.

Zanella, R., A. Sandoval, P. Santiago, V. A. Basiuk and J. M. Saniger. 2006. New preparation method of gold nanoparticles on SiO_2. J. Phys. Chem. B 110: 8559–8565.

Zhang, Z. and J. T. Yates. 2012. Band bending in semiconductors: Chemical and physical consequences at surfaces and interfaces. Chem. Rev. 112: 5520–5551.

Zhang, X., H. Shi and B. -Q. Xu. 2005. Catalysis by gold: Isolated surface Au^{3+} ions are active sites for selective hydrogenation of 1,3-butadiene over Au/ZrO_2 catalysts. Angew. Chem. Int. Ed. 44: 7132–7135.

Zhang, X., A. Du, H. Zhu, J. Jia, J. Wang and X. Ke. 2014. Surface plasmon-enhanced zeolite catalysis under light irradiation and its correlation with molecular polarity of reactants. Chem. Commun. 50: 13893–13895.

Zhang, Y., S. He, W. Guo, Y. Hu, J. Huang, J. R. Mulcahy et al. 2018. Surface-plasmon-driven hot electron photochemistry. Chem. Rev. 118: 2927–2954.

Zhao, J., S. C. Nguyen, R. Ye, B. Ye, H. Weller, G. A. Somorjai et al. 2017. A comparison of photocatalytic activities of gold nanoparticles following plasmonic and interband excitation and a strategy for harnessing interband hot carriers for solution phase photocatalysis. ACS Central Sci. 3: 482–488.

Zhou, Z., S. Kooi, M. Flytzani-Stephanopoulos and H. Saltsburg. 2008. The role of the interface in CO oxidation on Au/CeO_2 multilayer nanotowers. Adv. Funct. Mater. 18: 2801–2807.

Zhou, Z., S. Kooi, M. Flytzani-Stephanopoulos and H. Saltsburg. 2008. The role of the interface in CO oxidation on Au/CeO_2 multilayer nanotowers. Adv. Funct. Mater. 18: 2801–2807.

Chapter 7

Metallic Nanoparticles with Mesoporous Shells

Synthesis, Characterization and Applications

María Jazmín Penelas,[1] *Santiago Poklepovich-Caride*[2] *and Paula C. Angelomé*[2,*]

Introduction

Nanomaterials have received increasing attention since the end of the last century, mainly due to their unusual properties in comparison to bulk materials. The growth of the surface to volume ratio gives place to unique physical, mechanical, optical and magnetic properties, among others (Ozin et al. 2008). During the last decades, many researchers have made great efforts in the field, developing several synthetic procedures to obtain new materials such as quantum dots (Bera et al. 2010), magnetic (Reddy et al. 2012) and metallic nanoparticles (NPs) (Yang et al. 2016), etc. These nanomaterials could also act as building blocks, to produce new and more complex materials or nanocomposites.

Among these new classes of materials, the core-shell NPs have an expanding consideration. These nanoparticles are composed of a core, typically formed of a metallic or semiconductor material, covered with a shell, usually a polymer, metal or a metal oxide (El-Toni et al. 2016, Gawande et al. 2015). Core-shell nanoparticles are multifunctional platforms, where not only the core's composition, shape and size have an important impact on the material properties, but also the shell plays a key role by stabilizing the core and providing new properties to the composite (Hanske et al. 2018).

One of the most usual and useful cores are those based on noble metal NPs. These NPs have great appeal given their unique optical properties related to their Localized Surface Plasmon Resonance (LSPR), which has made them the subject of extensive research. LSPR consist of the coherent collective oscillations of delocalized electrons in the conduction band, and it is the result of the interaction between the noble metal NP and light with a specific frequency (Amendola et al. 2017). This frequency is directly related to the nanoparticle's shape, size and composition and also with the optical properties of the surrounding dielectric medium. In the case of gold, silver, copper and platinum LSPR gives place to strong extinction bands in the UV-Vis and NIR regions of the electromagnetic spectrum (Liz-Marzán 2006, Noguez 2007). Moreover, as a consequence of LSPR excitation, an electromagnetic field enhancement (Alvarez-Puebla et al. 2010), hot electrons

---

[1] Instituto de Nanosistemas, UNSAM, 25 de mayo 1021, San Martín (1650), Buenos Aires, Argentina.
[2] Gerencia Química & Instituto de Nanociencia y Nanotecnología, CAC, CNEA, CONICET, Av. General Paz 1499, San Martín (1650), Buenos Aires, Argentina.
* Corresponding author: angelome@cnea.gov.ar

production (Brongersma et al. 2015) and heat generation (Baffou and Quidant 2013) occur in the vicinity of the NPs. Therefore noble metal NPs have great relevance in areas such as sensing (Falahati et al. 2020) including Surface Enhanced Raman Spectroscopy (Langer et al. 2020, Reguera et al. 2017), biomedicine (Elahi et al. 2018, Kohout et al. 2018, Sharifi et al. 2019, Wu et al. 2019), catalysis (Linic et al. 2015, Liu and Corma 2018, Rodrigues et al. 2019) and sustainable energy (Choi et al. 2018, Jang et al. 2016), mentioning the more frequently reported.

Due to the great potential applications of noble metal NPs, many researchers have developed novel synthetic techniques, largely derived from colloid chemistry, to achieve good control over the NPs composition, shape and size. Since the pioneer Au NPs synthesis by reduction with citrate proposed by (Turkevich et al. 1951), many research groups have worked on the noble metal NPs synthesis. Nowadays a wide variety of procedures is available to obtain either isotropic or anisotropic NPs, such as spherical (Carbó-Argibay and Rodríguez-González 2016), polyhedral (Li et al. 2014, Lohse et al. 2014), rod-like (Vigderman et al. 2012), star-like (Guerrero-Martínez et al. 2011) and hollow nanoparticles (Genç et al. 2017), with different chemical compositions.

Although noble metal NPs are attractive building blocks for novel devices, they often show poor suitability for certain practical applications. The harsh conditions required in some environments can induce aggregation, sintering, thermal degradation, etching or other undesired processes (Davidson et al. 2018, Rai 2019). To avoid these disagreeable issues, a common pathway is to cover the metallic NPs with a shell, to give rise to core-shell particles. The most used kinds of shells are those composed by metal oxides prepared by the sol-gel approach. This technique offers a versatile method to obtain metal oxide shells with diverse composition, shape and thickness (Brinker and Scherer 1990). The most frequently used metal oxides are silica, titania and zirconia, chosen because of their physicochemical properties. In 1996 Liz-Marzán and coworkers developed, in a pioneering work, a method to cover gold and silver NPs with a SiO_2 coating using a slightly modified Stöber procedure (Stöber et al. 1968) to grow the oxide (Liz-Marzán et al. 1996). Over the last two decades, many researchers have focused on improving both the metal oxide shell formation and the stability of the synthesized nanocomposites (El-Toni et al. 2016, Gawande et al. 2015). Among the different kinds of metal oxide shells, it was found that those based on mesoporous (pore diameters between 2 and 50 nm, according to IUPAC) oxides are within the most promising. In 1992 Beck et al. designed and synthesized the first ordered mesoporous silica, called MCM-41. These oxide particles were prepared by the sol-gel method, combined with the use of a cationic surfactant as a supramolecular template, to give rise to a highly ordered hexagonal array of monodisperse pores (Beck et al. 1992). A few years later, Zhao et al. developed a synthetic method to obtain another kind of mesoporous silica: SBA-15 also with highly ordered porosity but with a bigger pore diameter and higher walls thickness (Zhao et al. 1998). The approach was immediately expanded to obtain other ordered mesoporous oxides (Yang et al. 1998), mainly TiO_2 and ZrO_2, using a similar synthetic approach. Since this innovative paper, there have been great improvements in the synthesis mesoporous oxides as powders (Kresge and Roth 2013), nanoparticles (Kankala et al. 2020) and thin films (Innocenzi and Malfatti 2013), achieving remarkable control over composition, pore size and pore ordering (Zou et al. 2020). The high specific surface of mesoporous oxides, combined with their versatile synthetic pathways confer them great potential in areas such as catalysis (Perego and Millini 2013), sensing (Melde and Johnson 2010), surface-enhanced Raman spectroscopy (Innocenzi and Malfatti 2019) and drug delivery (Florek et al. 2017), among others.

Thus, by combining noble metal NPs and Mesoporous Materials (MP), M@MP core-shell nanoparticles can be obtained. Advantages can be taken from both the metallic NPs unique optical and catalytical properties and the mesoporous oxides high and accessible area. Moreover a positive synergy between the two functional components can be expected.

This chapter focuses on the synthesis, characterization and application of these M@MP particles. First, the most common pathways and strategies to synthesize these materials will be described, with an emphasis in the most usual approaches (see Fig. 7.1). Later the wide range of characterization techniques used to understand the chemical and structural properties of these systems will be

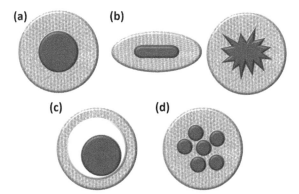

Figure 7.1: Main architecture reported for M@MP particles: (a) spherical cores, (b) anisotropical cores, (c) yolk-shell structure and (d) multiple cores.

described. Finally, the most relevant applications of M@MP in the areas of catalysis, sensing and biomedicine will be presented and discussed, using several key examples from the literature. The chapter will be mainly focus on core-shell nanoparticles containing a single metallic core, either isotropic (Fig. 7.1a) or anisotropic (Fig. 7.1b). However, some examples of yolk-shell (Fig. 7.1c) and multicore particles (Fig. 7.1d) will be also discussed.

Synthesis

The most well-established approach for obtaining M@MP NPs is the wet-chemical synthesis route since it allows nanoscale particle size, morphological control over the process and minimum energy loss compared with that of a top-down approach (Rai 2019). In general, M@MP particles are synthesized by a two-step process in which the metallic core formation first takes place, followed by the generation of the mesostructured oxide shell over the core. Likewise, the synthesis techniques of M@MP particles can be classified into two types: those in which the core particles are synthesized and separately incorporated into the system, and the one-pot synthesis approach, in which the core particles are synthesized *in situ* (Ghosh Chaudhuri and Paria 2012). Usually critical aspects during the synthesis of core-shell particles are: to avoid agglomeration of core particles in the reaction medium, to control the uniformity, thickness and mesostructure of the shell and to suppress the preferential formation of separate particles of the shell material rather than coating the core (Ghosh Chaudhuri and Paria 2012). The synthesis steps involved in the M@MP particles synthesis will be discussed using key successful approaches presented in the literature.

Two main methods for noble metal NPs synthesis are used for the production of M@MP particles: one-pot reduction methods and seed-mediated growth methods. The first family of methods, the one-pot reduction, are the simplest and well-established routes for the synthesis of metallic NPs, like Au, Ag and Pt (Xia et al. 2009). In this approach, a reducing agent is mixed with the metal precursor in the presence of stabilizing agents, such as ligands, polymers or surfactants. The size and shape of the NPs obtained via reduction methods depend on many factors, such as reducing agent, reaction rate, solvent, concentration, temperature and pH. It is important to note that bare metal NPs are unstable in solution hence stabilizing agents for electrostatic and/or steric stabilization are always required (Schärtl 2010). For instance, in their prominent work, Turkevich and co-workers used citrate both as a reducing agent and as a stabilizer during the synthesis of 15 nm Au NPs in an aqueous medium (Turkevich et al. 1951).

The second usual synthesis method is the seed-mediated growth one, which allows obtaining well-defined noble metal NPs of increased sizes and/or with anisotropic shapes (Ahmed et al. 2010, Sau and Rogach 2010). The method involves two steps: the seeds synthesis (generally small NPs with a spherical shape) by conventional chemical reduction method, followed by the addition of the seeds to a growth solution. During the second step, nucleation and growth take place via reduction of the metal ions by mild reducing agents, in the presence of auxiliary species for structure control. For example, Au nanorods (AuNRs) can be obtained by silver-assisted growth from spherical Au seeds, in the presence of cetyltrimethylammonium bromide (CTAB). During the process, a Ag^+-CTAB complex is formed, which is preferentially deposited on certain crystal facets of gold seeds, blocking the diffusion of the Au^+ and leading to a direction-fixed growth (Scarabelli et al. 2015). At the same time, bimetallic NPs (such as Au@Ag) can be obtained by galvanic replacement, possible when the metal in the growth solution has a higher reduction potential compared with the seeds material (Rodríguez-González et al. 2005).

Once the core NPs are formed, the mesoporous oxide shell building normally occurs through a heterogeneous nucleation process. The metal-oxide coating methods depend on the nature of both the noble metal cores and the oxide shells (Kankala et al. 2019). The sol-gel method is the most common method for metal oxide shell synthesis (Li and Zhao 2013). It is highly versatile and yet does not require complicated procedures or instruments for the synthesis of complex structures. The typical sol-gel process involves inorganic precursors (metal salts or alkoxides) that on reaction with water under basic conditions of hydrolysis and condensation, resulting in the formation of a 3D oxide network. The mesoporous texture of the shell is obtained by using surfactant molecules as organic templates, followed by hydrolysis and subsequent condensation of the oxide on them. Unsaturated shell material within the reaction mixture and low temperatures are generally used to favour coating on the core surface by heterogeneous nucleation (Ghosh Chaudhuri and Paria 2012). In general, the overall core-shell particle size and the shell thickness grow with increasing shell precursor concentration or decreasing seed content in the reaction mixture. Longer reaction times and thermal post-treatment at $T > 150°C$ were reported as strategies to improve mechanical and hydrolytic stability of the mesoporous oxide shell (Sanz-Ortiz et al. 2015).

Pioneer work in the field of wet-chemical M@MP nanoparticles synthesis was carried out by (Gorelikov and Matsuura 2008). In this work, CTAB-capped AuNRs were coated with a layer of mesoporous silica ($mSiO_2$) using a single-step procedure, in an aqueous medium containing NaOH, at room temperature. In this system, CTAB molecules stabilize the metallic NPs in the aqueous phase and serve as the organic template for the formation of the mesoporous coating by base-catalyzed hydrolysis of TEOS (see Fig. 7.2). In these reaction conditions, the CTAB surfactant molecules are strongly localized on the NPs surface in comparison with the solution, causing silica/CTAB primary particles to form and aggregate directly on the NPs surface. Thus, mesoporous shell growth is promoted, in comparison with free $mSiO_2$ particles formation. After CTAB removal by washing with methanol, ~ 15 nm thick coatings with poorly aligned pores of ~ 4 nm diameter separated by ~ 2 nm thick silica walls were obtained.

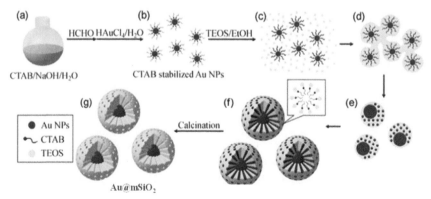

Figure 7.2: Scheme of the different stages of the proposed mechanism for the formation of the Au@mSiO₂ NPs by a one-pot approach: (a) solution of CTAB, NaOH and H₂O; (b) Au NPs stabilized with CTAB; (c) Au NPs assembled by silica micelles; (d) Au NPs coated by a thin SiO₂ layer; (e) Au cores encapsulated in mesostructured silica shells; (f) Au@mSiO₂ NPs with CTAB; (g) Au@mSiO₂ NPs without CTAB. Reproduced with permission from (Chen et al. 2013) Copyright 2013, Springer-Verlag.

The versatility of the methodology can be illustrated with the work of Joo and coworkers, who obtained Pt@mSiO$_2$ NPs from tetradecyltrimethylammonium bromide (TTAB)-stabilized Pt nanocubes (Joo et al. 2009). As earlier, the TTAB surfactant acts both as a stabilizer and as structure-directing template for the mSiO$_2$ shell formation.

Although the direct coating stands out for its simplicity and functionality, for certain applications it may not be the best option. For example, Fang and collaborators synthesized mesoporous silica-coated Pd@Ag nanoplates for which a direct contact of the core material with the external medium was not necessary, whereas a unique mesostructure with large accessible pore volume was a critical aspect (Fang et al. 2012). To this end, a dense silica spacer layer was grown on the Pd@Ag core using a modified Stöber method before mesoporous shell preparation through a two-step process. Cetyltrimethylammonium chloride (CTAC) micelles were first adsorbed on the surface of the Pd@ Ag@SiO$_2$ particles by electrostatic interaction, and then silica was co-assembled with CTAC to form the mesostructure under mild basic conditions (obtained using L-arginine), in the presence of mesitylene as a swelling agent. The as-prepared Pd@Ag@SiO$_2$@mSiO$_2$ particles present pores of ~ 10 nm diameter, much larger than the pores made under the Gorelikov's procedure. It is worth noting that the intermediate non-porous silica layer can also protect the core from etching in harsh conditions.

The direct coating method can be extended to NPs stabilized with other surfactants or capping ligands, if a previous replacement or overcoating with CTAB is performed. For example, Hernández Montoto et al. synthesized gold nanostars (AuNSs) coated with mSiO$_2$ by a multi-step procedure, that involves the synthesis of the AuNS cores via seeded growth using polyvinylpyrrolidone (PVP) as shape-directing agents in N,N-dimethylformamide, followed by the adsorption of the cationic surfactant and the formation of the mesoporous shell (Hernández Montoto et al. 2018). In this case, the mesoporous coating is a critical process because AuNSs can be easily reshaped due to oxidation of tips of gold atoms resulting in a loss of their optical properties. Fine adjustments of the synthesis parameters, such as CTAB/TEOS molar ratio, solvent composition, temperature, removal of oxygen from the reaction medium by purging with argon, etc., were required to successfully obtain the AuNS@mSiO$_2$ nanoparticles containing a multi-branched gold core.

Alternative approaches to obtain branched gold nanoparticles coated with mSiO$_2$ shells were reported by (Sanz-Ortiz et al. 2015) and (Zhao et al. 2017). In both cases, the radially oriented channels of the mesoporous shells were used as templates to grow spikes branching out from the core. The overgrowth of the tips was carried out using a method previously developed to grow surfactant-free NSs that does not require the use of capping agents (Yuan et al. 2012). The same methodology was applied to grow Au spheres, nanorods, and nanotriangles, after CTAB elimination by calcination at mild temperature ($\leq$ 300°C) to prevent reshaping of the anisotropic cores. Sanz-Ortiz et al. also reported that this treatment improves the shells stability against dissolution in water suspension. Furthermore, the accessibility, through the radial pores, of the Au NPs surface is an advantageous feature for catalysis and sensing applications (Sanz-Ortiz et al. 2015). An interesting aspect to mention is that, according to Sanz-Ortiz and Chen (Chen et al. 2019, Sanz-Ortiz et al. 2015) reports, during Ag coating of AuNRs@mSiO$_2$ without thermal treatment, no tips were obtained, but the resulting bimetallic core has a homogeneous silver shell. Chen and coworkers even reported that during dissolution and re-establishment of the Ag layer by redox cyclic reactions, the silica coating remained adhered to the metal core throughout the whole process. This behaviour has been attributed to an 'elastic' feature of the silica shell (Lin et al. 2015).

To obtain this particular mesostructure, slightly different conditions were used than those of Gorelikov's report: the synthesis was performed at relatively high reaction temperature (60°C), in a water/ethanol mixture, adjusted to pH 10 with ammonia, and using mechanical stirring. Zhao and coworkers further reported the addition of n-hexane as a pore-expanding agent, in order to enlarge the shell mesopores. Li et al. described a mechanism for this templating-Stöber process in which the presence of alcohol and ammonia play an active role in the formation of the parallel mesochannels (Li and Zhao 2013). However, according to Sanz-Ortiz et al., the organization of the pores and the quality of the obtained silica are easily affected by synthetic conditions and minor changes in the solvent, CTAB/TEOS molar ratio, temperature, stirring speed, pH, etc., may yield to either disordered pores or pores arranged in parallel or radial distributions (Sanz-Ortiz et al. 2015).

Sometimes however, such high-quality morphological control is not required, but other features of the synthesis, such as the use of inexpensive reagents, higher yields and adequate functionality, are prime aspects. Liong et al., for example, reported the encapsulation of silver nanocrystal obtained through an inexpensive non-hydrolytic method based on the reduction of silver acetate with oleylamine at high temperature (Liong et al. 2009). Since the resultant oleylamine-capped nanocrystals are unstable when transferred from the organic phase to the aqueous phase containing the CTAB, they must be coated with $mSiO_2$ quickly, using TEOS in a basic aqueous solution (pH = 11). With this synthesis, core-shell NPs are obtained, in which multiple silver nanocrystals are embedded at the centre of the spherical $mSiO_2$ structure (Fig. 7.1d). The dissolution of the core Ag NPs during the CTAB removal and the antimicrobial activity of the NPs due to the slow release of Ag^+ ions confirm the accessibility of the resultant mesostructured network.

The search for a simpler, faster and greener preparation of the M@MP NPs than the multistep procedures presented up to now, has motivated the development of one-pot synthesis processes. In this way, Han et al. synthesized M@MP spheres with Ag NPs as a core and an accessible and starburst ordered mesostructured silica as a shell, by an easy-handling one-pot synchronous approach (Han et al. 2011). The method combines several steps into one, including the *in situ* reduction of silver nitrate by formaldehyde, the $mSiO_2$ nanosphere generation by the assembly of CTAB-stabilized silver nanocrystals and silica-CTAB micelles, and the transfer, aggregation and growth of silver NPs to form a final single silver core in an incompact mesostructured SiO_2 framework. It is worth noting that, as in the already mentioned spatially confined Au@AgNR cores growth method, the 'soft' nature of the mesoporous shell in this synthesis medium plays an important role in obtaining the desired architecture (Lin et al. 2015).

Up to now, SiO_2 mesoporous shells were presented, but great efforts have also been invested in developing methods for the synthesis of other functional shells, such as mesoporous TiO_2, SnO_2 or ZrO_2. The main challenge to achieve heterogeneous nucleation and growth of TiO_2 with no aggregation of the core NPs is to control the hydrolysis and condensation kinetics of the highly reactive titanium alkoxide precursors during the sol-gel process (Li and Zhao 2013). A strategy usually adopted is to construct the mesoporous titania coating over a previously SiO_2 encapsulated metallic NPs, followed by a selective etching treatment to dissolve the silica middle layer. This approach renders a special class of core-shell nanostructures with an interstitial hollow space between the movable core and the porous shell, commonly known as nanorattle or yolk-shell NPs (Fig. 7.1c). Using this synthetic route (Zhang et al. 2014), constructed AuNR@void@m TiO_2 NPs applying a general method to coat colloidal particles with amorphous porous titania developed by (Demirörs et al. 2010) followed by the SiO_2 dissolving using NaOH. Later, Zhao et al. introduced the CTAB-templated method for the mesoporous titania direct coating of AuNRs in aqueous solutions (Zhao et al. 2016). Fine control of the reaction rate was achieved by careful addition of titanium diisopropoxide bis(acetylacetonate), a precursor with a slow hydrolysis rate, and the use of an optimized CTAB concentration. Calcination in air improved the crystallinity of the mesoporous TiO_2 shell, while it also resulted in the diminishing of the AuNRs aspect ratio.

Yolk–shell NPs were also obtained with porous ZrO_2, once again recurring to the use of silica sacrificial layers. With this approach, Arnal et al. synthesized a layer of porous zirconia on Au@SiO_2 through the hydrolysis of zirconium tetrabutoxide in the presence of a nonionic surfactant (Lutensol AO5) (Arnal et al. 2006), while Huang et al. followed a similar route using Brij 30 surfactant. In the latter case, the core consisted of thiol-capped sub-10 nm Au NPs (Huang et al. 2009).

Finally, another interesting example that deserves mention is the one reported by Zhou et al., in which the SnO_2 mesoporous coating of CTAB-stabilized noble metal NPs was achieved. Au nanospheres, AuNRs and AuNR@Ag covered with SnO_2 were obtained by a simple one-pot hydrothermal reaction, using Na_2SnO_3 as metal oxide precursor (Zhou et al. 2015). The heating, pH and stannate ion concentration were all adjusted to produce relatively homogeneous shell layers.

Characterization

It is well known that the properties of nanomaterials rely on their physicochemical and structural characteristics, which also define their possible applications (Gawande et al. 2015). In the case of the M@MP particles, a wide variety of parameters of the core, the shell, the particle and the suspension listed in Fig. 7.3, define their properties. Ideally, all these characteristics should be known to fully understand the structure-properties relationships of any newly developed material or synthetic path.

Most characterization techniques used for metallic or oxide NPs can be used to characterize M@MP particles, but one technique may not be sufficient to determine the wide variety of properties derived from the combination of two materials in a single particle (Ghosh Chaudhuri and Paria 2012). Thus, several standard and advanced characterization techniques should be used, whose results are complementary. Here the most common techniques applied to characterize M@MP particles will be described, showing some examples of the kind of results that can be expected.

The most widely used characterization techniques for M@MP particles are electronic microscopies and, among them, Transmission Electron Microscopy (TEM) stands out. Using this technique appropriately, both the core and the shell can be characterized in a single experiment. At a first glance, TEM allows the confirmation of the core-shell structure formation, through a contrast difference, and also gives information about the shell uniformity. Moreover, the overall particle size, core size and shape, shell thickness and shell's pore size can be determined. Some examples of the obtained images are presented in Fig. 7.4a and b, in which metallic AuNSs are seen in black, surrounded by a grey silica shell, whose porosity can be envisioned in higher magnification images. Traditional TEM measurements show a 2D projection of a 3D material, which might lead to erroneous interpretation of the particle morphology, especially in the case of highly anisotropic ones. In this way, Scanning Electron Microscopy (SEM) can provide useful information about the particle shape and the surface structural properties, including the size and order of the pores. SEM usually involves collecting secondary electrons and therefore generating only a surface image (Gawande et al. 2015), but carefully choosing the measurement conditions, both the shell and the core can be seen in a SEM image, as shown in Fig. 7.4c. This figure shows a Field Emission-SEM (FESEM) image in which the shells can be seen along with the AuNS core included within (Hernández Montoto et al. 2018). In this case, the metallic areas are brighter than the silica areas, due to the different interactions with the electrons. Furthermore, Scanning Transmission (STEM) experiments can be performed in adapted SEM or TEM equipment, giving rise to images like the one presented in Fig. 7.4d, in which the metallic core is clearly observed as a bright area, due to the use of dark field detection.

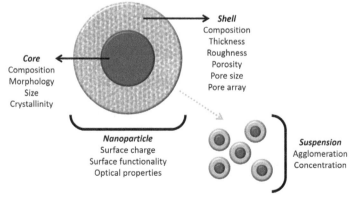

Figure 7.3: Properties and characteristics that can be determined in M@MP nanoparticles.

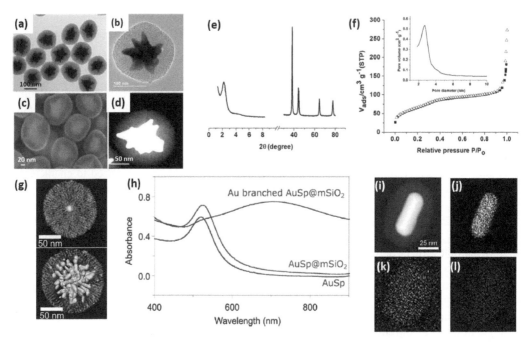

Figure 7.4: **(a)** TEM, **(b)** HR-TEM, **(c)** FE-SEM, and **(d)** DF-STEM images of AuNSs covered with mSiO$_2$. **(e)** PXRD patterns and **(f)** N$_2$ adsorption–desorption isotherms of the same particles. Reprinted with permission from (Hernández Montoto et al. 2018) Copyright 2018 American Chemical Society. **(g)** 3D TEM characterization of hybrid nanostructures containing Au spheres covered with mSiO$_2$ before (top) and after (bottom) core overgrowth, **(h)** evolution of the extinction spectra of 15 nm gold spheres (AuSp) stabilized in 0.1 M CTAB, after mSiO$_2$ shell growth, and after templated tip growth (Au branched AuSp@mSiO$_2$). All spectra were obtained using the same particle concentration. Source reference (Sanz-Ortiz et al. 2015) (pubs.acs.org/doi/10.1021/acsnano.5b04744, further permissions related to the material should be directed to the ACS). **(i)** HAADF-STEM image and HAADF-STEM-EDS mapping images of AuNR@mSiO$_2$ NPs; **(j)** Au mapping; **(k)** Si mapping; **(l)** O mapping. Reprinted with permission from (Liu et al. 2013). Copyright 2013 American Chemical Society.

As a step further to fully understand the tridimensional structure of M@MP NPs, several electron tomography techniques are available nowadays, offering the possibility to obtain accurate reconstructed 3D images of the particles (Bals et al. 2013, Modena et al. 2019). As an example on the use of such techniques, Fig. 7.4g shows the 3D reconstruction of a particle composed by a spherical Au core surrounded by mSiO$_2$ before and after the core overgrowth (Sanz-Ortiz et al. 2015). The images reveal the radial orientation of the pores and demonstrate the radial growth of the Au through them. More details about this technique can be found elsewhere (Sentosun et al. 2015).

Additionally, Energy Dispersive Spectrometry (EDS), which is usually attached as an accessory to electronic microscopes, represents a powerful methodology to determine the distribution of elements within a specified region (Gawande et al. 2015). This technique allows to qualitatively determine the presence of elements in a particular point or within a region, giving rise to composition maps. It can also be used to obtain the tridimensional elemental distribution when coupled with electron tomography measurements. Figure 7.4j, k and l present the elemental mapping results from a AuNR covered with chiral mesoporous SiO$_2$, obtained by HAADF-STEM-EDS. The image of the same NR obtained by HAADF-STEM is shown in Fig. 7.4i. The obtained results clearly reveal that the NPs are composed of gold cores and uniform mSiO$_2$ shells (Liu et al. 2013).

Another relevant technique to study M@MP NPs composition is Electron Energy Loss Spectroscopy (EELS), performed in TEM equipment (Gawande et al. 2015). This technique also gives information about single or small quantities of particles.

Another alternative for chemical characterization is the use of X-Ray Photoelectron Spectroscopy (XPS). This technique gives information about surface elemental composition and oxidation state, up to ~ 10 nm depth. When combined with etching experiments, depth analysis can be performed and the atomic composition of the M@MP NPs can be determined (Ghosh Chaudhuri and Paria 2012). Finally, when organic functional groups are included within the mesoporous shells, Fourier Transformed Infrared Spectroscopy (FTIR) based techniques are useful to determine the presence of their specific vibrational signals. The technique can also be used to follow changes in the surface functionalities during modification reaction steps.

Generally, a limited number of NPs can be visualized by electronic microscopies. To have more detailed information about the particles effective diameter and their aggregation state in solution, scattering techniques must be used. Moreover, these techniques allow following the growth of the mesoporous shells and determine their thickness evolution. Among them, the most commonly utilized are Small Angle X-Ray Scattering (SAXS) and Dynamic Light Scattering (DLS). Usually, no direct information on NPs shape can be obtained from these approaches, and an equivalent diameter, corresponding to that of a sphere behaving the same way as the sample under examination is usually returned as a characteristic size or a hydrodynamic radius (Modena et al. 2019). Visible light-scattering methods are commonly used for the routine analysis of colloidal suspensions, while SAXS requires less available equipment and more detailed data analysis. However, it is important to highlight that SAXS can provide accurate information on the inner diameter, outer diameter and size distribution, and it is usually more reliable for polydispersity determination (Mourdikoudis et al. 2018).

Surface charge, another key parameter with major implications on controlling the NPs interactions and the suspension stability, can be determined in an appropriately adapted DLS equipment. Zeta potential ζ is the key parameter used for such studies, defined as the electric potential difference between the stationary layer of charges surrounding the particles and the solution potential (Modena et al. 2019).

Crystalline structures of M@MP NPs is another key parameter. It can be elucidated using diffraction techniques based on electrons (Selected Area Electron Diffraction, SAED) or X-rays (XRD). SAED is performed in a TEM or HR-TEM and is useful to study the core and the shell crystallinity, in a local manner. XRD also allows the determination of crystalline structure for both the metallic core and the mesoporous shell in one diffraction experiment. XRD results are more representative since the data can be collected using a bigger quantity or even the whole sample. When one or both of the components of the M@MP particle are crystalline, the corresponding Bragg diffraction peaks appear in the diffractogram, allowing identification of the crystalline phase and quantitative assessment of mean crystallite size. Moreover, if the MP shell presents an ordered array of mesopores, a signal at low angles is expected, due to such pore ordering. Figure 7.4e shows an example of the use of the XRD technique to obtain information about AuNSs covered with ordered $mSiO_2$. The small-angle region exhibits a peak assigned to the hexagonal unit cell of the porous shell structure and in the wide-angle pattern, four diffraction peaks are present, attributed to the face-centred cubic gold lattice (Hernández Montoto et al. 2018).

For some applications, it is necessary to know not only the pore diameter and array of the porous shell but to have a proper determination of porosity and surface area. Gas adsorption/desorption techniques equipped with automated analyzers are the ones preferred for such determinations, usually utilizing N_2 as analysis gas. The Brunauer–Emmett–Teller (BET) method is commonly used for determination of the specific surface area of M@MP particles, whereas the Barrett–Joyner–Halenda (BJH) method can be employed to determine the pore volume and pore size of the materials (Gawande et al. 2015). As an example of the use of such techniques, Fig. 7.4f presents the results of N_2 adsorption–desorption isotherms of AuNSs covered with $mSiO_2$. A Type IV curve, typical for mesoporous materials, is clearly observed and the BET surface area and the average pore size can be obtained from it (Hernández Montoto et al. 2018).

Finally, in the case of Au, Ag and other plasmonic metallic cores, UV-vis spectroscopy can give valuable information about the particles optical properties. Moreover, the technique is also useful to follow the formation of the shell and to monitor changes in the metallic core size and shape (Ghosh Chaudhuri and Paria 2012). For such a determination, it is important to keep in mind that care should be taken in the spectra interpretation, due to scattering, particularly when the particles present diameters over 50 nm. A good example of the use of UV-vis spectroscopy to follow the formation and changes of M@MP NPs is presented in Fig. 7.4h, in which the spectrum of a Au spherical NP is compared with the one obtained after the silica shell growth. After this process, the LSPR band shift towards higher wavelengths is observed, due to the increase in the refractive index around the metallic surface. Moreover, when an overgrowth of the gold NPs is performed, a redshift and broadening of the LSPR band is observed, associated with the changes in the core shape (Sanz-Ortiz et al. 2015).

Applications

Catalysis

Metallic NPs have demonstrated to present, in comparison with bulk metals, improved efficacies for the catalysis of a wide variety of chemical reactions. This outstanding performance can be attributed to their high surface area, and also to the incremented reactivity of curved metal surfaces. In addition, metallic NPs can be designed to expose the crystal faces that present higher activities, an approach that is not easily achievable for bulk materials. As a consequence, a wide variety of catalytic applications of metallic NPs either in suspension or deposited on different supports have been presented to date (Liu and Corma 2018, Losch et al. 2019, Rodrigues et al. 2019).

Covering the metallic NPs with a shell can help to increment their long-term stability, preventing colloidal aggregation in solution and also fusion during high-temperature reactions. Thus, these shells can not only help to stabilize the particles in well-known catalytic applications but could also help to extend their use for reactions that work in extreme conditions. Additionally, the shells can help to improve the reusability of the catalysts, with the consequent environmental and economic benefits (Gawande et al. 2015, Lukosi et al. 2016, Wei et al. 2011). Among the available shell materials, mesoporous oxides stand out because they combine a high surface area and an adequate pore size to allow the diffusion of reactants and products. To date, several examples of the use of such composite materials have been presented and reviewed in the literature (Davidson et al. 2018, Hanske et al. 2018, Li and Tang 2014, Liu et al. 2012).

The improved performance of M@MP particles has been demonstrated in several works. An interesting example of such improvement was presented by Joo et al. using cubic/cuboctahedral Pt NPs (Fig. 7.5a) (Joo et al. 2009). The authors compared the performance for CO oxidation to CO_2 using these particles stabilized with a capping agent (TTAB) or covered with a $mSiO_2$ shell. As can be observed in Fig. 7.5b, the activity of the Pt@$mSiO_2$ catalyst was as high as that of TTAB-capped Pt NPs, indicating that the silica shell was porous enough to provide access to the metallic cores. In addition, the thermal stability of the $mSiO_2$ protected particles results much higher, allowing the study of catalytic reactions or surface phenomenon taking place on metallic surfaces at temperatures as high as 330°C.

Another example of the performance increment due to the mesoporous shell inclusion was presented by (Ndokoye et al. 2016). These authors synthesized AuNSs covered with $mSiO_2$ and applied them as catalysts for the well-known 4-nitrophenol reduction reaction (Hervés et al. 2012, Zhao et al. 2015). Reusability tests, that imply repeating the same reaction after the centrifugation and redispersion of the catalysts, demonstrated that the AuNS@$mSiO_2$ particles maintained the catalytic performance while AuNSs did not, due to the particle agglomeration.

Additionally, the shells can enhance the catalytic activity of the encapsulated metals. An example of such synergy was presented by (Zhou et al. 2015). In this work, Au NRs covered with $mSnO_2$ (Fig. 7.5c) were tested for the 4-nitrophenol reduction reaction. The incorporation of the SnO_2 layer resulted in a significantly enhanced catalytic activity in comparison with the naked Au rods, as shown in Fig. 7.5d. The effect was seen both in terms of reaction rate and in the shortening of the induction period. In the same work, the performance of Au spheres covered with $mSnO_2$ or $mSiO_2$ was compared. The $mSiO_2$ helped to reduce the induction time in comparison with the

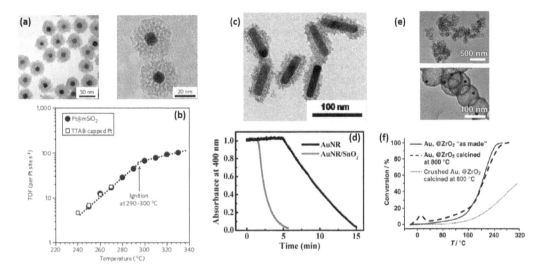

Figure 7.5: (a) TEM images of Pt nanocubes@mSiO$_2$ treated at 350°C, **(b)** CO oxidation activity of these particles compared with the one found on Pt nanoparticles. Reproduced with permission from (Joo et al. 2009), copyright 2008 Nature Publishing Group. **(c)** TEM image of AuNR@mSnO$_2$, **(d)** normalized absorbance of 4-nitrophenol absorption band at 400 nm as a function of time, in the presence of naked and mSnO$_2$ covered AuNRs. Reproduced with permission from (Zhou et al. 2015), copyright 2015 American Chemical Society. **(e)** TEM image of Au-mZrO$_2$ nanorattles, **(f)** CO oxidation catalytic performance of this sample, compared with the one treated at 800°C and the one treated at 800°C crushed, as indicated in the labels. Reproduced with permission from (Arnal et al. 2006), copyright 2006 Wiley.

pure metallic particles, but the changes in the reaction constant were minor. The mSnO$_2$ layer, on the other hand, generated a higher decrease in the induction time and a notable increase in the reaction constant. Since pure SnO$_2$ nanoparticles did not demonstrate to have any catalytic activity, the enhanced activity was attributed to a synergistic effect between the oxide and the metal. In particular, the authors explained the effect taking into account that SnO$_2$ is a semiconductor oxide, while SiO$_2$ is an insulator. As a semiconductor, SnO$_2$ forms a junction with the metal and thus allows the electron distribution within the material, increasing the chances for 4-nitrophenol to capture the electrons to be reduced.

Another very popular alternative to increase the catalytic activity of M@MP particles is the use of the so-called yolk-shell or nanorattles structures. In these kinds of particles, the metal core is not covalently connected to the shell, being able to move freely and allowing a higher metallic surface to be available for reaction. Additionally, the stabilization and high accessibility provided by the MPs shell is maintained (El-Toni et al. 2016, Hanske et al. 2018, Li and Tang 2014). The Schüth group was a pioneer in the production of yolk-shell particles and their application as catalysts. In one of their works, Au nanospheres encapsulated in a mZrO$_2$ shell (Fig. 7.5e), obtained after the dissolution of a sacrificial SiO$_2$ shell, were used to catalyze the oxidation of CO to CO$_2$ (Arnal et al. 2006). The performance of the encapsulated Au NPs was compared with an identical sample crushed and treated at the same temperature (800°C). The catalytic activity of the encapsulated particles was clearly superior, as can be seen in Fig. 7.5f, which presents the % of CO conversion as a function of temperature. The results can be explained taking into account that, due to the thermal treatment, substantial growth of the gold particles was observed for the crushed sample. The nanorattle particles, on the other hand, remained stable, proving again the stabilization effect offered by the shell.

Sensing

As it was mentioned above, noble metal NPs exhibit unique optical properties owing to the LSPR. This phenomenon provides these particles the potential to be used as sensors due to LSPR frequency sensitivity to changes in the NPs environment and the near-field enhancement in the NPs surface upon illumination, among other features (Alvarez-Puebla et al. 2010, Liz-Marzán 2006). Adding a MPs layer over the metallic NPs provides chemical stability to the system, especially when harsh sensing conditions are required (El-Toni et al. 2016, Hanske et al. 2018). Moreover, the porous layer could also improve the sensing performance in numerous ways such as facilitating the functionalization of the NPs or achieving high loads of a given analyte into the pores (Hanske et al. 2018, Talebzadeh et al. 2019).

One of the simplest types of sensors that can be prepared with metallic NPs are the LSPR based ones (Mayer and Hafner 2011, Sepúlveda et al. 2009). LSPR absorption band position shows a dependence not only with the shape, size and composition of the nanoparticle but also with the optical properties of the surrounding medium (Liz-Marzán 2006, Noguez 2007). Hence, M@MP particles can be used to detect changes in the refractive index in the vicinity of the metallic core. In this sense, Wu and coworkers demonstrated the high sensitivity of a LSPR sensor composed by AuNRs covered with a $mSiO_2$ shell. The plasmon shift of AuNR@$mSiO_2$ and AuNR@CTAB in water-diethylene glycol liquid mixtures with variable volume ratio (to give rise to different n) were measured (Wu and Xu 2009). As it can be observed in Fig. 7.6a, in the case of AuNR@CTAB a nonlinear response was obtained, that can be explained by the degradation of the CTAB bilayer due to the presence of organic solvents. Besides, a linear response was achieved with the AuNR@ $mSiO_2$NPs, yielding a sensitivity of 325 nm per Refractive Index Units (RIU) and proving that the mesoporous shell allows different liquid mixtures to diffuse and interact with the metallic core.

It is well known that silver LSPR frequency shows more sensitivity than the gold one against changes in the optical properties of the surrounding medium, thus the use of a silver coating often improves the performance of the sensors (Noguez 2007). In this regard, Chen and collaborators synthesized a core-shell system with AuNRs as a core with a first shell of silver and a second $mSiO_2$ shell (AuNR@Ag@$mSiO_2$) (Chen et al. 2019). The authors developed a sensing method involving two chemical reactions to determine the presence of Ascorbic Acid (AA). The procedure involves the addition of a certain amount of a strong oxidizing agent (permanganate, MnO_4^-) followed by the incorporation of different concentrations of AA. First, MnO_4^- produces the fast oxidation of

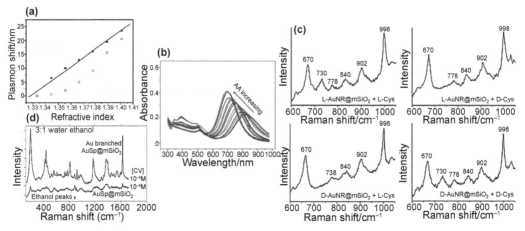

Figure 7.6: (a) Dependence of the longitudinal plasmon shifts of AuNR@CTAB (hollow circles) and AuNR@$mSiO_2$ (black circles) with the refractive index of different liquid mixtures. Reproduced with permission from (Wu and Xu 2009) Copyright 2009, American Chemical Society. (b) UV-Vis spectra of AuNR@Ag@$mSiO_2$ sensing platform with different AA concentrations. Reproduced from (Chen et al. 2019) with permission from the Centre National de la Recherche Scientifique and The Royal Society of Chemistry. (c) SERS spectra of D or L-Cys loaded onto L or D-AuNR@$mSiO_2$. Reprinted with permission from (Liu et al. 2013). Copyright 2013 American Chemical Society (d) SERS spectra of a 10^{-6} M crystal violet solution loaded into AuSp@$mSiO_2$ and AuNS@$mSiO_2$ particles. Source reference (Sanz-Ortiz et al. 2015) (pubs.acs.org/ doi/10.1021/acsnano.5b04744, further permissions related to the material should be directed to the ACS).

the silver shell, resulting in the redshift of the LSPR band. Later the AA reduces the silver ions and regrowth of the silver shell was observed, with the consequent LSPR band blue-shift. In Fig. 7.6b, this blue shift dependence on the AA concentration is presented.

Other kinds of sensors based on noble metal NPs are those in which the signal intensity is increased due to plasmonic features. Surface-Enhanced Raman spectroscopy (SERS) is a powerful technique to measure vibrational spectra of molecules adsorbed on plasmonic surfaces; which gives an improved detection limit compared to traditional Raman spectroscopy. The overall signal enhancement is due to a combination of an electromagnetic factor associated with the enhancement of the near-field in the vicinity of the NPs on illumination and a chemical factor due to the metal-target molecule bonding. Thus, the SERS enhancement mechanism requires the proximity of the target molecule and the metallic surface (Langer et al. 2020). Additionally, it is well-known that the anisotropic shapes such as rod-like or star-like NPs give a greater SERS signal enhancement, so this kind of architecture is usually preferred (Reguera et al. 2017). As discussed earlier, Sanz-Ortiz and coworkers synthesized AuNS@mSiO$_2$ particles (see Fig. 7.4g–h) with high connectivity of the radial pores (Sanz-Ortiz et al. 2015). This feature not only allows the controlled growth of the tips but also provides the ability for an analyte to interact with the metal core, by diffusing from the solution through the porous channels. In Fig.7.6d, the improvement in the crystal violet SERS sensing performance after the gold tips were grown is presented, compared with the system with spherical cores. This behaviour is attributed to the extraordinary enhancement of the near-field on the surface of the tips (Langer et al. 2020, Reguera et al. 2017). Additionally, these NPs were stable for long periods if stored correctly in a solution or as a powder, making them ideal for analytical applications. With a similar approach, Zou et al. developed mSiO$_2$ coated multi-branched gold NPs to quantify 4-bromomethcathinone, achieving a low detection limit and short measurement times (Zou et al. 2019).

The above-mentioned works have used the mSiO$_2$ shell as a protective layer and also as a template for branching gold cores. Moreover, the mesoporous shell can act as a substrate for anchoring functional groups. In this way, Carrasco et al. used the same strategy to synthesize AuNS@mSiO$_2$ with subsequent polymerization of a Molecularly Imprinted Polymer (MIP) (Chen et al. 2016) using the antimicrobial enrofloxacin (ENRO) as a template (Carrasco et al. 2016). The authors used ENRO as a target probe to test the AuNS@mSiO$_2$@MIP SERS sensing performance. A control probe in which the polymer was prepared using Boc-L-Phenylalanine as a template (CIP) was also prepared. A remarkably higher ENRO signal intensity was achieved when using AuNS@mSiO$_2$@MIP instead of AuNS@mSiO$_2$@CIP denoting not only the platform ability to recognize ENRO, but also the proximity of the molecule to the gold tips. Furthermore, the platform showed high selectivity for ENRO over other antimicrobials.

In a very interesting approach, Liu and co-workers developed a core-shell platform with potential application as chiral molecules sensor (Liu et al. 2013). The authors covered AuNRs with a mSiO$_2$ shell obtained using a chiral surfactant derived from phenylalanine (Phe) as template (see Fig. 7.4i–l). Later they evaluated the performance of the obtained NPs as a SERS sensor of Cysteine (Cys). As can be observed in Fig. 7.6c, when the shell-template and Cys molecules are of the same configuration (L-Cys and C16-L-Phe or D-Cys and C16-D-Phe) a new band at 730 cm^{-1} appears in the SERS spectra. This new band is related to a conformational change of the Cys molecule when it strongly interacts with the shell-template molecule. Moreover, the authors developed a semi-quantitative determination of the L-Cys percentage in a solution with a mixture of enantiomers.

Metal Enhancement Fluorescence (MEF) is another sensing technique obtained from the remarkable plasmonic properties of metallic NPs (Jeong et al. 2018, Li et al. 2017). Unlike SERS, here the contact between the analyte and the surface could promote signal loss, due to quenching mechanisms. Besides, the analyte must be close to the metal surface giving a plasmon coupling to escalate the fluorescence enhancement. Thus, the control over the distance between the analyte and the metal surface becomes decisive (Li et al. 2017). There are a few reported works in which M@MP platforms were used for sensing by MEF. For example, Yang et al. covered silver NPs with a mSiO$_2$ shell, using a dense silica layer with controlled thickness as a spacer located between the core and the porous layer (Yang et al. 2011). The authors loaded a fluorophore on the pores and studied the fluorescence enhancement capability of the system. Abadeer and coworkers prepared AuNR@mSiO$_2$NPs functionalized with an IR dye to study the dependence of the fluorescence enhancement on the distance between the dye and the metal surface (Abadeer et al. 2014). Luo et al. also synthesized AuNR@mSiO$_2$ embellished with a porphyrin derivative molecule and Cu^{2+} for using in *in vitro* sensing of H$_2$S (Luo et al. 2019). However, as of now the main benefits of the presence of a mesoporous shell instead of a dense one have not yet been exploited. Thus the development of MEF sensors with MPs shell with great pores connectivity and high achievable loadings is still a challenge in the field.

Nanotheranostics

Over the past few decades, nanotechnology has provided powerful tools for developing biocompatible multifunctional platforms for diagnosis and treatment in advanced biomedical applications, especially in cancer biology (Wagner et al. 2006). The most attractive feature of the use of NPs in biomedicine is the possibility of packing together several payloads to enable a comprehensive profile (area, type, stage) of a disease state, and simultaneously improve the outcome of that state (Farokhzad and Langer 2009, Jokerst and Gambhir 2011, Luo et al. 2011). In particular, the use of NPs for controlled drug delivery has sparked great interest due to their potential to enhance the efficacy and reduce the toxic side effects of the free-drug treatments. This NPs based approach allows the release of a high local concentration of anticancer or antibiotic active agents only where and when they are needed. Moreover, the integration of therapeutic agents and bioimaging, targeting and ablation components within one nanostructure gives rise to materials with concurrent and complementary diagnostic and therapeutic capabilities, known as 'theranostic' systems (Conde et al. 2014). The application of this class of nanosystems is promising for the development of synergistic chemo-photothermal combined treatments and specific imaging-guided cancer therapies (Huang et al. 2012).

M@MP particles occupy an outstanding place in the design and synthesis of these classes of nanomaterials since they allow the synergistic combination of different functional units in their structure at the nanoscale size (Girija and Balasubramanian 2019). On the one hand, the tunable plasmonic core can act as an imaging tag, thanks to the unique way in which it interacts with sound, visible light and other electromagnetic fields. Thus, the core can be used as an effective radiosensitizer, by enhancing the efficacy of physical radiation on tumour cells (Zhao et al. 2016), and as a multipurpose local nanoheater, by a photothermal effect (Melancon et al. 2011). AuNRs and other anisotropic Au NPs, such as stars, cages, etc., are usually preferred for therapeutic applications (Zhou et al. 2017). This is due to the combination of two features: their low toxicity and their optical properties. In particular, for these kinds of anisotropic NPs, the LSPR band is located in the near-infrared region (NIR, 700–1,000 nm range), a zone of low light absorption/scattering by biological tissues (i.e., haemoglobin and water), known as 'biological window'.

The imaging and thermal capabilities due to the photophysical properties of the metallic core can be successfully combined with the multifunctional features of the MPs oxide shell. The external mesoporous layer can provide protection and acts as a reservoir with high capacity, that allows incorporating drugs, dyes, contrast agents, etc. (Li et al. 2012). Silica has a prominent role compared to other oxides for nanotheranostic devices development, given its biocompatibility and the existence of well-established mild strategies for modifying the shell cavities and the external surface with a rich diversity of suitable (bio)organic functional moieties. The earlier characteristics, added to the relative simplicity of its synthesis, have determined that the most reported types of M@MP for theranostics applications are the ones with the AuNR@mSiO$_2$ structure. Figure 7.7 summarizes the therapeutic and diagnostic approaches based on these NPs that have been presented so far in the literature. Several key examples will be discussed next.

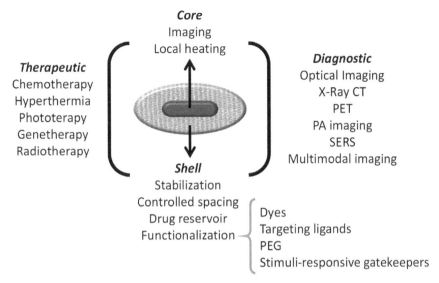

Figure 7.7: Scheme summarizing the main theranostic properties reported for M@MP NPs.

The development of contrast agents and imaging probes forms an essential research field in biomedical imaging since it improves the analysis of biological information and clinical diagnosis. Combining imaging techniques could overcome the possible limitations of using a single modality and thus improve diagnostic accuracy (Xu et al. 2018). X-ray Computed Tomography (CT) is a widely used anatomical technique based on the signal attenuation caused by a contrast agent accumulated in biological soft tissues. Gold has a higher atomic number and absorption coefficient than commercially available iodine-based contrast agents and thus, allows achieving better contrast and less bone and tissue interference, with a lower X-ray dose (Jackson et al. 2010). On the other hand, the use of NIR chromophores for *in vivo* optical imaging minimizes intrinsic background interference and provides relatively deep optical penetration (Lee et al. 2009). In this context, Luo et al. designed a dual-mode contrast agent for CT and NIR fluorescence imaging based on AuNR@mSiO$_2$ nanostructures loaded with the organic NIR dye indocyanine green (ICG) (Luo et al. 2011). The mSiO$_2$ shell provides spatially well-separated ICG molecules, preventing their aggregation and thus decreasing self-quenching fluorescence. In addition, the encapsulation of the ICG within the silica protects it against the external bioenvironment and prolongs its vascular half-life. This approach allowed observing ICG NIR fluorescence from nanocomposites up to 12 hours post intratumoral injection. In order to avoid fluorescent quenching due to molecule-metal contact, the chromophore was added during synthesis after the first TEOS injection. *In vivo* imaging results using these nanocomposites show the dual-mode imaging capability (CT and NIR fluorescence), which can be exploited to obtain quantifiable information of contrast agent accumulation, especially valuable in a preclinical setting.

The prospects of AuNR@mSiO$_2$ NPs for biomedical applications can be expanded by loading therapeutic components into the shell and exploiting the photothermal capabilities of the core (Fig. 7.8A). With this premise, Zhang et al. designed a theranostic nano-platform based on AuNR@mSiO$_2$ particles loaded with the well-known anticancer drug doxorubicin (DOX), and studied their performance in human lung cancer cells (A549) (Zhang et al. 2012). The Two-Photon Imaging (TPI) modality of the AuNRs was used to visualize the intracellular co-localization of the particles with DOX and some cellular compartments. By varying the concentrations of DOX and NPs, 14.5% of loading content was achieved in this system. The authors reported that about 5–10 times more DOX could be internalized when it is loaded in the nanocarrier. They attribute such behaviour to the strong interaction between the positive surface charge and the negatively charged cell membrane (Fig. 7.8C). Using low-intensity NIR irradiation, the release of stored drug was induced by

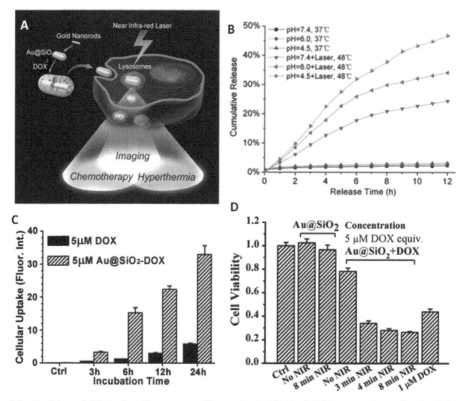

Figure 7.8: (A) Schematic illustration of mesoporous silica-coated Au NR (Au@SiO$_2$) as a novel multifunctional theranostic platform for cancer treatment. (B) DOX release profiles from Au@SiO$_2$–DOX with and without NIR laser irradiation at different pHs. (C) Cellular uptake of DOX and Au@SiO$_2$–DOX quantified by flow cytometry. (D) Differences in the viability of A549 cells irradiated by NIR laser for 8 minutes (Au@SiO$_2$) and 3, 4, and 8 minutes (Au@SiO$_2$–DOX), determined by CCK-8 assay. Reproduced with permission from (Zhang et al. 2012) Copyright 2012, John Wiley and Sons.

dissociation of the cationic DOX – surface silanolate groups interactions in both *in vitro* and *in vivo* assays (Fig. 7.8B, D). Excitation at a higher power, on the other hand, caused cell death by heat, through the photothermal therapy known as hyperthermia.

This scheme can be adapted to improve loading efficiency, surface features and to explore more powerful imaging modalities. For instance, Xu et al. obtained AuNR@mSiO$_2$ with enlarged pore size in the range of 4–8 nm to achieve 40,9% w/w of DOX content (Xu et al. 2018). The NPs were PEGylated to increase their colloidal stability, to reduce the adsorption of serum proteins in the physiological medium and to enhance the *in vivo* passive tumour targeting efficacy. Significantly enhanced chemo-photothermal combination therapy was observed due to photothermal effect and NIR light-triggered drug release of DOX at the tumour site. Positron Emission Tomography (PET) is a highly sensitive imaging method with very low background noise based on the detection of the gamma rays that are emitted by conjugated radioisotopes (Sun et al. 2015). The NPs described above were also radiolabelled with ^{89}Zr for PET imaging *in vivo* study by a chelator-free method, using the surface silanolate groups as O donors. Meanwhile, photoacoustic imaging, by converting laser into ultrasound (US) waves emission, combines the contrast of optical imaging with the high resolution and deep penetration of US imaging (Liu et al. 2016). This technique was used as a complementary *in vivo* imaging modality for PET taking advantage of the high absorption in the NIR range of the nanoplatforms. PET imaging showed that ^{89}Zr-labelled AuNR@mSiO$_2$(DOX)-PEG can passively target to the 4T1 murine breast cancer-bearing mice with high efficiency, based on enhanced permeability and retention effect.

In addition to cytotoxic drugs for chemotherapy, the mesoporous layer of M@mSiO$_2$ nanostructures can also be loaded with photosensitizers, for applications in photodynamic therapy (Khlebtsov et al. 2011). This therapy technique implies the excitation, using a tuned laser, of a photosensitizer that later transfers the energy to the nearby tissue oxygen. This process generates Reactive Oxygen Species (ROS) that can oxidize the cell's components and cause irreversible damage to tumour cells. In this frame, Zhao et al. designed a multifunctional nanoparticle composed of a AuNR core and a porphyrin-doped mSiO$_2$ shell (Zhao et al. 2010). The mesoporous layer allows the photosensitizers to be spatially well separated from each other, facilitating the generation and release of singlet oxygen. Moreover, the direct contact and energy transfer process from porphyrins (HP) to Au, which would quench the HP emission and reduce singlet oxygen generation efficiency, is prevented. The NPs display improved photosensitization effectiveness when irradiated at 633 nm compared with free HP. In addition, due to the AuNR core, the nanoparticles showed two-photon luminescence efficiency that allows optical monitoring of treated cells.

Nanocomposite based on mSiO$_2$-coated AuNRs can also be explored as non-viral vectors for gene therapy. This treatment modality consists of the delivery of small interfering RNA (siRNA) to silence specific proteins involved in apoptotic or drug resistance pathways (Kanasty et al. 2013). For example, Ni and coworkers developed AuNR@mSiO$_2$ particles using a large-pore mesoporous organosilica shell for gene and photothermal cooperative therapy of triple-negative breast cancer cells (TNBC) (Ni et al. 2017). The large and radially aligned pores guarantee capacity for large biomolecules, like nucleic acids, while ensuring gene protection. Thus, functional siRNA can be efficiently delivered into TNBC cells by the composite nanoparticle, causing much higher cell apoptosis by knocking down specific proteins. Simultaneously, the LSPR of the embedded AuNR is reserved for photothermal conversion. The results demonstrate that cooperative treatment could induce a 15-fold mice tumour inhibition rate than sole gene or photothermal therapy modality.

For both diagnostic and therapeutic purposes, selective accumulation inside the tissue of interest and greater cellular binding and uptake efficiency are desirable. This can be achieved by the immobilization of active targeting moieties on the NPs surface. A generalized strategy to accomplish such an effect is using ligands that present high affinity for the receptors that are typically overexpressed by target cells (Bazak et al. 2015), such as antibodies, aptamers, peptides or low molecular weight molecules. These ligands can act as transmembrane carriers, allowing the NPs internalization by receptor-mediated endocytosis. This strategy has been implemented on DOX loaded AuNR@mSiO$_2$ NPs to improve selectivity for chemo-photothermal therapy by (Liu et al. 2015). These authors labelled the mesoporous surface with folate moieties to specifically target KB cancer cells. In the same way, Shen et al. conjugated Arginine–glycine–aspartic acid (RGD) peptides on the terminal groups of PEG in order to target $\alpha_v\beta_3$ integrin extracellular receptors on several cancer cell-lines (Shen et al. 2013). The same recognition system was employed by Zhao et al. to obtain AuNRs@mSiO$_2$–RGD that could be internalized by TNBC by targeting tumour-specific antigens, leading to radiosensitization in combination with megavoltage RT both *in vitro* and *in vivo* (Zhao et al. 2016). The use of active targeting usually results in an increased cellular uptake level compared to the unconjugated NPs counterpart.

Although targeted NPs can deliver drugs to specific sites, another important property to be considered is the controlled release. With a focus on preventing drug leakage and optimizing the cargo release performance, active gating moieties can be incorporated into the porous host structure. This approach gives rise to hybrid nanosystems with the stimuli-responsive controlled release (Aznar et al. 2016). The 'on-command' release can be triggered by particular physical or biochemical internal conditions in the diseased tissue, such as pH, temperature or high concentration of certain intracellular biomolecule. For instance, Zong et al. developed an endogenous-responsive drug nanocarrier based on AuNR@Ag@mSiO$_2$ particles, in which the intracellular redox environment triggered drug delivery (Zong et al. 2013). The high level of cytosolic glutathione (GSH) compared with the extracellular medium (1–11 mM vs. 10 μM, respectively) induces a high reducing potential inside the cell. In the designed system, DOX is linked to the nanoparticle through disulphide bonds, which can be cleaved in the presence of GSH once the nanocarriers enter the cancer cells. On the other hand, disulphide bonds remain quite stable in the oxidative extracellular environment, avoiding premature DOX release. Furthermore, in this work the AuNR@Ag were tagged with a Raman reporter molecule DTNB, to obtain a SERS active cores which provide a highly distinguishable and narrow labelling signal of the nanocarrier. The silica layer effectively avoids the contact between DOX and the core and, as a result, its Raman signal is not enhanced. Thus, the dynamic process of the drug release inside living cells can be monitored by tracking the location of the nanoparticle by SERS signals and that of the drug through fluorescence.

Alternatively, the release in M@MP based-nanosystem can be induced remotely; for example, by externally applying a magnetic field, ultrasound or light of an adequate wavelength. Among various controlled release systems, light-triggered drug release has gained considerable attention owing to its non-invasiveness, remote responsiveness and controllable operation (Girija and Balasubramanian 2019). As one of the most studied strategies, thermo-sensitive gatekeepers can be attached to the pore opening of mesoporous shells loaded with chemotherapeutic agents. Then, by applying NIR irradiation, the local temperature can be raised by means of the photothermal conversion of the Au core, in order to trigger a specific behaviour in the capping group that unblocks the pores and allows cargo release. Following this plan, diverse smart sealing systems based on AuNR@mSiO$_2$ NPs have been designed for the thermal release of DOX. Some of the engineered strategies include the dehybridization of double-stranded oligonucleotides (Chang et al. 2012), phase transitions of biocompatible low melting point molecules like paraffin or 1-tetradecanol (Hernández Montoto et al. 2018, Liu et al. 2015), and the dissociation of supramolecular nanovalves (Li et al. 2014). One popular approach to control the release is the use of thermoresponsive polymers like poly(N-isopropylacrylamide) (PNIPAM) and derivatives, which can undergo a reversible temperature-dependent volume phase transition (VPTT) from a hydrophilic water-swollen state to a hydrophobic globular state (Bordat et al. 2019). For instance, Tang et al. coated AuNR@mSiO$_2$ NPs with poly(N-isopropylacrylamide-co-N-hydroxymethyl acrylamide), with tunable VPTT, obtained by varying the copolymer composition (Tang et al. 2012). Then the DOX was loaded at pH 8, near the pKa value of the drug (8.3), to minimize the electrostatic interactions while the molecules diffuse through the polymeric domain to adsorb into the silica mesostructure. Thermosensitive release experiments showed that the DOX released from the composite nanoparticles follows a distance-limiting mechanism. Therefore, the drug diffuses slowly through the polymeric layer when it is swollen at T < VPTT and more rapidly when it is shrunk at T > VPTT. *In vivo* assays using DOX-loaded composite nanoparticles after NIR laser irradiation suggest a synergistic effect of the chemo-photothermal therapy in cell viability reduction.

Conclusions and Future Perspectives

In this chapter, the current synthesis methods for M@MP nanoparticles have been reviewed, along with the most usual characterization techniques that allow understanding their structural and physicochemical properties. In addition, several applications of the composite NPs have been presented.

Some perspectives can be envisioned from the subjects that have been discussed. First, it is evident that most synthesis and applications so far are focused on Au NPs covered by mSiO$_2$. Thus, it is clear that incrementing the variety of cores and shells materials represents a synthetic challenge but could derive in uses that have not been tested so far. In the same direction, most of the published examples are focused on CTAB templated MPs shells. The increment of the pore size of the porous layer is also hard to achieve but could have an impact over the applications since bigger pores can incorporate a wider variety of cargoes. Moreover, the generation of architectures with incremented complexity could also impact on the future uses of M@MP particles. However, care should be taken in over-incrementing the synthetic steps, a fact that could result in difficult to scale-up processes (Cheng et al. 2012).

It is also interesting to note that, in literature, there are several examples on the uses of metallic core–dense oxide shell particles (Ghosh Chaudhuri and Paria 2012). Some of the applications of these particles, particularly in the area of energy (Rai 2019) and sensing (Hanske et al. 2018), may be improved by adding controlled porosity to the oxide shell.

Finally, it is important to highlight the fact that, to date, most of the applications of M@MP are far from being used at a commercial scale. Further experiments are needed to ensure safety and long-term stability of these composites before the promising results obtained at laboratory scale become a successful market product.

Acknowledgements

This work was partially funded by ANPCYT (PICT 2015-0351). MJP and SPC acknowledge CONICET for their postdoctoral and doctoral fellowship, respectively.

References

Abadeer, N. S., R. M. Brennan, L. W. Wilson and J. C. Murphy. 2014. Distance and plasmon wavelength dependent fluorescence of molecules bound to silica-coated gold nanorods. ACS Nano 8(8): 8392–8406.

Ahmed, W., E. Stefan Kooij, A. van Silfhout and B. Poelsema. 2010. Controlling the morphology of multi-branched gold nanoparticles. Nanotechnology 21(12): 125605.

Alvarez-Puebla, R., L. M. Liz-Marzán and F. J. García De Abajo. 2010. Light concentration at the nanometer scale. J. Phys. Chem. Lett. 1(16): 2428–2434.

Amendola, V., R. Pilot, M. Frasconi, O. M. Maragò and M. A. Iatì. 2017. Surface plasmon resonance in gold nanoparticles: A review. J. Phys. Condens. Matter 29(20).

Arnal, P. M., M. Comotti and F. Schüth. 2006. High-temperature-stable catalysts by hollow sphere encapsulation. Angew. Chemie - Int. Ed. 45(48): 8224–8227.

Aznar, E., M. Oroval, L. Pascual, J. R. Murguía, R. Martínez-Máñez and F. Sancenón. 2016. Gated materials for on-command release of guest molecules. Chem. Rev. 116(2): 561–718.

Baffou, G. and R. Quidant. 2013. Thermo-plasmonics: Using metallic nanostructures as nano-sources of heat. Laser Photonics Rev. 7(2): 171–187.

Bals, S., S. Van Aert and G. Van Tendeloo. 2013. High resolution electron tomography. Curr. Opin. Solid State Mater. Sci. 17(3): 107–114.

Bazak, R., M. Houri, S. El Achy, S. Kamel and T. Refaat. 2015. Cancer active targeting by nanoparticles: a comprehensive review of literature. J. Cancer Res. Clin. Oncol. 141(5): 769–784.

Beck, J. S., J. C. Vartuli, W. J. Roth, M. E. Leonowicz, C. T. Kresge, Schmitt et al. 1992. A new family of mesoporous molecular sieves prepared with liquid crystal templates. J. Am. Chem. Soc. 114(27): 10834–10843.

Bera, D., L. Qian, T. K. Tseng and P. H. Holloway. 2010. Quantum dots and their multimodal applications: A review. Materials (Basel). 3(4): 2260–2345.

Bordat, A., T. Boissenot, J. Nicolas and N. Tsapis. 2019. Thermoresponsive polymer nanocarriers for biomedical applications. Adv. Drug Deliv. Rev. 138: 167–192.

Brinker, C. J. and G. W. Scherer. 1990. Sol-Gel Science: The Physics and Chemistry of Sol-Gel Processing (1st ed.). Academic Press.

Brongersma, M. L., N. J. Halas and P. Nordlander. 2015. Plasmon-induced hot carrier science and technology. Nat. Nanotechnol. 10(1): 25–34.

Carbó-Argibay, E. and B. Rodríguez-González. 2016. Controlled growth of colloidal gold nanoparticles: Single-crystalline versus multiply-twinned particles. Isr. J. Chem. 56(4): 214–226.

Carrasco, S., E. Benito-Peña, F. Navarro-Villoslada, J. Langer, M. N. Sanz-Ortiz, J. Reguera et al. 2016. Multibranched gold-mesoporous silica nanoparticles coated with a molecularly imprinted polymer for label-free antibiotic surface-enhanced raman scattering analysis. Chem. Mater. 28(21): 7947–7954.

Chang, Y. T., P. Y. Liao, H. S. Sheu, Y. J. Tseng, F. Y. Cheng and C. S. Yeh. 2012. Near-Infrared light-responsive intracellular drug and sirna release using AU nanoensembles with oligonucleotide-capped silica shell. Adv. Mater. 24(25): 3309–3314.

Chen, J., R. Zhang, L. Han, B. Tu and D. Zhao. 2013. One-pot synthesis of thermally stable gold@mesoporous silica core-shell nanospheres with catalytic activity. Nano Res. 6(12): 871–879.

Chen, L., X. Wang, W. Lu, X. Wu and J. Li. 2016. Molecular imprinting: perspectives and applications. Chem. Soc. Rev. 45(8): 2137–2211.

Chen, L., M. Lin and P. Yang. 2019. Reproducible mesoporous silica-coated gold@silver nanoprobes for the bright colorimetric sensing of ascorbic acid. New J. Chem. 43(27): 10841–10849.

Cheng, Z., A. Al Zaki, J. Z. Hui, V. R. Muzykantov and A. Tsourkas. 2012. Multifunctional nanoparticles: Cost versus benefit of adding targeting and imaging capabilities. Science 338(6109): 903–910.

Choi, C. H., K. Chung, T. T. H. Nguyen and D. H. Kim. 2018. Plasmon-mediated electrocatalysis for sustainable energy: From electrochemical conversion of different feedstocks to fuel cell reactions. ACS Energy Lett. 3(6): 1415–1433.

Conde, J., J. T. Dias, V. Grazú, M. Moros, P. V. Baptista and J. M. de la Fuente. 2014. Revisiting 30 years of biofunctionalization and surface chemistry of inorganic nanoparticles for nanomedicine. Front. Chem. 2(JUL): 1–27.

Davidson, M., Y. Ji, G. J. Leong, N. C. Kovach, B. G. Trewyn and R. M. Richards. 2018. Hybrid mesoporous silica/noble-metal nanoparticle materials—synthesis and catalytic applications. ACS Appl. Nano Mater. 1(9): 4386–4400.

Demirörs, A. F., A. van Blaaderen and A. Imhof. 2010. A general method to coat colloidal particles with Titania. Langmuir 26(12): 9297–9303.

El-Toni, A. M., M. A. Habila, J. P. Labis, Z. A. Alothman, M. Alhoshan, A. A. Elzatahry and F. Zhang. 2016. Design, synthesis and applications of core-shell, hollow core, and nanorattle multifunctional nanostructures. Nanoscale 8(5): 2510–2531.

Elahi, N., M. Kamali and M. H. Baghersad. 2018. Recent biomedical applications of gold nanoparticles: A review. Talanta 184: 537–556.

Falahati, M., F. Attar, M. Sharifi, A. A. Saboury, A. Salihi, F. M. Aziz et al. 2020. Gold nanomaterials as key suppliers in biological and chemical sensing, catalysis, and medicine. Biochim. Biophys. Acta - Gen. Subj. 1864(1).

Fang, W., J. Yang, J. Gong and N. Zheng. 2012. Photo- and pH-triggered release of anticancer drugs from mesoporous silica-coated Pd@Ag nanoparticles. Adv. Funct. Mater. 22(4): 842–848.

Farokhzad, O. C. and R. Langer. 2009. Impact of nanotechnology on drug discovery & development pharmanext. ACS Nano 3(1): 16–20.

Florek, J., R. Caillard and F. Kleitz. 2017. Evaluation of mesoporous silica nanoparticles for oral drug delivery-current status and perspective of MSNs drug carriers. Nanoscale 9(40): 15252–15277.

Gawande, M. B., A. Goswami, T. Asefa, H. Guo, A. V. Biradar, Peng et al. 2015. Core-shell nanoparticles: synthesis and applications in catalysis and electrocatalysis. Chem. Soc. Rev. 44(21): 7540–7590.

Genç, A., J. Patarroyo, J. Sancho-Parramon, N. G. Bastús, V. Puntes and J. Arbiol. 2017. Hollow metal nanostructures for enhanced plasmonics: Synthesis, local plasmonic properties and applications. Nanophotonics 6(1): 193–213.

Ghosh Chaudhuri, R. and S. Paria. 2012. Core/shell nanoparticles: classes, properties, synthesis mechanisms, characterization, and applications. Chem. Rev. 112: 2373–2433.

Girija, A. R. and S. Balasubramanian. 2019. Theragnostic potentials of core/shell mesoporous silica nanostructures. Nanotheranostics 3(1): 1–40.

Gorelikov, I. and N. Matsuura. 2008. Single-step coating of mesoporous silica on cetyltrimethyl ammonium bromide-capped nanoparticles. Nano Lett. 8(1): 369–373.

Guerrero-Martínez, A., S. Barbosa, I. Pastoriza-Santos and L. M. Liz-Marzán. 2011. Nanostars shine bright for you. Colloidal synthesis, properties and applications of branched metallic nanoparticles. Curr. Opin. Colloid Interface Sci. 16(2): 118–127.

Han, L., H. Wei, B. Tu and D. Zhao. 2011. A facile one-pot synthesis of uniform core-shell silver nanoparticle@ mesoporous silica nanospheres. Chem. Commun. 47(30): 8536–8538.

Hanske, C., M. N. Sanz-Ortiz and L. M. Liz-Marzán. 2018. Silica-coated plasmonic metal nanoparticles in action. Adv. Mater. 30(27): 1–28.

Hernández Montoto, A., R. Montes, A. Samadi, M. Gorbe, J. M. Terrés, R. Cao-Milán et al. 2018. Gold nanostars coated with mesoporous silica are effective and nontoxic photothermal agents capable of gate keeping and laser-induced drug release. ACS Appl. Mater. Interfaces 10(33): 27644–27656.

Hervés, P., M. Pérez-Lorenzo, L. M. Liz-Marzán, J. Dzubiella, Y. Lub and M. Ballauff. 2012. Catalysis by metallic nanoparticles in aqueous solution: Model reactions. Chem. Soc. Rev. 41(17): 5577–5587.

Huang, X., C. Guo, J. Zuo, N. Zheng and G. D. Stucky. 2009. An assembly route to inorganic catalytic nanoreactors containing sub-10-nm gold nanoparticles with anti-aggregation properties. Small 5(3): 361–365.

Huang, Y., S. He, W. Cao, K. Cai and X. J. Liang. 2012. Biomedical nanomaterials for imaging-guided cancer therapy. Nanoscale 4(20): 6135–6149.

Innocenzi, P. and L. Malfatti. 2013. Mesoporous thin films: Properties and applications. Chem. Soc. Rev. 42(9): 4198–4216.

Innocenzi, P. and L. Malfatti. 2019. Mesoporous materials as platforms for surface-enhanced Raman scattering. TrAC – Trends Anal. Chem. 114: 233–241.

Jackson, P. A., W. N. W. A. Rahman, C. J. Wong, T. Ackerly and M. Geso. 2010. Potential dependent superiority of gold nanoparticles in comparison to iodinated contrast agents. Eur. J. Radiol. 75(1): 104–109.

Jang, Y. H., Y. J. Jang, S. Kim, L. N. Quan, K. Chung and D. H. Kim. 2016. Plasmonic solar cells: From rational design to mechanism overview. Chem. Rev. 116(24): 14982–15034.

Jeong, Y., Y. -M. Kook, K. Lee and W. -G. Koh. 2018. Metal enhanced fluorescence (MEF) for biosensors: General approaches and a review of recent developments. Biosens. Bioelectron. 111: 102–116.

Jokerst, J. V. and S. S. Gambhir. 2011. Molecular imaging with theranostic nanoparticles. Acc. Chem. Res. 44(10): 1050–1060.

Joo, S. H., J. Y. Park, C. -K. Tsung, Y. Yamada, P. Yang and G. A. Somorjai. 2009. Thermally stable Pt/mesoporous silica core-shell nanocatalysts for high-temperature reactions. Nat. Mater. 8(2): 126–131.

Kanasty, R., J. R. Dorkin, A. Vegas and D. Anderson. 2013. Delivery materials for siRNA therapeutics. Nat. Mater. 12(11): 967–977.

Kankala, R. K., H. Zhang, C. -G. C. Liu, K. R. Kanubaddi, C.-H. C. Lee, S. -B. S. Wang et al. 2019. Metal species–encapsulated mesoporous silica nanoparticles: current advancements and latest breakthroughs. Adv. Funct. Mater. 0(0): 1902652.

Kankala, R. K., Y. H. Han, J. Na, C. H. Lee, Z. Sun, Wang, S. Bin et al. 2020. Nanoarchitectured structure and surface biofunctionality of mesoporous silica nanoparticles. Adv. Mater. 1907035: 1–27.

Khlebtsov, B., E. Panfilova, V. Khanadeev, O. Bibikova, G. Terentyuk, A. Ivanov et al. 2011. Nanocomposites containing silica-coated gold–silver nanocages and Yb–2,4-dimethoxyhematoporphyrin: Multifunctional capability of IR-luminescence detection, photosensitization, and photothermolysis. ACS Nano 5(9): 7077–7089.

Kohout, C., C. Santi and L. Polito. 2018. Anisotropic gold nanoparticles in biomedical applications. Int. J. Mol. Sci. 19(11).

Kresge, C. T. and W. J. Roth. 2013. The discovery of mesoporous molecular sieves from the twenty year perspective. Chem. Soc. Rev. 42(9): 3663–3670.

Langer, J., D. Jimenez de Aberasturi, J. Aizpurua, R. A. Alvarez-Puebla, B. Auguié, J. J. Baumberg et al. 2020. Present and future of surface-enhanced raman scattering. ACS Nano 14(1): 28–117.

Lee, C. H., S. H. Cheng, Y. J. Wang, Y. C. Chen, N. T. Chen, J. Souris et al. 2009. Near-infrared mesoporous silica nanoparticles for optical imaging: Characterization and *in vivo* biodistribution. Adv. Funct. Mater. 19(2): 215–222.

Li, G. and Z. Tang. 2014. Noble metal nanoparticle@metal oxide core/yolk–shell nanostructures as catalysts: recent progress and perspective. Nanoscale 6(8): 3995–4011.

Li, H., L. -L. Tan, P. Jia, Q. -L. Li, Y. -L. Sun, J. Zhang et al. 2014. Near-infrared light-responsive supramolecular nanovalve based on mesoporous silica-coated gold nanorods. Chem. Sci. 5(7): 2804–2808.

Li, J. -F., C. -Y. Li and R. F. Aroca. 2017. Plasmon-enhanced fluorescence spectroscopy. Chem. Soc. Rev. 46(13): 3962–3979.

Li, N., P. Zhao and D. Astruc. 2014. Anisotropic gold nanoparticles: Synthesis, properties, applications, and toxicity. Angew. Chemie - Int. Ed. 53(7): 1756–1789.

Li, W. and D. Zhao. 2013. Extension of the stöber method to construct mesoporous SiO_2 and TiO_2 shells for uniform multifunctional core-shell structures. Adv. Mater. 25(1): 142–149.

Li, Z., J. C. Barnes, A. Bosoy, J. F. Stoddart and J. I. Zink. 2012. Mesoporous silica nanoparticles in biomedical applications. Chem. Soc. Rev. 41(7): 2590–2605.

Lin, M., Y. Wang, X. Sun, W. Wang and L. Chen. 2015. "Elastic" property of mesoporous silica shell: For dynamic surface enhanced Raman scattering ability monitoring of growing noble metal nanostructures via a simplified spatially confined growth method. ACS Appl. Mater. Interfaces 7(14): 7516–7525.

Linic, S., U. Aslam, C. Boerigter and M. Morabito. 2015. Photochemical transformations on plasmonic metal nanoparticles. Nat. Mater. 14(6): 567–576.

Liong, M., B. France, K. A. Bradley and J. I. Zink. 2009. Antimicrobial activity of silver nanocrystals encapsulated in mesoporous silica nanoparticles. Adv. Mater. 21(17): 1684–1689.

Liu, J., C. Detrembleur, M. C. De Pauw-Gillet, S. Mornet, C. Jérôme and E. Duguet. 2015. Gold nanorods coated with mesoporous silica shell as drug delivery system for remote near infrared light-activated release and potential phototherapy. Small 11(19): 2323–2332.

Liu, L. and A. Corma. 2018. Metal catalysts for heterogeneous catalysis: from single atoms to nanoclusters and nanoparticles. Chem. Rev. 118(10): 4981–5079.

Liu, S., S. Q. Bai, Y. Zheng, K. W. Shah and M. Y. Han. 2012. Composite metal-oxide nanocatalysts. ChemCatChem. 4(10): 1462–1484.

Liu, W., Z. Zhu, K. Deng, Z. Li, Y. Zhou, H. Qiu et al. 2013. Gold nanorod@chiral mesoporous silica core−shell nanoparticles with unique optical properties. J. Am. Chem. Soc. 135: 9659−9664.

Liu, Y., L. Nie and X. Chen. 2016. Photoacoustic molecular imaging: from multiscale biomedical applications towards early-stage theranostics. Trends Biotechnol. 34(5): 420–433.

Liz-Marzán, L. M., M. Giersig and P. Mulvaney. 1996. Synthesis of nanosized gold–silica core–shell particles. Langmuir 12(18): 4329–4335.

Liz-Marzán, L. M. 2006. Tailoring surface plasmons through the morphology and assembly of metal nanoparticles. Langmuir 22(1): 32–41.

Lohse, S. E., N. D. Burrows, L. Scarabelli, L. M. Liz-Marzán and C. J. Murphy. 2014. Anisotropic noble metal nanocrystal growth: The role of halides. Chem. Mater. 26(1): 34–43.

Losch, P., W. Huang, E. D. Goodman, C. J. Wrasman, A. Holm, A. R. Riscoe et al. 2019. Colloidal nanocrystals for heterogeneous catalysis. Nano Today 24: 15–47.

Lukosi, M., H. Zhu and S. Dai. 2016. Recent advances in gold-metal oxide core-shell nanoparticles: Synthesis, characterization, and their application for heterogeneous catalysis. Front. Chem. Sci. Eng. 10(1): 39–56.

Luo, T., P. Huang, G. Gao, G. Shen, S. Fu, D. Cui et al. 2011. Mesoporous silica-coated gold nanorods with embedded indocyanine green for dual mode X-ray CT and NIR fluorescence imaging. Opt. Express 19(18): 17030–17039.

Luo, Y., Y. Song, C. Zhu, S. Li, M. Xian, C. M. Wai et al. 2019. Visualization of endogenous hydrogen sulfide in living cells based on Au nanorods@silica enhanced fluorescence. Anal. Chim. Acta 1053: 81–88.

Mayer, K. M. and J. H. Hafner. 2011. Localized surface plasmon resonance sensors. Chem. Rev. 111(6): 3828–3857.

Melancon, M. P., M. Zhou and C. Li. 2011. Cancer theranostics with near-infrared light-activatable multimodal nanoparticles. Acc. Chem. Res. 44(10): 947–956.

Melde, B. J. and B. J. Johnson. 2010. Mesoporous materials in sensing: Morphology and functionality at the meso-interface. Anal. Bioanal. Chem. 398(4): 1565–1573.

Modena, M. M., B. Rühle, T. P. Burg and S. Wuttke. 2019. Nanoparticle characterization: What to measure? Adv. Mater. 31(32): 1–26.

Mourdikoudis, S., R. M. Pallares and N. T. K. Thanh. 2018. Characterization techniques for nanoparticles: Comparison and complementarity upon studying nanoparticle properties. Nanoscale 10(27): 12871–12934.

Ndokoye, P., Q. Zhao, X. Li, T. Li, M. O. Tade and S. Wang. 2016. Branch number matters: Promoting catalytic reduction of 4-nitrophenol over gold nanostars by raising the number of branches and coating with mesoporous SiO$_2$. J. Colloid Interface Sci. 477: 1–7.

Ni, Q., Z. Teng, M. Dang, Y. Tian, Y. Zhang, P. Huang et al. 2017. Gold nanorod embedded large-pore mesoporous organosilica nanospheres for gene and photothermal cooperative therapy of triple negative breast cancer. Nanoscale 9(4): 1466–1474.

Noguez, C. 2007. Surface Plasmons on metal nanoparticles: The influence of shape and physical environment. J. Phys. Chem. C 111(10): 3806–3819.

Ozin, G. A., A. Arsenault and L. Cademartiri. 2008. Nanochemistry: A Chemical Approach to Nanomaterials (2nd ed.). The Royal Society of Chemistry.

Perego, C. and R. Millini. 2013. Porous materials in catalysis: Challenges for mesoporous materials. Chem. Soc. Rev. 42(9): 3956–3976.

Rai, P. 2019. Plasmonic noble metal@metal oxide core-shell nanoparticles for dye-sensitized solar cell applications. Sustain. Energy Fuels 3(1): 63–91.

Reddy, L. H., J. L. Arias, J. Nicolas and P. Couvreur. 2012. Magnetic nanoparticles: Design and characterization, toxicity and biocompatibility, pharmaceutical and biomedical applications. Chem. Rev. 112(11): 5818–5878.

Reguera, J., J. Langer, D. Jiménez De Aberasturi and L. M. Liz-Marzán. 2017. Anisotropic metal nanoparticles for surface enhanced Raman scattering. Chem. Soc. Rev. 46(13): 3866–3885.

Rodrigues, T. S., A. G. M. Da Silva and P. H. C. Camargo. 2019. Nanocatalysis by noble metal nanoparticles: Controlled synthesis for the optimization and understanding of activities. J. Mater. Chem. A 7(11): 5857–5874.

Rodríguez-González, B., A. Burrows, M. Watanabe, C. J. Kiely and L. M. Liz Marzán. 2005. Multishell bimetallic AuAg nanoparticles: synthesis, structure and optical properties. J. Mater. Chem. 15(17): 1755–1759.

Sanz-Ortiz, M. N., K. Sentosun, S. Bals and L. M. Liz-Marzán. 2015. Templated growth of surface enhanced raman scattering-active branched gold nanoparticles within radial mesoporous silica shells. ACS Nano 9(10): 10489–10497.

Sau, T. K. and A. L. Rogach. 2010. Nonspherical noble metal nanoparticles: Colloid-chemical synthesis and morphology control. Adv. Mater. 22(16): 1781–1804.

Scarabelli, L., A. Sánchez-Iglesias, J. Pérez-Juste and L. M. Liz-Marzán. 2015. A "tips and tricks" practical guide to the synthesis of gold nanorods. J. Phys. Chem. Lett. 6(21): 4270–4279.

Schärtl, W. 2010. Current directions in core-shell nanoparticle design. Nanoscale 2(6): 829–843.

Sentosun, K., M. N. Sanz Ortiz, K. J. Batenburg, L. M. Liz-Marzán and S. Bals. 2015. Combination of HAADF-STEM and ADF-STEM tomography for core-shell hybrid materials. Part. Part. Syst. Charact. 32(12): 1063–1067.

Sepúlveda, B., P. C. Angelomé, L. M. Lechuga and L. M. Liz-Marzán. 2009. LSPR-based nanobiosensors. Nano Today 4(3): 244–251.

Sharifi, M., F. Attar, A. A. Saboury, K. Akhtari, N. Hooshmand, A. Hasan et al. 2019. Plasmonic gold nanoparticles: Optical manipulation, imaging, drug delivery and therapy. J. Control. Release 311–312(June): 170–189.

Shen, S., H. Tang, X. Zhang, J. Ren, Z. Pang, D. Wang et al. 2013. Targeting mesoporous silica-encapsulated gold nanorods for chemo-photothermal therapy with near-infrared radiation. Biomaterials 34(12): 3150–3158.

Stöber, W., A. Fink and E. Bohn. 1968. Controlled growth of monodisperse silica spheres in the micron size range. J. Colloid Interface Sci. 62–69.

Sun, X., W. Cai and X. Chen. 2015. Positron emission tomography imaging using radiolabeled inorganic nanomaterials. Acc. Chem. Res. 48(2): 286–294.

Talebzadeh, S., C. Queffélec and D. A. Knight. 2019. Surface modification of plasmonic noble metal–metal oxide core–shell nanoparticles. Nanoscale Adv. 1(12): 4578–4591.

Tang, H., S. Shen, J. Guo, B. Chang, X. Jiang and W. Yang. 2012. Gold nanorods@mSiO$_2$ with a smart polymer shell responsive to heat/near-infrared light for chemo-photothermal therapy. J. Mater. Chem. 22(31): 16095–16103.

Turkevich, J., P. C. Stevenson and J. Hillier. 1951. A study of the nucleation and growth processes in the synthesis of colloidal gold. Discuss. Faraday Soc. 11(0): 55–75.

Vigderman, L., B. P. Khanal and E. R. Zubarev. 2012. Functional gold nanorods: Synthesis, self-assembly, and sensing applications. Adv. Mater. 24(36): 4811–4841.

Wagner, V., A. Dullaart, A. K. Bock and A. Zweck. 2006. The emerging nanomedicine landscape. Nat. Biotechnol. 24(10): 1211–1217.

Wei, S., Q. Wang, J. Zhu, L. Sun, H. Lin and Z. Guo. 2011. Multifunctional composite core-shell nanoparticles. Nanoscale 3(11): 4474–4502.

Wu, C. and Q. -H. Xu. 2009. Stable and functionable mesoporous silica-coated gold nanorods as sensitive localized surface plasmon resonance (LSPR) nanosensors. Langmuir 25(16): 9441–9446.

Wu, Y., M. R. K. Ali, K. Chen, N. Fang and M. A. El-Sayed. 2019. Gold nanoparticles in biological optical imaging. Nano Today 24: 120–140.

Xia, Y., Y. Xiong, B. Lim and S. E. Skrabalak. 2009. Shape-controlled synthesis of metal nanocrystals: Simple chemistry meets complex physics? Angew. Chemie - Int. Ed. 48(1): 60–103.

Xu, C., F. Chen, H. F. Valdovinos, D. Jiang, S. Goel, B. Yu et al. 2018. Bacteria-like mesoporous silica-coated gold nanorods for positron emission tomography and photoacoustic imaging-guided chemo-photothermal combined therapy. Biomaterials 165: 56–65.

Yang, J., F. Zhang, Y. Chen, S. Qian, P. Hu, W. Li et al. 2011. Core-shell Ag@SiO$_2$@mSiO$_2$ mesoporous nanocarriers for metal-enhanced fluorescence. Chem. Commun. 47(42): 11618–11620.

Yang, P., T. Deng, D. Zhao, P. Feng, D. Pine, B. F. Chmelka et al. 1998. Hierarchically ordered oxides. Science 282(5397): 2244–2246.

Yang, P., J. Zheng, Y. Xu, Q. Zhang and L. Jiang. 2016. Colloidal synthesis and applications of plasmonic metal nanoparticles. Adv. Mater. 28(47): 10508–10517.

Yuan, H., C. G. Khoury, H. Hwang, C. M. Wilson, G. A. Grant and T. Vo-Dinh. 2012. Gold nanostars: surfactant-free synthesis, 3D modelling, and two-photon photoluminescence imaging. Nanotechnology 23(7): 75102.

Zhang, W., Y. Wang, X. Sun, W. Wang and L. Chen. 2014. Mesoporous titania based yolk–shell nanoparticles as multifunctional theranostic platforms for SERS imaging and chemo-photothermal treatment. Nanoscale 6(23): 14514–14522.

Zhang, Z., L. Wang, J. Wang, X. Jiang, X. Li, Z. Hu et al. 2012. Mesoporous silica-coated gold nanorods as a light-mediated multifunctional theranostic platform for cancer treatment. Adv. Mater. 24(11): 1418–1423.

Zhao, D., J. Feng, Q. Huo, N. Melosh, G. H. Fredrickson, B. F. Chmelka et al. 1998. Triblock copolymer syntheses of mesoporous silica with periodic 50 to 300 angstrom pores. Science 279(5350): 548–552.

Zhao, J., P. Xu, Y. Li, J. Wu, J. Xue, Q. Zhu et al. 2016. Direct coating of mesoporous titania on CTAB-capped gold nanorods. Nanoscale 8(10): 5417–5421.

Zhao, J., L. Long, G. Weng, J. Li, J. Zhu and J. W. Zhao. 2017. Multi-branch Au/Ag bimetallic core-shell-satellite nanoparticles as a versatile SERS substrate: The effect of Au branches in a mesoporous silica interlayer. J. Mater. Chem. C 5(48): 12678–12687.

Zhao, N., Z. Yang, B. Li, J. Meng, Z. Shi, P. Li et al. 2016. RGD-conjugated mesoporous silica-encapsulated gold nanorods enhance the sensitization of triplenegative breast cancer to megavoltage radiation therapy. Int. J. Nanomedicine 11: 5595–5610.

Zhao, P., X. Feng, D. Huang, G. Yang and D. Astruc. 2015. Basic concepts and recent advances in nitrophenol reduction by gold- and other transition metal nanoparticles. Coord. Chem. Rev. 287: 114–136.

Zhao, T., H. Wu, S. Q. Yao, Q. H. Xu and G. Q. Xu. 2010. Nanocomposites containing gold nanorods and porphyrin-doped mesoporous silica with dual capability of two-photon imaging and photosensitization. Langmuir 26(18): 14937–14942.

Zhou, J., Z. Cao, N. Panwar, R. Hu, X. Wang, J. Qu et al. 2017. Functionalized gold nanorods for nanomedicine: Past, present and future. Coord. Chem. Rev. 352: 15–66.

Zhou, N., L. Polavarapu, Q. Wang and Q. H. Xu. 2015. Mesoporous SnO_2-coated metal nanoparticles with enhanced catalytic efficiency. ACS Appl. Mater. Interfaces 7(8): 4844–4850.

Zong, S., Z. Wang, H. Chen, J. Yang and Y. Cui. 2013. Surface enhanced Raman scattering traceable and glutathione responsive nanocarrier for the intracellular drug delivery. Anal. Chem. 85(4): 2223–2230.

Zou, Y., H. Chen, Y. Li, X. Yuan, X. Zhao, W. Chen et al. 2019. Synthesis of mesoporous-silica coated multi-branched gold nanoparticles for surface enhanced Raman scattering evaluation of 4-bromomethcathinone. J. Saudi Chem. Soc. 23(3): 378–383.

Zou, Y., X. Zhou, J. Ma, X. Yang and Y. Deng. 2020. Recent advances in amphiphilic block copolymer templated mesoporous metal-based materials: Assembly engineering and applications. Chem. Soc. Rev. 49(4): 1173–1208.

CHAPTER 8

Nanoparticles Based Composites and Hybrids

Functionalities and Synthetic Methods and Study Cases

Victoria Benavente Llorente, Antonella Loiácono and *Esteban A. Franceschini**

Introduction

Metallic, ceramic and polymeric composite materials have been widely used as structural materials in different fields such as aerospace, energy, construction, etc., resulting in excellent mechanical properties and lighter materials. Despite their enormous potential, composite materials are usually expensive compared to other simpler materials such as ceramics and polymers. However, this higher cost can be compensated since composites can be modified to obtain unique properties and additional functions such as self-healing, energy storage and harvesting, electric/thermal conductivity and biocompatibility. Thus, multifunctional composites are how they become profitable for mass applications.

In the last decades, the increasing demand in the miniaturization of devices, as well as the increase in the energy consumption of electronic devices, required an increase in the development of multifunctional compounds, which are not only structural, but also comply with magnetic, electrical, thermal, chemical, optical and biological functionalities, considering future specific applications. Economic savings are achieved when two or more individual systems are replaced by a system with multiple functions, increasing efficiency and, for example, reducing energy consumption or increasing the autonomy time.

In this first part of the chapter, a few of the most common functionalities sought in composite-based multifunctional material systems (MFMS) will be analyzed and some of the conventional and unconventional synthetic methods will be explored.

Composite Structural Materials

As mentioned in the introduction (Chapter 1), structural properties such as stiffness, strength, ductility, energy absorption, damping, thermal stability, among others, are part of the set of functions that give multifunctionality to MFMS. One of the leitmotifs in the development of MFMS with structural properties is the reduction of weight, particularly for construction, transport in all its forms

INFIQC, Facultad de Ciencias Químicas, Universidad Nacional de Córdoba-CONICET. 5000 Córdoba, Argentina.
* Corresponding author: esteban.franceschini@mi.unc.edu.ar

(air, sea and land), clothing and sports, portable electronics, among an infinity of other applications (Gibson et al. 2001, Chan et al. 2018a, Chung 2019, Asp and Greenhalgh 2014).

Moreover, considering that the synergistic effect between materials is increased by reducing the size of the dispersed phase, the use of nano-reinforcements in polymers greatly improves the mechanical properties. Koratkar and Suhr (2005) measured increases of more than 1,000% in the polycarbonate loss modulus without detriment to other properties, such as the storage modulus, when 2% by weight of Single-Walled Carbon Nanotubes (SWCNT) are added, probably due to frictional sliding at the nanotube/polycarbonate interfaces which led to enhanced energy dissipation (Zhou et al. 2004). A similar effect was found by Rajoria and Jalili (2005) for Multi-Walled Carbon Nanotubes (MWNT) modified epoxy, although no significant effect on the storage modulus was observed.

Another illustrative example was proposed by Teh et al. (2007) where it is possible to simultaneously increase the resistance and elastic modulus of the epoxy compound by the addition of silica and epoxy microparticles and greatly reducing the Coefficient of Thermal Expansion (CTE) of the MFMS. It is important to note that in all these examples an optimization of the synthesis parameters was necessary to maximize the desired properties since an amount that was too large of the dispersed phase would detract from the mechanical properties of the matrix, due to agglomeration and reduced Disperse Phase (DP)/matrix interfacial strength. Moreover, in many cases, it is not possible to improve different properties simultaneously. Usually the improvement of some properties causes reductions in others, so, often it is a compromising situation between achieving the desired effects at the detriment of other properties that are not of interest. A clear example was presented by Manjunatha et al. (2009) for a glass fibre reinforced epoxy composite, the epoxy matrix was modified by the incorporation of rubber particles improving the tensile fatigue life by three, while causing a reduction of 12.7 and 5.2% in the elastic modulus and tensile strength, respectively.

Hence, in an ideal case, it is important to establish which are the properties of primary interest and which can be surrendered to obtain the desired multifunctionality. Often, this is not possible, especially when dealing with structural properties, where small changes in the matrix/disperse phase ratio can generate huge losses in them.

Integrated Structural and Non-structural Functions

The development of multi-scale nanocomposites containing nano-reinforcements has enabled simultaneous improvements not only in structural functions but also in the incorporation of other non-structural functions. Different cases in which one or multiple non-structural functions are introduced will be analyzed next.

Self-healing Function

A very interesting property, particularly for materials that must perform their functions in isolated areas, where maintenance is difficult, is self-healing when damaged. At the same time this type of function, even for conventional systems, allows reducing the cost of maintenance, which is why a great effort has been made in recent years to include this capability to a wide variety of other functions, particularly structural, electricity and heat conduction functions, etc. (Wu et al. 2008).

One of the strategies used for the development of self-healing MFMS is to create systems using microcapsules of a healing agent and a catalyst to polymerize it when necessary. White et al. (2001) designed self-healing materials (mainly polymers and composites) based on this concept. An example of this behaviour is shown in Fig. 8.1. When a fracture is generated and expands through the material, in this case, a polymer, breaks the microcapsules of the healing agent which leak through the fracture due to capillary pressure. There is a probability that some of these agents will come into contact with the catalyst which initiates the reaction and blocks the fracture repairing the polymer. Thus, the efficiency of healing is studied by analyzing the mechanical properties of

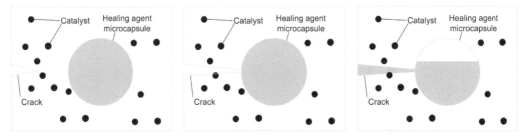

Figure 8.1: Illustration of self-healing of cracks in polymers by the use of a microencapsulated healing agent and a catalyst for polymerizing the healing agent.

the material, specifically the fracture toughness fatigue loading (Brown et al. 2005a,b, 2006) and low-velocity impact loading (Patel et al. 2010), and making the ratio between the properties of the recovered material and the original. The more similar the properties of the recovered to the original material, the greater the healing efficiency. White and co-workers (White et al. 2001) analyzed the mechanical properties of a virgin and a self-healed epoxy and showed that fracture load and toughness of the self-healed samples reached 75% of the values obtained for the virgin epoxy.

These values can be improved by optimizing parameters such as the concentration of the microcapsule and the choice of catalyst, reaching efficiency values of 90% (Brown et al. 2002). Even 100% of healing values have been obtained (Caruso et al. 2008) by using a mixture of solvent and epoxy monomer.

Other strategies for achieving self-healing composites consist in the application of different microencapsulation methods as the use of mesoporous silica (Kirk et al. 2009), self-healing polymers employed as a matrix in fibre-reinforced carbon composites (Williams et al. 2007), and the utilization of three-dimensional microvascular networks in the substrate under an epoxy layer to allow continuous administration of healing agents for self-healing of damage from repeated cracking in the coating (Toohey et al. 2007).

Electrical and/or Thermal Conductivity

These properties are among the most sought after in composites, but most of them have polymeric matrixes which are usually poor conductors, although they have the advantage of being very light. This combination of properties is important in applications such as aircraft structures, where electrically isolated structures may be damaged by lightning strikes (Gibson et al. 2007).

On the other hand, high thermal conductivity is particularly important for electronic circuits and propulsion cooling systems. Thus, there are many examples in the literature of simultaneous improvements of mechanical and electrical/thermal properties of nanocomposites (Allaoui et al. 2002, Qiu et al. 2007, Kalaitzidou et al. 2007, Cebeci et al. 2009).

One of the most common techniques found in the literature to improve the structural properties of nanocomposites and at the same time the electrical/thermal properties is the incorporation of carbon nanotubes. A small concentration of carbon nanotubes (CNT) in polymers boosts the electrical conductivity of the nanocomposite. An example is given by Bauhofer and Kovacks (2009) in Fig. 8.2 where the electrical conductivity of CNT/epoxy nanocomposites increases by nearly six orders when the CNT concentration is increased by only two.

In this case, the CNT concentration in the polymer that characterizes the insulator–conductor transition, described as 'percolation threshold', is only 0.04 wt.%. An ultra-low percolation threshold of 0.0025 wt.% for aligned MWNT/epoxy nanocomposites is presented in the literature (Sandler et al. 2003). The ultra-low value is due to the extremely high aspect ratio of CNTs that makes the continuous conducting relatively easy to form a CNTs network in the insulating epoxy matrix. Thus, it is important to use high aspect ratio CNTs and select processing methods that preserve the structures avoiding ruptures (Li et al. 2007, Thosthenson et al. 2009).

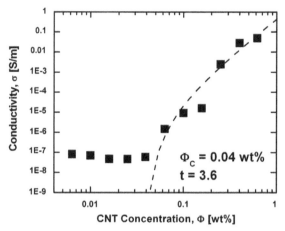

Figure 8.2: The electrical conductivity of CNT/epoxy nanocomposites at various CNT concentrations. The percolation threshold is 0.04 wt.% (Reprinted from Bauhofer and Kovacs (2009), with permission from Elsevier).

On the other hand, unlike the electrical conductivity, the thermal conductivity increases gradually with CNT concentration on CNT-polymer nanocomposites and no sharp percolation threshold is observed (Shenogina et al. 2005). The difference is that while the electrical transport is mainly due to the percolation of the CNT network, thermal transport is dominated by the matrix (Thosthenson et al. 2009). In this case, CNT-polymer composites do not show a percolation threshold, small amounts of nanotubes lead to high increases in composite thermal conductivity, although higher CNTs concentration are required. Thus, 7 wt.% CNT were used to reach between 55–57% of thermal conductivity increase using both SWNT and MWNT, respectively (Bonnet et al. 2007, Kim et al. 2007). However, growth of up to 125% have been reported using as little as 1 wt.% SWNT by optimizing synthesis parameters (Biercuk et al. 2002). Nevertheless, it is not possible to include a large amount of dispersed phase without generating structural changes in the matrix or complicating the synthesis process. For instance, Ganguli et al. (2008) found that adding graphite by more than 4 wt.% increased the viscosity of the synthetic mixture beyond the working window for the moulding process, whereas when chemically functionalized and exfoliated graphite flakes were able to reach loads as high as 20 wt.%, thus generating a huge increase in thermal conductivity.

Energy Harvesting/Storage

The concept of energy collection and storage is simple, but it is interesting to analyze how to make multifunctional structures and which are the functionality options, in addition to the energy function. One of the ideas of multifunctionality in energy harvesting/storage systems is to extract small amounts of energy from the movement and/or deformation of a host structure and convert it into electrical energy. Another very popular strategy is to use a system with defined structural properties to generate or store energy reducing the size of the battery required by a device.

The multifunctional structure/energy harvesting is extremely important when the size and weight of the device are limiting factors, such as wireless sensors for monitoring health (Park et al. 2008, Sodano et al. 2004, Anton and Sodano 2007, Cook-Chennault 2008) and aviation, particularly for Unmanned Aerial Vehicles (UAVs) where the weight is a determinant factor in the device autonomy.

One of the most common ways for energy collection uses piezoelectric materials to transduce mechanical deformations of structures that undergo continuous vibrations such as beams into electrical energy (Sodano et al. 2005).

Some examples of multifunctionality, battery/structure were discussed in Chapter 1. Although there are a large number of examples that can be analyzed, some of the most interesting are presented below.

Kim et al. (2009) used a copper inkjet-printed electronic circuit on a foil to connect a Li-ion battery to a solar thin-film panel. The resulting device was embedded within carbon/epoxy layers to develop an energy harvesting/storage laminate. The obtained MFMS was subjected to different mechanical loading techniques and it was found that when the electrodes are thicker than 4 μm, the mechanical properties are maintained.

Liu et al. (2009) developed a structural battery where the cathode in the Li-ion polymer battery was modified using a high molecular weight polymer reinforced with carbon nanofibres (Fig. 8.3). In order to increase the structural resistance of the battery, the liquid electrolyte was changed by a solid polymer and the electrode separator was reinforced with non-conductive fibres. This is an interesting design, although the battery traction modulus was only 3 GPa, similar to that found in polymethyl methacrylate (PMMA), and the energy density obtained was low.

However as mentioned above, it is very difficult to achieve good structural properties and good battery efficiency. In Fig. 8.4 a graph of conductivity versus compressive modulus constructed by Snyder et al. (2007) for different formulations of polymeric electrolytes for structural-batteries was presented. It can be seen how increasing the properties towards one of the functions decreases the properties of the other. That is, by improving efficiency as a battery, the structural properties decline and vice versa. This behaviour makes it difficult to optimize both properties and only a compromise situation can be reached (Snyder et al. 2009).

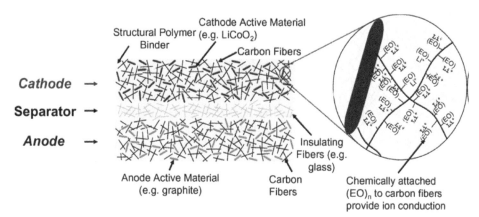

Figure 8.3: Construction of structural battery (Reprinted from Liu et al. (2009) with permission from Elsevier).

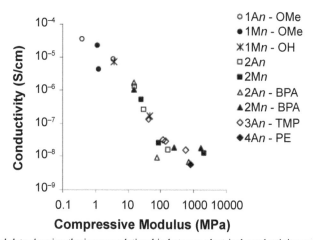

Figure 8.4: Experimental data showing the inverse relationship between electrical conductivity and compressive modulus for several polymer electrolytes for use in structural batteries (Reprinted from Snyder et al. (2007) with permission from American Chemical Society).

Thus although the integrated structural batteries without undermining the structural properties, currently allow little charge to help in reducing the necessary battery volume and therefore the weight of the prototype making them useful for situations where a small amount of energy is required for a long time. In cases where a large discharge of energy is required, structural capacitors can provide rapid discharge. Some model cases are presented in the literature by O'Brien et al. (2006). It is important to note that they found the same limitations discussed for structural batteries in terms of the properties of each of its functions. However, in the case of structural capacitors, it is possible to improve efficiency (Baechle et al. 2007).

Lin and Sodano (2009a) found that a SiC/BaTiO$_3$ piezoelectric structural fibre (Lin and Soldano 2009b) can constitute a structural capacitor by using the dielectric nature of the BaTiO$_3$ coating.

The ones presented above are some of the multiple examples found in the literature on composites with applications in energy storage/harvesting, this being an area where more and more materials are required to respond to the growing demand for systems with unique properties.

Biocompatibility

Nanomaterials, particularly those capable of multifunctionality, offer tremendous potential for many medical uses, like prosthetics, imaging, therapy, etc. One of the classic examples is MFMS based on Ti and its alloys for applications in prostheses. This is because they meet many of the requirements for these purposes, acceptable structural properties and good bone integration. A great deal of work has been done to modify the surface thus modifying the chemical properties of the bone-surface interface to achieve faster osseointegration and a focus on the response of osteoblasts to the implant surface.

There are many situations where rapid healing is needed or the quality and/or quantity of bones are poor, requiring osteoinductive behaviour of the surface that is not detrimental to the mechanical properties of the prosthesis, given by bulk. The main approaches to solve these problems can be divided into two categories: (1) Bioactive approach (*in vivo* induced apatite precipitation) or (2) Topographical stimulus of the osteoblasts through *ad-hoc* tailored roughness.

The bioactive approach is usually accomplished by applying an apatite or bioactive glass coating, using electrochemical processes or chemical surface treatments; and the mechanism of bioactivity of the surfaces can be related to ion exchange with the body fluids (Valente and Andreana 2016) and/or to surface charge effects and micro or nano-scale topography (Yang et al. 2008).

In the topographical stimulus approach, the use of composites has two fundamental objectives, to increase the structural stability of the materials or to facilitate osseointegration. Ouyang et al. (2019) presented a metal-metal Ti-Mg composite fabricated by spark plasma sintering. *In vitro* experiments were conducted (cells proliferation and differentiation) and a rat model with femur condyle defect was employed for *in vivo* tests. The results showed that these composites exhibited a greater cytocompatibility than pure Ti, while the results of the microcomputed tomography showed that the volume of the bone trabecula was significantly more abundant around the Ti-Mg implants than around the Ti implants, which indicates that more active new bones were formed around the composite implants and significantly greater osseointegration around the Ti-Mg implants than around the Ti implants was achieved.

Chan et al. (2018) presented three bioactive glass fibre reinforced composite implants and the *in vitro* biocompatibility and *in vivo* osseointegration were analyzed.

In vivo osseointegration performance was studied using rabbit femurs and the results were evaluated using microcomputed tomography, histology and histomorphometric analysis after 1, 2, 4 and 8 weeks of healing. *In vitro* results indicate that the cell cultures carried out on the composites showed better growth and exhibited a more differentiated phenotype than the cells grown on a titanium alloy substrate (Ti$_6$Al$_4$V). During *in vivo* studies, the histological evaluation showed more bone regeneration within composite implants during the initial healing period. In addition, the composite implants showed better bone volume/tissue volume values at 4 weeks and this was accompanied by contact with the bone-implant values at 8 weeks compared with the Ti$_6$Al$_4$V group.

These are just some examples where the use of composites improves the biocompatibility properties, compared to the results obtained using alloys and simple materials.

Sensing and Actuation

Among non-structural functions, sensing and actuating are two closely related functions and often in the literature it is possible to use the same material for both functions. There are numerous articles in the literature that have covered much of the recent research related to multifunctional structures that can be used as sensors and actuators. Composites based on CNT (Li et al. 2008, Gibson et al. 2007) can have these functions and even shape memory polymers (Ratna and Karger-Kocsis 2008) and piezoelectric materials are capable of presenting sensing and actuation functions. Materials featuring electromechanical couplings such as lead zirconatetitanate (PZT), polyvinylidene fluoride (PVDF) and aluminium nitride (AlN) can be integrated into structures for detection and the effectiveness with which a piezoelectric material converts the applied mechanical energy into electrical energy is characterized by the piezoelectric coupling coefficient (Lesieutre and Davis 1997). Piezoelectric transducers can be configured as sensors when the design involves the generation of an electrical signal from the application of stress or strain of the material, using direct piezoelectric effect as actuators when the design of the device is optimized to generate stress or strain using the reverse piezoelectric effect (Tadigadapa and Mateti 2009). Normally, piezoelectric based sensors are configured as direct mechanical transducers or as resonators, where the measured resonance frequency and amplitude are modified by external mechanical inputs.

Figure 8.5 illustrates the most widely used piezoelectric sensor configurations. The most common forms of piezoelectric materials are wafers (Giurgiutiu and Zagrai 2011) and thin films (Muralt et al. 2009). Both materials have been scaled and miniaturized during the last few decades which makes them excellent candidates to be incorporated into microelectromechanical systems (MEMS). Despite this, most piezoelectric materials are usually made of relatively inert nitrides and metal oxides and semiconductors, some of these elements are incompatible with standard Complementary Metal-Oxide-Semiconductor (CMOS) technology.

In addition to films, composite piezoelectric fibres (PFCs) have been studied as possible active components in MFMS. Hagood and Bent (1993) and Bent et al. (1995), incorporated micron-sized piezoelectric fibres into an epoxy matrix to which PZT powder had been added to reduce dielectric fibre/matrix mismatch. To obtain this material, PFC sheets were embedded between graphite/epoxy sheets and electrodes in which the electric field required for the action is applied, obtaining a composite that shows a good relationship between the measured deformation and the applied electric field.

There is a great variety of structures that can be used, each with its particular properties. Unlike piezoelectric fibres with a solid cross-section, hollow piezoelectric fibres (Fernandez et al. 1996, Brei and Cannon 2004) offer the advantage of lower operating voltage and a wider choice of possible matrix materials. This is because the design parameters in hollow fibres are more varied than in dense fibres and other tunable parameters exits such as the matrix/fibre ratio, Young's modulus, an aspect ratio of the individual fibres and overall active composite volume fraction.

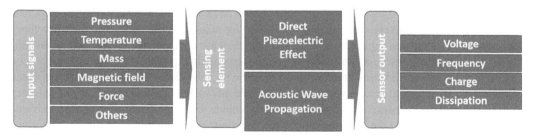

Figure 8.5: Different operation modes of piezoelectric sensors (adapted from Tadigadapa and Mateti 2009 with permission from IOP Publishing).

Lin and Sodano (2008, 2009c) developed piezoelectric structural fibres consisting of carbon fibres coated with a piezoelectric layer and an external electrode. As earlier, radial poling results in a longitudinal drive of the fibre, obtaining electromechanical coupling coefficients as high as 65–70%. These piezoelectric compounds are suitable for vibration control, damping, energy harvesting or monitoring of structural health.

Another interesting application for composites with sensor/actuation functions is the monitoring of structural health (Montalvao et al. 2006, Zou et al. 2000, Wu et al. 2008, Park et al. 2008), which are used for the detection of damage in multifunctional composite structures. Wu et al. (2009) used PZT actuators and Bragg Fibre Grating (FBG) fibre optic sensors to form a network of actuators/sensors for damage detection in composite laminates. The advantage of this approach, compared to those that use PZT as simultaneous actuators and sensors (Lin and Chang 2002), is that with separate systems it is possible to decouple the mechanisms of signal transmission and the elimination of crosstalk between the actuator signals and the sensor in the network. This type of MFMS opens the way to a much more comprehensive approach in the repair of structural damage using artificial neural networks to analyze data from piezoelectric sensor networks and classify and locate damage in composite structures (Watkins et al. 2007, Haywood et al. 2005, Yu et al. 2007). Thus, the use of Layer-by-Layer assembly (LbL) (Srivastava et al. 2005) allows the sequential deposition of different thin films at the nanoscale, and consequently the development of multipurpose laminar sensors. Some examples of multipurpose sensors designed using LbL technology include carbon nanotube-polyelectrolyte multilayer composite material for monitoring strain and corrosion (Loh et al. 2007). To do this, the CNT concentration determines the sensitivity to strain and the type of polyelectrolyte determines the sensitivity to pH. Other examples include manufacturing passive wireless sensor that does not require a battery power supply (Loh et al. 2008), high-strength multifunctional compounds for biological implants, anti-corrosive coatings and thermal/electrical interface materials (Olek et al. 2004, Shim et al. 2006).

Synthesis of Nanoparticles-based Composites and Hybrids

The specific properties of a nanoparticle-based composite or hybrid can be finely tuned by the control of the synthetic method used. Therefore, the main fundamental and experimental aspects of conventional preparation methods, as polymerization and powder metallurgy, will be summarized next. Electrochemical composite plating as an unconventional pathway for the preparation of metallic matrix composites will also be described.

Polymerization

Nanoscale building blocks in a polymeric matrix are an important category of organic-inorganic hybrid materials with novel and controllable functionalities. Numerous procedures for the preparation of nanostructures-based Polymer Matrix Composites (PMC) have been proposed. Here different approaches for nanomaterials-based PMC preparation are described considering a simple classification reported (Zhao et al. 2011): (1) direct compounding and (2) *in situ* synthesis. In literature, the polymer matrix is described as the polymer 'host', and the nanostructures as the 'filler'.

The direct compounding approach presents several advantages such as very simple equipment requirements, comparatively low cost and suitability for massive production. The nanostructures and polymer are prepared separately at first, and then they are compounded by solution, emulsion, fusion or mechanical forces. As an example, solution compounding involves the simple mixing of the prepared polymer and nanostructure in a solvent with subsequent evaporation of the solvent (Coleman et al. 2006). Melt compounding presents the advantage of avoiding the use of solvents and is commonly used when polymers and fillers cannot be solubilized in the same solvent. Nanostructures are mixed with the molten polymer for direct compounding.

Although direct compounding is a very simple method, it presents a limited success for numerous systems because the nanostructures usually exhibit a large tendency to form aggregates during the process. The formation of aggregates greatly diminishes the advantages of their small dimensions. Polymer degradation on melt compounding is also sometimes severe (Zhao et al. 2011). These limitations should be considered when using the direct compounding method to prepare nanomaterials-based PMC. To decrease the aggregation of nanostructures during direct compounding, various surface treatments have been adopted for the nanostructure (Punetha et al. 2017). The grafting of organic molecules with functional groups in the polymeric matrix also serve as an anchorage point aiding the dispersion of nanoparticles (Belhout et al. 2019). In addition,the compounding conditions such as temperature and time, the configuration of the reactor and solvent-free routes can also be adjusted to achieve a good dispersion of nanoparticles in polymer matrices (Ning et al. 2019, Noh et al. 2015).

The *in situ* synthesis approach is also widely used to prepare nanomaterials-based PMC. According to different starting materials and fabrication processes, *in situ* synthesis can involve the *in situ* synthesis of nanostructures inside a polymer matrix or the *in situ* polymerization starting from a mixture of the monomers and the nanostructures. Each route is described and examples are given for a better illustration of the processes involved.

For the *in situ* synthesis of metal or metal oxide nanoparticles in a polymer matrix, ions are preloaded within a polymer matrix to serve as nanoparticle precursors, where ions could be distributed uniformly. Zhao et al. (2019) prepared Cu_2O nanoparticles embedded in a hyper-cross-linked polymer matrix. For this purpose, the polymeric matrix was mixed with a $CuCl_2$ ethanolic solution under stirring and ultra-sonication. The growth of Cu_2O nanoparticles was induced on the addition of an NH_2OH/HCl solution to the mixture. In this way, they reported the preparation of this PMC with well-dispersed Cu_2O nanoparticles, as a photocatalyst with improved stability and activity.

The other route uses the monomers as precursors of the polymeric hosts and the nanoparticles as starting materials. First, the nanoparticles are dispersed into the monomers or precursors of the polymeric hosts, and the mixture is then polymerized on addition of a suitable initiator. An adequate dispersion of the nanostructures into the liquid monomers or precursors would avoid their agglomeration in the polymer matrix and thereafter improve the interfacial interactions between both nanostructure and polymer matrix. As for direct compounding, the surface functionalization of nanostructures was also reported as a strategy to improve their dispersion in the polymer matrix for the *in situ* polymerization procedure (Wang et al. 2015).

Shokry et al. (2019) reported the *in situ* oxidative polymerization of polyaniline (PANi) to obtain well-dispersed Ag nanoparticles and graphene oxide quantum dots in a PANi matrix, as a highly stable and luminescent ternary composite.

Interestingly, the surface of nanoparticles can be exploited as a provider of the initiator for *in situ* polymerization in certain cases. Fe_3O_4 nanoparticles were used for a surface-initiation polymerization method. After surface treatment, Fe_3O_4 nanoparticles could catalyze the H_2O_2 decomposition to produce hydroxyl radical. These hydroxyl radicals were used to initiate the polymerization of the monomer in the surface of Fe_3O_4 nanoparticles, leading to the encapsulation of the nanoparticles in a polymer microsphere (Wang et al. 2020). The surface-initiation of polymerization was also achieved by UV irradiation of TiO_2 or TiO_2 covered nanoparticles, in the presence of different monomers. When TiO_2 absorbs a photon, electrons and holes generated in the surface drive the free radical polymerization nearby, producing core-shell nanocomposites with a polymer shell of polystyrene, poly(methyl methacrylate) or poly(N-isopropylacrylamide) (Wang et al. 2016).

As an alternative method, *in situ* interfacial polymerization has also been used for the preparation of nanomaterials-based PMC. Zhu et al. (2018) used interfacial polymerization to prepare sulphur covered carbon nanofibres embedded in a PANi matrix, as a high-performance cathode for Li-S batteries. Interfacial polymerization involves the contact of two immiscible phases, containing an initiator and monomer separately. On contact of the two immiscible phases, the initiator will start the polymerization in the interface, suppressing secondary growth and allowing it to obtain nanofibres.

Powder Metallurgy

Powder Metallurgy (PM) comprises a wide range of conventional processes that are used to produce dense 3-dimensional parts starting from metal powder (Upadhyaya 2014). When the metal powder is mixed with nanostructured materials, a composite powder is obtained and a Metal Matrix Composite (MMC) can be acquired. The PM approach has a major advantage of the ease for combining dissimilar materials that are difficult to obtain by other composite production methods (Chang and Zhao 2013).

PM can be performed through a large number of different processes. Here selected techniques for preparation of nanomaterials based MMC will be briefly described.

Among all the PM processes, the most common is the press-and-sinter process, and it is better than the conventional melting and casting for producing nanoparticles based MMC (Joshua et al. 2018). This PM route allows distribution of the dispersed phase or reinforcement nanoparticles without the segregation phenomenon, typical of casting processes (Chang and Zhao 2013). The press-and-sinter process involves three basic steps: powder mixing, compaction and sintering. Each step comprises different alternatives that are further described below.

The metal or alloy powder is mechanically mixed with the desired nanostructured material in order to obtain specific functionality. When a further reaction of the powders is required before the press and sinter steps, mechanical alloying is used. Mechanical Alloying (MA) has emerged in the last decades as an important step in the powder optimization process. Using a room temperature reactor known as a ball mill, a mechanically induced solid-state reaction can be accomplished between the powders of the reactant material. MA is a powerful tool for preparing a wide range of nanocomposite materials, characterized as a top-down approach (Eskandarany 2015).

The compaction step involves the application of high pressure, commonly using a rigid dye and special mechanical or hydraulic presses. The metal powder compaction has several functions such as the consolidation of the powder in a desired shape and dimensions, and to impart the level and type of porosity (Thümmler and Oberacker 1993). When uniform powders are placed together, the particles pack the space in between. As an example, a random packing of monodispersed spheres is usually taken to fill 64% of the available volume. The empty volume will decrease when the powder is pressed to form a compact, usually called the 'green compact' or 'green body'. The composite powders can be shaped by various PM techniques, including uniaxial pressing, cold isostatic pressing, metal injection moulding and powder rolling. Selection of the shaping process for consolidating the green body depends on the powder composition and the specifically required geometry of the part. For example, the simplest uniaxial pressing is selected for simple dye forms, while powder rolling is selected for sheets and strips (Chen et al. 2016).

Sintering consists of the process by which an assembly of particles chemically bond themselves into a coherent body (Fang 2010). The driving force of solid-state sintering is the excess surface energy. Smaller particles have larger surface energy, and in the consequence, they sinter faster than larger particles. Since atomic motion increases with temperature, sintering is accelerated by high temperatures. However, the temperature must be maintained below the melting point of the major constituent. From a microscopic point of view, the first stage of sintering corresponds to the neck growth between contacting particles. An intermediate stage is identified with pore rounding and the onset of grain growth. The final stage of sintering occurs when the pores collapse into closed spheres, giving a reduced impediment to grain growth. During all three stages, the microstructural changes are associated with the motion of atoms, that are transported by mechanisms such as surface diffusion and grain boundary diffusion. When sintered, the particles will bond and the empty volume fraction will be reduced. Therefore, densification or shrinkage of the sintered part is very often associated with all types of sintering.

Hot compaction processes are PM routes involving simultaneous pressing and sintering, such as Hot Isostatic Pressing (HIP or HIPing), hot extrusion, hot rolling and Field Activated Sintering Technique (FAST) (Upadhyaya 2014). The hot compaction processes aim to acquire a fully dense part. The HIP involves the application of high temperature and pressure and is suitable for materials that would not otherwise compact even at considerably higher pressures. Several reports of nanomaterials based MMC are prepared through HIP under vacuum (Zhang et al. 2018, Cui et al. 2020). The Field Activated Sintering Technique (FAST) also known as Sparking Plasma Sintering (SPS) has also been used as an alternative for the sintering of nanomaterials based MMC (Chen et al. 2019, Abe et al. 2020). The FAST process combines the simultaneous application of a high-density pulsed current, uniaxial pressure and low temperature (Fang 2010). It is stated that the FAST activates the particle's surface and therefore increases the rate of the sintering process, although the fundamental aspects related to this process remain a subject of debate. The key advantages of SPS include very high densification, minimal grain size growth, short sintering time and low sintering temperatures.

Electrochemical Composite Plating

The electrochemical composite plating is an unconventional pathway for the preparation of nanomaterials-based composites and hybrids. This method is used for electrodeposition of composites formed by an organic or inorganic dispersed phase within a metal matrix. In brief, a metal salt is reduced on a substrate in an electrochemical cell by applying a current, in the presence of dispersed particles. The nanoparticles are co-deposited during the plating, yielding a composite coating on the substrate. The main uses of electro deposited composites have been the improvement of wear resistance, corrosion resistance and lubrication of automotive parts. In recent years it was also used intensively in the area of electrocatalysis.

The properties of an electrodeposited composite depend mainly on the deposition conditions. Hence, it is important to understand the co-deposition mechanism. In summary, the incorporation of nanoparticles in the metal matrix can be described in a four-step process as follows: (1) Surface charge accumulation on the suspended nanoparticles (2) Mass transport from suspension bulk to electrode-electrolyte interface (3) Electrode and particle interaction (4) Incorporation of the particle and irreversible entrapment simultaneous with the growth of the metal layer (See Fig. 8.6) (Low et al. 2005).

The embedding of nanoparticles in a metal coating depends on the different parameters of electroplating, which includes: characteristics of the substrate, properties of nano/microparticles (concentration, surface charge, size, shape, composition), the composition of the electrolyte

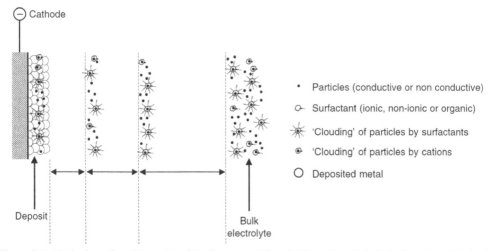

Figure 8.6: Mechanism of particle co-deposition into a metal deposit. The regions include the formation of ionic clouds around the particles (bulk electrolyte, typical length in cm); convective movement toward the cathode (convection layer, typical length < 1 mm); diffusion through a concentration boundary layer (diffusion layer, typical dimensions of hundreds of μm); electrical double layer (typical dimensions of nm) followed by adsorption and entrapment of particles (Reprinted from Low et al. (2006) with permission from Elsevier).

(concentration, presence of additives, temperature, pH), applied current program (direct current, pulsed current) and hydrodynamics (laminar, turbulent regime) together with the electrode geometry (rotating disc electrode, plate-in-tanks). There is a great deal of information about the relation of the above-mentioned parameters with the electrocatalyst properties (Gómez et al. 2019a,b, Loiácono et al. 2020). However, it is difficult to obtain an exact trend for all experimental parameters. Most recent research has suggested the effect of three global factors which should be considered: First, the applied current density, second, the type of particle and concentration, last, the agitation of the bath or movement of the electrode (Shim et al. 2006).

The most commonly used method is direct current. In this case, nanoparticles are embedded simultaneously with the reduction reaction of the ionic species that form the metal layer (Gómez et al. 2018). Only current density can be varied when using this method. The use of pulsed current has allowed the incorporation of higher concentrations of nanoparticles. In this case, other variables can be tuned such as frequency and duty cycle, providing a wider range of possibilities.

The nanoparticles concentration in the electroplating bath significantly affects the number of nanoparticles embedded, and this adjusts for a Langmuir adsorption isotherm. Therefore, the nanoparticle fraction in the composite increases substantially with the concentration of nanoparticles in the bath. There is a saturation of the embedded nanoparticle fraction for large concentrations of nanoparticles. The fraction of co-deposited nanoparticles has been shown to affect both, on the surface roughness and the microstructure of the deposit. This observation has been reported in several studies; thus, the presence of nanoparticles disturbs the crystalline metal growth. In this way, the number of crystal defects increases, favouring a nanocrystalline structure with a higher micro-hardness.

Bath stirring has two purposes: first, to keep micro/nanoparticles suspended in the electrolyte and second to transport particles from the bulk to the cathode surface. It is reported that increased agitation usually allows embedding larger amounts of nanoparticles in the deposit. However, excessive agitation can lead to a low number of deposited particles. This is explained in relation to vigorous hydrodynamic forces removing particles from the cathode surface, thus, preventing their entrapment in the metallic deposit.

Even when only three main aspects were described, as explained earlier, there are many variables affecting the co-deposition process. Therefore, it is important to consider them for a rational design of an MMC with optimal properties.

Inorganic NPs in Metallic Matrix Composites (MMC) Obtained by Electroplating

Electroplating can be suitably combined with improvement obtained from nanostructured materials. Since electrochemical processes are largely controlled by the chemical composition, crystalline phases and surface properties of the electrode, electrodeposition is a convenient method of regulating electrode properties. Moreover, with the emergence of nanostructured materials in the past few decades, there is a great need for different inorganic nanomaterials with tunable physicochemical properties. Therefore, considering the variety of metals that can be electrodeposited and the development of inorganic nanostructured materials, co-deposition of metal with inorganic nanoparticles enables the production of a wide range of composite materials.

The co-deposition of inorganic particles during the electroplating of a metallic matrix has found application in various fields and currently comprises an area of intense study. At the beginning of the 20th century, electrodeposited nickel films that supported sand particles were used as non-slip coatings on ship ladders (Hovestad and Janssen 2006). In the 80s, the development was focused on coatings with improved resistance to corrosion and wear. This was applied in the automotive industry, e.g., composite materials such as Ni|SiC served as abrasive protection coatings in car engines (Hovestad and Janssen 1995). At present, composites based on inorganic nanoparticles in a metallic matrix can greatly contribute to the area of renewable energies, as it will be described next.

Nowadays, there is a large number of research oriented to the efficient exploitation of renewable and clean energies. In this context, hydrogen emerges as an interesting energy vector, which can be obtained from intermittent renewable energy sources such as solar or wind energy. Particularly, water electrolysis in an alkaline medium presents great potential, allowing hydrogen production with high purity. In this case, electrocatalysis plays a major role. This has promoted, in recent years, an increasing number of studies on electrocatalysts based on non-noble metals, which can reduce production costs compared to platinum (Cao et al. 2018, Darband et al. 2019, Togharei et al. 2020, Li et al. 2018). The growing demand for new materials featuring defined properties for Hydrogen Evolution Reaction (HER) offers a promising prospect for composite materials applications.

According to current literature, various non-noble metals such as Ni, Co and Fe are active for HER catalysis in basic media (Inamuddin 2020). However, their catalytic activity is low, and therefore it is necessary to improve it. This can be achieved either by means of an alloy or by increasing its surface area (Navarro-Flores et al. 2005). When large-scale production is planned, the improvement of catalytic properties can reduce the final cost of hydrogen production, so it is a crucial aspect to be considered.

Non-noble metal electrodes with various oxide nanoparticles have shown synergistic effects on improving catalytic properties towards HER (Łosiewicz et al. 2004, Lačnjevac et al. 2012, Elias and Hedge 2016). Although there are different methods to produce metal matrix composites, the most widely used are electroplating, powder metallurgy, metal spraying and internal oxidation (Hovestad and Janssen 2006). The last three techniques require high temperatures, which is a great disadvantage because it increases production costs. Furthermore, these methods do not generally offer the possibility of preparing coatings or thin films, which is of utmost importance for electrocatalysis. On the other hand, electroplating involves simple equipment and does not require high temperatures. These are great advantages, that allow reducing production costs and making scale-up easier.

The Catalytic Activity of MMC

Some relevant results on composites prepared by electrodeposition are discussed below. In particular, papers concerning the embedding of metallic oxides nanoparticle in non-noble metals were selected. They were also oriented to the effect of the incorporation of such nanoparticles in the catalytic properties towards HER.

A commonly used transition metal oxide for the preparation of composites is TiO_2, because of its low cost, abundance and stability in basic media.

Danilov et al. (2016), reported the co-deposition of TiO_2 nanoparticles (25 nm) on Fe electrodes using direct current in a methanesulphonate bath. In this work, the concentration of TiO_2 in the electrodeposition bath at various current densities was studied. The number of nanoparticles in the composite was larger with increasing nanoparticle concentration. As mentioned above, this behaviour is explained based on a Langmuir adsorption isotherm. A decrease in nanoparticle incorporation was also observed for higher current densities. The electrodes presented a composition ranging from 2 wt. to 5 wt.% of TiO_2 relative to Fe. TiO_2 was found as nanoparticle agglomerates and the Fe surface structure did not vary significantly.

The electrochemical characterization of the prepared electrodes indicated an improvement of catalytic properties by increasing the amount of TiO_2 in this composition range (NaOH 1 M, 298 K). Larger values of j_0 (exchange current density) were found for composites, increasing by up to 50% with the embedding of TiO_2. On the other hand, R_{ct}[1] (Charge transfer resistance) also decreased significantly due to the presence of nanoparticles. The main factor affecting the catalysis was attributed to the appearance of active surface sites towards HER in the Fe|TiO_2 composites.

Franceschini et al. (2019) reported an improvement in the catalytic activity of Ni|TiO_2 composites with respect to Ni. This work compared the use of two different nanostructured TiO_2:

---

[1] Randles-CPE model.

commercial nanoparticles (25 nm) and mesoporous nanoparticles (500 nm). Both were dispersed in a traditional Watts bath (same g/L concentration) and prepared by a direct current. The resulting electrodes had a ratio of 2% at TiO_2, uniformly distributed in the surface as nanoparticle aggregates. The catalytic activity (KOH 1 M, 298 K) was improved in the case of mesoporous TiO_2, indicated by an increase in j_0 of about 50%. However, it did not improve significantly for the embedding of TiO_2 nanoparticles. The greater electrocatalytic activity of the Ni|TiO_2 composite is attributed to a synergistic effect related to a greater number of electroactive centres affected by the incorporation of mesoporous nanoparticles.

Electrodes were aged to establish their stability. The ageing process consisted of a 4-hour chronoamperometry at hydrogen generation overpotential. After ageing, the absence of nickel hydrides on composites was confirmed by Raman spectroscopy. This accounts for both composites, with TiO_2 nanoparticles and mesoporous TiO_2. Furthermore, activation of the electrode was observed, since j_0 evaluated after the ageing process was several orders of magnitude greater than a nickel for mesoporous TiO_2 composite. Therefore, this work reports not only an increase of catalytic activity but also of stability by including TiO_2 mesoporous nanostructures.

Rare earth oxide nanoparticles as CeO_2 have also been used in the development of electrocatalysts for HER. Zheng et al. (2012), studied the electrodeposition of Ni electrodes with CeO_2 particles in a sulphamate bath. Their efforts were focused on the embedding of different particle sizes: CeO_2 microparticles (7 µm) and nanoparticles (20 nm). Under the conditions specified in their research, a composition between 3.5 to 14% at was found for the CeO_2 using different microparticles concentrations. On the other hand, compositions of 3 to 7% were found for CeO_2 nanoparticles prepared in the same range of concentration.

It was found that nanoparticle embedding decreases the preferential orientation of the (200) planes. The change in the (1 1 1)/(2 0 0) ratio with respect to Ni is attributed to the formation of different chemical species adsorbed on the metal electrolyte interface during the electrodeposition process, altering the growth of Ni crystal planes.

J_0 increases an order of 72 and 42 times for composites with microparticles and nanoparticles, respectively. R_{ct}^2 decreases from 1 to 2 orders of magnitude in composites compared to nickel. Unlike the work of Danilov et al. (2016), the catalytic activity reaches a maximum value for an intermediate value of nanoparticle fraction in the composite. In other words, an increase in the number of nanoparticles embedded does not guarantee a consequent better catalytic activity in every material. However, j_0 remains several times above pure nickel for the largest CeO_2 content in composite.

In this case, it was proposed that the CeO_2 particles are active for HER, boosting the number of active sites. The authors discussed the possibility of an adsorption interaction between Ce (IV) and H. This interaction could explain a synergistic effect between the particles and the nickel matrix that favours migration of H to CeO_2, giving rise to a greater number of sites for progressive dissociation of water molecules.

According to the literature reviewed, it can be concluded that co-deposition of nanoparticles in a non-noble metal matrix constitutes a versatile strategy for improvement of catalytic properties for HER in basic media. Higher catalytic activity for micro/nanoparticle embedding may be due to an increase in the surface area or an increase in the intrinsic catalytic activity. This is why it is important to measure the electroactive area for a appropriate debate.

It was found that the preparation of these composites by electrodeposition not only favours the catalytic properties of the electrodes but their stability in ageing processes. On the other hand, there might be optimal ranges of nanoparticle fraction in the deposit, which is why exploration is essential in the preparation of these composite.

---

[2] Armstrong and Henderson with two CPE model.

Co-deposition of Metallic Oxide Nanoparticles in Metallic Alloys

Alloys are suitable for the preparation of composite coatings. They constitute an excellent starting point for the improvement of HER catalysts (Gomez et al. 2018). Therefore, two studies about composites based in alloys are presented here.

Sheng et al. (2018) prepared CoW|CeO$_2$ composites with nanoparticles (0.2–2 µm) suspended in a citrate bath. First, it was found that the amount of CeO$_2$ particles in the composite increases for higher nanoparticle concentration in the bath; however, this simultaneously affected the W to Co ratio. As CeO$_2$ increased, the W to Co ratio decreased. This also influenced the crystalline arrangement that depends on the alloy composition. The same electroplating bath in the absence of nanoparticles produced an amorphous CoW alloy, according to the Co to W ratio (50 wt.%). It is well known that for this composition range CoW alloy is amorphous. The inclusion of the nanoparticles, in this particular case, modified the alloy composition, and therefore, changed the crystal structure from amorphous to polycrystalline. CeO$_2$ nanoparticles significantly affected the electrodeposition process of the CoW alloy. This was explained by the authors in terms of the presence of CeO$_2$ hindering the diffusion of ternary tungsten complexes from the bulk to the electrolyte electrode interface. On the other hand, it was observed that the composites have a more compact shape with fewer cracks or defects compared to the CoW alloy. The surface structure of the composite shows a nodular shape with some aggregates of particles (See Fig. 8.7).

J$_0$ in the composite increases about 15 times for electrodes having 4 wt.% CeO$_2$, 22 wt.% W and 74 wt.% Co, in relation to the prepared alloy with a 50 wt.% W and 50 wt.% Co. The data obtained from Electrochemical Impedance spectroscopy (EIS) showed that R$_{ct}$[3] decreases significantly when the particles of CeO$_2$ are co-deposited.

The intrinsic catalytic activity value (j$_0$/R$_f$) was discussed in order to discriminate surface area contributions. Since the intrinsic catalytic activity was found to be higher in the composite, authors conclude that the incorporation of CeO$_2$ particles has a synergistic effect on improving the catalytic properties towards HER. However, as can be deduced from data, comparison of the composite to alloy involves changes in the alloy composition and crystalline structure. As embedding of nanoparticles might affect the alloy composition, it is a challenge to prepare comparable samples in all senses.

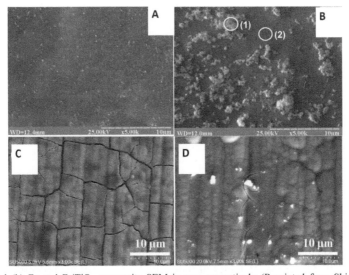

Figure 8.7: (a) and (b) Fe and Fe|TiO$_2$ composite SEM images respectively (Reprinted from Shim et al. 2006 with permission from Elsevier). SEM images of (c) CoW alloy and (d) CoW|CeO$_2$ composite (Reprinted from Sheng et al. (2018) with permission from Elsevier).

---

[3] Randles-CPE model.

Zhang et al. (2018) conducted research in this direction. They were trying to discern the role of nanoparticles in alloys composites. A composite and the alloy counterpart with the same composition was prepared.

Co-deposition of NiW alloy with TiO_2 (25 nm) nanoparticles was studied. A series of electrocatalysts was prepared using a constant TiO_2 bath concentration, but different current densities. This allowed them to obtain composites and the alloy counterpart with comparable surface area and also W to Ni composition.

In all the composites prepared at different current densities, the composition was around 1 wt.% TiO_2. However, for different current densities, different compositions of W with respect to Ni were obtained. In this case, depending on the composition of the alloy, the addition of TiO_2 affected or not. For low ratios of W to Ni, the incorporation of nanoparticles did not modify significantly j_0 (see Fig. 8.8). On the other hand, composites with higher W ratio presented an improvement of catalytic properties due to the presence of TiO_2 nanoparticles (see Fig. 8.8).

Comparing NiW alloys with the same composition, and with the same surface area (measured with ferrocene redox couple), an increase in intrinsic catalytic activity for HER is observed, therefore the authors concluded that the increase in catalytic activity is inherent to the incorporation of TiO_2.

As a conclusion, co-deposition of metallic oxide nanoparticles can also improve the catalytic properties of an alloy. Since embedding of nanoparticles might modify the composition of the alloy, it is interesting to find conditions where it is possible to compare the composite with the corresponding alloy counterpart.

It is clear from the different literature described, that nanoparticle properties play an essential role in the improvement of catalytic properties, either due to their size, surface and chemical properties.

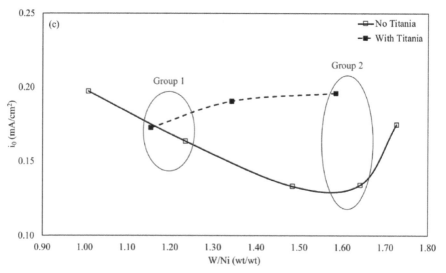

Figure 8.8: Exchange current density of NiW alloy deposits (solid line) and NiW|TiO_2 composites (dash line) (Reprinted from Zhang et al. (2018) with permission from Elsevier).

Organic Nanostructures in a Metallic Matrix

The concept of hybrid materials (otherwise hybrid organic/inorganic materials) developed in the 80s with the advances in deposition strategies of inorganic films, as sol-gel processes, where synthetic conditions allow the combination of inorganic domains with organic counterparts by chemical reactions (Azadmanjiri et al. 2018, Meng 2017). Thus, each of the parts had particular properties, and while the inorganic part contributed to its electrical resistivity, excellent thermal and chemical stability, the organic part provided tuneable chemical reactivity, adjustable optical and electronic properties, low density, between others (Franceschini et al. 2016, Franceschini and Lacconi 2018, Gómez et al. 2019a).

Besides the above-mentioned sol-gel approach (Tahara et al. 2005), numerous bottom-up approaches for materials deposition were developed for many applications including Layer–by–Layer (LbL) assembly (Keller et al. 1994), Chemical Vapour Deposition (CVD) method (Spee et al. 2015), Atomic (or Molecular) Layer Deposition (ALD or MLD) techniques (George et al. 1996, Ritala et al. 1999, Yoshimura et al. 2001, Kubono et al. 1996).

Moreover, two or three-dimensional hybrid structures, such as mesoporous hybrid films, particles and core-shells, allows the synergistic combination of many structural, chemical, optical and electric properties presented in the components (Lee et al. 2012).

Superficial Modification with Organic Molecules

In recent years, a great effort has been made to develop ways to anchor organic functional groups in inorganic matrices, giving space to a large range of hybrid materials with adjustable functionality. Calvo et al. (2009) performed superficial modifications on mesoporous silica thin films produced by a one-pot sol-gel method by growing zwitterionic poly (methacryloyl-L-lysine) (PML) brushes. The modification process was carried out by surface-initiated radical polymerization of the methacryloyl-L-lysine monomers using an initiator in the proper conditions (Yameen et al. 2009) and the mesoporous film modification was corroborated by X-ray Photoelectron Spectroscopy (XPS) and Diffuse Reflectance Infrared Fourier Transform Spectroscopy (DRIFTS). Thus, the presence of a brush with particular properties, in this case, a polyzwitterionic brush although they could be others, introduced great changes in the surface properties of the material. The combination of different mesoporous films and polymer brushes allows the manipulation of nanoscale topological and chemical aspects of the materials resulting in a synergistic effect between the inorganic and organic building blocks.

In a more recent work, Franceschini et al. (2016) developed a three-stage method to carry out the selective functionalization of mesoporous films. This method makes it easy to select organic functionality to be placed either on the outer surface or the inner surface.

This agreeable method allows the fast and selective functionalization of mesoporous thin films to be carried out by controlled deposition of alkoxide vapours, without affecting the stability of the material and leaving the internal surface of the pore intact. For this, during the first step of silane steam functionalization, the internal surface of the pore is kept blocked using the pore template. This template is removed through successive washes during the second step, thus releasing the inner surface of the pore. In the third step, a conventional silane anchoring surface modification method is used to modify the inner surface of the pores. In this case, two different functional groups were used on the inner and outer surface of the pores, to determine the location of the same using surface characterization techniques such as XPS and bulk such as EDS. The selected groups were trimethylchlorosilane (TMCS) and (3-aminopropyl) triethoxysilane (APTES).

In this way, the ability to select the spatial location of functional groups in 3D structures enables the rational development of a huge variety of MFMS.

Metal Matrix Growth Around Organic Nanostructures

Another way to build a hybrid nanomaterial is to use an organic nanomaterial as a scaffold (dispersed phase), such as carbon nanotubes, graphene, etc. And grow a second material around it (matrix). There are different methods to do this, particularly when metal matrices are used, chemical or electrochemical reductions are the most apt. This is because these organic materials have enormous stability under reduced conditions, maintaining their properties unchanged after the deposition of the matrix.

In addition to coating them, CNTs can be filled with metals, semiconductors, by different methods (Leonhardt et al. 2006, Hampel et al. 2006, Monthiux 2002, Tasis et al. 2006). Thus, the rational combination of CNTs with metal structures leads to new structural, optical, magnetic and electronic functionalities (Cleuziou et al. 2011, Garcia-Vidal et al. 1998, Borowiak-Palen et al. 2006, Tripathi and Bholanth 2010).

CNT reinforced materials have been shown to have a high Young modulus and tremendous tensile strength. This gives rise to the fact that in recent decades a large number of studies to grant multifunctionality to materials have been carried, such as metals (Bakshi et al. 2010, Choi and Bae 2011) and ceramics (Peigney et al. 2000, Sharma and Kothiyal 2015), through the inclusion of different types of CNT.

One of the main drawbacks in preparing CNT composites with metals is found when trying to distribute them homogeneously, since CNT normally agglomerates, so the mechanical properties obtained are not uniform. However, agglomeration does not appear to be a drawback for other applications, for example, catalysis where a marked synergy between CNT and metals is usually reported (Gao et al. 2019, Cheng and Jiang 2015). A similar effect can be found in materials where the incorporation of reduced graphene oxide or its derivatives are incorporated into metal matrices to decrease corrosion of non-noble metals and increase catalytic efficiency (Gómez et al. 2019a, Franceschini and Lacconi 2018).

Organic Nanostructures in an Organic Matrix

The development of carbon/carbon composites or C/C composites dates back to the late 1960s in response to the structural limitations introduced in graphite, which until then was the material typically used for ultra-high temperature applications (even 3,000°C) (Thomas 1993, Fitzer and Manocha 1998). Thus, carbonaceous matrix composites normally reinforced with carbon fibre began to develop. These materials have an extraordinarily high tenacity and chemical resistance, added to the inherent high thermal stability at more than 3,000°C in an oxygen-free environment, high thermal conductivity, low thermal expansion coefficient, high thermal shock resistance and low recession in high-pressure environments, high fracture resistance and a low density, even below 2 g/cm^3, make these compounds the most suitable candidates for a large number of applications. Initial developments, due to the high costs of these compounds, were restricted to space and military applications, such as rocket nozzles and thermal protection system for re-entry space vehicles (Fitzer and Manocha 1998, Krenkel 2008, Schmidt et al. 1999a,b,c, Evans and Zok 1994, Aly-Hassan et al. 2003a,b, Hatta et al. 2005a,b,c, 2004).

However, one of the most interesting properties of C/C compounds is that mechanical resistance increases with operating temperature as opposed to the response found in metals and ceramics, where an opposite trend is observed (Goto et al. 2003, Sauder et al. 2004, Aoki et al. 2007), which makes C/C composites ideal for high-temperature applications, not only in aerospace and military applications but also for other applications, such as automotive (on brake discs).

As discussed in Chapter 1, the reduction of vehicle weight is of particular interest, particularly in the aerospace industry, since this generates a considerable increase in autonomy, reducing the amount of fuel required and therefore costs.[4] For this reason, the use of C/C composites appears as excellent alternatives to replace structural materials in prototypes, particularly those that have multifunctional properties, for example, the thermal conductivity (working as a heat exchanger) or the energy harvesting.

Here some of the more common C/C compounds, particularly those considered to be MFMS will be discussed.

---

[4] Conventional rocket designs cost approximately $ 20,000/kg to transport payloads into Earth orbit and increase to $ 200,000/kg in the case of deep space.

Expanded Graphite and Graphene

Materials with a hierarchical structure have been intensively studied in various fields of artificial photosynthesis, photo and electrocatalysis, dye-sensitized solar cells and fuel cells, among a wide range of applications (Wang et al. 2008, Soller-Illia et al. 2002, Huang et al. 2001, Guo et al. 2008, Qin et al. 2019), because this large surface area provides a large specific surface area for reactions, mass transport or dispersal of active sites (Li et al. 2012, Magasinski et al. 2010). However, another possible and less obvious application is that of thermal energy storage, since as mentioned earlier, heat transfer occurs mainly through the matrix and the porous structure allows the storage of thermal energy (Li et al. 2018, Tang et al. 2017).

Expanded graphite is a material with a porous structure commonly used for the manufacture of phase change material composites to improve thermal conductivity (Zhang and Fang 2006, Sari and Kairapleki 2007). However, its porosity is not high enough to obtain good performance for the storage of thermal energy, largely limiting its wide application. Thus, the expanded graphite can be mixed with another porous material to increase the porous volume and improve its efficiency.

One option to improve the properties of expanded graphite composites is the use of Hyper-Crosslinked Polymer (HCP), which not only displays a large specific surface area but also exhibits porous structure with good stability (Tan and Tan 2017, Liu et al. 2009). These polymers have many advantages, among them an easy preparation process but perhaps the most important, they can be designed to have some specific functionalities (Yang et al. 2008, Ouyang et al. 2019, Chan et al. 2018b).

Liu et al. (2020) developed a hierarchical structured Expanded Graphite (EG)/HCP material in a 'knitting' HCP protocol. They used it as a superior holding material to impregnate PCM with the aim of addressing the problem of low thermal conductivity and poor encapsulation rate of a phase change material composite. This is due to the combination of the properties of EG and HCP, which demonstrated high thermal conductivity and large specific surface area respectively.

Moreover, there are numerous works where hierarchical 3D reduced graphene porous carbon-based or graphene foam-based composites for superior thermal energy storage performance are reported (Li et al. 2018, Qi et al. 2016), but also for many other applications.

It is reported in the literature that 5 wt.% graphene (Ren et al. 2018) in a composite improves the thermal conductivity of the resulting nanocomposite reaching a maximum value of 1.18 Wm^{-1}K^{-1}, which added to the effectiveness of electromagnetic interference shielding (EMI) of 38 dB makes it an excellent candidate for being used in electronic applications such as electronic packaging. These properties, and others such as strain sensing and Joule heating (Kernin et al. 2019), originate from the possibility that graphene presents to generate interconnected networks when used as a dispersed phase in a polymer matrix. The possibility of these interconnections, added to the electronic and optical properties of graphene transfer multifunctionality to their nanocomposites, for example in the fields of optoelectronics, energy storage, electronics as well as biomedical applications (Ma et al. 2018, Chen et al. 2018, Olszowska et al. 2017).

With the rapid advancement of high-resolution 3D printing technology, particularly the so-called Additive Manufacturing (AM) technologies, it is believed to have been the perfect complement for the use of nanocomposites in prototypes and mass industrial production, particularly for graphene nanocomposites, which has great flexibility and thermal resistance.

The 3D printing process by AM allows the manufacture of a part with complex 3D geometry designed with Computer-Aided Design (CAD) software without any mould and without any intermediate step, such as joining or assembling, allowing to overcome many limitations in the use of graphene (and many other materials). Nevertheless, significant progress has been made in the development of 2D nanofilms (with techniques such as direct writing, soft lithography and photolithography), the manufacture of multifunctional 3D structures still remains a challenge.

Some examples of printed graphene/polymer resin nanocomposites with a complex 3D structure are shown in Fig. 8.9, demonstrating the great potential of this technique, particularly when combined with multifunctional materials such as graphene-based or expanded graphite particles.

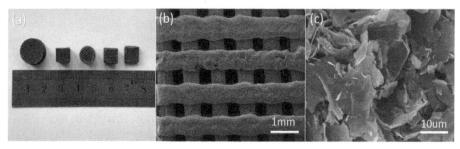

Figure 8.9: Morphology of three-dimensional graphene-based structures. (a) Photograph of three-dimensional graphene-based structures. (b) and (c) SEM micrographs of the surface of the structures with different magnification (Reprinted from You et al. (2018) with permission from Elsevier).

CNT and Carbon Fibres

One of the most common forms of C/C composites is the inclusion of various types of nanofillers in traditional polymeric matrices to improve their mechanical, thermal, electrical and other barrier properties, adding multifunctionality to the compounds. Examples include 2D nanomaterials (e.g., graphene flakes), 1D (e.g., single-walled (SWCNT), multiple-walled (MWCNT), cup-stacked carbon nanotubes (CSCNT), and produced carbon fibre Vapour (VGCF)) or 0D (e.g., Fullerene) (Ajayan et al. 2003, Kojima et al. 1993, Gojny et al. 2005, Delozier et al. 2005, Zhu et al. 2006, Liu et al. 2005, Kim et al. 2005, Ogasawara et al. 2006). Since their publication (Iijima 1991), carbon nanotubes (CNT) have been the subject of extensive academic and applied research. Numerous investigations in the last 30 years have reported on extraordinary thermal and electrical behaviour, stiffness, strength and mechanical resistance for CNT (Thostenson et al. 2001).

In order to improve the properties of carbon fibre reinforced polymers, different techniques have been designed to achieve the hybridization of carbon fibres with carbon nanotubes (CNT) (Zhao et al. 2008, Hassanzadeh-Aghdam et al. 2018, Dong et al. 2014, Kamae and Drzal 2012, Xiao et al. 2018, Zhou et al. 2017), thus combining the extraordinary physical and mechanical properties of CNTs by coating the surface of the carbon fibre with the CNT to provide a stronger interfacial adhesion between the carbon fibre and the polymer matrix. Thus, a system with a hierarchical structure with carbon fibres at the microscale and CNT at the nanoscale is used. This structure where the dispersed phase has this hierarchical structure forms a composite system with a 3D network structure with unique characteristics.

Research on hybrid CF-CNT has become increasingly popular. CNT/carbon fibre hybridization is performed by directly depositing or growing CNTs on the surface of the carbon fibres. Thus, CF-CNT hybridization is considered one of the best approaches to improve interfacial adhesion between fibre and polymer matrix. Some of the methods used include Chemical Vapour Deposition (CVD), electrophoretic deposition (EPD), electrospray deposition (ESD) and chemical functionalization.

Two examples of this hybridization, shown by the electrospray deposition technique, are shown in Fig. 8.10. It can be seen how the presence of CNTs considerably increases the area/volume ratio, thus explaining the increased interaction between the dispersed phase and the matrix in a composite.

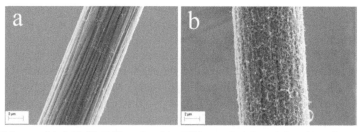

Figure 8.10: SEM image of hybrid CF-CNT produced by ESD as a function of deposition time: (a) 2 minutes and (b) 10 minutes (Reprinted from Li et al. (2016) with permission from Elsevier).

Conclusion

The synthesis of nanostructured composites is currently one of the most explored ways to obtain materials for multiple applications. In this chapter, different combination alternatives between dispersed phases and organic and inorganic matrices have been explored by analyzing various synthesis pathways and presenting some representative examples. The use of nanostructured dispersed phases maximizes the synergistic effect between the materials that compose the composite. In this way, a rational choice of structure, chemical composition, and concentration of the dispersed phase and matrix will allow obtaining materials with unique functionalities adaptable to different applications.

Acknowledgements

The authors thank the financial support from Agencia Nacional de Promoción Científica y Tecnológica (PICT 2017-0250), SECyT-UNC and CONICET (PUE2017). AL thanks for her Scholarships to stimulate scientific vocation granted by CIN and YPF foundation. VBL thanks CONICET for her postdoctoral fellowship. EAF is a permanent research fellow of CONICET. The authors thank Dr Ana Spitale for her assistance in translation.

References

Abe, J. O., A. P. I. Popoola and O. M. Popoola. 2020. Consolidation of Ti6Al4V alloy and refractory nitride nanoparticles by spark plasma sintering method: Microstructure, mechanical, corrosion and oxidation characteristics. Mater. Sci. Eng. A. 774(1): 138920.

Ajayan, P. A, L. S. Schadler and P. V. Braun. 2003. Polymer-based and polymer-filled nanocomposites. Nanocomposite Science and Technology. Weinheim: Wiley-VCH.

Allaoui, A., S. Bai, H. M. Cheng and J. B. Bai. 2002. Mechanical and electrical properties of a MWNT/epoxy composite. Compos. Sci. Technol. 62: 1993–8.

Aly-Hassan, M. S., H. Hatta and S. Wakayama. 2003a. Effect of zigzag damage extension mechanism on fracture toughness of cross-ply laminated carbon/carbon composites. Adv. Compos. Mater. 12(2–3): 223–36.

Aly-Hassan, M. S., H. Hatta, S. Wakayama, M. Watanabe and K. Miyagawa. 2003b. Comparison of 2D and 3D carbon/carbon composites with respect to damage and fracture resistance. Carbon 41(5): 1069–78.

Anton, S. R. and H. A. Sodano. 2007. A review of power harvesting using piezoelectric materials (2003–2006). Smart Mater Struct. 16(3): R1–21.

Aoki, T., Y. Yamane, T. Ogasawara, T. Ogawa, S. Sugimoto and T. Ishikawa. 2007. Measurements of fiber bundle interfacial properties of three-dimensionally reinforced carbon/carbon composites up to 2273 K. Carbon 45: 459–67.

Asp, L. E. and E. S. Greenhalgh. 2014. Structural power composites. Compos. Sci. Technol. 101: 41–61.

Azadmanjiri, J., J. Wang, C. C. Berndt and A. Yu. 2018. 2D layered organic–inorganic heterostructures for clean energy applications. J. Mater. Chem. A. 6: 3824–3849.

Baechle, D. M., D. J. O'Brien and E. D. Wetzel. 2007. Design and processing of structural composite capacitors. *In*: Proc. Int. SAMPE Symposium and Exhibition, M and P – from Coast to Coast and Around the World.

Bakshi, S. R., D. Lahiri and A. Agarwal. 2010. Carbon nanotube reinforced metal matrix composites—a review. Int. Mater. Rev. 55: 41–64.

Bauhofer, W. and J. Z. Kovacs. 2009. A review and analysis of electrical percolation in carbon nanotube polymer composites. Compos. Sci. Technol. 69(10): 1486–98.

Belhout, S. A., F. R. Baptista, S. J. Devereux, A. W. Parker, A. D. Ward and S. J. Quinn. 2019. Preparation of polymer gold nanoparticle composites with tunable plasmon coupling and their application as SERS substrates. Nanoscale 11(42): 19884–19894.

Bent, A. A., N. W. Hagood and J. P. Rogers. 1995. Anisotropic actuation with piezoelectric fiber composites. J. Intell. Mater. Syst. Struct. 6: 338–49.

Biercuk, M. J., M. C. Llaguno, M. Radosavljevic, J. K. Hyun and A. T. Johnson. 2002. Carbon nanotube composites for thermal management. Appl. Phys. Lett. 80(15): 2767–9.

Bonnet, P., D. Sireude, B. Garnier and O. Chauvet. 2007. Thermal properties and percolation in carbon nanotube–polymer composites. Appl. Phys. Lett. 91: 201910.

Borowiak-Palen, E., M. H. Ruemmeli, T. Gemming, T. Pichler, R. J. Kalenczuk and S. R. Silva. 2006. Silver filled single-wall carbon nanotubes synthesis, structural and electronic properties. Nanotech. 17: 2415–9.

Brei, D. and B. J. Cannon. 2004. Piezoceramic hollow fiber active composites. Compos. Sci. Technol. 64: 245–61.

Brown, E. N., N. R. Sottos and S. R. White. 2002. Fracture testing of a self-healing polymer composite. Exp. Mech. 42: 372–9.

Brown, E. N., S. R. White and N. R. Sottos. 2005a. Retardation and repair of fatigue cracks in a microcapsule toughened epoxy composite—part I: manual infiltration. Compos. Sci. Technol. 65: 2466–73.

Brown, E. N., S. R. White and N. R. Sottos. 2005b. Retardation and repair of fatigue cracks in a microcapsule toughened epoxy composite—part II: *in situ* self-healing. Compos. Sci. Technol. 65: 2474–80.

Brown, E. N., S. R. White and N. R. Sottos. 2006. Fatigue crack propagation in microcapsule toughened epoxy. J. Mater. Sci. 41(19): 6266–73.

Calvo, A., B. Yameen, F. J. Williams, Galo J. A. A. Soler-Illia and O. Azzaroni. 2009. Mesoporous films and polymer brushes helping each other to modulate ionic transport in nanoconfined environments. An interesting example of synergism in functional hybrid assemblies. J. Am. Chem. Soc. 131: 10866–10868.

Cao, X., D. Jia, D. Li, L. Cui and J. Liu. 2018. One-step co-electrodeposition of hierarchical radial NixP nanospheres on Ni foam as highly active flexible electrodes for hydrogen evolution reaction and supercapacitor. Chem. Eng. J. 348: 310–318.

Caruso, M. M., B. J. Blaiszik, S. R. White, N. R. Sottos and J. S. Moore. 2008. Full recovery of fracture toughness using a nontoxic solvent-based self-healing system. Adv. Funct. Mater. 18(13): 1898–904.

Cebeci, H., R. Guzman de Villoria, A. J. Hart and B. L. Wardle. 2009. Multifunctional properties of high volume fraction aligned carbon nanotube polymer composites with controlled morphology. Compos. Sci. Technol. 69: 2649–56.

Chan, K. -Y., B. H. Lin, N. Hameed, J. H. Lee and K. T. Lau. 2018a. A critical review on multifunctional composites as structural capacitors for energy storage. Compos. Struct. 188: 126–142.

Chan, Y. -H., W. -Z. Lew, E. Lu, T. Loretz, L. Lu, C. -T. Lin et al. 2018b. An evaluation of the biocompatibility andosseointegration of novel glass fiber reinforced composite implants: *In vitro* and *in vivo* studies. Dental Mater. 34: 470–485.

Chang, I. and Y. Zhao. 2013. Advances in Powder Metallurgy. Woodhead Publishing Limited.

Chen, B., K. Kondoh, J. Umeda, S. Li, L. Jia and J. Li. 2019. Interfacial *in-situ* Al_2O_3 nanoparticles enhance load transfer in carbon nanotube (CNT)-reinforced aluminum matrix composites. J. Alloys Compd. 789: 25–29.

Chen, L., H. Yu, J. Zhong, J. Wu and W. Su. 2018. Graphene based hybrid/composite for electron field emission: a review. J. Alloy Compd. 749: 60–84.

Chen, Y., X. Zhang, E. Liu, C. He, C. Shi, Jiajun Li et al. 2016. Fabrication of *in-situ* grown graphene reinforced Cu matrix composites. Sci. Rep. 6(1): 1–9.

Cheng, Y. and S. P. Jiang. 2015. Advances in electrocatalysts for oxygen evolution reaction of water electrolysis-from metal oxides to carbon nanotubes. Prog. Nat. Sci. Mat. Int. 25(6): 545–553.

Choi, H. J. and D. H. Bae. 2011. Strengthening and toughening of aluminum by single-walled carbon nanotubes. Mat. Sci. Eng. A-Struct. 528: 2412–7.

Chung, D. D. L. 2019. A review of multifunctional polymer-matrix structural composites. Compos. B Eng. 160: 644–660.

Cleuziou, J. -P., W. Wernsdorfe, T. Ondarçuhu and M. Monthioux. 2011. Electrical detection of individual magnetic nanoparticles encapsulated in carbon nanotubes. ACS Nano 5(3): 2348–55.

Coleman, J. N., U. Khan, W. J. Blau and Y. K. Gun'ko. 2006. Small but strong: A review of the mechanical properties of carbon nanotube-polymer composites. Carbon N. Y. 44(9): 1624–1652.

Cook-Chennault, K. A., N. Thambi and A. M. Sastry. 2008. Powering MEMS portable devices—a review of non-regenerative and regenerative power supply systems with special emphasis on piezoelectric energy harvesting systems. Smart Mater. Struct. 17(4): 043001.

Cui, G., Y. Liu, S. Li, H. Liu, G. Gao and Z. Kou. 2020. Nano-TiO_2 reinforced CoCr matrix wear resistant composites and high-temperature tribological behaviors under unlubricated condition. Sci. Rep. 10(1): 1–12.

Danilov, F. I., A. V. Tsurkan, E. A. Vasil'Eva and V. S. Protsenko. 2016. Electrocatalytic activity of composite Fe/TiO_2 electrodeposits for hydrogen evolution reaction in alkaline solutions. Int. J. Hydrogen Energy. 41(18): 7363–7372.

Darband, G. B., M. Aliofkhazraei, A. S. Rouhaghdam and M. A. Kiani. 2019. Three-dimensional Ni-Co alloy hierarchical nanostructure as efficient non-noble-metal electrocatalyst for hydrogen evolution reaction. Appl. Surf. Sci. 465: 846–862.

Delozier, D. M., K. A. Watson, J. G. Smith and J. W. Connell. 2005. Preparation and characterization of space durable polymer nanocomposite films. Compos. Sci. Technol. 65: 749–55.

Dong, L., F. Hou, Y. Li, L. Wang, H. Gao and Y. Tang. 2014. Preparation of continuous carbon nanotube networks in carbon fiber/epoxy composite. Compos. Appl. Sci. Manuf. 56: 248–55.

El-Eskandarany, M. S. 2015. Mechanical Alloying. Elsevier: pp. 1–12.

Elias, L. and A. C. Hegde. 2016. Modification of Ni-P alloy coatings for better hydrogen production by electrochemical dissolution and TiO$_2$ nanoparticles. RSC Adv. 6(70): 66204–66214.

Evans, A. G. and F. W. Zok. 1994. Review: the physics and mechanics of fibre-reinforced brittle matrix composites. J. Mater. Sci. 29: 3857–96.

Fang, Z. Z. 2010. Sintering of Advanced Materials. Elsevier.

Fernandez, J. F., A. Dogan, Q. M. Zhang, J. F. Tressler and R. E. Newnham. 1996. Hollow piezoelectric composites. Sens. Actuators 51: 183–92.

Fitzer, E. and L. M. Manocha. 1998. Carbon Reinforcements and Carbon/Carbon Composites. Berlin: Springer Verlag GmbH & Co.

Franceschini, E. A., E. de la Llave, F. J. Williams and Galo J. A. A. Soler-Illia. 2016. A simple three step method for selective placement of organic groups in mesoporous silica thin films. Mater. Chem. Phys. 169: 82–88.

Franceschini, E. A. and Gabriela I. Lacconi. 2018. Synthesis and performance of nickel/reduced graphene oxide hybrid for hydrogen evolution reaction. Electrocatalysis 9: 47–58.

Franceschini, E. A., M. J. Gomez and G. I. Lacconi. 2019. One step synthesis of high efficiency nickel/mesoporous TiO$_2$ hybrid catalyst for hydrogen evolution reaction. J. Energy Chem. 29: 79–87.

Ganguli, S., A. K. Roy and D. P. Anderson. 2008. Improved thermal conductivity for chemically functionalized exfoliated graphite/epoxy composites. Carbon 46(5): 806–17.

Gao, W., D. Wen, J. C. Ho and Y. Qu. 2019. Incorporation of rare earth elements with transition metal–based materials for electrocatalysis: a review for recent progress. Mater. Today Chem. 12: 266–281.

García-Vidal, F. J., J. M. Pitarke and J. B. Pendry. 1998. Silver-filled carbon nanotubes used as spectroscopic enhancers. Phys. Rev. B. 58(11): 6783–6.

George, S. M., A. W. Ott and J. W. Klaus. 1996. Surface chemistry for atomic layer growth. J. Phys. Chem. 100: 13121–13131.

Gibson, R. F., Y. Chen and H. Zhao. 2001. Improvement of vibration damping capacity and fracture toughness in composite laminates by the use of polymeric interleaves. J. Eng. Mater. Technol. 123: 309–14.

Gibson, R. F., E. O. Ayorinde and Y. -F. Wen. 2007. Vibrations of carbon nanotubes and their composites: a review. Compos. Sci. Technol. 67: 1–28.

Gibson, T., S. Putthanarat, J. C. Fielding, A. Drain, K. Will and M. Stoffel. 2007. Conductive nanocomposites: focus on lightning strike protection. In: Proc. Int. SAMPE Tech Conf. and Exhibition—From Art to Science: Advancing Mater and Proc. Eng.

Giurgiutiu, V. and A. N. Zagrai. 2001. Characterization of piezoelectric wafer active sensors. J. Intell. Mater. Syst. Struct. 11(12): 959–76.

Gojny, F. H., M. H. G. Wichmann, B. Fiedler and K. Schulte. 2005. Influence of different carbon nanotubes on the mechanical properties of epoxy matrix composites—a comparative study. Compos. Sci. Technol. 65: 2300–13.

Gomez, M. J., E. A. Franceschini and G. I. Lacconi. 2018. Ni and NixCoy alloys electrodeposited on stainless steel AISI 316L for hydrogen evolution reaction. Electrocatalysis 9(4): 459–470.

Gómez, M. J., A. Loiácono, L. A. Pérez, E. A. Franceschini and G. I. Lacconi. 2019a. Highly efficient hybrid Ni/ nitrogenated graphene electrocatalysts for hydrogen evolution reaction. ACS Omega 4(1): 2206–2216.

Gómez, M. J., L. A. Diaz, E. A. Franceschini, G. I. Lacconi and G. C. Abuin. 2019b. 3D nanostructured NiMo catalyst electrodeposited on 316L stainless steel for hydrogen generation in industrial applications. J. Appl. Electrochem. 49: 1227–1238.

Goto, K., H. Hatta, M. Oe and T. Koizumi. 2003. Tensile strength and deformation of a two-dimensional carbon–carbon composite at elevated temperature. J. Am. Ceram. 86(12): 2129–35.

Guo, Y. -G., J. -S. Hu and L. -J. Wan. 2008. Nanostructured materials for electrochemical energy conversion and storage devices. Adv. Mater. 20(15): 2878–2887.

Hagood, N. W. and A. A. Bent. 1993. Development of piezoelectric fiber composites for structural actuation. In: Proc. 34th AIAA Structures, Structural Dynamics and Mater Conference; AIAA Paper No. 93-1717, La Jolla, CA.

Hampel, S., A. Leonhardt, D. Selbmann, K. Biedermann, D. Elefant, C. Müller et al. 2006. Growth and characterization of filled carbon nanotubes with ferromagnetic properties. Carbon 44: 2316–22.

Hassanzadeh-Aghdam, M. K., R. Ansari and A. Darvizeh. 2018. Micromechanical analysis of carbon nanotube-coated fiber-reinforced hybrid composites. Int. J. Eng. Sci. 130: 215–29.

Hatta, H., L. Denk, T. Watanabe, I. Shiota and M. S. Aly-Hassan. 2004. Fracture behavior of carbon/ carbon composites with cross-ply lamination. Compos. Mater. 38(17): 1479–94.

Hatta, H., K. Goto and T. Aoki. 2005a. Strengths of C/C composites under tensile, shear, and compressive loading: role of interfacial shear strength. Compos. Sci. Technol. 65(15–16): 2550–62.

Hatta, H., K. Goto, S. Ikegaki, I. Kawahara, M. S. Aly-Hassan and H. Hamada. 2005b. Tensile strength and fiber/ matrix interfacial properties of 2D- and 3D-carbon/carbon composites. Eur. Ceram. Soc. 25(4): 535–42.

Hatta, H., M. S. Aly-Hassan, Y. Hatsukade, S. Wakayama, H. Suemasu and N. Kasai. 2005c. Damage detection of C/C composites using ESPI and SQUID techniques. Compos. Sci. Technol. 65(7–8): 1098–106.

Haywood, J., P. T. Coverley, W. J. Staszewski and K. Worden. 2005. An automated impact monitor for a composite panel employing smart sensor technology. Smart Mater. Struct. 14: 265–71.

Hovestad A. and L. J. J. Janssen. 1995. Electrochemical codeposition of inert particles in a metallic matrix. J. Appl. Electrochem. 25: 519–527, doi: 10.1007/BF00573209.

Hovestad, A. and L. J. J. Janssen. 2006. Electroplating of metal matrix composites by codeposition of suspended particles. Mod. Asp. Electrochem. 38: 475–532.

Huang, Y., X. F. Duan, Q. Q. Wei and C. M. Lieber. 2001. Directed assembly of one-dimensional nanostructures into functional networks. Science 291(5504): 630–633.

Iijima, S. 1991. Helical microtubules of graphite carbon. Nature 354: 56–8.

Inamuddin, Boddula, R. and A. M. Asiri. 2020. Methods for Electrocatalysis. Advanced Materials and Allied Applications. Berlin, Springer Verlag GmbH & Co.

Joshua, K. J., S. J. Vijay and D. P. Selvaraj. 2018. Effect of nano TiO_2 particles on microhardness and microstructural behavior of AA7068 metal matrix composites. Ceram. Int. 44(17): 20774–20781.

Kalaitzidou, K., H. Fukushima and L. T. Drzal. 2007. Multifunctional polypropylene composites produced by incorporation of exfoliated graphite nanoplatelets. Carbon 45: 1446–52.

Kamae, T. and L. T. Drzal. 2012. Carbon fiber/epoxy composite property enhancement through incorporation of carbon nanotubes at the fiber–matrix interphase–Part I: the development of carbon nanotube coated carbon fibers and the evaluation of their adhesion. Compos. Appl. Sci. Manuf. 43: 1569–77.

Keller, S. W., H. Kim and T. E. Mallouk. 1994. Layer–by–layer assembly of intercalation compounds and heterostructures on surfaces: toward molecular "beaker" epitaxy. J. Am. Chem. Soc. 116: 8817–8818.

Kernin, A., K. Wan, Y. Liu, X. Shi, J. Kong, E. Bilotti et al. 2019. The effect of graphene network formation on the electrical, mechanical, and multifunctional properties of graphene/epoxy nanocomposites. Compos. Sci. Technol. 169: 224–31.

Kim, H. S., J. S. Kang, J. S. Park, H. T. Hahn, H. C. Jung and J. W. Joung. 2009. Inkjet printed electronics for multifunctional composite structure. Compos. Sci. Technol. 69(7–8): 1256–64.

Kim, J. K., C. Hu, R. S. C. Woo and M. -L. Sham. 2005. Moisture barrier characteristics of organoclay epoxy nanocomposites. Compos. Sci. Technol. 65: 805–13.

Kim, Y. A., S. Kamio S. Tajiri, T. Hayashi, S. M. Song, M. Endo et al. 2007. Enhanced thermal conductivity of carbon fiber/phenolic resin composites by the introduction of carbon nanotubes. Appl. Phys. Lett. 90: 093125.

Kirk, J. G., S. Naik, J. C. Moosbrugger, D. J. Morrison, D. Volkov and I. Sokolov. 2009. Self-healing epoxy composites based on the use of nanoporous silica capsules. Int. J. Fract. 159(1): 101–2.

Kojima, Y., A. Usuki, M. Kawasaki, A. Okada, Y. Fukushima, T. Kurauchi et al. 1993. Mechanical properties of nylon 6-clay hybrid. J. Mater. Res. 8: 1185–9.

Koratkar, N. A. and J. Suhr. 2005. Characterizing energy dissipation in single-walled carbon nanotube polycarbonate composites. Appl. Phys. Lett. 87: 063102.

Krenkel, W. [ed.]. 2008. Ceramic Matrix Composites: Fiber Reinforced Ceramics and Their Applications. Weinheim, Germany: John Wiley & Sons.

Kubono, A., N. Yuasa, H. Shao, S. Umemoto and N. Okui. 1996. *In-situ* study on alternating vapor deposition polymerization of alkyl polyamide with normal molecular orientation. Thin Solid Films 289: 107–111.

Lačnjevac, U. Č., B. M. Jović, V. D. Jović and N. V. Krstajić. 2012. Determination of kinetic parameters for the hydrogen evolution reaction on the electrodeposited $Ni-MoO_2$ composite coating in alkaline solution. J. Electroanal. Chem. 677–680: 31–40.

Lee, B. H., B. Yoon, A. I. Abdulagatov, R. A. Hall and S. M. George. 2012. Growth and properties of hybrid organic–inorganic metalcone films using molecular layer deposition techniques. Adv. Funct. Mater. 23: 532–546.

Leonhardt, A., S. Hampel, C. Müller, I. Mönch, R. Koseva, M. Ritschel et al. 2006. Synthesis, properties, and applications of ferromagnetic-filled carbon nanotubes. Chem. Vap. Dep. 12(6): 380–7.

Lesieutre, G. A. and C. L. Davis. 1997. Can a coupling coefficient of a piezoelectric device be higher than those of its active material? J. Intell. Mater. Syst. Struct. 8(10): 859–67.

Li, A., C. Dong, W. Dong, D. G. Atinafu, H. Gao, X. Chen et al. 2018. Hierarchical 3D reduced graphene porous-carbon-based PCMs for superior thermal energy storage performance. ACS Appl. Mater. Interfaces 10(38): 32093–32101.

Li, C., E. T. Thostenson and T. -W. Chou. 2008. Sensors and actuators based on carbon nanotubes and their composites: a review. Compos. Sci. Technol. 68: 1227–49.

Li, D., X. Cheng, Y. Li, H. Zou, G. Yu, G. Li et al. 2018. Effect of MOF derived hierarchical Co_3O_4/expanded graphite on thermal performance of stearic acid phase change material. Sol. Energy 171: 142–149.

Li, J., P. C. Ma, W. S. Chow, C. K. To, B. Z. Tang and J. -K. Kim. 2007. Correlations between percolation threshold, dispersion state, and aspect ratio of carbon nanotubes. Adv. Funct. Mater. 17: 3207–15.

Li, Q., J. S. Church, M. Naebe and B. L. Fox. 2016. A systematic investigation into a novel method for preparing carbon fibre–carbon nanotube hybrid structures. Compos. Appl. Sci. Manuf. 90: 174–85.

Li, Y., Z. -Y. Fu and B. L. Su. 2012. Hierarchically structured porous materials for energy conversion and storage. Adv. Funct. Mater. 22(22): 4634–4667.

Li, Y., X. Zhang, A. Hu and M. Li. 2018. Morphological variation of electrodeposited nanostructured Ni-Co alloy electrodes and their property for hydrogen evolution reaction. Int. J. Hydrog. Energy 22012–22020.

Lin, M. and F. -K. Chang. 2002. The manufacture of composite structures with a built-in network of piezoceramics. Compos. Sci. Technol. 62: 919–39.

Lin, Y. and H. A. Sodano. 2008. Concept and model of a piezoelectric structural fiber for multifunctional composites. Compos. Sci. Technol. 68: 1911–8.

Lin, Y. and H. A. Sodano. 2009a. Characterization of multifunctional structural capacitors for embedded energy storage. J. Appl. Phys. 106: 114108.

Lin, Y. and H. A. Sodano. 2009b Fabrication and electromechanical characterization of a piezoelectric structural fiber for multifunctional composites. Adv. Funct. Mater. 19: 592–8.

Liu, C., Y. Song, Z. Xu, J. Zhao and Z. Rao. 2020. Highly efficient thermal energy storage enabled by a hierarchical structured hypercrosslinked polymer/expanded graphite composite. Int. J. Heat Mass Tran. 148: 119068–79.

Liu, P., E. Sherman and A. Jacobsen. 2009. Design and fabrication of multifunctional structural batteries. J. Power Sources 189(1): 646–50.

Liu, W., S. V. Hoa and M. Pugh. 2005. Fracture toughness and water uptake of high-performance epoxy/nanoclay nanocomposites. Compos. Sci. Technol. 65: 2364–73.

Loh, K. J., J. Kim, J. P. Lynch, N. W. S. Kam and N. Kotov. 2007. Multifunctional layer-by-layer carbon nanotube-polyelectrolyte thin films for strain and corrosion sensing. Smart Mater. Struct. 16: 429–38.

Loh, K. J., J. P. Lynch and N. Kotov. 2008. Passive wireless sensing using SWMT-based multifunctional thin film patches. Int. J. Appl. Electromag. Mech. 28: 887–94.

Loiácono, A., M. Gómez, E. Franceschini and G. Lacconi. 2020. Enhanced hydrogen evolution activity of $Ni[MoS_2]$ hybrids in alkaline electrolyte. Electrocatalysis 11. 10.1007/s12678-020-00588-w.

Łosiewicz, B., A. Budniok, E. Rówiński, E. Łagiewka and A. Lasia. 2004. The structure, morphology and electrochemical impedance study of the hydrogen evolution reaction on the modified nickel electrodes. Int. J. Hydrog. Energy 29(2): 145–157.

Low, C. T. J., R. G. A. Wills and F. C. Walsh. 2006. Electrodeposition of composite coatings containing nanoparticles in a metal deposit. Surf. Coatings Technol. 201(1–2): 371–383.

Ma, Y., J. Han, M. Wang, X. Chen and S. Jia. 2018. Electrophoretic deposition of graphene-based materials: a review of materials and their applications. J. Mater. 4(2): 108–20.

Magasinski, A., P. Dixon, B. Hertzberg, A. Kvit, J. Ayala and G. Yushin. 2010. High-performance lithium-ion anodes using a hierarchical bottom-up approach. Nat. Mater. 9(4): 353–358.

Manjunatha, C. M., A. C. Taylor, A. J. Kinloch and S. Sprenger. 2009. The tensile fatigue behavior of a GFRP composite with rubber particle modified epoxy matrix. J. Reinf. Plast. Compos. 29(14): 2170–2183.

Meng, X. 2017. An overview of molecular layer deposition for organic and organic–inorganic hybrid materials: mechanisms, growth characteristics, and promising applications. J. Mater. Chem. A 5: 18326–18378.

Montalvao, D., N. M. M. Maia and A. M. R. Ribeiro. 2006. A review of vibration-based structural health monitoring with special emphasis on composite materials. Shock Vib. Digest. 38(4): 295–324.

Monthioux, M. 2002. Filling single-wall carbon nanotubes. Carbon 40: 1809–23.

Muralt, P., R. G. Polcawich and S. Troiler-McKinstry. 2009. Piezoelectric thin films for sensors, actuators and energy harvesting. MRS Bull. 34(9): 658–64.

Navarro-Flores, E., Z. Chong and S. Omanovic. 2005. Characterization of Ni, NiMo, NiW and NiFe electroactive coatings as electrocatalysts for hydrogen evolution in an acidic medium. J. Mol. Catal. A Chem. 226(2): 179–197.

Ning, X., A. M. Jimenez, J. Pribyl, S. Li, B. Benicewicz, S. K. Kumar et al. 2019. Nanoparticle organization by growing polyethylene crystal fronts. ACS Macro Lett. 8(10): 1341–1346.

Noh, Y. J., H. -I. Joh, J. Yu, S. H. Hwang, S. Lee, C. H. Lee et al. 2015. Ultra-high dispersion of graphene in polymer composite via solvent free fabrication and functionalization. Sci. Rep. 5: 1–7.

O'Brien, D. J., D. M. Baechle and E. D. Wetzel. 2006. Multifunctional structural composite capacitors for US Army applications. *In*: Proc. Int. SAMPE Tech. Conf., 38th Fall Tech Conf: Global Adv. Mater. Proc. Eng.

Ogasawara, T., Y. Ishida, T. Ishikawa, T. Aoki and T. Ogura. 2006. Helium gas permeability of montmorillonite/epoxy nanocomposites. Compos. Part A Appl. Sci. Manuf. 37(12): 2236–40.

Olek, M., J. Ostrander, S. Jurga, H. Mohwald, N. Kotov, K. Kempa et al. 2004. Layer-by-layer assembled composites from multiwall carbon nanotubes with different morphologies. Nano Lett. 4(10): 1889–95.

Olszowska, K., J. Pang, P. S. Wrobel, L. Zhao, H. Q. Ta, Z. Liu et al. 2017. Three-dimensional nanostructured graphene: synthesis and energy, environmental and biomedical applications. Synth. Met. 234: 53–85.

Ouyang, S., Q. Huang, Y. Liu, Z. Ouyang and L. Liang. 2019. Powder metallurgical Ti-Mg metal-metal composites facilitate osteoconduction and osseointegration for orthopedic application. Bioact. Mater. 4: 37–42.

Park, G., T. Rosing, M. D. Todd, C. R. Farrar and W. Hodgkiss. 2008. Energy harvesting for structural health monitoring sensor networks. J. Infrastruct. Syst. 14(1): 64–79.

Patel, A. J., N. R. Sottos, E. D. Wetzel and S. R. White. 2010. Autonomic healing of low-velocity impact damage in fiber-reinforced composites. Compos. Part A: Appl. Sci. Manuf. 41(3): 360–8.

Peigney, A., C. Laurent, E. Flahaut and A. Rousset. 2000. Carbon nanotubes in novel ceramic matrix nanocomposites. Ceram. Int. 26: 677–83.

Punetha, V. D., S. Rana, H. -J. Yoo, A. Chaurasia, J. T. McLeskey, M. S. Ramasamy et al. 2017. Functionalization of carbon nanomaterials for advanced polymer nanocomposites: A comparison study between CNT and graphene. Prog. Polym. Sci. 67: 1–47.

Qi, G., J. Yang, R. Bao, D. Xia, M. Cao, W. Yang et al. 2016. Hierarchical graphene foam-based phase change materials with enhanced thermal conductivity and shape stability for efficient solar-to-thermal energy conversion and storage. Nano Res. 10(3): 802–813.

Qin, L., H. Yi, G. Zeng, C. Lai, D. Huang, P. Xu et al. 2019. Hierarchical porous carbon material restricted Au catalyst for highly catalytic reduction of nitroaromatics. J. Hazard. Mater. 380: 120864.

Qiu, J., C. Zhang, B. Wang and R. Liang. 2007. Carbon nanotube integrated multifunctional composites. Nanotechnology 18: 275708.

Rajoria, H. and N. Jalili. 2005. Passive vibration damping enhancement using carbon nanotube–epoxy reinforced composites. Compos. Sci. Technol. 65: 2079–93.

Ratna, D. and J. Karger-Kocsis. 2008. Recent advances in shape memory polymers and composites: a review. J. Mater. Sci. 43(1): 254–69.

Ren, F., D. Song, Z. Li, L. Jia, Y. Zhao, D. Yan et al. 2018. Synergistic effect of graphene nanosheets and carbonyl iron–nickel alloy hybrid filler on electromagnetic interference shielding and thermal conductivity of cyanate ester composites. J. Mater. Chem. C. 6(6): 1476–86.

Ritala, M., M. Leskela, J. Dekker, C. Mutsaers, P. J. Soininen and J. Skarp. 1999. Perfectly conformal tin and Al_2O_3 films deposited by atomic layer deposition. Chem. Vap. Depos. 5: 99–101.

Sandler, J. K. W., J. E. Kirk, I. A. Kinloch. M. S. P. Shaffer and A. H. Windle. 2003. Ultra-low electrical percolation threshold in carbon-nanotube–epoxy composites. Polymer 44: 5893–9.

Sarı, A. and A. Karaipekli. 2007. Thermal conductivity and latent heat thermal energy storage characteristics of paraffin/expanded graphite composite as phase change material. Appl. Therm. Eng. 27(8–9): 1271–1277.

Sauder, C., J. Lamon and R. Pailler. 2004. The tensile behavior of carbon fibers at high temperatures up to 2400°C. Carbon 42: 715–25.

Schmidt, D. L., K. E. Davidson and S. Theibert. 1999a. Unique applications of carbon–carbon composite materials (Part One). SAMPE J. 35(3): 27–39.

Schmidt, D. L., K. E. Davidson and S. Theibert. 1999b. Unique applications of carbon–carbon composite materials (Part Two). SAMPE J. 35(4): 51–63.

Schmidt, D. L., K. E. Davidson and S. Theibert. 1999c. Unique applications of carbon–carbon composite materials (Part Three). SAMPE J. 35(5): 47–55.

Sharma, S. and N. C. Kothiyal. 2015. Synergistic effect of zero-dimensional spherical carbon nanoparticles and one-dimensional carbon nanotubes on properties of cement-based ceramic matrix: microstructural perspectives and crystallization investigations. Compos. Interfaces 22: 899–921.

Sheng, M., W. Weng, Y. Wang, Q. Wu and S. Hou. 2018. Co-W/CeO_2 composite coatings for highly active electrocatalysis of hydrogen evolution reaction. J. Alloys Compd. 743: 682–690.

Shenogina, N., S. Shenogin, L. Xue and P. Keblinski. 2005. On the lack of thermal percolation in carbon nanotube composites. Appl. Phys. Lett. 87: 133106.

Shim, B. S., J. Starkovich and N. Kotov. 2006. Multilayer composites from vapor-grown carbon nano-fibers. Compos. Sci. Technol. 66: 1174–81.

Shokry, A., M. M. A. Khalil, H. Ibrahim, M. Soliman and S. Ebrahim. 2019. Highly luminescent ternary nanocomposite of polyaniline, silver nanoparticles and graphene oxide quantum dots. Sci. Rep. 9(1): 1–12.

Snyder, J. F., R. H. Carter and E. D. Wetzel. 2007. Electrochemical and mechanical behavior in mechanically robust solid polymer electrolytes for use in multifunctional structural batteries. Chem. Mater. 19(15): 3793–801.

Snyder, J. F., E. L. Wong and C. W. Hubbard. 2009. Evaluation of commercially available carbon fibers, fabrics and papers for potential use in multifunctional energy storage applications. J. Electrochem. Soc. 156(3): A215–24.

Sodano, H. A., D. J. Inman and G. Park. 2004. A review of power harvesting from vibration using piezoelectric materials. Shock Vib. Digest. 36(3): 197–205.

Sodano, H. A., D. J. Inman and G. Park. 2005. Generation and storage of electricity from power harvesting devices. J. Intell. Mater. Syst. Struct. 16(1): 67–75.

Soler-illia, G. J. D., C. Sanchez, B. Lebeau and J. Patarin. 2002. Chemical strategies to design textured materials: from microporous and mesoporous oxides to nanonetworks and hierarchical structures. Chem. Rev. 102(11): 4093–4138.

Spee, D. A., J. K. Rath and R. E. I. Schropp. 2015. Using hot wire and initiated chemical vapor deposition for gas barrier thin film encapsulation. Thin Solid Films 575: 67–71.

Srivastava, A., A. Agarwal, D. Chakraborty and A. Dutta. 2005. Control of smart laminated FRP structures using artificial neural networks. J. Reinf. Plast. Compos. 24(13): 1353–64.

Tadigadapa, S. and K. Mateti. 2009. Piezoelectric MEMS sensors: state-of-the-art and perspectives. Meas. Sci. Technol. 20: 092001.

Tahara, S., Y. Takeda and Y. Sugahara. 2005. Preparation of organic-inorganic hybrids possessing nanosheets with perovskite-related structures via exfoliation during a sol–gel process. Chem. Mater. 17: 6198–6204.

Tan, L. and B. Tan. 2017. Hypercrosslinked porous polymer materials: design, synthesis, and applications. Chem. Soc. Rev. 46(11): 3322–3356.

Tang, J., M. Yang, F. Yu, X. Chen, L. Tan and G. Wang. 2017. 1-Octadecanol@hierarchical porous polymer composite as a novel shape-stability phase change material for latent heat thermal energy storage. Appl. Energy. 187: 514–522.

Tasis, D., N. Tagmatarchis, A. Bianco and M. Prato. 2006. Chemistry of carnon nanotubes. Chem. Rev. 106: 1105–36.

Teh, P. L., M. Mariatti, H. M. Akil, C. K. Yeoh, K. N. Seetharama, A. N. R. Wagiman et al. 2007. The properties of epoxy resin coated silica fillers composites. Mater. Lett. 61: 2156–8.

Thomas, C. R. [ed.]. 1993. Essentials of carbon/carbon composites. pp. 1–36. *In*: C. R. Thomas [ed.]. Overview— What are Carbon-Carbon Composites and What Do They Offer? Great Britain: Royal Society of Chemistry.

Thostenson, E. T., Z. Ren and T. W. Chou. 2001. Review: advances in the science and technology of carbon nanotubes and their composites. Compos. Sci. Technol. 61: 1899–912.

Thostenson, E. T., S. Ziaee and T. -W. Chou. 2009. Processing and electrical properties of carbon nanotube/vinyl ester nanocomposites. Compos. Sci. Technol. 69: 801–4.

Thümmler, F. and R. Oberacker. 1993. An Introduction to Powder Metallurgy (1st ed.). London, The Institute of Materials.

Toghraei, A., T. Shahrabi and G. BaratiDarband. 2020. Electrodeposition of self-supported Ni-Mo-P film on Ni foam as an affordable and high-performance electrocatalyst toward hydrogen evolution reaction. Electrochim. Acta. 335: 135643.

Toohey, K. S., N. R. Sottos, J. A. Lewis, J. S. Moore and S. R. White. 2007. Self-healing materials with microvascular networks. Nat. Mater. 6: 581–5.

Tripathi, S. M. and T. S. Bholanath. 2010. Synthesis and study of applications of metal coated carbon nanotubes. Int. J. Adv. Sci. Technol. 16: 31–42.

Upadhyaya, G. S. 2014. Powder Metallurgy Technology. Cambridge Int. Sci. Publ. 1: 1–5.

Valente, N. A. and S. Andreana. 2016. Peri-implant disease: what we know and what we need to know. J. Periodontal Implant. Sci. 46: 136–151.

Wang, D. W., F. Li, M. Liu, G. Q. Lu and H. -M. Cheng. 2008. 3D aperiodic hierarchical porous graphitic carbon material for high-rate electrochemical capacitive energy storage. Angew. Chem. 47(2): 373–376.

Wang, X., E. N. Kalali and D. Y. Wang. 2015. An *in-situ* polymerization approach for functionalized MoS_2/nylon-6 nanocomposites with enhanced mechanical properties and thermal stability. J. Mater. Chem. A. 3(47): 24112–24120.

Wang, X., Q. Lu, X. Wang, J. Joo, M. Dahl, B. Liu et al. 2016. Photocatalytic surface-initiated polymerization on TiO_2 toward well-defined composite nanostructures. ACS Appl. Mater. Interfaces 8(1): 538–546.

Wang, Y., L. Shi, Y. Jin, S. Sun, P. Gao, Y. Wei et al. 2020. Surface-initiated polymerization for the preparation of magnetic polymer composites. Polym. Chem. 11(10): 1797–1805.

Watkins, S. E., F. Akhavan, R. Dua, K. Chandrashekhara and D. C. Wunsch. 2007. Impact induced damage characterization of composite plates using neural networks. Smart Mater. Struct. 16: 515–24.

White, S. R., N. R. Sottos, P. H. Geubelle, J. S. Moore, M. R. Kessler, S. R. Sriram et al. 2001. Autonomic healing of polymer composites. Nature 409: 794–7.

Williams, G., R. Trask and I. Bond. 2007. A self-healing carbon fibre reinforced polymer for aerospace applications. Compos. Part A: Appl. Sci. Manuf. 38(6): 1525–32.

Wu, D. Y., S. Meure and D. Solomon. 2008. Self-healing polymeric materials: a review of recent developments. Prog. Polym. Sci. 33(5): 479–522.

Wu, Z., X. P. Qing and F. -K. Chang. 2009. Damage detection for composite laminate plates with a distributed hybrid PZT/FBG sensor network. J. Intell. Mater. Syst. Struct. 20: 1069–77.

Xiao, C., Y. Tan, X. Wang, L. Gao, L. Wang and Z. Qi. 2018. Study on interfacial and mechanical improvement of carbon fiber/epoxy composites by depositing multi-walled carbon nanotubes on fibers. Chem. Phys. Lett. 703: 8–16.

Yameen, B., M. Ali, R. Neumann, W. Ensinger, W. Knoll and O. Azzaroni. 2009. Single conical nanopores displaying ph-tunable rectifying characteristics. manipulating ionic transport with zwitterionic polymer brushes. J. Am. Chem. Soc. 131: 2070–2071.

Yang, Z., S. Si, X. Zeng, C. Zhang and H. Dai. 2008. Mechanism and kinetics of apatite formation on nanocrystalline TiO_2 coatings: a quartz crystal microbalance study. Acta Biomater. 4: 560–568.

Yoshimura, T., S. Tatsuura and W. Sotoyama. 1991. Polymer films formed with monolayer growth steps by molecular layer deposition. Appl. Phys. Lett. 59: 482–484.

You, X., J. Yang, Q. Feng, K. Huang, H. Zhou, J. Hu and S. Dong. 2018. Three-dimensional graphene-based materials by direct ink writing method for lightweight application. Int. J. Lightweight Mat. Manuf. 1(2): 96–101.

Yu, L., L. Cheng, L. H. Yam, Y. J. Yan and J. S. Jiang. 2007. Online damage detection for laminated composite shells partially filled with fluid. Compos. Struct. 80: 334–42.

Zhang, J., S. Yang, Z. Chen, H. Wu, J. Zhao and Z. Jiang. 2018. Graphene encapsulated SiC nanoparticles as tribology-favoured nanofillers in aluminium composite. Compos. Part B Eng. 162(11): 45–453.

Zhang, Y., H. K. Bilan and E. Podlaha. 2018. Enhancing the hydrogen evolution reaction with Ni-W-TiO_2 composites. Electrochem. commun. 96: 108–112.

Zhang, Z. and X. Fang. 2006. Study on paraffin/expanded graphite composite phase change thermal energy storage material. Energy Convers. Manag. 47(3): 303–310.

Zhao, J., L. Liu, Q. Guo, J. Shi, G. Zhai, J. Song et al. 2008. Growth of carbon nanotubes on the surface of carbon fibers. Carbon 2: 380–3.

Zhao, Q., K. Wang, J. Wang, Y. Guo, A. Yoshida, A. Abudula et al. 2019. Cu_2O nanoparticle hyper-cross-linked polymer composites for the visible-light photocatalytic degradation of methyl orange. ACS Appl. Nano Mater. 2(5): 2706–2712.

Zhao, X., L. Lv, B. Pan, W. Zhang, S. Zhang and Q. Zhang. 2011. Polymer-supported nanocomposites for environmental application: A review. Chem. Eng. J. 170(2–3): 381–394.

Zheng, Z., N. Li, C. Q. Wang, D. Y. Li, Y. M. Zhu and G. Wu. 2012. Ni-CeO_2 composite cathode material for hydrogen evolution reaction in alkaline electrolyte. Int. J. Hydrog. Energy. 37(19): 13921–13932.

Zhou, H., X. Du, H. Y. Liu, H. Zhou, Y. Zhang and Y. W. Mai. 2017. Delamination toughening of carbon fiber/epoxy laminates by hierarchical carbon nanotube-short carbon fiber interleaves. Compos. Sci. Technol. 140: 46–53.

Zhou, X., E. Shin, K. W. Wang and C. E. Bakis. 2004. Interfacial damping characteristics of carbon nanotube-based composites. Compos. Sci. Technol. 64: 2425–37.

Zhu, B. -K., S. -H. Xie, Z. -K. Xu and Y. -Y. Xu. 2006. Preparation and properties of polyimide/multi-walled carbon nanotubes (MWCTs) nanocomposites. Compos. Sci. Technol. 66: 54854.

Zhu, P., J. Zhu, C. Yan, M. Dirican, J. Zang, H. Jia et al. 2018. *In situ* polymerization of nanostructured conductive polymer on 3D sulfur/carbon nanofiber composite network as cathode for high-performance lithium–sulfur batteries. Adv. Mater. Interfaces 5(10): 1–10.

Zou, Y., L. Tong and G. P. Steven. 2000. Vibration-based model dependent damage identification and health monitoring for composite structures—a review. J. Sound. Vib. 230(2): 357–78.

CHAPTER 9

Effect of Functionalization on Graphenic Surfaces in their Properties

María del Carmen Rojas,[2] M. Victoria Bracamonte,[2] Martín Zoloff Michof,[1] Patricio Vélez,[1] Fernanda Stragliotto[2] and Guillermina L. Luque[1],*

Introduction

Graphene is a single-atom-thick planar sheet of hexagonally arranged sp^2 bonded carbon atoms, with a C-C bond length of 0.142 nm. It is the based element of some carbon allotropes, such as graphite, carbon nanotubes and fullerenes (Singh et al. 2011, Bazylewski and Fanchini 2019, Rodríguez-Pérez et al. 2013). Graphene is the building block of graphite (one of the most abundant material on Earth) and even though it was theoretically established in 1940 (Wallace 1946), it was not until 2004 that Geim and Novoselov, two Russian scientists at the University of Manchester, successfully identified single layers of graphene in a simple table top experiment (Novoselov et al. 2004) and six-years later were awarded the Nobel Prize in physics for their discovery. This material, the thinnest one found to date exhibits great electrical and thermal conductivity, mechanical stability, flexibility and low coefficient of thermal expansion (Novoselov et al. 2005, Zhong et al. 2017, Bazylewski and Fanchini 2019, Balandin et al. 2008). The incredibly higher conductivity is due to the presence of an electron cloud above and below each carbon ring, which overlaps to create a continuous pi orbital across the whole graphene layer, leading to higher electric conductivity in comparison with other materials (Saba and Jawaid 2018).

All the great properties graphene presents make this material a very interesting one to be applied in the field of physics, chemistry, material science and biotechnology, among others (Zhong et al. 2017, Bazylewski and Fanchini 2019), and attracted tremendous interest not only in the academic world but also in industry. Nevertheless, in general, the progress of technology from the discovery to real application in a product is slow and tortuous, and graphene is no exception in this.

In order to implement the large potential applications, the functionalization of graphene it is often necessary for its practical use (Thakur and Thakur 2015, Dubey and Oh 2012). And this is the reason why there is continuous and important research for chemical modification of graphene obtained in this way, graphene derivatives, which are being increasingly studied nowadays. The principal types of functionalization as are summarized in Fig. 9.1 are covalent and non-covalent methods as can be seen in (Raji et al. 2018). The first one is based on the formation of covalent

[1] INFIQC, Departamento de Química Teórica y Computacional, Facultad de Ciencias Químicas, Universidad Nacional de Córdoba-CONICET. 5000 Córdoba, Argentina.
[2] IFEG, Facultad de Matemática Astronomía y Física, Universidad Nacional de Córdoba-CONICET. 5000 Córdoba, Argentina.
* Corresponding author: guillerminaluque@unc.edu.ar

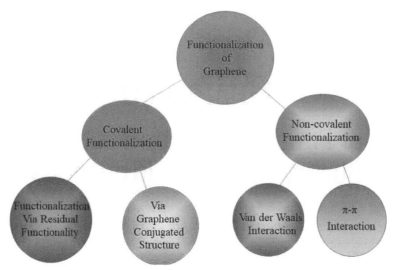

Figure 9.1: Types of graphene surface functionalization.

bonds, and the last one on the interactions between molecules and graphene surface (such as van der Waals or pi interactions). In general, the covalent functionalization may take place either by the formation of covalent bonds between free radicals or electrophiles with the carbon-carbon double bond of pristine graphene on the edge carbons that are more reactive than those located in the inner pi surface or on defects on the surface that are prone to functionalization (Yang et al. 2018). This type of functionalization generally involves the rehybridization of carbon atoms from sp² to sp³ and tailors the chemical, thermal and mechanical properties of graphene, but at the same time produces the loss of the electronic properties in it. Another way is by a covalent bond between oxygen groups of one of the most used and famous graphene derivative, Graphene Oxide (GO), and different organic functional groups (Eigler and Hirsch 2014). GO has a history of its own, and extends decades back with some studies involved in the chemistry of graphite. Even though the exact structure of GO is difficult to determine, after oxidation of graphene, the aromatic lattice is broken up with oxidized functional groups such as epoxides, alcohols, ketones, carbonyl and carboxyl groups (Erickson et al. 2010).

The non-covalent functionalization of graphene or its derivate, such as GO, is mainly based on physical interactions, such as pi interactions between counter molecules and the surface of graphene or GO (Georgakilas et al. 2016).

Experimental Studies of the Effects of Covalent and Non-Covalent Functionalization of Graphene

The functionalization of pristine graphene sheets and its derivatives with different functional groups has been developed for several purposes, namely: (i) to allow their dispersibility in specific solvents, (ii) to introduce a band gap in their electronic structure for electronic and optical applications, (iii) to anchor a specific molecule to recognize a target in sensors and/or (iv) to introduce, in general, new properties combined with the ones from graphene, among others. This functionalization can be performed by several methods, principally, into two categories: covalent and non-covalent functionalization.

Covalent Functionalization

Covalent functionalization means to perform an irreversible process that will change the pristine structure of graphene and its derivatives through the formation of new covalent bonds. In general, the functionalization reaction between graphene and organic molecules includes two main routes as stated above: the formation of covalent bond (i) breaking the C=C unions of pristine graphene and (ii) using the oxygen groups present in graphene derivatives (Georgakilas et al. 2016). Despite the methodology used; the change in the original structure of graphene impacts the final properties of the functionalized material. Huang et al. (2013) have studied the reaction of diazonium salt to produce free radicals, which attacks the sp^2 carbon atoms of graphene forming a covalent bond, Fig. 9.2.

These authors used a nitrophenyldiazonium functionalized graphene (DFG) as a study case and they made a systematic analysis to characterize both, the micro and electronic structures and transport properties. Raman spectroscopy, Atomic Force Microscopy (AFM) and Transmission Electron Microscopy (TEM) confirmed not only the inhomogeneous distribution of the diazonium functionalization on graphene but also how the sp^3 hybridization induced shrinking of lattice parameters and the loosening of the conjugated pi-bond. From the transport properties characterization, Huang et al. 2013 observed that graphene with a low amount of modifying molecules exhibited higher resistivity than pristine one; while extending the reaction, the resistivity was suppressed to smaller values than the pristine graphene. These results implied that DFG was heavily p-type doped, as was expected considering the electron-withdrawing character of the nitrophenyl group used as a modifier. As a result, an energy gap was opened and a spin polarization was produced in graphene, giving it a ferromagnetic semiconductive character which could be applied especially in spintronics. Using the same approach, Farquhar et al. (2016) modified a few layers of graphene using a nitrobenzenediazonium salt (NBD). The synthesis produces a nitrophenyl modified few-layer graphene (FLG_{NP}) which was mounted onto an electrode and used for the study of its topography, morphology and electrochemical properties. Raman spectroscopy confirmed the functionalization of the surface showing an increase in the average D/G ratio from (0.14 ± 0.03) to (0.26 ± 0.04) for FLG and FLG_{NP}, respectively. The covalent nature of the bond was proved by XPS and infrared spectroscopy. In both cases, the apparition of both –C-C– and –N=N– linkages in the film demonstrate at least two grafting mechanisms operating for spontaneously formed films. The electrochemical characterization of these samples showed an electrochemical discontinuity. However, the FLG's functionalization enhanced the permeability of the multilayered systems giving them a distinct advantage for capacitor applications. In a later work, Farquhar et al. (2018) applied the diazonium functionalization of graphene to prevent restacking of FLG under conditions where the per-sheet capacitance is not diminished by the spacer groups or the stacking arrangements. Following this objective, they investigated both, the effect of spacer layer thickness and also the chemical nature of spacer-groups by modifying FLG with thin films of aminophenyl (AP) and carboxyphenyl (CP). The capacitance determinations were made by Electrochemical Impedance Spectroscopy (EIS) and Cyclic Voltammetry (CV) measurements. The results showed that independently on the type of functionalization, the modified FLGs with diazonium salts are porous to the electrolyte solution and exhibit an increased capacitance due to the fast ion diffusion and full double layer formation at both FLG-solution interfaces.

Another type of reaction used for the functionalization of graphene is the so-called Diels-Alder. Georgakilas et al. (2010) were one of the first in the use of the 1,2 cycloaddition reaction to modify the dispersibility properties of graphene in different solvents. They used azomethineylide

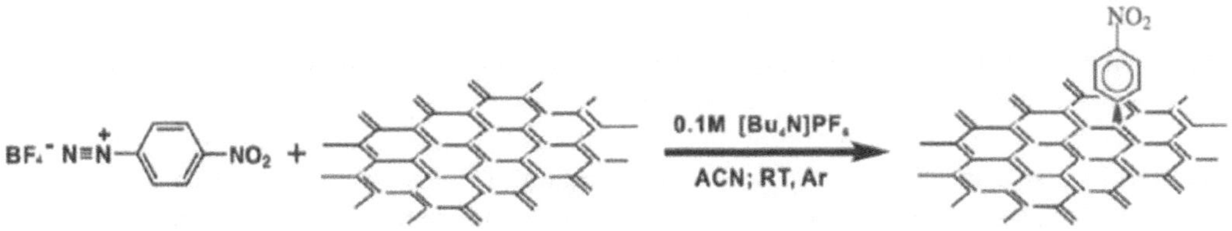

Figure 9.2: Scheme of the reaction processes of aryl diazonium functionalization on graphene (Reprinted from Huang et al. 2013 with permission from American Chemical Society).

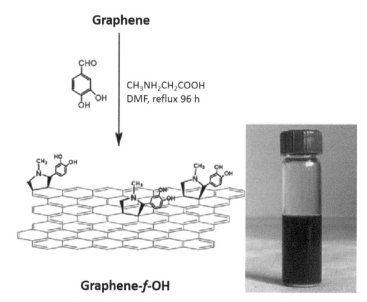

Graphene

CH₃NH₂CH₂COOH
DMF, reflux 96 h

Graphene-*f*-OH

Figure 9.3: Schematic representation of the 1,3 dipolar cycloaddition of azomethineylide on graphene (Reprinted from Georgakilas et al. 2010 with the permission from Royal Society of Chemistry).

as a reactive specie which reacts with C=C bond-forming a functionalized graphene, Fig. 9.3, with augmented dispersibility in organic and aqueous solvents.

The increment in dispersibility is a goal pursued in many publications due to its importance in other processes such as ink production (Saidina et al. 2019), corrosion protection application (Ramezanzadeh et al. 2016), batteries production (YongJian et al. 2020), among others. Oh et al. (2019) applied the functionalization of graphene to improve the scratch healing and reprocessing ability of thermosetting polymers. Using furfurylamine as the active reactant for the cycloaddition reaction, these authors functionalized graphene with epoxy groups. The functionalization was confirmed through FT-IR spectroscopy, DSC thermogravimetry and rheological measurements. The scratch healing and reprocessing ability were reached even at 150°C, improving the physical-chemical properties of these thermosetting polymers and their potential application in the plastic industry. Feng et al. (2019) also used the Diels-Alder (DA) reaction to modified graphene. Briefly, Expanded Graphite (EG) was used as a diene and Maleic Anhydride (MA) was chosen as a dienophile. The DA reaction between EG and MA was conducted at 160°C in a mixer. During the DA reaction, the solvent transfers a strong shearing force, exfoliating the EG into MA-functionalized graphene nanosheets (MAG). While Raman spectroscopy indicated that melt-blending can effectively promote dissociation of the π–π stacking, functionalization and subsequent exfoliation of EG into MAG; XPS and FTIR exhibited the formation of hydrolyzed carboxylic acids on the surface of MAG. This functionalization of graphene not only allowed its dispersion in hydrogenated nitrile butadiene rubber, but also improved the mechanical—and electrical—properties of the composite.

Another type of functionalization of graphene is the substitution of some C atoms with other elements, such as nitrogen, phosphorus or boron. In this case, the generated materials are no longer known as functionalized, they are called doped graphene. Several examples on the effect of this doping on the properties of the pristine graphene can be found in the bibliography. Guo et al. (2010) chose N as a natural candidate for doping graphene to create an n-type semiconductor. To synthesize the material, they used a two steps procedure consisting of N⁺-ion irradiation of a graphene sample followed by NH₃ annealing. Although the incorporation of N into graphene was corroborated by various techniques—Auger electron spectroscopy, XPS and Raman, the most interesting results were obtained during the electrical characterization. A Field-Effect Transistor (FET) was fabricated on a 300 nm $SiO_2/p^{2+}Si$ substrate using N-doped graphene and the source-drain conductance and back-gate voltage (Gsd-Vg) curves were measured. The presented results show how the transport property changed compared to that of the FET made by intrinsic graphene, that is, the Dirac point position moved from positive Vg to negative Vg, indicating the transition of graphene from p-type to n-type after annealing in NH_3, Fig. 9.4.

Using another strategy, also prepared n-type semiconductor graphene using phosphorus doping. The synthesis was done by mixing sodium hypophosphite monohydrate and sodium carbonate anhydrous and calcinating the mixture at 950°C for 10 minutes with Ar flow. According to their results, the obtained lamellar microstructure is composed of carbon in a honeycomb structure and phosphorus having a thickness of 0.58 nm, on average. The apparition of peaks a 132.8 eV and 284.7 eV in the C 1s and P 2p XPS spectra, respectively, shows that the phosphorus atoms were doped into graphene rather than adsorbed onto the surface of graphene. Besides optical and electrical results obtained by the authors, demonstrated changes in the work function and Fermi level of doped graphene with respect to the pure one, confirmed the n-type semiconductor character of the phosphorus-doped sample. Besides n-type, p-type semiconductor doped graphene samples can also be synthesized. Sheng et al. (2012) synthesized boron-doped graphene through thermal annealing of graphite oxide in the presence of boron oxide, Fig. 9.5. In this way, boron appearing from B_2O_3 vapor replaces carbon atoms within graphene structures forming a planar structure with an average thickness of ca. 2 nm. XPS analysis of the samples, showed a shift in the B 1s peak to a higher energy binding in respect to pure boron, suggesting that not only boron atoms are surrounded by carbon and oxygen atoms, but also indicating the presence of oxidized boron atoms in this sample. The as-prepared boron-doped graphene presented a p-type character which improves its electrocatalytic performance towards the oxygen reduction reaction.

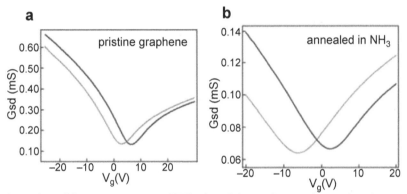

Figure 9.4: Comparison of the transport property of FET using pristine graphene graphene and graphene annealed in NH_3 after irradiation (black and grey curves were measured in air and vacuum, respectively) (Reprinted from Guo et al. 2010 with permission from American Chemical Society).

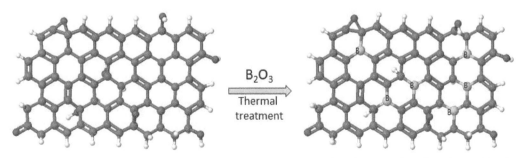

Figure 9.5: Scheme of the incorporation of B atoms on a graphene layer through the reaction of graphene oxide with vapored B_2O_3 (Reprinted from Sheng et al. 2012 with permission from Royal Society of Chemistry).

Non-covalent Functionalization

The non-covalent functionalization of graphene is a very important task for most applications because the extended π system of graphenic nanostructures is not interrupted, and in consequence, important properties of graphene—such as electric conductivity or mechanical strength—are not affected (Karousis et al. 2010, Georgakilas et al. 2012). This approach is very attractive in order to induce additional properties of graphene without losing its intrinsic chemical and physical characteristics. When one refers to non-covalent functionalization it usually relates to Van der Waals forces or π–π interactions between graphene or graphene oxide with organic molecules or polymers. Usually, van der Waals forces are developed when the organic molecules or polymers have hydrophobic character, π–π interactions are common when molecules present short to highly extended π system (Karousis et al. 2010). In all the instances, noncovalent functionalization leads to enhanced dispersibility, biocompatibility, reactivity, binding capacity or sensing properties (Georgakilas et al. 2016).

π–π Interactions

In order to have these types of interactions, there is a need to have π systems and consider the geometry of the interacting species. It is important to have an overlap between the two components in order to have a good interaction among them and this is favored with the planarity of the components. Graphene normally has a planar structure that usually interacts with small aromatics molecules. These interactions are influenced by different factors related to the molecules, their donating or withdrawing ability, its substituents, size and polarity (Georgakilas et al. 2016).

Using π–π interactions, Cheng et al. (2011), developed a method based on photolithography to create graphene p-n junctions preserving the unique band structure of graphene. To do this, they used a two steps procedure. First, the graphene device was fabricated using a shadow mask technique and the subsequent vacuum and thermal annealing to avoid the residual resist polymer on the graphene surface and achieve high carrier mobility. Second, a well-designed scheme utilizing a microfluidic channel technique was used to realize the spatially selective functionalization and solution-based chemical doping of graphene. The chemically modified graphene samples were systematically characterized by optical microscopy, surface topography, potential measurements and spatially resolved Raman spectroscopic imaging. While the pi-pi interaction of the modifier molecule was confirmed by atomic force microscopy, the high quality of the graphene p-n junctions was supported by the observations of the high carrier mobility, the Fermi energy difference, and the unconventional measured quantum hall effect. On other hand, Liu and coworkers (Gmbh and Technologies 2010), also used π–π interactions to prepare a thermosensitive graphene/poly(N-isopropylacrylamide) (PNIPAAm) composite without affecting the electrical properties of graphene. To prepare the material, they simply mixed dissolved PNIPAAm and graphene in water and sonicate them for 20 minutes. Then the mixture was centrifuged and redispersed several times in order to remove any free PNIPAAm from the composite. The π–π interactions of PNIPAAm and graphene were confirmed by attenuated total reflection infrared (ATR-IR) spectroscopy with the relative intensity of the C=O absorption to the aromatic stretch for the composite is reduced in respect to the pure polymer consistent with contributions from the underlying graphene. While thermogravimetric and XPS analysis also confirmed the formation of the composite, AFM revealed a planar structure of 5 nm thickness corresponding to a sandwich-like nanostructure.

Van der Waals Interactions

Apart from the aromaticity that is involved in the π–π interactions, graphene is also characterized by its hydrophobic character, which promotes interactions of graphene with hydrophobic or partially hydrophobic-organic molecules such as surfactants, ionic liquids or macromolecules. These interactions have been mostly used to incorporate graphene into the polymer's matrix and/or to disperse it in both, aqueous and organic media. Depending on the case, a hydrophobic interaction of graphene with the aliphatic part of surfactants; or a hydrophilic reaction between ionic, zwitterionic or highly polar group—from the surfactant with functional groups in graphene will take place (Fernández-Merino et al. 2012). Some of the amphiphilic organic molecules and macromolecules that have been used successfully as graphene dispersants and stabilizers are cellulose derivatives (Bourlinos et al. 2009), lignin (Saiful Badri et al. 2017), albumin (Puglia et al. 2020) and sodium dodecyl benzenesulfonate (Lotya et al. 2009), among others. Lotya et al. (2009) also proposed an organic molecule -sodium cholate- as a surfactant for the exfoliation of graphene. Water-based 0.3 mg mL^{-1} dispersions of four-layer graphene were obtained. The transmission electron microscopic analysis of the samples confirmed the presence of graphene multilayers with a characteristic aspect ratio length/width fairly constant between 2.0 and 2.7; suggesting asymmetric flakes. Lastly, the authors applied the exfoliated graphene to prepare free-standing films with electrical and mechanical properties relevant for many applications. Bourlinos et al. (2009) described an example using polyvinylpyrrolidone (PVP) to interact with graphene layers to obtain in this way, stable colloidal high-quality graphene protected with a polymer with enhanced mechanical properties. The exfoliation of graphene without affection of the sp^2 character of the carbon core was proved by AFM, TEM and Raman studies. As shown in Fig. 9.6, the spectrum of graphite and the dispersed fraction are identical in position. However, the relatively low intensity and broadness of the D band coupled with the much higher intensity and sharpness of the G band strongly suggests low defect content, virtually no oxidization and the presence of multilayer flakes in the as-dispersed graphitic solids.

Phillipson et al. (2016) reported tunable n-type doping of graphene using self-assembled networks of aliphatic amines which differ only in the length of the alkyl chain. Octadecylamine (ODA) and nonacosylamine (NCA) were used as alkylated amine molecules because their length determines the density of the amine groups at the interface—acting as a spacer—as revealed by Scanning Tunnelling Microscopy (STM) and atomic force microscopy. Raman spectroscopy measurements carried out on SiO$_2$ are shown in Fig. 9.7a where higher n-type doping can be seen from ODA compared to NCA molecules, based on a larger redshift of the G peak position and a higher increase in the I(2D)/I(G) ratio. A charge carrier concentration 1.59 times higher with ODA compared to NCA was found, which is in good agreement with the ratio of alkyl chain length of NCA and ODA (1.6 times longer). The doping mechanism was explained by the formation of a dipole layer between graphene and the NH$_2$ groups of ODA and NCA molecules. The transport measurements by FET, Fig. 9.7b, also shows a shift corresponding to an injection of 11.2 ×

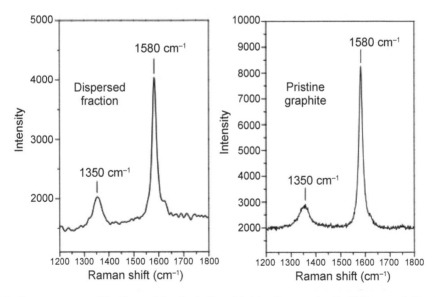

Figure 9.6: Raman spectrum of the dispersed fraction in the solid-state after removing the PVP excess (left) and starting graphite powder (right) (Reprinted from Bourlinos et al. 2009 with permission from Elsevier).

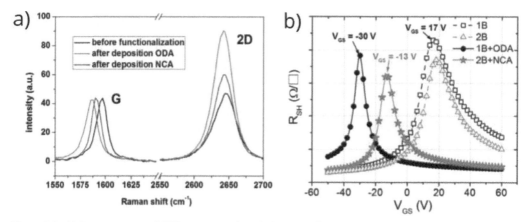

Figure 9.7: (a) Raman spectra of CVD grown graphene before and after deposition of ODA and NCA. (b) Transfer characteristics of devices before and after functionalization with ODA and NCA obtained by electrical characterization of FET devices (Reprinted from Phillipson et al. 2016 with permission from Royal Society of Chemistry).

10^{12} cm^{-2} electrons from the ODA self-assembled networks and of 7.4×10^{12} cm^{-2} for NCA, reflecting a 1.52 times higher doping for ODA functionalized devices compared to NCA functionalized ones.

Theoretical Studies of the Effects of Covalent and Non-Covalent Functionalization of Graphene

Here the focus will be on the effect of covalent and non-covalent functionalization of graphene surfaces on their electronic structure as determined by computational modeling methods, based mainly on the Density Functional Theory (DFT).

Covalent Functionalization

Elemental Functionalization (Heteroatom Doped Graphene)

General Description

Heteroatom-doped Graphene (HG) refers to a new structure where carbon atoms from the hexagonal honeycomb lattice of graphene are substituted by N, B, S, O, P, Al, Si, among other atoms. Several theoretical studies on these systems comprise DFT calculations.

Currently numerous studies can be found, where the geometric, energetic, electrical, optical and vibrational properties of many HG systems are comprehensively characterized (Denis et al. 2009, Denis 2010, Gholizadeh and Yu 2014, Thakur et al. 2018, Dai and Yuan 2010, Pramanik and Kang 2011, Chen et al. 2012, Shao et al. 2013). The main effects of heteroatom-doping in graphene are addressed here, as well as some selected applications.

Denis et al. (2009), studied the doping of a graphene sheet with S atoms. Their results suggest that the synthesis of graphene doped with S is possible since the energy of formation of sulfur doping is not too high. In addition, they studied the energy and electrical contributions of a monolayer and a bilayer of graphene doped with Al, Si, P and S (Denis 2010). Their results showed that phosphorus is the best choice to open a band gap in graphene and also to create interlayer bonds.

Gholizadeh and Yu (2014) performed theoretical investigations of differentiating the work function of graphene monolayers doped with different kinds of atoms from the IIA–VIA groups (N, B, P, O, S, Si, As, Se, Ge, Al, Ga) of the Periodic Table. The calculations demonstrate that the work functions of all HG are a linear function of the intensity of the applied external electric field. The slopes of the lines deviate from the ideal value to a different extent, depending mainly on the polarization of the heteroatom–carbon bonds and the production of induced dipole moments on the HG.

Dobrota et al. (2017), studied the oxidative capacity and reactivity of the graphene basal plane with vacancies or with substitutional doping by B, N, P and S. Their results showed that the presence of these defects enhances the reactivity of graphene. In particular, these sites act as strong attractors for hydroxyl groups.

Thakur et al. (2018) investigated how structural, electronic and optical properties of pristine graphene changes when it is doped with Al, Al-N, Al-P and Al-S dopants. Al, Al-P and Al-S doping improve the graphene surface as a more electron-rich system. On the other hand, the absorption spectra of Al-P and Al-N co-doped graphene are shifted towards the lower wavelength in comparison to Al-doped graphene. For each doping system, the authors carried out a careful study of natural population analysis, HOMO-LUMO orbital analysis, DOS analysis and Electrostatic potential. Using the Time-Dependent Density Functional Theory (TD-DFT), they simulated the UV–visible absorption spectrum.

Gas Sensing Applications

The possibility of tuning the chemical and electronic properties of graphene through heteroatom-doping is of great importance and has attracted attention for application in gas sensors. DFT calculations can contribute to a more fundamental understanding of the chemical properties of HG systems that are valuable for gas sensing. The performance of electrochemical sensors can be improved by the use of doped graphene materials because the electrochemically active sites introduced in HG facilitate charge transfer, adsorption and activation of analytes and anchoring of molecules.

Dai and Yuan (2010) studied the adsorption of molecular oxygen (O_2) on graphene doped with the following heteroatoms: B, N, Al, Si, P, Cr and Mn. O_2 is physisorbed on B- and N-doped graphene with small adsorption energy. Chemisorption was observed on Al-, Si-, P-, Cr- and Mn-doped graphene. Later Pramanik and Kang (2011), studied the O_2 and the nitrogen monoxide (NO) adsorption on pristine, N-doped, and P-doped graphene including the van der Waals interaction. The van der Waals interaction makes an important contribution to the physisorption energy and the adsorption geometry of these gases in pristine and N-doped graphene.

Chen et al. (2012) studied the adsorption of NO, N_2O, and NO_2 on pristine and silicon (Si)-doped graphene using DFT methods. Their results indicate that, while adsorption of the three molecules on pristine graphene is very weak, Si-doping enhances the interaction of these molecules with a graphene sheet. Based on these results, Si-doped graphene can be expected to be a good sensor for NO and NO_2 detection, as well as a metal-free catalyst for N_2O reduction.

Shao et al. (2013), studied the adsorption of sulfur dioxide (SO_2) on pristine graphene and heteroatom-doped (B, N, Al, Si, Cr, Mn, Ag, Au, and Pt) graphene from DFT calculations. SO_2 molecule is adsorbed weakly on pristine graphene and B-, N-doped graphene. Strong chemisorption is observed on Al-, Si-, Cr-, Mn-, Ag-, Au- and Pt-doped graphene. Their results show that HG could be used as a gas sensor for SO_2. The adsorption energy showed that Cr and Mn may be the best choices among all the dopants.

Applications in Li-S Batteries

Lithium-sulfur (Li-S) batteries have been intensively researched in the last decade as a promising option for electrochemical energy storage systems owing to its high energy density, low cost and environmental friendliness. The cost-effective application of lithium-sulfur batteries is still hampered by the rapid decrease in their capacity, the self-discharge caused by the dissolution and the shuttle effect of soluble lithium polysulfides (LiPS) in the electrolyte. Therefore, it is imperative to control the solubility and transport of the polysulfide species to increase the stability of the Li-S batteries. In order to overcome the shuttle mechanism, from a theoretical point of view, one of the proposals is the use of functionalized graphene sheets that are capable of retaining long-chain LiPS.

Different studies of modified graphene and their interaction with polysulfides can be found in the literature. Jand et al. (2016) analyzed the effect of the adsorption of lithium polysulfides with different chain lengths (Li_2S_x) on pristine and defective graphene. Hou et al. (Hou et al. 2016), studied different HG systems in the interaction with polysulfides. Velez et al. (2019) used first-principle calculations to characterize the interaction between LiPS and different HG. The evaluation of residence time for LiPS on the surface of HG was performed in accordance with the transition state theory. The results revealed that Al or Si atoms were the most effective heteroatoms for LiPSs retention.

Oxidized Graphene: Covalent Functionalization by Epoxy, Carboxylic and Hydroxyl Groups

As seen above, graphene oxide can be characterized as a single graphitic monolayer with randomly distributed aromatic regions (sp^2 carbon atoms) and oxygenated aliphatic regions (sp^3 carbon atoms) containing hydroxyl, epoxy, carbonyl and carboxyl functional groups. The epoxy and hydroxyl groups lie above and below each graphene layer and the carboxylic groups exist usually at the edges of the layers; however, it should be noted that the chemistry and heterogeneity of graphene oxide are still greatly debated. The presence of oxygen groups on the surface of GO provides a remarkable hydrophilic character and analogous chemical reactivity. Therefore, GO has attracted tremendous interest over the past decade due to their unique and excellent electronic, optical, mechanical and chemical properties.

The theoretical approach of GO comes mainly from DFT calculations. The first theoretical works of GO with an ordered and amorphous structure dates from the beginning of the last decade. Yan et al. (Yan et al. 2009, Yan and Chou 2010) presented a detailed study of the oxidized functional groups (epoxide and hydroxyl) on graphene, based on DFT calculations. They showed the effects of single functional groups and their various combinations on the electronic and structural properties. It was found that single functional groups can induce interesting electronic bound states in graphene. The results suggested that functionalization of graphene by oxidation will significantly alter the electronic properties of graphene. Investigations on possible ordered structures with different compositions of epoxy and hydroxyl groups show that the hydroxyl groups could form chain-like structures stabilized by the hydrogen bonding between these groups, close to the epoxy groups.

Using genetic algorithm and first-principles approaches (Xiang et al. 2010), searched the most stable structure of oxidized graphene by considering various epoxy groups (normal epoxy, unzipped epoxy and epoxy pair), and found that phase separation between bare graphene and fully oxidized graphene is thermodynamically favorable in partially oxidized graphene. Liu et al. (2012) based on first-principles calculations and experimental observations, constructed amorphous structural models of GO with different coverage and hydroxyl/epoxy ratios randomly adding epoxy and hydroxyl groups onto a perfect graphene super cell.

The thermodynamically most favorable amorphous GO models always contain some locally ordered structures in the short-range, due to a compromise of the formation of hydrogen bonds, the existence of dangling bonds and the retention of the p bonds. Compared to the ordered counterparts, these amorphous GO structures possess good stability at low oxygen coverage. Varying the oxygen coverage and the ratio of epoxy and hydroxyl groups provides an efficient way to tune the electronic properties of the GO-based materials.

On the other hand, and in the direction of generating a GO model that takes into account different degrees of oxidation (Šljivančanin et al. 2013), used ab initio calculations based on Density Functional Theory (DFT) to investigate the binding of atomic oxygen on graphene, considering adsorption structures with sizes varying from sub-nanometric clusters to infinite layers. They studied the high, medium and low O coverage regime. From DFT results obtained for small clusters, they constructed a simple model able to describe the energetics of the O islands with sizes beyond those that can be directly treated by first-principles methods. The works selected comprise only a small look at the pioneering studies from the computational theoretical point of view that contributed to the elucidation of the properties and potentialities of the GO. Currently, the use of GO in materials science is widespread. In the context of theoretical studies, some are related to Li-S batteries (Wasalathilake et al. 2018) and metal-ion batteries (Moon et al. 2015, Dobrota et al. 2015).

In literature, there are numerous studies of the reactivity of GOs against the possible adsorptions of H, Cl, Pt, OH (Dobrota et al. 2016) and of different alkali metals (Boukhvalov et al. 2012) as well as studies of the interaction of S8 and long-chain polysulfides with GO (Wasalathilake et al. 2018).

Non-covalent Functionalization

The non-covalent functionalization of graphene and its derivatives can be achieved by means of the adsorption of an atomic species (usually a transition metal atom), a small isolated molecule or many molecules forming an interfacial layer. In the first two cases, the focus of the studies is mostly on the type of bond formed and its strength, whereas for the latter the aim is usually set at the modification and tuning of the electronic properties of graphene. The type of molecules that can be adsorbed on graphene range from small gaseous molecules, such as H_2 or O_2, to aromatic molecules, biomolecules (e.g., DNA bases, oligosaccharides, etc.), and polymeric chains. The adsorption is mostly based on van der Waals forces, electrostatic interactions or π–π interactions.

Given the size of the systems that are of interest from the application point of view, theoretical studies usually rely on DFT calculations. It is well known that DFT has serious difficulties in accounting for long-range interactions, such as the non-covalent interactions mentioned above. Therefore, in the present-day state-of-the-art computer modeling calculations the inclusion of some type of correction to describe these types of interactions is mandatory. Two of the most popular approaches are adding an empirical correction to the Kohn-Sham energy, such as Grimme's DFT-D method (Grimme et al. 2010, Caldeweyher et al. 2017) or using a purpose-designed functional, such as some of the highly parameterized hybrid Minnesota functionals (Zhao et al. 2006, Zhao and Truhlar 2008). In fact, the graphene surface has become the prototypical testing ground for van der Waals correction methods.

The interaction of graphene sheets with individual Transition Metal (TM) atoms was studied a while ago by DFT methods. This is relevant for the storage of H_2 on graphene. Pristine graphene displays a poor interaction with H_2, but the addition of dopants or the functionalization with TM atoms can greatly enhance the H_2 storage capacity of the material. Because metals tend to diffuse on surfaces and form clusters, studies of the various preferential coordination sites and the activation barriers that separate sites are important for predicting the behavior of single adsorbed metal atoms on such surfaces. Although in some of the earliest studies van der Waals corrections were not included, a few interesting trends were established. For instance, it was found that 3d TM atoms, with the exception of Cr, Mn, and Cu, are chemisorbed onto graphene with an η6 hollow geometry with binding energies ranging between 1.09–1.74 eV. On the other hand, half-filled Cr and Mn and filled Cu 3d atoms are physisorbed on these surfaces, with no preferred adsorption site. These trends were explained in terms of the Density Of States (DOS), Bader charge and molecular orbital diagram analysis (Valencia et al. 2010). The adsorption of H_2 on Pd-decorated graphene sheets was also studied and it could be verified that the bonding between the carbon-supported Pd particles and H_2 molecules is based on the Kubas coordination model and leads to a strengthened H-H bond with respect to the free Pd-H_2 complex (López-Corral et al. 2011). In a more recent study, the adsorption of H_2 on pristine, B-doped and with a vacancy defect graphene surfaces decorated with Rh atoms was examined by DFT corrected by Grimme's D2 scheme (Ambrusi et al. 2016). DOS and Bader charge analysis showed that there is a transfer in the range of 0.38–0.73 e⁻ from Rh to the graphene surface with and without defects. At the same time, overlap population and bond order analysis were useful to determine the reasons as to why the graphene with a single vacancy defect is the best choice for hydrogen storage purpose. It was found that the vacancy not only helps to stabilize the dispersion of individual Rh atoms but also displays H_2-molecular adsorption between the physisorbed and chemisorbed states, avoiding the hydrogen dissociation.

Another interesting application of graphenic nanostructures is their utility as platforms for the stabilization and the delivery of organic molecules that are used as drugs in therapeutic protocols. Regarding the immobilization of these molecules, noncovalent interactions such as π–π stacking are preferable in most cases because the release of the molecules is much easier to manipulate in comparison with that of covalently bonded molecules on the graphenic surface. In a non-periodic hybrid, DFT with dispersion corrections study the energetics of the adsorption of the anticancer chemotherapeutic camptothecin (CPT) on graphene, Boron Nitride (BN), modified graphene and Graphene Oxide (GO) clusters were determined (Saikia and Deka 2013). Global reactivity descriptors such as the electronic chemical potential (η) and hardness (μ) were used to characterize the interaction. It was demonstrated that the noncovalent functionalization of graphene with CPT is mediated by π–π stacking interaction whereas in the case of GO it is through the polar functional groups. The same authors also studied the noncovalent functionalization of graphene with pyrazinamide (PZA), another chemotherapeutic drug (Saikia and Deka 2014). It was shown that the adsorption of PZA on graphene does not affect the planarity of the sheet suggesting that the adsorption is governed by weak dipole-dipole interactions. The increase in the dipole moment of pristine graphene from zero up to ca. 2.7 D on functionalization with PZA, shows that this molecule can induce significant polarizability within graphene.

The embellishment of the graphene surface with aromatic molecules is probably one of the most obvious ways to functionalize this material and some of its derivatives due to favorable π–π stacking interactions. Some of the early works based on LDA calculations showed that by adsorbing simple aromatic molecules such as borazine ($B_3N_3H_6$), triazine ($C_3N_3H_3$) and benzene (C_6H_6) it was possible to tune the band gap of graphene (Chang et al. 2012). Functionalization of graphene with the chromophore imidazophenazine and its derivatives (Zarudnev et al. 2016) with molecules that possess a large conjugated π-system of bonds, such as porphyrins (Karachevtsev et al. 2018) has been investigated by DFT using the hybrid M05-2X functional. The hybrid meta-GGA M05-2X and M06-2X functionals are capable of predicting the structures and the interaction energies of noncovalent π–π stacked complexes formed by nucleotides (Gu et al. 2011), and even between Watson-Crick base pairs and carbon surfaces such as nanotubes and graphene (Stepanian et al. 2014), in close agreement with the much more costly post-Hartree–Fock MP2 method. In the case of the work on imidazophenazine and its derivatives, benzene, tetracene and imidazole were also tested. It was found that symmetrical molecules such as benzene and tetracene give rise to a uniform distribution of the electrostatic potential on graphene. On the other hand, an asymmetrical potential distribution is generated by the imidazole molecule. The asymmetry is caused by two nitrogen atoms in this molecule which possess small negative partial charges. The calculations also show that for asymmetrical linear molecules such as imidazophenazines, there is a difference in the binding energies with graphene along the zigzag and armchair direction and that the electrostatic potentials on the surfaces of these chromophores, as well as of their hybrids with graphene, have different signs that are mainly caused by the presence of nitrogen and hydrogen atoms in their structure. In the case of the adsorption of porphyrin derivatives on graphene, it was shown that the interaction of the porphyrin core is mainly due to π–π stacking, while for the tetraphenylporphyrin the interaction with the carbonaceous surface leads to a distortion of the flat structure of the porphyrin core. The decrease of the binding energy caused by this effect is compensated by the additional van der Waals interactions of the side-chain phenyl groups with graphene. Another interesting and quite recent theoretical study explores the impact of surface irregularities and chemical functionalization of graphene-based nanostructures on the electronic stabilities of diverse graphene-polypyrrolenanocomposites (Samanta and Das 2019). In this work, ab initio-based DFT calculations using M06-2X functional were performed. Pristine graphene, two different graphene surfaces with defects and three types of graphene oxides were represented using a finite graphene sheet. The pristine surface consists of 20 hexagonal rings capped with 20 terminating C-H bonds satisfying the residual valence. A multilayer graphene model consisting of two graphene layers was also considered. Polypyrrole was modeled by means of a finite oligomer of three pyrrole molecules. The calculated interaction energies of the oligomer with the different surfaces range between 0.6–1.2 eV. Different types of analysis were performed to characterize and dissect the molecule-surface interaction, such as DOS spectra, IR frequencies and global chemical reactivity descriptors.

The different types of noncovalent interactions, such as π–π, lone pair-π and H-bond and their specific role in stabilizing the graphene-polymer composites were further elucidated by analyzing electrostatic potential maps, Beckeiso surface maps and reduced density gradient isosurfaces. It was found that, the adsorption of polypyrrole on the graphene with defects is facilitated by both π-π stacking and lone pair-π interactions, whereas the enhancement in the interaction of the polymer with the graphene oxide surface appears from the H-bond interactions between pyrrole -NH bonds and the superficial epoxy functional groups.

Noncovalent functionalization of graphene by carbohydrates has the potential to improve graphene dispersibility and its use in biomedical applications. The interaction of graphene surfaces with the cyclic oligosaccharide β-cyclodextrin has been studied from a theoretical point of view (Jaiyong and Bryce 2017). In this work, the interaction energies computed at M06-2X/def2-TZVPP level were compared with those obtained from semi empirical methods, such as PM6-DH2, PM7 and DFTB3. It was found that for non-covalent intermolecular interactions, methods such as PM6-DH2 and PM7 reproduce high-level quantum mechanical approaches with good accuracy. DFTB3 provided the best computational approach of the approximate quantum methods considered for the evaluation of interaction energies in large graphene-carbohydrate systems.

Possible Applications of Functionalized Graphenic Surfaces

Graphene and functionalized graphenic surfaces have many possible applications; nevertheless, its commercialization to be used at an industrial scale is in the early stages. This is mainly due to the technical difficulty to produce graphene on an industrial level, which impacts on its price leading to an increment in the final price of the products that contain graphene (Tiwari et al. 2020).

In general, the applications that require high-quality graphene would take longer since it is more difficult to produce, making it more expensive. That is why the applications that in the beginning were thought to be primary applications, are these days left aside, and others are getting more attention.

Here some of the main applications of graphene and its derivatives are described. When graphene was first discovered, one of the main applications was thought to be electronic devices due to its high electrical conductivity, charge carrier and transparency (Tiwari et al. 2020, Wei and Kivioja 2013, Randviir et al. 2014) however these days it is not the case. This is mainly because the application of graphene in electronic devices needs high quality and uniformity of graphene. Atomic features, such as defects, impurities, disorders and anchored groups will lead to the alteration of these properties. In this regard, the controllable synthesis is the most challenging issue. The way of producing high-quality graphene is by chemical vapor deposition, but this technique is mainly of a disadvantage because of its high cost.

Graphene is useful in electronic devices, but also in photonics and optoelectronics that combine the optical and electronic properties (Wang et al. 2020). In optoelectronic devices graphene functions as a transparent conductor due to its low sheet resistance and high transparency, nevertheless the resistance is higher than the conventionally used of Indium Tin Oxide.

As can be seen, the great versatility of the graphene surface towards functionalization makes it a great material to be used in a wide number of applications (Dasari et al. 2017). As a result, the functionalized graphene monolayers find potential applications in diverging fields as an active and/or passive component (Fig. 9.8). here the main applications will be discussed.

These days one of the main applications of graphene or its derivatives is in the field of composites. Composite materials are those formed by combining two or more materials with different properties to produce a new material with better and unique properties than the materials separated alone, taking advantage of the best properties of each (Mohan et al. 2018). In this application, the oxidizing form of graphene, graphene oxide or reduced graphene oxide that is produced by exfoliation of graphite oxide is more popular due to its better dispersibility in aqueous solutions (Dasari et al. 2017). The incorporation of graphene and its derivatives in composites could result in better properties

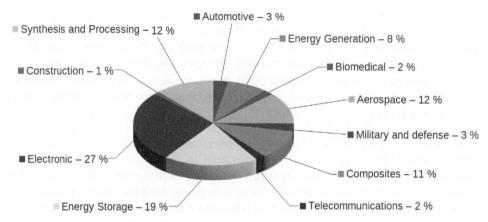

Figure 9.8: Applications based on graphenic structures.

leading to newer applications. The manufacture of such composites requires not only the graphene production on a sufficient scale but also that it could be incorporated and homogeneously distributed in the composite. Graphene can be added as a reinforcing agent in a polymer matrix improving its mechanical, thermal, electrical or other properties (Lawal 2019). The composites have been produced by covalent or no covalent functionalization of graphene, such as *in situ* polymerization or other techniques such as solution blending or melt blending (Mohan et al. 2018, Jinhong Du and Cheng 2012) and used later in several applications. Wu et al. (2020) reviewed the commercially available materials as well as research activities related to recent progress on high-performance polymer nanocomposites that are being used in various additive manufacturing techniques.

Many researchers have used GO for the fabrication of graphene nanocomposites. Song et al. (2019) reported the preparation and application of GO/waterborne polyurethane (WPU) composite paste. They analyzed the rheological properties of GO/WPU composite paste, which were greatly affecting the printing pattern definition. They also investigated the influences of reduced GO (rGO)/ WPU composite paste printed fabrics with various GO contents on the UV Protection Factor (UPF) values and UV transmittance. The presence of a polymeric solution during the possible reduction step prevents the agglomeration of the sheets.

Also inorganic nanostructures such as gold or silver can form a composite with graphene using different methods in order to enhance the electrical, thermal and mechanical properties. Sanes et al. (2020) in their work have summarized the recent development of nanocomposite materials that combine a thermoplastic matrix with different forms of graphene or graphene oxide nanofillers. Darabdhara et al. (2019) provided insights into the methods of decorations of graphene and GO with Ag and Au nanoparticles and their uses. The biocompatibility of both Au/Ag NPs and GO with biomolecules is used for their potent application in several biomedical fields such as drug delivery, photothermal therapy and biosensors.

Another main application based on graphene is in the field of energy storage and conversion. The energy crisis has given way to constant research in alternative sources of energy and in enhancing the capabilities of the existing ones. In this manner graphene is a promising material to be used in many energy conversion devices, such as fuel cells that have the advantage of having high conversion efficiency, low operating temperature and ease of handling (Fan et al. 2020, Saba and Jawaid 2018, Hossain et al. 2020, Farooqui et al. 2018). It is challenging to develop a metal-free Oxygen Reduction Reaction (ORR) catalyst with high current density and durable material to replace or to reduce the amount of Pt needed. In this way (Choi et al. 2011), reported the potentiality of Pt deposited on the surface of GNS using a surfactant as a cathode material for fuel cells. In their work, it can be noted that the current density was increased to three times that of the Pt/c. Graphene and its derivatives have also been used in photovoltaic cells (devices that convert light into electrical energy) (Tiwari et al. 2020) and solar thermal fuel cells (Choi 2011, Sani et al. 2018) due to the capacity of changing graphene band gap by doping or functionalization.

In the field of energy storage, electrochemical capacitors based on double-layer capacitance are good electrical energy storage devices that offer high energy density, high power capability and fast charge/discharge ability without degradation, long life and weight in comparison with the conventional dielectric capacitors. Graphene with its high specific area and great electron transport properties, chemical and mechanical stability, is a very promising material to be applied in energy storage devices (Tao et al. 2020, Grace and Rajendran 2016, Ke and Wang 2016).

Concerning batteries, graphene can be used as part of the anode or cathode material and modifying the separator (*Advanced Battery Materials* 2019). Regarding lithium-ion batteries and sodium-ion batteries, graphene has been used as an active material, additive, support or as part of a composite for anodic active materials (Kumar et al. 2019). Graphene as an additive helps to prevent the agglomeration of active material nano/microparticles used in anodes, improving the electron conductivity, Li^+ mobility and enhancing the mechanical stability of the whole electrode. In the case of alloying based anodes, such as silicon and tin, it also helps to buffer the changes in the volume suffered by these materials during charge and discharge cycles. As for cathodes, graphene has been used as an additive during the synthesis of the cathodic active materials following the same strategies as for anode materials.

In the case of lithium-sulfur batteries, graphene could help not only to diminish the impact of sulfur's insulating nature but also to buffer the volume changes (Zhang et al. 2018). Furthermore, if graphene is functionalized it could prevent the occurrence of the shuttle mechanism, by the interaction of the functional groups with lithium polysulfides as pointed out recently in several reports. All these features improved the usage of the active material and enhanced the overall performance of the battery. For separators, graphene and its derivatives are widely used to impede or reduce the migration of soluble polysulfides, reducing in this way the shuttle mechanism and improving the electrochemical behaviour. Recently, graphene derivatives were also used to protect lithium metal anodes preventing the Li dendrites growth and its consequent degradation (Hu et al. 2020).

In the field of water desalination, the functionalization of graphene is applied in different ways. Currently, ~ 1.2 billion people around the world are suffering from a lack of water and its adverse consequences on health, food and energy. Seawater comprises a vast supply of water (97.5% of all the water on the planet), the population growth, increased industrialization and greater energy needs and, on the other hand, loss of snowmelt, shrinkage of glaciers and so on will worsen this situation in the coming years (Nicolaï et al. 2014, Surwade et al. 2015). Seawater desalination has been mainly performed by multistage flash distillation and Reverse Osmosis (RO) (Spiegler and El-Sayed 2001). This kind of distillation needs desalination plants that are based on heating and then condensing seawater that consumes large amounts of thermal and electric energy, thus emitting greenhouse gases extensively (Shannon et al. 2008). The core of the RO process is a semi permeable membrane that separates pure water from seawater (Lee et al. 2011, Van Der Bruggen et al. 2003). For the RO desalination process, the development of more permeable membranes can reduce energy consumption and thus related costs. Although the necessary application of pressure in advanced RO systems is nearly equal to the thermodynamic limit, any extra reduction can notably affect the membrane performance (Hegab and Zou 2015). Classic semipermeable RO membrane is still based on the same polyamide thin-film composite design that was developed three decades ago (Xu et al. 2012, Cohen-Tanugi and Grossman 2012). Graphene can be regarded as an 'ultimate' RO membrane because it is stronger, thinner and more chemically resistant than the polyamide active layers in thin-film composite RO membranes (Cohen-Tanugi and Grossman 2015). The atomic thinness of graphene ($d \approx 0.34$ nm) can lead to larger water permeability than the polyamide active layer in thin-film composite membranes ($d \approx 100$ nm) and shows better tolerance to chlorine than polyamide (Sun et al. 2013). Graphene as Graphene Oxide (GO) shows antimicrobial properties, thus lowering membrane biofouling, i.e., improving the membrane lifetime and energy consumption of the water purification processes (Mahmoud et al. 2015, Goh and Ismail 2015). Graphene can be used for the construction of desalination membranes in various forms, such as pristine graphene, GO and reduced GO (rGO) (Zhao et al. 2014, He et al. 1998). Despite the significant advantages of monolayer graphene membranes, especially in terms of water permeability, the fabrication of leak-free, large-area monolayer graphene membranes with controlled pore density and size on the industrial scale is challenging (Cohen-Tanugi et al. 2016). One solution for this challenge is the fabrication of desalination membranes based on stacked GO nanosheets (Werber et al. 2016). GO nanosheets can be produced on a large scale with low cost by chemical oxidation and the ultrasonic exfoliation of graphite. This method promises the cost-efficient and industrially applicable fabrication of stacked membranes (Dreyer et al. 2010, Geim and Novoselov 2007). In addition, the presence of oxygen-containing functional groups, such as carboxyl groups, on the GO nanosheets enables functionalization and thus allows related charge-based interactions with water pollutants (An et al. 2016).

Another type of material which is gaining importance these days due to its versatility in synthesis, biocompatibility and biodegradability are the three-dimensional Graphene Foams (GFs). Graphene-based 3D foam structures (3D Gfs) have fascinating properties such as enormous specific surface area (Zhao et al. 2017), high pore volume (Huang et al. 2012), good structural integrity (Jinlong et al. 2017), thermal conductivity (Nieto et al. 2015), super-strong adsorption capacity (Yang et al. 2014), great porous structure (Chen et al. 2011), compressibility (Zhang et al. 2015), low mass density (Chi et al. 2017), high energy dissipation (Wang et al. 2018), controllable degradation rate, and improvement in electrical conductivity without agglomeration (Liu et al. 2018). The graphene 3D foam structure has a wide range of applications including catalyst supports (Long et al. 2011), water-purification (Tao et al. 2016), lithium-ion batteries (Xu et al. 2017), thermal energy storage (Lu et al. 2017), electrochemical sensors and biosensors (Yue et al. 2014), tissue engineering and regenerative medicine (Crowder et al. 2013). Regarding materials, they can be used in tissue engineering approaches with various advantages. Conventional implants are usually based on metal and silicon materials, which possess two major disadvantages: (i) lack of flexibility and (ii) poor chemical stability *in vivo* (Reina et al. 2017). Recent advances in nanotechnology have revolutionized the fields of tissue engineering and regenerative medicine. Among these promising biomaterials, 2D graphene showed excellent potential use in various aspects of nanotechnology-based tissue engineering (Akhavan et al. 2013). Nanosized GO sheets, as an allotrope of carbon, have received a lot of attention in the design of tissue engineering strategies for functional myocardium regeneration (McLaughlin et al. 2016). Assemblies of 2D graphene sheets into 3D architecture is required because a combination of the unique intrinsic properties of graphene and versatility of 3D porous structures which will pave a way towards high-performance graphene-based materials in tissue engineering and regenerative medicine (Li and Shi 2012). Tissue engineering is a field that benefited enormously from these unique structures. Li et al. (2013) reported the first use of these 3D porous structures in nerve tissue engineering Zhang et al. (2018), described the fabrication of GO-3D mesh as tissue scaffolds demonstrating that changing the ratio of NaCl to GO, produced changes in the porosity and strength of the final 3D architecture. Graphene foams are also used in nerve tissue regeneration. Recently, a large list of neurodegenerative disorders increasingly necessitated the design of powerful scaffolds with high efficiency in nerve tissue engineering in this regard, graphene foams serves thanks to their 3D microenvironments and highly conductive pathways for charge transport, in which cells are able to resemble their *in vivo* counterparts (Pazoki-Toroudi et al. 2016, Ajami et al. 2017, Ghadernezhad et al. 2016).

The nature and large specific area of 3D GFs makes them suitable materials for the loading and controlled released of essential biomaterials such as hydroxyapatite, the most widely accepted biomaterial for the repair and reconstruction of bone tissue defects (Nie et al. 2017). In the field of bone tissue regeneration, the unique and highly favorable properties of porous rGO 3D scaffolds make them a suitable option for bone tissue regeneration. For example (Tian et al. 2017), prepared a nanohydroxyapatite incorporated porous rGO 3D scaffold by *in situ* self-assembly in a hydrothermal reaction kettle for bone tissue regeneration. The advent of 3D graphene architecture has opened up new horizons for the development of stem cell-based tissue regeneration. Many research studies are still required for better understanding the impact of 3D graphene architecture on modern and progressive tissue engineering.

Conclusions and Remarks

In this chapter, the effect of functionalization of graphene surfaces and their possible applications is viewed. The main types of functionalization and the effect of these in the final properties of the achieved material are described from an experimental and theoretical point of view. Finally, some of the main applications of graphene and its functionalized derivatives are discussed.

It can be concluded that the great versatility of the graphene surface towards covalent and no-covalent functionalization makes it a very interesting material to be applied in a wide range of applications. And the interest in this material has not diminished over the years, since it was first isolated. Each functionalization is worth studying and this has been revealed in the case of graphene oxide where a lot of research was conducted. This could occur for other graphene-based materials. The fact that graphene can be easily functionalized, has expanded the number of potential applications for graphene-based materials.

Despite the significant advances in the synthesis and the many ways of functionalization of graphene, many challenges need to be addressed to implement graphene and its derivatives in the industrial scale. There is still the need to lower the cost and perform large scale functionalization processes in order to achieve large scale production to implement graphene and its derivatives at an industrial scale and not only in research. In this case, the improvement in production and processing of graphene-based materials are required to get more reproducibility and better performance of the material in different application fields.

Acknowledgements

The authors would like to acknowledge partial financial support from grants PIO Conicet-YPF 3855/15, PID Conicet-11220150100624, Program BID-Foncyt (PICT-2015-1605), and SeCyT of the Universidad Nacional de Córdoba and YPF-Tecnología (Y-TEC), Argentina.

References

Advanced Battery Materials. 2019. Advanced Battery Materials. https://doi.org/10.1002/9781119407713.

Ajami, Marjan, Hamidreza Pazoki-Toroudi, Hamed Amani, Seyed Fazel Nabavi, Nady Braidy, Rosa Anna Vacca et al. 2017. Therapeutic role of sirtuins in neurodegenerative disease and their modulation by polyphenols. Neurosci. Biobehav. Rev. 73: 39–47. https://doi.org/10.1016/j.neubiorev.2016.11.022.

Akhavan, Omid, Elham Ghaderi and Mahla Shahsavar. 2013. Graphene nanogrids for selective and fast osteogenic differentiation of human mesenchymal stem cells. Carbon 59: 200–211. https://doi.org/10.1016/j.carbon.2013.03.010.

Ambrusi, Rubén E., C. Romina Luna, Alfredo Juan and María E. Pronsato. 2016. DFT study of Rh-decorated pristine, B-doped and vacancy defected graphene for hydrogen adsorption. RSC Adv. 6(87): 83926–41. https://doi.org/10.1039/c6ra16604k.

An, Di, Ling Yang, Ting Jie Wang and Boyang Liu. 2016. Separation performance of graphene oxide membrane in aqueous solution. Ind. Eng. Chem. Res. 55(17): 4803–10. https://doi.org/10.1021/acs.iecr.6b00620.

Balandin, Alexander A., Suchismita Ghosh, Wenzhong Bao, Irene Calizo, Desalegne Teweldebrhan, Feng Miao et al. 2008. Superior thermal conductivity of single-layer graphene. Nano Lett. 8(3): 902–7. https://doi.org/10.1021/nl0731872.

Bazylewski, Paul and Giovanni Fanchini. 2019. Graphene: Properties and applications. Comprehen. Nanosci. Nanotechnol. 1–5(i): 287–304. https://doi.org/10.1016/B978-0-12-803581-8.10416-3.

Boukhvalov, Danil W., Daniel R. Dreyer, Christopher W. Bielawski and Young Woo Son. 2012. A computational investigation of the catalytic properties of graphene oxide: exploring mechanisms by using DFT methods. ChemCatChem. 4(11): 1844–49. https://doi.org/10.1002/cctc.201200210.

Bourlinos, Athanasios B., Vasilios Georgakilas, Radek Zboril, Theodore A. Steriotis, Athanasios K. Stubos and Christos Trapalis. 2009. Aqueous-phase exfoliation of graphite in the presence of polyvinylpyrrolidone for the production of water-soluble graphenes. Solid State Commun. 149(47–48): 2172–76. https://doi.org/10.1016/j.ssc.2009.09.018.

Bruggen, Bart Van Der, Carlo Vandecasteele, Tim Van Gestel, Wim Doyen and Roger Leysen. 2003. A review of pressure-driven membrane processes in wastewater treatment and drinking water production. Environ. Prog. 22(1): 46–56. https://doi.org/10.1002/ep.670220116.

Caldeweyher, Eike, Christoph Bannwarth and Stefan Grimme. 2017. Extension of the D3 dispersion coefficient model. J. Chem. Phys. 147(3). https://doi.org/10.1063/1.4993215.

Chang, Chung Huai, Xiaofeng Fan, Lain Jong Li and Jer Lai Kuo. 2012. Band gap tuning of graphene by adsorption of aromatic molecules. J. Chem. Phys. C 116(25): 13788–94. https://doi.org/10.1021/jp302293p.

Chang, Jingbo, Guihua Zhou, Erik R. Christensen, Robert Heideman and Junhong Chen. 2014. Graphene-based sensors for detection of heavy metals in water: A review chemosensors and chemoreception. Anal. Bioanal. Chem. 406(16): 3957–75. https://doi.org/10.1007/s00216-014-7804-x.

Chen, Ying, Bo Gao, Jing Xiang Zhao, Qing Hai Cai and Hong Gang Fu. 2012. Si-doped graphene: An ideal sensor for NO- or NO_2-detection and metal-free catalyst for N_2O-reduction. J. Mol. Model. 18(5): 2043–54. https://doi.org/10.1007/s00894-011-1226-x.

Chen, Zongping, Wencai Ren, Libo Gao, Bilu Liu, Songfeng Pei and Hui Ming Cheng. 2011. Three-dimensional flexible and conductive interconnected graphene networks grown by chemical vapour deposition. Nat. Mater. 10(6): 424–28. https://doi.org/10.1038/nmat3001.

Cheng, Hung Chieh, Ren Jye Shiue, Chia Chang Tsai, Wei Hua Wang and Yit Tsong Chen. 2011. High-quality graphene p-n junctions via resist-free fabrication and solution-based noncovalent functionalization. ACS Nano 5(3): 2051–59. https://doi.org/10.1021/nn103221v.

Chi, Kai, Zheye Zhang, Qiying Lv, Chuyi Xie, Jian Xiao, Fei Xiao et al. 2017. Well-ordered oxygen-deficient $CoMoO_4$ and Fe_2O_3 nanoplate arrays on 3D graphene foam: toward flexible asymmetric supercapacitors with enhanced capacitive properties. ACS Appl. Mater. Interfaces 9(7): 6044–6053. doi: 10.1021/acsami.6b14810.

Choi, Sung Mook, Min Ho Seo, Hyung Ju Kim and Won Bae Kim. 2011. Synthesis of surface-functionalized graphene nanosheets with high pt-loadings and their applications to methanol electrooxidation. Carbon 49(3): 904–9. https://doi.org/10.1016/j.carbon.2010.10.055.

Choi, Sung. 2011. Synthesis of surface-functionalized graphene nanosheetswith high Pt-loadings and their applications to methanolelectrooxidation. Carbon 11(1): 904–9. https://doi.org/10.1038/s41467-020-15116-z.

Cohen-Tanugi, David and Jeffrey C. Grossman. 2012. Water desalination across nanoporous graphene. Nano Lett. 12(7): 3602–8. https://doi.org/10.1021/nl3012853.

Cohen-Tanugi, David and Jeffrey C. Grossman. 2015. Nanoporous graphene as a reverse osmosis membrane: Recent insights from theory and simulation. Desalination 366: 59–70. https://doi.org/10.1016/j.desal.2014.12.046.

Cohen-Tanugi, David, Li Chiang Lin and Jeffrey C. Grossman. 2016. Multilayer nanoporous graphene membranes for water desalination. Nano Lett. 16(2): 1027–33. https://doi.org/10.1021/acs.nanolett.5b04089.

Crowder, Spencer W., Dhiraj Prasai, Rutwik Rath, Daniel A. Balikov, Hojae Bae, Kirill I. Bolotin et al. 2013. Three-dimensional graphene foams promote osteogenic differentiation of human mesenchymal stem cells. Nanoscale 5(10): 4171–76. https://doi.org/10.1039/c3nr00803g.

Dai, Jiayu and Jianmin Yuan. 2010. Adsorption of molecular oxygen on doped graphene: atomic, electronic and magnetic properties. April. https://doi.org/10.1103/PhysRevB.81.165414.

Darabdhara, Gitashree, Manash R. Das, Surya P. Singh, Aravind K. Rengan, Sabine Szunerits and Rabah Boukherroub. 2019. Ag and Au nanoparticles/reduced graphene oxide composite materials: synthesis and application in diagnostics and therapeutics. Adv. Colloid Interface Sci. https://doi.org/10.1016/j.cis.2019.101991.

Dasari, Bhagya Lakshmi, Jamshid M. Nouri, Dermot Brabazon and Sumsun Naher. 2017. Graphene and derivatives—synthesis techniques, properties and their energy applications. Energy 140(December): 766–78. https://doi.org/10.1016/j.energy.2017.08.048.

Denis, Pablo A., Ricardo Faccio and Alvaro W. Mombru. 2009. Is it possible to dope single-walled carbon nanotubes and graphene with sulfur? ChemPhysChem. 10(4): 715–22. https://doi.org/10.1002/cphc.200800592.

Denis, Pablo A. 2010. Band gap opening of monolayer and bilayer graphene doped with aluminium, silicon, phosphorus, and sulfur. Chem. Phys. Lett. 492(4–6): 251–57. https://doi.org/10.1016/j.cplett.2010.04.038.

Dobrota, Ana S., Igor A. Pašti, Slavko V. Mentus and Natalia V. Skorodumova. 2016. A general view on the reactivity of the oxygen-functionalized graphene basal plane. Phys. Chem. Chem. Phys. 18(9): 6580–86. https://doi.org/10.1039/c5cp07612a.

Dobrota, Ana S., Igor A. Pašti, Slavko V. Mentusab and Natalia V. Skorodumova. 2017. A DFT study of the interplay between dopants and oxygen functional groups over the graphene basal plane—implications in energy-related applications. Phys. Chem. Chem. Phys. 19(12): 8530–40. https://doi.org/10.1039/c7cp00344g.

Dobrota, Ana S., Igor A. Pašti and Natalia V. Skorodumova. 2015. Oxidized graphene as an electrode material for rechargeable metal-ion batteries—a DFT point of view. Electrochim. Acta 176(August): 1092–99. https://doi.org/10.1016/j.electacta.2015.07.125.

Dreyer, Daniel R., Sungjin Park, Christopher W. Bielawski and Rodney S. Ruoff. 2010. The chemistry of graphene oxide. Chem. Soc. Rev. 39(1): 228–40. https://doi.org/10.1039/b917103g.

Dubey, Pawan K. and Il-kwon Oh. 2012. Review on functionalized graphenes and their applications review on functionalized graphenes and their applications. No. May 2014.

Eigler, Siegfried and Andreas Hirsch. 2014. Chemistry with graphene and graphene oxide—challenges for synthetic chemists. Angew. Chem. Int. Ed. Engl. 53(30): 7720–38. https://doi.org/10.1002/anie.201402780.

Erickson, Kris, Rolf Erni, Zonghoon Lee, Nasim Alem, Will Gannett and Alex Zettl. 2010. Determination of the local chemical structure of graphene oxide and reduced graphene oxide. Adv. Mater. 22(40): 4467–72. https://doi.org/10.1002/adma.201000732.

Fan, Youjun, Bo Yang and Chuyan Rong. 2020. Functionalized carbon nanomaterials for advanced anode catalysts of fuel cells. Advanced Nanomaterials for Electrochemical-Based Energy Conversion and Storage. Elsevier Inc. https://doi.org/10.1016/b978-0-12-814558-6.00007-1.

Farooqui, U. R., A. L. Ahmad and N. A. Hamid. 2018. Graphene oxide: a promising membrane material for fuel cells. Renew. Sustain. Energy Rev. 82(August 2016): 714–33. https://doi.org/10.1016/j.rser.2017.09.081.

Farquhar, Anna K., Haidee M. Dykstra, Mark R. Waterland, Alison J. Downard and Paula A. Brooksby. 2016. Spontaneous modification of free-floating few-layer graphene by aryldiazonium ions: electrochemistry, atomic force microscopy, and infrared spectroscopy from grafted films. J. Phys. Chem. C 120(14): 7543–52. https://doi.org/10.1021/acs.jpcc.5b11279.

Farquhar, Anna K., Paula A. Brooksby and Alison J. Downard. 2018. Controlled spacing of few-layer graphene sheets using molecular spacers: capacitance that scales with sheet number. ACS Appl. Nano Mater. 1(3): 1420–29. https://doi.org/10.1021/acsanm.8b00280.

Feng, Zhanbin, Hongli Zuo, Jing Hu, Bing Yu, Nanying Ning, Ming Tian et al. 2019. *In situ* exfoliation of graphite into graphene nanosheets in elastomer composites based on diels-alder reaction during melt blending. Ind. Eng. Chem. Res. 58(29): 13182–89. https://doi.org/10.1021/acs.iecr.9b02580.

Fernández-Merino, M. J., J. I. Paredes, S. Villar-Rodil, L. Guardia, P. Solís-Fernández, D. Salinas-Torres et al. 2012. Investigating the influence of surfactants on the stabilization of aqueous reduced graphene oxide dispersions and the characteristics of their composite films. Carbon 50(9): 3184–94. https://doi.org/10.1016/j.carbon.2011.10.039.

Geim, A. K. and K. S. Novoselov. 2007. The rise of graphene. Nat. Mater. 6: 183–91.

Georgakilas, Vasilios, Athanasios B. Bourlinos, Radek Zboril, Theodore A. Steriotis, Panagiotis Dallas, Athanasios K. Stubos et al. 2010. Organic functionalisation of graphenes. Chem. Comm. 46(10): 1766–68. https://doi.org/10.1039/b922081j.

Georgakilas, Vasilios, Michal Otyepka, Athanasios B. Bourlinos, Vimlesh Chandra, Namdong Kim, K. Christian Kemp et al. 2012. Functionalization of graphene: covalent and non-covalent approach. Chem. Rev. 112(11): 6156–6214. https://doi.org/10.1021/cr3000412.

Georgakilas, Vasilios, Jitendra N. Tiwari, K. Christian Kemp, Jason A. Perman, Athanasios B. Bourlinos, Kwang S. Kim et al. 2016. Noncovalent functionalization of graphene and graphene oxide for energy materials, biosensing, catalytic, and biomedical applications. Chem. Rev. 116(9): 5464–5519. https://doi.org/10.1021/acs.chemrev.5b00620.

Ghadernezhad, Negar, Leila Khalaj, Hamidreza Pazoki-Toroudi, Masoumeh Mirmasoumi and Ghorbangol Ashabi. 2016. Metformin pretreatment enhanced learning and memory in cerebral Forebrain Ischaemia: The role of the AMPK/BDNF/P70SK signalling pathway. Pharm. Biol. 54(10): 2211–19. https://doi.org/10.3109/13880209.2016.1150306.

Gholizadeh, Reza and Yang Xin Yu. 2014. Work functions of pristine and heteroatom-doped graphenes under different external electric fields: An ab initio DFT study. J. Phys. Chem. C 118(48): 28274–82. https://doi.org/10.1021/jp5095195.

Gmbh, Evonik Degussa and Creavis Technologies. 2010. Cm 2/Vs and an on/off ratio >10 6, with negligible hysteresis, on standard silicon. Polymer 48: 1973–78. https://doi.org/10.1002/POLA.

Goh, P. S. and A. F. Ismail. 2015. Graphene-based nanomaterial: the state-of-the-art material for cutting edge desalination technology. Desalination 356: 115–28. https://doi.org/10.1016/j.desal.2014.10.001.

Grace, A. Nirmala and Ramachandran Rajendran. 2016. Advanced materials for supercapacitors. Eco-Friendly Nano-Hybrid Materials for Advanced Engineering Applications. https://doi.org/10.1201/b19935-7.

Grimme, Stefan, Jens Antony, Stephan Ehrlich and Helge Krieg. 2010. A consistent and accurate ab initio parametrization of density functional dispersion correction (DFT-D) for the 94 elements H-Pu. J. Phys. Chem. 132(15). https://doi.org/10.1063/1.3382344.

Gu, Jiande, Jing Wang and Jerzy Leszczynski. 2011. Stacking and H-bonding patterns of DGpdC and DGpdCpdG: Performance of the M05-2X and M06-2X minnesota density functionals for the single strand DNA. Chem. Phys. Lett. 512(1–3): 108–12. https://doi.org/10.1016/j.cplett.2011.06.085.

Guo, Beidou, Qian Liu, Erdan Chen, Hewei Zhu, Liang Fang and Jian Ru Gong. 2010. Controllable N-doping of graphene. Nano Lett. 10(12): 4975–80. https://doi.org/10.1021/nl103079j.

He, Heyong, Jacek Klinowski, Michael Forster and Anton Lerf. 1998. A new structural model for graphite oxide. Chem. Phys. Lett. 287(1–2): 53–56. https://doi.org/10.1016/S0009-2614(98)00144-4.

Hegab, Hanaa M. and Linda Zou. 2015. Graphene oxide-assisted membranes: fabrication and potential applications in desalination and water purification. J. Membr. Sci. 484: 95–106. https://doi.org/10.1016/j.memsci.2015.03.011.

Hossain, Shahzad, Abdalla M. Abdalla, Suleyha B. H. Suhaili, Imtiaz Kamal, Shabana P. S. Shaikh, Mohamed K. Dawood et al. 2020. Nanostructured graphene materials utilization in fuel cells and batteries: a review. J. Energy Storage 29(November 2019): 101386. https://doi.org/10.1016/j.est.2020.101386.

Hou, Ting Zheng, Xiang Chen, Hong Jie Peng, Jia Qi Huang, Bo Quan Li, Qiang Zhang et al. 2016. Design principles for heteroatom-doped nanocarbon to achieve strong anchoring of polysulfides for lithium–sulfur batteries. Small, June, 3283–91. https://doi.org/10.1002/smll.201600809.

Hu, Yin, Wei Chen, Tianyu Lei, Yu Jiao, Hongbo Wang, Xuepeng Wang et al. 2020. Graphene quantum dots as the nucleation sites and interfacial regulator to suppress lithium dendrites for high-loading lithium-sulfur battery. Nano Energy. https://doi.org/10.1016/j.nanoen.2019.104373.

Huang, Ping, Long Jing, Huarui Zhu and Xueyun Gao. 2013. Diazonium functionalized graphene: microstructure, electric, and magnetic properties. Acc. Chem. Res. 46(1): 43–52. https://doi.org/10.1021/ar300070a.

Huang, Xiaodan, Kun Qian, Jie Yang, Jun Zhang, Li Li, Chengzhong Yu et al. 2012. Functional nanoporous graphene foams with controlled pore sizes. Adv. Mater. 24(32): 4419–4423. doi: 10.1002/adma.201201680.

Jaiyong, Panichakorn and Richard A. Bryce. 2017. Approximate quantum chemical methods for modelling carbohydrate conformation and aromatic interactions: β-cyclodextrin and its adsorption on a single-layer graphene sheet. Phys. Chem. Chem. Phys. 19(23): 15346–55. https://doi.org/10.1039/c7cp02160g.

Jand, Sara Panahian, Yanxin Chen and Payam Kaghazchi. 2016. Comparative theoretical study of adsorption of lithium polysulfides (Li2Sx) on pristine and defective graphene. J. Power Sources 308(March): 166–71. https://doi.org/10.1016/j.jpowsour.2016.01.062.

Jinhong Du and Hui-Ming Cheng. 2012. The fabrication, properties, and uses of graphene/polymer composites. Macromol. Chem. Phys. 213: 1060–1077. https://doi.org/10.1002/macp.

Jinlong, Lv, Yang Meng, Ken Suzuki and Hideo Miura. 2017. Fabrication of 3D graphene foam for a highly conducting electrode. Mater. Lett. 196: 369–372. doi: 10.1016/j.matlet.2017.03.079.

Karachevtsev, V. A., S. G. Stepanian, M. V. Karachevtsev and L. Adamowicz. 2018. Graphene induced molecular flattening of meso-5,10,15,20-tetraphenyl porphyrin: DFT calculations and molecular dynamics simulations. Comput. Theor. Chem. 1133(June): 1–6. https://doi.org/10.1016/j.comptc.2018.04.009.

Karousis, Nikolaos, Nikos Tagmatarchis and Dimitrios Tasis. 2010. Current progress on the chemical modification of carbon nanotubes. Chem. Rev. 110(9): 5366–97. https://doi.org/10.1021/cr100018g.

Ke, Qingqing and John Wang. 2016. Graphene-based materials for supercapacitor electrodes—a review. J. Materiomics. 2(1): 37–54. https://doi.org/10.1016/j.jmat.2016.01.001.

Kumar, Rajesh, Sumanta Sahoo, E. Joanni, Rajesh Kumar Singh, Wai Kian Tan, Kamal Krishna Kar et al. 2019. Recent progress in the synthesis of graphene and derived materials for next generation electrodes of high performance lithium ion batteries. Progress in Energy and Combustion Science. https://doi.org/10.1016/j.pecs.2019.100786.

Lawal, Abdulazeez T. 2019. Graphene-based nano composites and their applications. A review. Biosensors and Bioelectronics 141(June): 111384. https://doi.org/10.1016/j.bios.2019.111384.

Lee, Changgu, Xiaoding Wei, Jeffrey W. Kysar and James Hone. 2008. Of monolayer graphene. Science 321(July): 385–88.

Lee, Kah Peng, Tom C. Arnot and Davide Mattia. 2011. A review of reverse osmosis membrane materials for desalination-development to date and future potential. J. Membr. Sci. 370(1–2): 1–22. https://doi.org/10.1016/j.memsci.2010.12.036.

Li, Chun and Gaoquan Shi. 2012. Three-dimensional graphene architectures. Nanoscale 4(18): 5549–63. https://doi.org/10.1039/c2nr31467c.

Li, Ning, Qi Zhang, Song Gao, Qin Song, Rong Huang, Long Wang et al. 2013. Three-dimensional graphene foam as a biocompatible and conductive scaffold for neural stem cells. Sci. Rep. 3: 1604. doi: 10.1038/srep01604.

Liu, Feng, Chao Wang and Qiheng Tang. 2018. Conductivity maximum in 3D graphene foams. Small 14(32): 1801458. https://doi.org/10.1002/smll.201801458.

Liu, Lizhao, Lu Wang, Junfeng Gao, Jijun Zhao, Xingfa Gao and Zhongfang Chen. 2012. Amorphous structural models for graphene oxides. Carbon 50(4): 1690–98. https://doi.org/10.1016/j.carbon.2011.12.014.

Long, Ying, Congcong Zhang, Xingxin Wang, Jianping Gao, Wei Wang and Yu Liu. 2011. Oxidation of SO_2 to SO_3 catalyzed by graphene oxide foams. J. Mater. Chem. 21(36): 13934–13941. doi: 10.1039/c1jm12031j.

López-Corral, Ignacio, Estefanía Germán, Alfredo Juan, María A. Volpe and Graciela P. Brizuela. 2011. DFT study of hydrogen adsorption on palladium decorated graphene. J. Phys. Chem. C 115(10): 4315–23. https://doi.org/10.1021/jp110067w.

Lotya, Mustafa, Yenny Hernandez, Paul J. King, Ronan J. Smith, Valeria Nicolosi, Lisa S. Karlsson et al. 2009. Liquid phase production of graphene by exfoliation of graphite in surfactant/water solutions. J. Am. Chem. Soc. 131(10): 3611–20. https://doi.org/10.1021/ja807449u.

Lu, Yanying, Ning Zhang, Shuang Jiang, Yudong Zhang, Meng Zhou, Zhanliang Tao et al. 2017. High-capacity and ultrafast na-ion storage of a self-supported 3d porous antimony persulfide–graphene foam architecture. Nano Lett. 17(6): 3668–3674. doi: 10.1021/acs.nanolett.7b00889.

Mahmoud, Khaled A., Bilal Mansoor, Ali Mansour and Marwan Khraisheh. 2015. Functional graphene nanosheets: the next generation membranes for water desalination. Desalination 356: 208–25. https://doi.org/10.1016/j.desal.2014.10.022.

McLaughlin, Sarah, James Podrebarac, Marc Ruel, Erik J. Suuronen, Brian McNeill and Emilio I. Alarcon. 2016. Nano-engineered biomaterials for tissue regeneration: what has been achieved so far? Front. Mater. 3(June). https://doi.org/10.3389/fmats.2016.00027.

Mohan, Velram Balaji, Kin tak Lau, David Hui and Debes Bhattacharyya. 2018. Graphene-based materials and their composites: a review on production, applications and product limitations. Compos. Part B: Eng. 142(December 2017): 200–220. https://doi.org/10.1016/j.compositesb.2018.01.013.

Moon, Hye Sook, Ji Hye Lee, Soonchul Kwon, Il Tae Kim and Seung Geol Lee. 2015. Mechanisms of Na adsorption on graphene and graphene oxide: density functional theory approach. Carbon Lett. 16(2): 116–20. https://doi.org/10.5714/CL.2015.16.2.116.

Nicolaï, Adrien, Bobby G. Sumpter and Vincent Meunier. 2014. Tunable water desalination across graphene oxide framework membranes. Phys. Chem. Chem. Phys. 16(18): 8646–54. https://doi.org/10.1039/c4cp01051e.

Nie, Wei, Cheng Peng, Xiaojun Zhou, Liang Chen, Weizhong Wang, Yanzhong Zhang et al. 2017. Three-dimensional porous scaffold by self-assembly of reduced graphene oxide and nano-hydroxyapatite composites for bone tissue engineering. Carbon 116: 325–37. https://doi.org/10.1016/j.carbon.2017.02.013.

Nieto, Andy, Rupak Dua, Cheng Zhang, Benjamin Boesl, Sharan Ramaswamy and Arvind Agarwal. 2015. Three dimensional graphene foam/polymer hybrid as a high strength biocompatible scaffold. Adv. Funct. Mater. 25(25): 3916–3924. doi: 10.1002/adfm.201500876.

Novoselov, K. S., A. K. Geim, S. V. Morozov, D. Jiang, Y. Zhang, S. V. Dubonos et al. 2004. Electric field effect in atomically thin carbon films. Science 306(5696): 666–69.

Novoselov, K. S., A. K. Geim, S. V. Morozov, D. Jiang, M. I. Katsnelson, I. V. Grigorieva et al. 2005. Two-dimensional gas of massless dirac fermions in graphene. Nature 438(7065): 197–200. https://doi.org/10.1038/nature04233.

Oh, Cho-Rong, Dae-Il Lee, Jun-Hong Park and Dai-Soo Lee. 2019. Thermally healable and recyclable graphene-nanoplate/epoxy composites via an *in-situ* diels-alder reaction on the graphene-nanoplate surface. Polymers 11(6). https://doi.org/10.3390/polym11061057.

Pazoki-Toroudi, Hamidreza, Hamed Amani, Marjan Ajami, Seyed Fazel Nabavi, Nady Braidy, Pandima Devi Kasi et al. 2016. Targeting MTOR signaling by polyphenols: a new therapeutic target for ageing. Ageing Res. Rev. 31: 55–66. https://doi.org/10.1016/j.arr.2016.07.004.

Phillipson, Roald, César J. Lockhart De La Rosa, Joan Teyssandier, Peter Walke, Deepali Waghray, Yasuhiko Fujita et al. 2016. Tunable doping of graphene by using physisorbed self-assembled networks. Nanoscale 8(48): 20017–26. https://doi.org/10.1039/c6nr07912a.

Pramanik, Anup and Hong Seok Kang. 2011. Density functional theory study of O_2 and NO adsorption on heteroatom-doped graphenes including the van der waals interaction. J. Phys. Chem. C 115(22): 10971–78. https://doi.org/10.1021/jp200783b.

Puglia, Megan K., Sohan Aziz, Kevin M. Brady, Mark O'Neill and Challa V. Kumar. 2020. Stirred not shaken: facile production of high-quality, high-concentration graphene aqueous suspensions assisted by a protein. ACS Appl. Mater. Interfaces 12(3): 3815–26. https://doi.org/10.1021/acsami.9b15121.

Raji, Marya, Nadia Zari, Abou el Kacem Qaiss and Rachid Bouhfid. 2018. Chemical preparation and functionalization techniques of graphene and graphene oxide. Functionalized Graphene Nanocomposites and Their Derivatives: Synthesis, Processing and Applications. Elsevier Inc. https://doi.org/10.1016/B978-0-12-814548-7.00001-5.

Ramezanzadeh, B., S. Niroumandrad, A. Ahmadi, M. Mahdavian and M. H. Mohamadzadeh Moghadam. 2016. Enhancement of barrier and corrosion protection performance of an epoxy coating through wet transfer of amino functionalized graphene oxide. Corros. Sci. 103: 283–304. https://doi.org/10.1016/j.corsci.2015.11.033.

Randviir, Edward P., Dale A. C. Brownson and Craig E. Banks. 2014. A decade of graphene research: production, applications and outlook. Mater. Today 17(9): 426–32. https://doi.org/10.1016/j.mattod.2014.06.001.

Reina, Giacomo, José Miguel González-Domínguez, Alejandro Criado, Ester Vázquez, Alberto Bianco and Maurizio Prato. 2017. Promises, facts and challenges for graphene in biomedical applications. Chem. Soc. Rev. 46(15): 4400–4416. https://doi.org/10.1039/c7cs00363c.

Rodríguez-Pérez, Laura, Ángeles Herranz and Nazario Martín. 2013. The chemistry of pristine graphene. Chem. Comm. 49(36): 3721–35. https://doi.org/10.1039/c3cc38950b.

Saba, N. and M. Jawaid. 2018. Energy and Environmental Applications of Graphene and Its Derivatives. Polymer-Based Nanocomposites for Energy and Environmental Applications: A Volume in Woodhead Publishing Series in Composites Science and Engineering. Elsevier Ltd. https://doi.org/10.1016/B978-0-08-102262-7.00004-0.

Saidina, D. S., N. Eawwiboonthanakit, M. Mariatti, S. Fontana and C. Hérold. 2019. Recent development of graphene-based ink and other conductive material-based inks for flexible electronics. J. Electron. Mater. 48(6): 3428–50. https://doi.org/10.1007/s11664-019-07183-w.

Saiful Badri, Muhammad Ashraf, Muhamad Mat Salleh, Noor Far ain Md Noor, Mohd Yusri Abd Rahman and Akrajas Ali Umar. 2017. Green synthesis of few-layered graphene from aqueous processed graphite exfoliation for graphene thin film preparation. Mater. Chem. Phys. 193: 212–19. https://doi.org/10.1016/j.matchemphys.2017.02.029.

Saikia, Nabanita and Ramesh C. Deka. 2013. Ab initio study on the noncovalent adsorption of camptothecin anticancer drug onto graphene, defect modified graphene and graphene oxide. J. Comput. Aid. Mol. Des. 27(9): 807–21. https://doi.org/10.1007/s10822-013-9681-3.

Saikiaa, Nabanita and Ramesh C. Deka. 2014. Density functional study on noncovalent functionalization of pyrazinamide chemotherapeutic with graphene and its prototypes. New J. Chem. 38(3): 1116–28. https://doi.org/10.1039/c3nj00735a.

Samanta, Pabitra Narayan and Kalyan Kumar Das. 2019. Deciphering the impact of surface defects and functionalization on the binding strength and electronic properties of graphene-polypyrrole nanocomposites: a first-principles approach. J. Phys. Chem. C 123(9): 5447–59. https://doi.org/10.1021/acs.jpcc.8b11173.

Sanes, José, Cristian Sánchez, Ramón Pamies, María Dolores Avilés and María Dolores Bermúdez. 2020. Extrusion of polymer nanocomposites with graphene and graphene derivative nanofillers: an overview of recent developments. Materials 13(3). https://doi.org/10.3390/ma13030549.

Sani, Elisa, Javier P. Vallejo, David Cabaleiro and Luis Lugo. 2018. Functionalized graphene nanoplatelet-nanofluids for solar thermal collectors. Sol. Energy Mater. Sol. Cells 185(April): 205–9. https://doi.org/10.1016/j.solmat.2018.05.038.

Shannon, Mark A., Paul W. Bohn, Menachem Elimelech, John G. Georgiadis, Benito J. Marĩas and Anne M. Mayes. 2008. Science and technology for water purification in the coming decades. Nature 452(7185): 301–10. https://doi.org/10.1038/nature06599.

Shao, Li, Guangde Chen, Honggang Ye, Yelong Wu, Zhijuan Qiao, Youzhang Zhu et al. 2013. Sulfur dioxide adsorbed on graphene and heteroatom-doped graphene: a first-principles study. Eur. Phys. J. B. 86(2). https://doi.org/10.1140/epjb/e2012-30853-y.

Sheng, Zhen Huan, Hong Li Gao, Wen Jing Bao, Feng Bin Wang and Xing Hua Xia. 2012. Synthesis of boron doped graphene for oxygen reduction reaction in fuel cells. J. Mater. Chem. 22(2): 390–95. https://doi.org/10.1039/c1jm14694g.

Singh, Virendra, Daeha Joung, Lei Zhai, Soumen Das, Saiful I. Khondaker and Sudipta Seal. 2011. Graphene based materials: past, present and future. Prog. Mater. Sci. 56(8): 1178–1271. https://doi.org/10.1016/j.pmatsci.2011.03.003.

Šljivančanin, Željko, Aleksandar S. Milošević, Zoran S. Popović and Filip R. Vukajlović. 2013. Binding of atomic oxygen on graphene from small epoxy clusters to a fully oxidized surface. Carbon 54: 482–88. https://doi.org/10.1016/j.carbon.2012.12.008.

Song, Weihua, Bo Wang, Lihua Fan, Fangqing Ge and Chaoxia Wang. 2019. Graphene oxide/waterborne polyurethane composites forfine patternfabrication and ultrastrong ultraviolet protection cotton fabric via screenprinting. App. Surf. Sci. 463: 403–411. https://doi.org/10.1016/j.apsusc.2018.08.167.

Spiegler, K. S. and Y. M. El-Sayed. 2001. The energetics of desalination processes. Desalination 134(1–3): 109–28. https://doi.org/10.1016/S0011-9164(01)00121-7.

Stepanian, S. G., M. V. Karachevtsev, V. A. Karachevtsev and L. Adamowicz. 2014. Interactions of the watson-crick nucleic acid base pairs with carbon nanotubes and graphene: DFT and MP2 study. Chem. Phys. Lett. 610–611(August): 186–91. https://doi.org/10.1016/j.cplett.2014.07.035.

Sun, Pengzhan, Miao Zhu, Kunlin Wang, Minlin Zhong, Jinquan Wei, Dehai Wu et al. 2013. Selective ion penetration of graphene oxide membranes. ACS Nano 7(1): 428–37. https://doi.org/10.1021/nn304471w.

Surwade, Sumedh P., Sergei N. Smirnov, Ivan V. Vlassiouk, Raymond R. Unocic, Gabriel M. Veith, Sheng Dai et al. 2015. Water desalination using nanoporous single-layer graphene. Nat. Nanotechnol. 10(5): 459–64. https://doi.org/10.1038/nnano.2015.37.

Tao, Hengcong, Qun Fan, Tao Ma, Shizhen Liu, Henry Gysling, John Texter et al. 2020. Two-dimensional materials for energy conversion and storage. Prog. Mater. Sci. 111: 100637. https://doi.org/10.1016/j.pmatsci.2020.100637.

Tao, Hua-Chao, Shou-Chao Zhu, Xue-Lin Yang, Lu-Lu Zhang and Shi-Bing Ni. 2016. Systematic investigation of reduced graphene oxide foams for high-performance supercapacitors. Electrochim. Acta. 190: 168–177. doi: 10.1016/j.electacta.2015.12.179.

Thakur, Samir, Sankar M. Borah and Nirab C. Adhikary. 2018. A DFT study of structural, electronic and optical properties of heteroatom doped monolayer graphene. Optik 168(September): 228–36. https://doi.org/10.1016/j.ijleo.2018.04.099.

Thakur, Vijay Kumar and Manju Kumari Thakur. 2015. Chemical functionalization of carbon nanomaterials: chemistry and applications. Chemical Functionalization of Carbon Nanomaterials: Chemistry and Applications 1–1061.

Tian, Zizhu, Lixun Huang, Xibo Pei, Junyu Chen, Tong Wang, Tao Yang et al. 2017. Electrochemical synthesis of three-dimensional porous reduced graphene oxide film: preparation and *in vitro* osteogenic activity evaluation. Colloids and Surf. B: Biointerfaces 155: 150–58. https://doi.org/10.1016/j.colsurfb.2017.04.012.

Tiwari, Santosh K., Sumanta Sahoo, Nannan Wang and Andrzej Huczko. 2020. Graphene research and their outputs: status and prospect. J. Sci. Adv. Mater. Dev. 5(1): 10–29. https://doi.org/10.1016/j.jsamd.2020.01.006.

Tiwari, Sourabh, Anushka Purabgola and Balasubramanian Kandasubramanian. 2020. Functionalised graphene as flexible electrodes for polymer photovoltaics. J. Alloys Compd. 825: 153954. https://doi.org/10.1016/j.jallcom.2020.153954.

Valencia, Hubert, Adrià Gil and Gilles Frapper. 2010. Trends in the adsorption of 3d transition metal atoms onto graphene and nanotube surfaces: A DFT study and molecular orbital analysis. J. Phys. Chem. C 114(33): 14141–53. https://doi.org/10.1021/jp103445v.

Vélez, Patricio, M. Laura Para, Guillermina L. Luque, Daniel Barraco and Ezequiel P. M. Leiva. 2019. Modeling of substitutionally modified graphene structures to prevent the shuttle mechanism in lithium-sulfur batteries. Electrochim. Acta 309(June): 402–14. https://doi.org/10.1016/j.electacta.2019.04.062.

Wallace, P. R. 1946. The band theory of graphite. Phys. Rev. B 71: 622–634. http://journals.aps.org/pr/pdf/10.1103/PhysRev.71.622.

Wang, Chao, Douxing Pan and Shaohua Chen. 2018. Energy dissipative mechanism of graphene foam materials. Carbon 132: 641–650. doi: 10.1016/j.carbon.2018.02.085.

Wang, J., J. Song, X. Mu and M. Sun. 2020. Optoelectronic and photoelectric properties and applications of graphene-based nanostructures. Mater. Today Phys. 13. https://doi.org/10.1016/j.mtphys.2020.100196.

Wang, Zhijie, Hong Gao, Qing Zhang, Yuqing Liu, Jun Chen and Zaiping Guo. 2019. Recent advances in 3D graphene architectures and their composites for energy storage applications. Small 15(3): 1–21. https://doi.org/10.1002/smll.201803858.

Wasalathilake, Kimal Chandula, Md Roknuzzaman, Kostya Ostrikov, Godwin A. Ayoko and Cheng Yan. 2018. Interaction between functionalized graphene and sulfur compounds in a lithium-sulfur battery-a density functional theory investigation. RSC Adv. 8(5): 2271–79. https://doi.org/10.1039/c7ra11628d.

Wei, Di and Jani Kivioja. 2013. Graphene for energy solutions and its industrialization. Nanoscale 5(21): 10108–26. https://doi.org/10.1039/c3nr03312k.

Werber, Jay R., Chinedum O. Osuji and Menachem Elimelech. 2016. Materials for next-generation desalination and water purification membranes. Nat. Rev. Mater. 1. https://doi.org/10.1038/natrevmats.2016.18.

Wu, H., W. P. Fahy, S. Kim, H. Kim, N. Zhao, L. Pilato et al. 2020. Recent developments in polymers/polymer nanocomposites for additive manufacturing. Prog. Mater. Sci. 111(November 2019). https://doi.org/10.1016/j.pmatsci.2020.100638.

Wu, Yujun, Chuanbao Cao, Chen Qiao, Yu Wu, Lifen Yang and Waqar Younas. 2019. Bandgap-tunable phosphorus-doped monolayer graphene with enhanced visible-light photocatalytic H2-production activity. J. Mater. Chem. C 7(34): 10613–22. https://doi.org/10.1039/c9tc03539g.

Xiang, H. J., Su Huai Wei and X. G. Gong. 2010. Structural motifs in oxidized graphene: a genetic algorithm study based on density functional theory. PPhys. Rev. B Condens. Matter Mater. Phys. 82(3). https://doi.org/10.1103/PhysRevB.82.035416.

Xu, Peng Tao, Ji Xiang Yang, Ke Sai Wang, Zhen Zhou and Pan Wen Shen. 2012. Porous graphene: properties, preparation, and potential applications. Chin. Sci. bull. 57(23): 2948–55. https://doi.org/10.1007/s11434-012-5121-3.

Xu, Yunhe, Lun Li and Wenxin Huang. 2017. Porous graphene oxide prepared on nickel foam by electrophoretic deposition and thermal reduction as high-performance supercapacitor electrodes. Materials 10(8): 936. doi: 10.3390/ma10080936

Yan, Jia An, Lede Xian and M. Y. Chou. 2009. Structural and electronic properties of oxidized graphene. Phys. Rev. Lett. 103(8). https://doi.org/10.1103/PhysRevLett.103.086802.

Yan, Jia An and M. Y. Chou. 2010. Oxidation functional groups on graphene: structural and electronic properties. Phys. Rev. B Condens. Matter Mater. Phys. 82(12). https://doi.org/10.1103/PhysRevB.82.125403.

Yang, Gao, Lihua Li, Wing Bun Lee and Man Cheung Ng. 2018. Structure of graphene and its disorders: a review. Sci. Technol. Adv. Mater. 19(1): 613–48. https://doi.org/10.1080/14686996.2018.1494493.

Yang, Sudong, Lin Chen, Lei Mu and Peng-Cheng Ma. 2014. Magnetic graphene foam for efficient adsorption of oil and organic solvents. J. Colloid Interface Sci. 430: 337–344. doi: 10.1016/j.jcis.2014.05.062.

YongJian, W. U., Tang RenHeng, L. I. WenChao, Wang Ying, Huang Ling and Ouyang LiuZhang. 2020. A high-quality aqueous graphene conductive slurry applied in anode of lithium-ion batteries. J. Alloys Compd. 830: 154575. https://doi.org/10.1016/j.jallcom.2020.154575.

Yue, Hong Yan, Shuo Huang, Jian Chang, Chaejeong Heo, Fei Yao, Subash Adhikari et al. 2014. ZnO nanowire arrays on 3D hierachical graphene foam: biomarker detection of Parkinson's disease. ACS Nano. 8(2): 1639–1646. doi: 10.1021/nn405961p.

Zarudnev, Eugene, Stepan Stepanian, Ludwik Adamowicz and Victor Karachevtsev. 2016. Noncovalent interaction of graphene with heterocyclic compounds: benzene, imidazole, tetracene, and imidazophenazines. ChemPhysChem. 17(8): 1204–12. https://doi.org/10.1002/cphc.201500839.

Zhang, Yan, Xinhui Xia, Bo Liu, Shengjue Deng, Dong Xie, Qi Liu et al. 2019. Multiscale graphene-based materials for applications in sodium ion batteries. Adv. Energy Mater. 9(8): 1–35. https://doi.org/10.1002/aenm.201803342.

Zhang, Yi., Yi Huang, Tengfei Zhang, Huicong Chang, Peishuang Xiao, Honghui Chen et al. 2015. Broadband and tunable high-performance microwave absorption of an ultralight and highly compressible graphene foam. Adv. Mater. 27(12): 2049–2053. doi: 10.1002/adma.20140578.

Zhang, Ying, Xiao Liu, Kayla Michelson, Rachana Trivedi, Xu Wu, Eric Schepp et al. 2018. Graphene oxide-based biocompatible 3D mesh with a tunable porosity and tensility for cell culture. ACS Biomater. Sci. Eng. 4(5): 1505–17. https://doi.org/10.1021/acsbiomaterials.8b00190.

Zhang, Yunya, Zan Gao, Ningning Song, Jiajun He and Xiaodong Li. 2018. Graphene and its derivatives in lithium–sulfur batteries. Mater. Today Energy 9: 319–35. https://doi.org/10.1016/j.mtener.2018.06.001.

Zhao, Daoli, Lu Zhang, David Siebold, DeArmond, Noe Alvarez, Vesselin Shanov et al. 2017. Electrochemical studies of three dimensional graphene foam as an electrode material. Electroanal. 29(6): 1506–1512. doi: 10.1002/elan.201700057.

Zhao, Jian, Zhenyu Wang, Jason C. White and Baoshan Xing. 2014. Graphene in the aquatic environment: adsorption, dispersion, toxicity and transformation. Environ. Sci. Technol. 48(17): 9995–10009. https://doi.org/10.1021/es5022679.

Zhao, Yan, Nathan E. Schultz and Donald G. Truhlar. 2006. Design of density functionals by combining the method of constraint satisfaction with parametrization for thermochemistry, thermochemical kinetics, and noncovalent interactions. J. Chem. Theory Comput. 2(2): 364–82. https://doi.org/10.1021/ct0502763.

Zhao, Yan and Donald G. Truhlar. 2008. The M06 suite of density functionals for main group thermochemistry, thermochemical kinetics, noncovalent interactions, excited states, and transition elements: two new functionals and systematic testing of four M06-class functionals and 12 other functionals. Theor. Chem. Acc. 120(1–3): 215–41. https://doi.org/10.1007/s00214-007-0310-x.

Zhong, Yujia, Zhen Zhen and Hongwei Zhu. 2017. Graphene: fundamental research and potential applications. FlatChem. 4: 20–32. https://doi.org/10.1016/j.flatc.2017.06.008.

CHAPTER 10
Two-Dimensional Hybrid Nanomaterials

Luis A. Pérez,[1] *Federico Fioravanti,*[1] *Diana M. Arciniegas Jaimes,*[2]
Noelia Bajales Luna[2,*] *and Gabriela I. Lacconi*[1,*]

Introduction

Today graphene is a two-dimensional reference system of a thin material with atomic thickness and carbon atoms positioned in a perfect hexagonal network, exhibiting a very stable behaviour when interacting with any external agents (Geim and Novoselov 2007). This is the result of its exceptional properties such as high thermal and electrical conductivity, zero bandgaps, high optical transmittance and its superior mechanical resistance in comparison to that of steel (Pimenta et al. 2007, Jorio et al. 2011, Bao and Loh 2012, Zhang 2015, Panwar et al. 2019). All kinds of two-dimensional materials have high surface areas through which they are chemically and physically very reactive, exhibiting quantum confinement effects, mainly at the edges and surface defects where they concentrate the particular electronic, magnetic, photonic and catalytic characteristics.

Nanoscience and nanotechnology have focused on basic developments for their use in specific-objectives to develop novel materials of interest in medicine, energy, agriculture, engineering, physic, optoelectronics, etc. (Bonaccorso et al. 2010, Han et al. 2019, Shaohua et al. 2020, Galeotti et al. 2020). However, certain applications require some particular characteristics that even graphene or two-dimensional nanomaterials cannot supply by themselves. Thus, in order to explain the new and future requirements, intensive research of specific synthesis methods, behaviour and advanced characterization of two-dimensional materials such as Graphene Oxide (GO), reduced Graphene Oxide (rGO) and Transition Metal Dichalcogenides (TMDC), enhanced with nanoparticles in their structures, have been carried out (Siamaki et al. 2011, Dragoman and Dragoman 2017, Li et al. 2018, Chen et al. 2018, Wen et al. 2018, Guan and Han 2019, Cong et al. 2020, Cuniberto et al. 2020, Zhao et al. 2020, Wang et al. 2020). Some of the present reasons that justify the effort for seeking the modification of these materials depend on the desired functionalities as well as on the limitations to resolve interdisciplinary situations in optoelectronics, spintronic, energy storage, biosensors, etc.

Among many two-dimensional carbonaceous materials, GO is nowadays one of the most used to obtain hybrid systems since it provides low-cost routes and easy implementation in the laboratory. GO is obtained by extreme oxidation of graphite in order to incorporate oxygen functional groups (mainly hydroxyl, epoxy and carboxyl) in the carbon matrix.

Depending on the oxidation conditions, the size of flakes and the content of Csp^2 and Csp^3 domains in the flakes can be different (Suk et al. 2010, Marquardt et al. 2011). Generally, aqueous dispersions of GO are brown, contain highly hydrophilic flakes and their presence in the hybrid material can be

[1] INFIQC, Dpto. Fisicoquímica, Facultad de Ciencias Químicas, UNC, 5000 Córdoba, Argentina.
[2] IFEG, Facultad de Matemática, Astronomía, Física y Computación, UNC, 5000 Córdoba, Argentina.
* Corresponding authors: Noelia.bajales.luna@unc.edu.ar; glacconi@fcq.unc.edu.ar

recognized through a vibrational characterization when the extinction UV-vis spectrum in respect to the initial carbon signals evidence $\pi \rightarrow \pi^*$ and n $\rightarrow \pi^*$ transitions corresponding to C-C and C=O bonds, respectively.

On the one hand, a key tool to design and control the physical properties of carbonaceous hybrid materials to develop high-impact technology appears when GO flakes are embedded with magnetic nanostructures such as nickel, iron or cobalt nanoparticles (Zhang et al. 2019, Behera et al. 2019, Sebastian et al. 2019). Thus large magnetic permeability, enhanced sensitivity as well as low coercivity and remanence for magnetic shielding are attributes that can be multiplexed with those of the GO matrix. It is well-known that a behaviour closer to that of the graphene monoatomic film can be obtained through the progressive partial reduction of the oxygen groups of GO. Therefore chemical, thermal, optical and electrochemical reduction methods have been extensively used in order to obtain rGO flakes (Stankovich et al. 2007, Murphy et al. 2013, Zhong et al. 2013, Huang et al. 2015, Badhulika et al. 2015, Naqvi et al. 2019). The latter are used as building blocks of hybrid nanomaterials due to their properties and particular active sites that enable the possibility of obtaining novel materials with functionalities. In addition, two-dimensional Transition Metal Dichalcogenides (TMDC), such as MoS_2, WS_2, TeS_2, SeS_2 have received great attention for academic research and technological applications since they could play a fundamental role as elements in sensors, photoelectronic devices, photocatalysts, elements of clinical diagnosis, etc. (Pumera and Huiling Loo 2014, Biscaras et al. 2015, Donarelli and Ottaviano 2018, Saifur et al. 2018, Lu et al. 2018), in compliance with the respective requirements based on their special optical, electrical and structural properties.

It has been extensively demonstrated with achievements how composites of micro- and macroscopic dimensions can contribute to improving the properties of host materials (Telesio et al. 2018, Khan et al. 2020). In this way, it is worth noting that one of the aspects that restricts the use of two-dimensional nanomaterial sheets is the weak absorption of radiation due to the thin thickness to be specifically applied in optical sensors, optoelectronics and photovoltaic devices (Fai Makand Shan 2016, Wang et al. 2019). One strategy to increase light absorption efficiency is its use as one component or block in the hybrid platform building. For example, by the assembly of two-dimensional flakes with plasmonic nanoparticles, enhancement of light absorption is achieved and results in an integral improvement of the electrical and optical properties. At the same time, hybrids with AuNPs and rGO flakes produce an increase in the efficiency of organic photovoltaic devices for energy storage, which is caused by functionalities of Localized Surface Plasmon Resonance (LSPR) to generate higher photocurrents (Ou 2015, Che et al. 2020). Furthermore, the presence of AuNPs on MoS_2 monolayers induces photoluminescence quenching (Yin et al. 2019). As a consequence, different multifunctional hybrids such as Ni-rGO (Franceschini and Lacconi 2017, Lemes et al. 2019) and Ni-MoS_2 (Jin et al. 2018, Gómez et al. 2019) are used in electrocatalysts, heterostructures with two-dimensional flakes in photocatalysts (Luo et al. 2016), biosensors (Yin et al. 2015, Segovia et al. 2019), light-emitting diodes (Choi and Kamat 2013), robust and mechanically resistant materials to friction or erosion, superconductors (Biscaras et al. 2015, Lu et al. 2018), etc.

The above-mentioned examples show that both GO and rGO flakes are fundamental blocks for building hybrids nanomaterials since they have many different sites suitable to form heterojunctions. The latter is possible due to the differences in hydrophilicity, electrical and thermal conductivities, optical and mechanical properties, chemical reactivity and electrochemical-catalytic activity. In general, all two-dimensional nanomaterials have proved to be advantageous in regard to mechanical, optical, chemical and physical properties for their use as support flakes for low dimensional metal and oxide particles, to build multifunctional hybrid platforms.

Despite these advantages, many aspects still remain unresolved. Due to this, an incessant search towards a strategical optimization of methods of tuning properties by combining materials of a different chemical nature is the current goal to further meet the demand of reaching a diversity of applications that the world needs today. Moreover, the notable permanent requests within the field of nanotechnology need the participation of extremely small systems with satisfactory multifunctionalities. Such demands on the new generation of promising nanomaterials are motivated by the claims of lower costs, higher versatility and increased efficiency performed in smaller volumes to encourage and promote their scalability and manufacturing. In order to achieve these special needs, new developments for hybrids materials with highly controlled synthesis and manipulation by means of advanced high-resolution methods for characterization are eagerly awaited, to ensure stability and low toxicity, among other required conditions.

This chapter describes the partial current state of the art on synthesis studies, type of assemblies, chemical and morphological nature, advanced characterization and the direct application of hybrid nanomaterials with two-dimensional flakes and metallic nanoparticles. In fact, the partiality of such a description is due to the accelerated growth in the development of different complex nanosystems as a limitless challenge for solving multidisciplinary problems.

Experimental Strategies to Prepare and Characterize Hybrid Nanomaterials

Here the recent progress related to the preparation and characterization methods of two-dimensional hybrid nanomaterials are summarized. Various factors that can affect their chemical nature, stability, composition or structure to enhance, favour or supply different behaviours in respect to those coming from the individual components are detailed. As well, some challenging issues that need to be solved for future research are discussed.

Since the rise of graphene as a leading reference two-dimensional material (Geim and Novoselov 2007), multiple protocols and sharp strategies for the synthesis of multifunctional carbon-based nanohybrids have been developed accurately (Zhang 2015, Panwar et al. 2019, Han et al. 2019), and further extended to other two-dimensional materials. In addition, the experimental pathway and methodologies used in the preparation of two-dimensional carbon-based (graphene, GO or rGO) samples depend on not only the surface properties of the two-dimensional source but also on the desired functionalities.

Effective and low-cost bottom-up routes designed to create carbon-based two-dimensional hybrids have been encouraged in the last years, and supported through the wide variety of the latest developed methods of synthesis that lead to single GO sheets and GO thin films as well (Brodie 1859, Hummers and Offeman 1958). One of the most used methods to produce huge amounts of GO samples (flakes and films) is driven by the oxidation with $KMnO_4$ in H_2SO_4 and $NaNO_3$ (Brodie 1859, Hummers and Offeman 1958). Thus, through the hydrolysis of these intercalated compounds, hydroxyl, ether and carboxyic groups are included in the carbon matrix. Besides, the modified Hummers method can provide GO sheets with high yield, other authors (Marcano et al. 2010) improved this last method by means of centrifugation and dialysis steps in order to increase the number of oxygenated groups, which have an essential purpose in the preparation of hybrid materials. Accordingly, among the most current developments, a remarkable one has shown an effective pathway to obtain graphene oxide sheets of high quality through accurate control of the synthesis parameters (Ranjan et al. 2018). These latter have strongly impacted in the increasing development of a new generation of advanced hybrid nanomaterials created from graphene oxide flakes and films that also contain other bi-dimensional flakes of MoS_2, WS_2 or AuNPs for a wide range of applications, as it will be later described in the 'Multifunctional Applications' section.

In the case of GO, its hydrophilicity and abundant surface moieties enable a strong interaction with different materials and substances in aqueous media. On the other hand, graphene cannot be dispersed in water due to its hydrophobicity. Therefore, an organic solvent can be used to obtain graphene dispersions. In this case, the chemical properties of each two-dimensional material should be taken into account when preparing the composites. Recent reports have demonstrated that electrochemical formation of PdNPs leads to a nanomaterial that exhibits a highlighted role as a catalyst for carbon-carbon cross-coupling and other organic reactions (Uberman et al. 2014). Simultaneous chemical reduction with polyethylene glycol of Ag^+ ions and GO flakes facilitates the creation of two-dimensional flakes with plasmonic NPs in solution and provides new SERS active substrates to be used as ultrasensitive sensors of analytes (Dalfovo et al. 2014, Fioravanti et al. 2020). At the same time, UV-irradiation facilitates the formation of heterojunctions with graphene and nanomaterials (Badhulika et al. 2015).

When GO flakes or films are intentionally reduced in a controllable way, following some specific pattern, a carbon-based hybrid nanomaterial takes place. In fact, many studies have reported that GO can be reduced to graphene-like sheets by removing the oxygen-containing groups, revealing how versatile the resultant hybrid material is as an elemental component of novel nanodevices. Thus, as GO is generally considered an insulating material, a reduction process is required to modify its electrical properties. These reduction procedures are often chemically aggressive or require high-temperature annealing (Becerril et al. 2008). Additionally, electrical and thermal properties can be also modified depending on the oxygen and carbon atoms ratio on the surface. Some authors have obtained an almost complete reduction from GO to reduced GO (rGO), through chemical conversion by sodium borohydride in a strongly acidic medium, followed by thermal annealing (Gao et al. 2009). This pathway leads to the restoration of the π-conjugated structure for highly soluble in organic medium and conductive graphene materials. At the same time, thermal reduction of GO by laser irradiation involves different stages that give progressive elimination of intercalated water molecules and oxygen functional groups (Schniepp et al. 2006).

Some investigations have shown that GO films can overcome a reduction process, generating damage in the GO lattice to reach the graphene oxide ablation at certain laser doses irradiation (Mehta et al. 2017). Depending on the direction of the work, this latter phenomenon could become a complementary technological approach for designing two-dimensional hybrid nanodevices. Some authors (Pérez et al. 2019) succeeded to turn this outward problem into a versatile and outstanding solution to perform GO-rGO hybrid nanomaterials. In order to get such a hybrid, the authors first prepared clean coverages of GO flakes on silicon oxide (SiOx) surfaces to obtain two kinds of samples, GO flakes and GO films. These GO coatings were performed inside GO dispersions, carrying out a fairly simple but controllable immersion method that consists of dipping the SiOx substrates in a vertical position, during 600 seconds at 25°C. Therefore, two different concentrations of GO dispersions could be prepared that lead to two different configurations, layered GO films (Fig. 10.1a) and layered GO flakes (Fig. 10.1b).

After obtaining both configurations, an experimental approach by combining Raman spectroscopy and Atomic Force Microscopy (AFM) scanning was developed, to tailor hybrid conducting-insulating rGO-GO microchannelled patterns of nanometric depths (Pérez et al. 2019). Thusly, a detailed analysis of Raman measurements effects can allow controlling laser-induced damage on both GO surface films and GO individual flakes to reach hybrid nanomaterials.

Magnetic hybrid composites represent another wide increasing branch of the material science field with important contributions to the development of two-dimensional hybrid nanomaterials. In particular, when low dimensional carbon-based materials are combined with Ni structures, a larger strength compared to that of pure Ni could be obtained (Zhang et al. 2019, Behera et al. 2019, Sebastian et al. 2019). Ni-rich systems are widely used as magnetic core materials in diverse technological applications, such as magnetic recording heads, magneto-resistive random-access

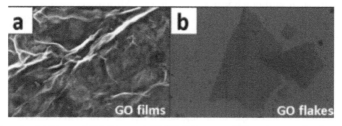

Figure 10.1: SEM images of (a) layered GO films and (b) GO flakes, onto SiOx substrate after immersion (600 seconds) in 1.0 mg mL⁻¹ and 0.08 mg mL⁻¹ GO dispersions, respectively (Reprinted from Pérez et al. (2019), with permission from Elsevier).

memories, among many others. Besides, the high Ni content provides excellent corrosion resistance to the alloy. When the dimension of these Ni-rich structures are in the nanoscale, such as nanowires or nanotubes (Escrig et al. 2008, Palma et al. 2018) and are mixed with carbon-based materials like graphene, GO or rGO can provide novel properties compared to 3-D atomic arrangements (Li et al. 2016). On the one hand, the strategy of embedding GO films or GO flakes with such magnetic nanostructures could provide a key tool to design, tune and control the physical properties of the hybrid nanomaterials for high-impact technology in order to obtain enhanced multiplexed attributes. On the other hand, the thermal reduction of graphene oxide flakes is also one of the most efficient steps to fabricate stable hybrids with Ni and Co NPs (Huang et al. 2015).

Among all the available structural characterization techniques for low dimensional materials, Raman spectroscopy appears as a simple, fast and reliable tool, especially for carbon-based systems. In fact, since the isolation of graphene (Geim and Novoselov 2007), Raman spectroscopy has been one of the main techniques used for a qualitative structural description such as determination of the number of layers, identification of defect density and its nature (D, D' and 2D), grain boundaries, thickness, roughness, doping, strain and thermal conductivity of graphene films and related materials (Ferrari and Robertson 2000). It is worth noting here that a proper selection of experimental conditions must be carefully performed to prevent or enhance the damage effects (Guo et al. 2012, Nikolenko 2013), as it was described through the rGO-GO example (Pérez et al. 2019) since high-quality Raman spectra of bi-dimensional materials need long acquisition times of scattered signals when measurements are performed at low power incident radiation. To illustrate the influence of these latter factors, Fig. 10.2 shows the dependence of D band frequency (Figs. 10.2a and 10.2b) and intensity (Figs. 10.2c and 10.2d) in Raman spectra of GO films and individual GO flakes, recorded with 488, 514.5 and 632.8 nm laser at different powers (Pérez et al. 2019).

In addition to Raman spectroscopy, another characterization technique that is especially relevant for determining the properties of two-dimensional hybrid nanomaterials is the so-called Surface-Enhanced Raman Spectroscopy, SERS. In fact, SERS combined with Raman spectroscopy has been

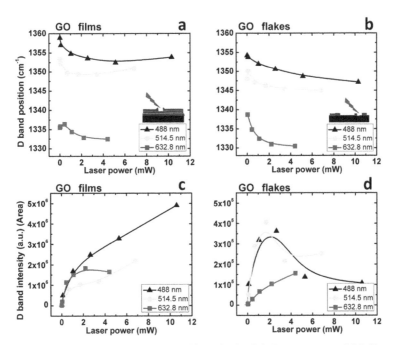

Figure 10.2: Dependence of D band frequency (a, b) and intensity (c, d) in Raman spectra of GO films and individual GO flakes, recorded with 488, 514.5 and 632.8 nm laser at different powers. All data are shown after the baseline spectral discount. Acquisition time was 10 seconds and average 10 times. Intensity values were established by the area of the Raman signal (Reprinted from Pérez et al. (2019), with permission from Elsevier).

extensively used for performing the characterization of graphenoids. This is because, as seen earlier, Raman spectroscopy provides significant information, such as the intensity (area) relation and the bands position, as fingerprint parameters that are helpful during the reduction of GO to rGO, besides other analyses. In two-dimensional hybrids, especially when a component of the composite is plasmonic active, the enhanced near fields produces an amplification of the spectroscopic signal of the other components. Thus, the enhancements can be used for sensing purposes, i.e., the detection of analytes adsorbed on the two-dimensional hybrid surface, but also for increasing the detection limit of the composite. A clear example of the latter is given by electrophoretic deposition of AuNPs on graphene-coated ITO electrodes, which can be used to obtain transparent and efficient detection of methylene blue dye through the sensitive increase of Raman signal from the plasmonic particles and graphene hybrid structures (Fioravanti et al. 2020).

In a step further, Tip-Enhanced Raman Spectroscopy (TERS) is a technique that combines all the advantages of SERS spectroscopy with surface topography characterization. In this advanced technique, the surface sample is scanned by a plasmonic tip, measuring the surface structure and at the same time, provides enhanced Raman signal due to the tip-sample interaction when it is approached. Consequently, TERS can retrieve the differential moieties from the edges, the defects or the flake basal domain. Moreover, TERS can be performed on two-dimensional hybrids to identify each component with notable high spatial resolution. When any plasmonic components in the composite are present, a strong optical coupling leads to the origin of a hot spot. A current work conducted by Pérez et al. showed that when SERS and TERS measurements were performed on heterojunctions of graphene with AuNPs (Pérez et al. 2016), the optical coupling produces spectral changes in the hybrid that are related with the interaction of the tip. In addition, the authors concluded that a shift in frequency of D and D' bands of the doped graphene is observed when the gold tip approaches the film surface, i.e., induction of changes in the electrical properties in graphene is promoted by the presence of AuNPs. Hence, such techniques enable the correlation between the topographical features with the spectroscopic ones. This remarkable advanced facility is undoubtedly a great advantage of such techniques for performing reliable measurements on two-dimensional hybrid nanomaterials since it allows collecting fresh and correlated data, therefore avoiding other possible effects that appear when sequential measurements are carried out. In fact, factors such as sample ageing or scanning damages that take place due to sequential and accumulative measurements, are usually consequences that create difficultly in the reproducing the results. At the same time, when a sample is first measured through Raman spectroscopy to determine the carbonaceous nature and then, the topographical characterization of the same sample is performed, which is a direct and clear correlation between Raman spectra and topography images, obtained by scanning probe microscopies, for example, is not a simple task and many times, is not attainable.

Another main complementary technique that gives strong support to Raman spectroscopy, as well as AFM, SERS and TERS, is X-ray Photoelectron Spectroscopy (XPS), since these measurements provide information about the surface chemical composition of the components in the hybrid composite in order to establish not only the chemical nature of the individual constituents but also to detect and monitor their potential changes after forming the hybrid nanomaterial (Pérez et al. 2019). In the case of magnetic hybrid nanomaterials, magnetic measurements should be performed by different kinds of magnetometres, such as Vibrating Sample Magnetometre (VSM) or longitudinal Magneto-Optical Kerr Effect Magnetometre (MOKE), to compare the magnetic properties of the composite with those exhibited by the individual components. Regarding the substrates, paramagnetic (Al) and diamagnetic (Au) contributions should be removed in all the cases.

Novel Multiplexed Properties

A two-dimensional hybrid system emerged as a new class of nanomaterial engineered by assembling two or more different components into a novel non conventional one. The latter offers better performance since it is designed and built to strengthen attributes that are inherent to individual components as well as they are also bottom-up created to increase the potential applications of the new generated materials. Scientific findings have indicated that two-dimensional hybrid nanomaterials can replace the standard ones since they provide more advantages especially in the areas of alternative energies, health, environmental remediation, optoelectronics, spintronics, etc. (Franceschini and Lacconi 2017, Ou 205, Yin et al. 2019, Che et al. 2020). Here the most significant attributes are highlighted that come for hybrid nanomaterials when they are built up, pointing out the differences with these properties that appear from individual components.

Physical and chemical properties of two-dimensional hybrids can be substantially different from those of the isolated two-dimensional effect. Thus, chemical reactivity can be lowered or tuned by the addition of a thin layer of an inert material, as well as thermal, magnetic and electrical properties could have been modified from those inherent of the individual components. At the same time, changes in the characteristic morphology and topography of the isolated components, with respect to those observed from the obtained hybrid, can be clear fingerprints of the emergent properties assigned to the new nanomaterial.

An advantage related to the rising properties reported for hybrid nanomaterials formed in solution is the availability to be used under different configurations. For example, dispersions of two-dimensional plasmonic-nanoparticle hybrids are very useful tools for the detection of species in solution, with very high sensitivity, by SERS spectroscopy. They can also be used as catalysts for organic chemistry reactions (Uberman et al. 2014). On the other hand, by drop-casting of hybrid dispersions or by its incorporation into the matrix of metallic coatings, their functionalities can be extended as an electrode to study various processes (Franceschini and Lacconi 2017).

In the case of GO-rGO hybrid nanomaterials, recent studies have demonstrated how the topography of GO films can be modified during laser irradiation, performed within certain experimental conditions (Pérez et al. 2019). The remarkable result of the latter appeared when GO surface regions with lateral dimensions of some micrometres were laser-irradiated in a controlled XY direction, at a constant Z, by using the AFM scan head without a cantilever (Pérez et al. 2019). In fact, the notable presence of damaged GO films was detected by performing AFM images before and after the irradiation (Fig. 10.3a). After identifying the damaged region, depth profiles on the irradiated GO films were measured in tapping mode AFM imaging, as it is depicted in Fig. 10.3b. The topography of GO shows a thickness decrease of about 20 nm on the irradiated area due to the loss of functional groups, an effect that was assigned to the reduction of GO. In order to monitor the same effect of the irradiation on the topography of GO individual flakes, AFM characterization was also performed (Pérez et al. 2019).

It is worth noting that very special care of the flake position should be taken in such kind of experiments, to allow the comparison between the morphology before and after the irradiation

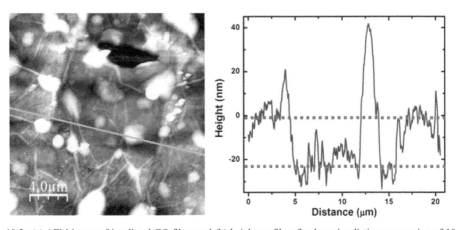

Figure 10.3: (a) AFM image of irradiated GO films and (b) height profile, after laser irradiation on a region of 10 μm × 10 μm at 633 nm and 5 mW of power (Reprinted from Pérez et al. (2019), with permission from Elsevier).

(Fig. 10.4). In this latter image, Fig. 10.4b shows evidence of the presence of a circular hole produced by the laser spot on the flake. The thickness of the flake was around 2 nm (Fig. 10.4c). After identifying the damage on the GO flake, the authors performed Raman mapping to monitor the intensity changes of the D signal (Fig. 10.4d). Therefore, the differences in the laser-induced effect for both configurations, GO films and GO flakes, could be explained by the predominant interaction among layers of GO flakes. Thus, the enhanced heat transfer was supported the modification of GO films.

The evolution of D and G bands for GO flakes as a function of the spectra acquisition time (Fig. 10.5) has been also investigated (Pérez et al. 2019). In this latter case, the optical image of one GO flake on the SiOx surface is shown (488 nm laser at 4.75 mW, after 600 seconds irradiation, Fig. 10.5a). The notable effect appears when the Raman intensity of D and G bands (Fig. 10.5e) dropped off very sharply in the first 80 seconds, at which time the hole in the optical image is clearly distinguished. An increasing D and G intensity decay joined with a fast increase of the SiOx signal can be correlated to the decrease of disorder-induced states that occurs during the reduction of GO, even in solution. Furthermore, an increase in the D/G ratio intensities during the first 100 seconds (Fig. 10.5d) was also observed as a remarkable change.

Chemical differences that take place during the formation of the hybrid nanomaterials can be monitored for advanced surface techniques, such as XPS. For example, Fig. 10.6 shows a comparison among the spectra of a non-irradiated (Fig. 10.6a) and irradiated GO films (Fig. 10.6b) (Pérez et al. 2019). Thus, the composition of present oxygen groups can be determined by deconvolution of the C 1 s core level peak.

A suitable comparison between the relative amounts of single and double C-O bonds indicates that the observed diminution of oxygenated groups in the irradiated GO leads to the restoration of sp^2 carbon domains (Figs. 10.6a and 10.6b). Therefore, the irradiated zone on the surface has been photo-thermally reduced from GO to rGO (reduced graphene oxide), giving rise in this way to a two-dimensional hybrid nanomaterial.

In the case of magnetic hybrid nanomaterials, oxidation reactions that can overcome the magnetic nanostructures that interact with the functional groups of GO can be accurately detected by XPS measurements.

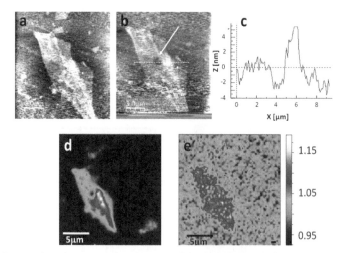

Figure 10.4: AFM images show the topography of one GO flake (a) before and (b) after recording a Raman spectrum obtained at 488 nm, ~ 5 mW and 600 s. (c) Height profile depicted from AFM image of the damaged GO flake. (d) Raman map with the intensity of D band at 1330 cm^{-1} in a 20 × 20 μm^2 framework irradiated area, measured at 633 nm at 0.45 mW, using a 100x (09 NA) objective. Integration time: 1 second with 1 average. (e) Raman image of D/G intensities ratio with the same experimental parameters than used in (d) (Reprinted from Pérez et al. (2019), with permission from Elsevier).

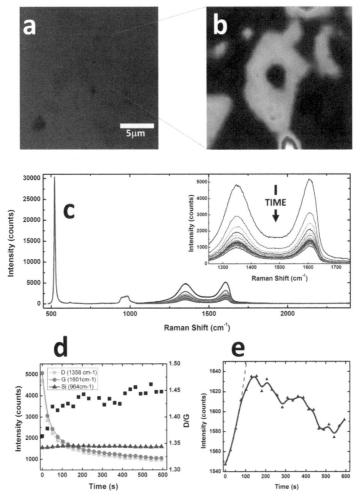

Figure 10.5: (a) Optical image of a single GO flake on SiOx after irradiation with 488 nm at 4 mW power laser, (b) D intensity Raman imaging from the optical image in (a). (c) Raman spectra evolution that took place at the same point on the flake, recording each spectrum every 25.8 seconds (5 seconds acquisition, 5 times average); the black arrow indicates the diminution of the intensities. (d) Dependence of intensity from D, G and SiOx signals with the time during laser irradiation on the left Y-axis and ratio D/G on the right Y-axis; (e) signal from Si (980 cm^{-1}) intensity evolution (Reprinted from Pérez et al. (2019), with permission from Elsevier).

Multifunctional Applications

The extraordinary properties resulting from these novel hybrid nanomaterials allow expanding the capabilities as part of a new generation of high-impact devices as indispensable and non-replaceable tools, very exclusive in structure and multifunctionalities, provide certainly benefits to their further potential commercial production. Indeed, the evidence of collective properties described above allows one to understand the reason for the current scientific and technological impulse to promote their development for overcoming the challenges necessary in their use, controlled production, high quality and efficiency for the functions particularly designed. In fact, diverse two-dimensional hybrids are applied for imaging analysis, disease diagnosis, therapy and prevention in biomedicine; active electrocatalysts; antibacterial agent; drug-delivery; lubricant; stiffness or flexibility reinforcement and components of optoelectronics, plasmonics, magnetic and optical devices (Bonaccorso et al. 2010, Wen et al. 2018, Panwar et al. 2019, Han et al. 2019, Guan and Han 2019, Galeotti et al. 2020, Cuniberto et al. 2020, Cong et al. 2020, Zhao et al. 2020, Shaohua et al. 2020). In addition, diverse

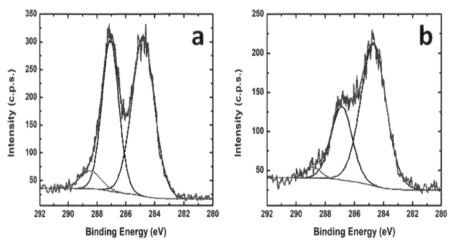

Figure 10.6: XPS spectrum from C 1 s of GO film (a) before and (b) after the sample exposure to the 633 nm laser radiation at ~ 5 mW power (Reprinted from Pérez et al. (2019), with permission from Elsevier).

types of systems to be used like portable sensors, biosensors, Field-Effect Transistors (FET) and pH controllers with flexible and transparent MoS_2 flakes have been performed (Pumera and Huiling 2014, Wen et al. 2018, Donarelli and Ottaviano 2018, Saifur Rahman et al. 2018, Dasari Shareena et al. 2018). This is possible because the building blocks are integrated hybrid nanomaterials with wide versatility and certain changes in their physicochemical properties, which allows modular integration among them. Here some outstanding examples have been selected where collective properties are useful for different applications.

On the one hand, doping of graphene films can be induced by the presence of Au NPs incorporated into its structure. Recent investigations reported how to transfer graphene without polymer assistance, but only with Au NPs that were finally demonstrated to be ideal SERS and TERS platforms (Pérez et al. 2016). TERS and imaging of SERS spectra were recorded on heterojunctions of graphene and Au NPs, concluding that Au removed electrons from graphene. Changes in the position and intensity of phonon Raman bands were also observed (Dalfovo et al. 2014). Therefore, doping of graphene is feasible with hybrid systems since they promote changes in the electrical properties of graphene. As it was mentioned earlier, ultrasensitive platforms obtained using electrophoretic deposition of Au NPs on graphene-coated ITO electrodes for SERS detection of analytes, can be effectively achieved by depositing AgNPs or AuNPs on ITO electrodes, where the increase of the electromagnetic field around the plasmonic particles in the graphene hybrid structures leads to sensitive detection of adsorbed analytes (Fioravanti et al. 2020, Muñetón Arboleda et al 2020). An interesting pathway to obtain rGO-NPs hybrid solutions is based on electrochemical reduction of GO flakes aqueous dispersions in the presence of metallic precursors. This scalable process allows one to obtain large volumes of highly stable two-dimensional hybrids in only a single step procedure, even under high ionic force media.

Depending on the metal precursor used in the synthesis, different capabilities can be accomplished: catalytic activity, optical absorption and spectroscopic signal enhancing, high electrical conductivity, etc. In addition, the possibility of more complex structures such us bimetallic ones can be easily reached during posterior stage-reactions as it is described below.

As can be seen in Fig. 10.7 (unpublished data), electrochemical synthesis aimed to obtain dispersions constituted by rGO-PdNPs hybrid nanomaterials (Fig. 10.7a) has been performed in a single-step procedure. Both GO and $PdCl_4^{2-}$ ions are electrochemically reduced, and their concentrations can be changed tuning the coverage degree. The resulting nanostructures lead to the formation of nanoparticles with an average diameter of ~ 35 nm onto the rGO flake surface (STEM image, Fig. 10.7b). PdNPs have been extensively used in organic catalysts (i.e., Heck and

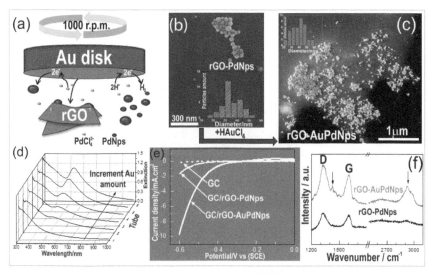

Figure 10.7: (a) Scheme of synthesis for rGO-PdNPs hybrid nanomaterials in solution by electrochemical reduction. STEM characterization (b) rGO-PdNPs, (c) rGO-AuPdNPs. (d) UV-Visible spectral evolution during the $AuCl_4^-$ ions addition. (e) HER performance in acid medium. (f) Raman/SERS spectra of rGO-PdNPs and rGO-AuPdNPs colloids (Unpublished data).

Suzuki cross-coupling reactions) (Uberman et al. 2014). RGO-PdNPs two-dimensional hybrids performed to attain the hydrogen electrochemical reduction (HER) are also promising for catalytic applications as it is shown in Fig. 10.7e. Other functionalities, such as optical response, can be developed by further modifications of the hybrid composition. Thus, when equal volumes of the hybrid dispersion are added to different amounts of $HAuCl_4$ solutions, the formation of bimetallic complex nanoparticles supported on rGO flakes is achieved. In the PdAuNPs hybrid, the galvanic replacement takes place in the PdNPs surface, generating gold deposition over the rGO-PdNPs hybrids. Figure 10.7c shows the result of STEM characterization, giving evidence to a large number of particles with an average diameter between ~ 40 and 50 nm, which is in good agreement with UV-Visible (D) spectra, where the appearance of the LSPR of the AuNPs at ~ 545 nm is expected. The rGO-supported bimetallic hybrid is a Raman enhancer (Fig. 10.7f), allowing the detection of the presence of rGO and the SERS spectra of PVP molecules, used as a stabilizer. The performance of the bimetallic hybrid in the HER in an acid medium (Fig. 10.7e) was evaluated, demonstrating a higher catalytic activity compared to the rGO-PdNPs hybrid.

Another interesting low-cost and versatile application of two-dimensional hybrid nanomaterials is focused on GO-rGO systems. Such systems can be achieved once reproducible experimental conditions necessary to reach the reduction of GO have been carefully determined. In fact, as it was mentioned earlier, some authors (Pérez et al. 2019) reported how the development of a simple lithographic method for tailoring and patterning conducting microchannels of rGO among GO films using a strategic combination of Raman spectroscopy and AFM is feasible (Fig. 10.8). The hybrid microchanenelled pattern supported on SiOx was accomplished using 632.8 nm of laser wavelength and 5 mW of power during 600 seconds, by means of the combination of Raman spectroscopy and the AFM scan head. AFM images measured before and after the irradiation were reproducible with respect to that obtained in Fig. 10.3 shown earlier.

Figure 10.9 develops other lithographic examples that were achieved using a technological approach. The Tic-Tac-Toe pattern that is shown in Fig. 10.9a was obtained by performing rGO microchannels, using the proposed controlled framework. Institutional logo patterns were also successfully imprinted on the GO films (Fig. 10.9b), at 632.8 nm and 5 mW of power. In this way, controlled laser-induced structural and chemical changes in multilayered GO films and individual GO flakes supported on SiOx substrates were produced, using a novel technical approach that

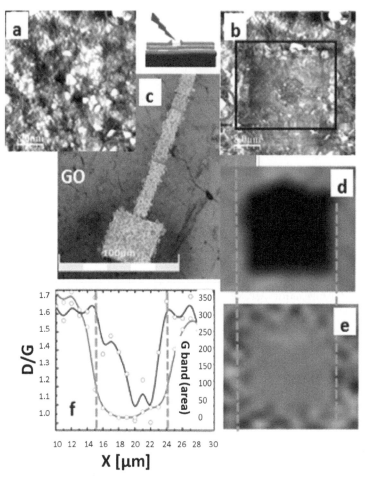

Figure 10.8: Microchannelled pattern: (a) AFM image of GO films before irradiation, (b) AFM image of irradiated GO employing the same experimental conditions used in Fig. 10.3. (c) Optical image of one patterned zone obtained from irradiation of GO films prepared in (a). (d) and (e) Raman images of D and G bands, respectively; and (f) Raman spectra of D/G intensity ratio (black line) and G band intensity (grey line), both corresponding to the irradiated region (Reprinted from Pérez et al. (2019), with permission from Elsevier).

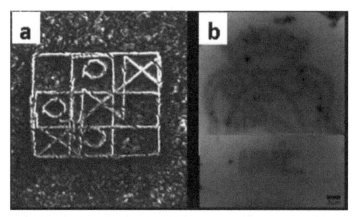

Figure 10.9: rGO-based pattern of (a) Tic-Tac-Toe on GO films covering a glass substrate (SEM image) and (b) National University of Cordoba logos (optical image of GO-rGO hybrid) (Reprinted from Pérez et al. (2019), with permission from Elsevier).

combines Raman spectroscopy synchronized with X-Y scanning for tailoring hybrid conducting-insulating GO-rGO microchannelled patterns of nanometric depths (Pérez et al. 2019).

In addition to the examples described earlier, magnetic hybrid systems also appear as novel advanced nanomaterials. Thus, a kind of hybrid composites constituted by nanowires (NWs) of Permalloy (Py), an alloy with about 20% of iron and 80% of nickel, and multi-layered GO flakes can be developed, as it is presented in Fig. 10.10 (Arciniegas Jaimes et al. 2020). Figure 10.10a discloses well-separated GO films that evidence lateral dimensions of several micrometres of extension. Such films can be achieved by random piling of individual GO flakes that work as building blocks to create the carbonaceous matrix of the hybrid nanomaterial. On the other hand, decoupled magnetic nanowires, which have been synthesized inside AAO membranes, are shown in Fig. 10.10b, where a high density of isolated unidimensional nanostructures is reachable by dissolving the AAO template (Raviolo et al. 2020, Meneses et al. 2018a, b, Riva et al. 2016). A lower density of decoupled PyNWs with serrated morphology is observed in the inset in Fig. 10.10b. PyNWs of shorter lengths are also present since they are crushed during the decoupling process. It is worth noting that the decoupling procedure is generally conducted after an exhaustive structural characterization performed on several samples in order to determine the appropriate alumina-dissolution times that lead to a considerable amount of nanostructures (nanowires or nanotubes), but preserving the morphology and magnetic nature of them (Escrig et al. 2008, Meneses et al. 2018a, b, Riva et al. 2016).

Figure 10.10c displays a SEM image of the magnetic composite sample, where the GO film surface has been embedded with some of the described above decoupled Py nanowires, most of them oriented parallel to the GO surface and probably overlapped among some GO layers. Figure 10.10d also gives evidence of the presence of alumina-embedded Py NWs in the composite sample. These Py agglomerates are located around the edges of the multi-layered GO films and intercalated among

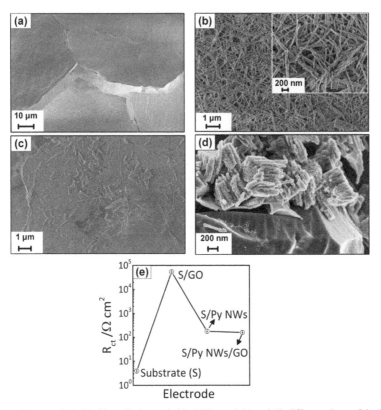

Figure 10.10: SEM images of (a) GO films, (b) decoupled Py NWs, and (c) and (d) different views of the Py NWs on GO flakes. (e) Charge-transfer resistance of the different electrodes. Inset in (b) lower density of Py NWs decoupled from AAO template (Reprinted from Arciniegas Jaimes et al. (2020), with permission from Springer Nature).

their layers as well. By means of electrochemical measurements performed on such kinds of novel magnetic hybrid composites, it is possible to determine the charge transfer resistance (R_{ct}) of each component as well as the one corresponding to the composite, as it is shown in Fig. 10.10e. This striking behaviour indicates that, despite the surface inhomogeneities that can appear because of the preparation method or partial surface oxidation of the magnetic structures, a clear R_{ct} trend can be found. In this particular case, it is worth noting that a marked improvement of the conductivity has been obtained for electrodes made of hybrid composites constituted by a low density of magnetic nanowires and large two-dimensional films of GO. Therefore, two-dimensional magnetic hybrid composites are promising candidates in the development of attractive and stable molecular sensing nanostructured devices.

The examples described above, show how advanced applications of two-dimensional hybrid nanomaterials are widely diverse since they exhibit multiple cooperative functionalities. Therefore, they have become attractive candidates to be used as part of a new generation of optical sensors, electrochemical systems, photovoltaic and optoelectronic devices, biomimetic and biosensor, active light-absorbing materials, textiles and polymers, nanotribological tools, advanced medical instruments, flexible and electronic devices.

Outlook and Future Perspectives

Advanced nanomaterials obtained from hybrids heterostructures containing NPs and two-dimensional components have revealed through evidence and different developments, the fundamental role that they can play to tackle many challenges that lead to solutions in contributing to several current problems concerning health, energy, technology and environment.

Therefore, the efforts made to search and reach synergy between attributes from each component are strongly motivated by the goal of leading to the activation of capacities and efficiency able to promote a new generation of advanced hybrid nanomaterials. For this reason, active and extensive multidisciplinary research related to two-dimensional hybrids materials is dynamically increasing and advancing to resolve diverse applied situations in all disciplinary areas of science, pursuing the fundamental aim of bringing a better quality of life from the nanoscience and nanotechnology.

Acknowledgements

The authors gratefully acknowledge PhD Esteban Franceschini for his diligent editing and mainly the invitation to contribute to the present chapter of the book "Nanostructured Multifunctional Materials: Synthesis, Characterization, Applications and Computational Simulation", which tackles pioneering research fields of nanoscience and nanotechnology knowledge. Most of the measurements here were performed by using Raman microscopy and AFM facilities, supported by Laboratorio de Nanoscopía y Nanofotónica LANN-SNM of MINCyT (PME1544) at INFIQC-CONICET. LAMARX also provided technical assistance to perform the SEM micrographies corresponding to the cited works. CONICET and SECYT-UNC gave financial support to carry out many of the experiments that were cited in the present chapter.

References

Arboleda, D. M. and F. J. Ibañez. 2020. Improved electrocatalysis and electrophoretic deposition due to the strong synergy between Au and Ag Nanparticles. Chem. Select 5: 9839.

Arciniegas Jaimes, D. M., P. Márquez, A. Ovalle, J. Escrig, O. Linarez Pérez and N. Bajales. 2020. Permalloy nanowires/graphene oxide composite with enhanced conductive properties. Sci. Rep. 10: 13742. 1–13.

Badhulika, S., T. Terse-Thakoor, C. Villarreal and A. Mulchandani. 2015. Graphene hybrids: synthesis strategies and applications in sensors and sensitized solar cells. Front Chem. 3(38): 1–19.

Bao, Q. and K. P. Loh. 2012. Graphene photonics, plasmonics, and broadband optoelectronic devices. ACS Nano 6: 3677–3694.

Becerril, H. A., J. Mao, Z. Liu, R. M. Stoltenberg, Z. Bao and Y. Chen. 2008. Evaluation of solution-processed reduced graphene oxide films as transparent conductors. ACS Nano 2: 463–470.

Behera, A., S. Mansingh, K. K. Das and K. Parida. 2019. Synergistic $ZnFe_2O_4$-carbon allotropes nanocomposite photocatalyst for norfloxacin degradation and Cr (VI) reduction. J. Colloid Interface Sci. 544: 96–111.

Biscaras, J., Z. Chen, A. Paradisi and A. Shukla. 2015. Onset of two-dimensional superconductivity in space charge doped few-layer molybdenum disulfide. Nat. Commun. 6: 8826. 1–8.

Bonaccorso, F., Z. Sun, T. Hasan and A. C. Ferrari. 2010. Graphene photonics and optoelectronics. Nat. Photonics 4: 611–622.

Brodie, B. C. 1859. On the atomic weight of graphite. Philos. Trans. R. Soc. Lond. Ser. B Biol. Sci. 149: 249–259.

Che, Y., Q. Liu, B. Lu, J. Zhai, K. Wang and Z. Liu. 2020. Plasmonic ternary hybrid photocatalyst based on polymeric g-C_3N_4 towards visible light hydrogen generation. Sci. Rep. 10: 721. 1–12.

Chen, Y., Z. Fan, Z. Zhang, W. Niu, C. Li, N. Yang et al. 2018. Two-dimensional metal nanomaterials: synthesis, properties, and applications. Chem. Rev. 118: 6409–6455.

Choi, H. and P. V. Kamat. 2013. CdS nanowire solar cells: dual role of squaraine dye as a sensitizer and a hole transporter. J. Phys. Chem. Lett. 4: 3983–3991.

Cong, X., X. Liu, M. Lin and P. Tan. 2020. Application of Raman spectroscopy to probe fundamental properties of two-dimensional materials. 2D Mater. Appl. 4: 1–12.

Cuniberto, E., A. Alharbi, T. Wu, Z. Huang, K. Sardashti, K. You et al. 2020. Nano-engineering the material structure of preferentially oriented nano-graphitic carbon for making high-performance electrochemical micro-sensors. Sci. Rep. 10: 1–11.

Dalfovo, M. C., G. I. Lacconi, M. Moreno, M. C. Yappert, G. U. Sumanasekera, R. C. Salvarezza et al. 2014. Synergy between graphene and Au nanoparticles (Heterojunction) towards quenching. Improving Raman signal, and UV light sensing. ACS Appl. Mat. Interfaces 6: 6384–6391.

DasariShareena, Th. P., D. McShan, A. K. Dasmahapatra and P. B. Tchounwou. 2018. A review on graphene-based nanomaterials in biomedical applications and risks in environment and health. Nano-Micro Lett. 10: 53. 1–34.

Donarelli, M. and L. Ottaviano. 2018. 2D materials for gas sensing applications: a review on graphene oxide, MoS_2, WS_2 and phosphoren. Sensors 18: 3638. 1–45.

Dragoman, M. and D. Dragoman. 2017. 2D Nanoelectronics: Physics and Devices of Atomically Thin Materials. Springer Books.

Escrig, J., R. Lavin, J. L. Palma, J. C. Denardin, D. Altbir, A. Cortes et al. 2008. Geometry dependence of coercivity in Ni nanowire arrays. Nanotechnol. 19: 075713. 1–6.

Fai Mak, K. and J. Shan. 2016. Photonics and optoelectronics of 2D semiconductor transition metal dichalcogenides. Nat. Photonics 10: 216–226.

Ferrari, A. C. and J. Robertson. 2000. Interpretation of Raman spectra of disordered and amorphous carbon. Phys. Rev. B. 61: 14095–14107.

Fioravanti, F., D. MuñetónArboleda, G. I. Lacconi and F. J. Ibañez. 2020. Characterization of SERS platforms designed by electrophoretic deposition on CVD graphene and ITO/glass. Mater. Advances RSC (in press).

Franceschini, E. A. and G. I. Lacconi. 2017. Synthesis and properties of nickel-graphene hybrid for hydrogen evolution reaction. Electrocatalysis 9: 47–58.

Galeotti, G., F. De Marchi, E. Hamzehpoor, O. MacLean, M. Rajeswara Rao, Y. Chen et al. 2020. Synthesis of mesoscale ordered two-dimensional π-conjugated polymers with semiconducting properties. Nat. Mat. 1–7.

Gao, W., L. B. Alemany, L. Ci and P. M. Ajayan. 2009. New insights into the structure and reduction of graphite oxide. Nat. Chem. 1: 403–408.

Geim, A. and K. Novoselov. 2007. The rise of graphene. Nature Mater. 6: 183–191.

Gómez, M. J., A. Loiácono, L. Pérez, E. Franceschini and G. I. Lacconi. 2019. Highly efficient hybrid Ni/nitrogenated-graphene electrocatalysts for hydrogen evolution reaction. ACS Omega 4: 2206–2216.

Guan, G. and M. -Y. Han. 2019. Functionalized hybridization of 2D nanomaterials. Adv. Sci. 6: 1901837. 1–32.

Guo, L., R. -Q. Shao, Y. -L. Zhang, H. -B. Jiang, X. -B. Li, S. -Y. Xie et al. 2012. Bandgap tailoring and synchronous microdevices patterning of graphene oxides. J. Phys. Chem. C 116: 3594–3599.

Han, S. A., J. Lee, W. Seung, J. Lee and S. Kim. 2019. Patchable and implantable 2D nanogenerator. Small: 1903519. 1–14.

Huang, X., G. Zhao and X. Wang. 2015. Fabrication of reduced graphene oxide/metal (Cu, Ni, Co) nanoparticle hybrid composites via a facile thermal reduction method. RSC Adv. 5: 49973–49978.

Hummers, W. S. and R. E. Offeman. 1958. Preparation of graphitic oxide. J. Am. Chem. Soc. 80: 1339–1339.

Jin, H., C. Guo, X. Liu, J. Liu, A. Vasileff, Y. Jiao et al. 2018. Emerging two-dimensional nanomaterials for electrocatalysis. Chem. Rev. 118: 6337–6408.

Jorio, A., R. Saito, M. S. Dresselhaus and G. Dresselhaus. 2011. Raman Spectroscopy in Graphene Related Systems. Wiley VchVerlag Gmbh, Weinheim, Germany.

Khan, K., A. Khan Tareen, M. Aslam, R. Wang, Y. Zhang, A. Mahmood et al. 2020. Recent developments in emerging two-dimensional materials and their applications. J. Mater. Chem. C 8: 387–440.

Lemes P. G., D. Sebastián del Río, E. Pastor Tejera, L. Elorri and M. Jesús. 2019. N-doped graphene catalysts with high nitrogen concentration for the oxygen reduction reaction. J. Power Sources 438: 227036. 1–10.

Li, H., L. Xu, H. Sitinamaluwa, K. Wasalathilake and C. Yang. 2016. Coating Fe_2O_3 with graphene oxide for high-performance sodium-ion battery anode. Compos. Commun. 1: 48–53.

Li, H., Y. Li, A. Aljarb, Y. Shi and L. Li. 2018. Epitaxial growth of two-dimensional layered transition-metal dichalcogenides: growth mechanism, controllability, and scalability. Chem. Rev. 118: 6134–6150.

Lu, J., O. Zheliuk, Q. Chen, I. Leermakers, N. E. Hussey, U. Zeitler et al. 2018. Full superconducting dome of strong Ising protection in gated monolayer WS_2. Proc. Nat. Acad. Sci. USA 115: 3551–3556.

Luo, B., G. Liu and L. Wang. 2016. Recent advances in 2D materials for photocatalysis. Nanoscale 8: 6904–6920.

Marcano, D. C., D. V. Kosynkin, J. M. Berlin, A. Sinitskii, Z. Sun, A. Slesarev et al. 2010. Improved synthesis of graphene oxide. ACS Nano 4: 4806–4814.

Marquardt, D., C. Vollmer, R. Thomann, P. Steurer, R. Mulhaupt, E. Redel et al. 2011. The use of microwave irradiation for the easy synthesis of graphene-supported transition metal nanoparticles in ionic liquids. Carbon 49: 1326–1332.

Mehta, J. S., A. C. Faucett, A. Sharma and J. M. Mativetsky. 2017. How reliable Are Raman spectroscopy measurements of graphene oxide? J. Phys. Chem. C 121: 16584–16591.

Meneses, F., A. Pedernera, C. Blanco, N. Bajales, S. E. Urreta and P. G. Bercoff. 2018a. L10-FeNi ordered phase in AC electrodeposited iron-nickel biphasic nanowires. J. Alloys Compd. 766: 373–381.

Meneses, F., S. E. Urreta, J. Escrig and P. G. Bercoff. 2018b. Temperature dependence of the effective anisotropy in Ni nanowire arrays. Curr. Appl. Phys. 18: 1240–1247.

Murphy, S., L. B. Huang and O. V. Kamat. 2013. Reduced graphene oxide-silver nanoparticle composite as an active SERS material. J. Phys. Chem. C 117: 4740–4747.

Naqvi, T. K., A. K. Srivastava, M. M. Kulkarni, A. M. Siddiqui and P. K. Dwivedi. 2019. Silver nanoparticles decorated reduced graphene oxide (rGO) SERS sensor for multiple analytes. Appl. Surf. Sci. 478: 887–895.

Nikolenko, A. S. 2013. Laser heating effect on Raman spectra of Si nanocrystals embedded into SiOx matrix. Semicond. Phys. Quant. Electr. & Optoelectr. 16: 86–90.

Ou, C. F. 2015. The effect of graphene/Ag nanoparticles addition on the performances of organic solar cells. J. Mat. Sci. Chem. Eng. 3: 30–35.

Palma, J. L., A. Pereira, R. Alvaro, J. M. García-Martín and J. Escrig. 2018. Magnetic properties of Fe_3O_4 antidot arrays synthesized by AFIR: Atomic layer deposition, focused Ion beam and thermal reduction. Beilstein J. Nanotechnol. 9: 1728–1734.

Panwar, N., A. M. Soehartono, K. K. Chan, S. Zeng, G. Xu, J. Qu et al. 2019. Nanocarbons for biology and medicine: sensing, imaging, and drug delivery. Chem. Rev. 119, 16: 9559–9656.

Pérez, L. A., M. C. Dalfovo, H. Troiani, A. Soldati, G. I. Lacconi and F. J. Ibañez. 2016. CVD graphene transferred with Au nanoparticles: An ideal platform for TERS and SERS on a single triangular nanoplate. J. Phys. Chem. C. 120: 8315–8322.

Pérez, L. A., N. Bajales and G. I. Lacconi. 2019. Raman spectroscopy coupled with AFM scan head: A versatile combination for tailoring graphene oxide/reduced graphene oxide hybrid materials. Appl. Surf. Sci. 495: 143539. 1–11.

Pimenta, M. A., G. Dresselhaus, M. S. Dresselhaus, L. G. Cancado, A. Jorio and R. Saito. 2007. Studying disorder in graphite-based systems by Raman spectroscopy. Phys. Chem. Chem. Phys. 9: 1276–1291.

Pumera, M. and A. Huiling Loo. 2014. Layered transition-metal dichalcogenides (MoS_2 and WS_2) for sensing and biosensing. TrAC Trends in Anal. Chem. 61: 49–53.

Ranjan, P., S. Agrawal, A. Sinha, T. R. Rao, J. Balakrishnan and A. D. Thakur. 2018. A low-cost non-explosive synthesis of graphene oxide for scalable applications. Sci. Rep. 8: 12007–12020.

Raviolo, S., A. Pereira, D. M. ArciniegasJaimes, J. Escrig and N. Bajales. 2020. Angular dependence of the magnetic properties of permalloy nanowire arrays: A comparative analysis between experiment and simulation. J. Mag. Mag. Mat. 499: 166240. 1–21.

Riva, S., G. Pozo-López, A. M. Condó, M. S. Viqueira, S. E. Urreta, D. R. Cornejo et al. 2016. Biphasic FeRh nanowires synthesized by AC electrodeposition. J. Alloys Compd. 688: 804–813.

Saifur Rahman, M., M. R. Hasan, K. A. Rikta and M. S. Anower. 2018. A novel graphene coated surface plasmon resonance biosensor with tungsten disulfide (WS_2) for sensing DNA hybridization. Opt. Mat. 75: 567–573.

Schniepp, H. C., J. L. Li, M. J. McAllister, H. Sai, M. Herrera-Alonso, D. H. Adamson et al. 2006. Functionalized single graphene sheets derived from splitting graphite oxide. J. Phys. Chem. B 110: 8535–8539.

Sebastian, N., W. -C. Yu, Y. -C. Hu, D. Balram and Y. -H. Yu. 2019. Sonochemical synthesis of iron-graphene oxide/honeycomb-like ZnO ternary nanohybrids for sensitive electrochemical detection of antipsychotic drug chlorpromazine. Ultrason. Sonochem. 59: 104696. 1–14.

Segovia, M., M. Alegría, J. Aliaga, S. Celedon, L. Ballesteros, C. Sotomayor-Torres et al. 2019. Heterostructured 2D ZnO hybrid nanocomposites sensitized with cubic Cu_2O nanoparticles for sunlight photocatalysis. J. Mat. Sci. 54: 13523–13536.

Shaohua, C., L. Qiu and H. Cheng. 2020. Carbon-based fibers for advanced electrochemical energy storage devices. Chem. Rev. 120(5): 2811–2878.

Siamaki, A. R., A. E. Rahman, S. Khder, V. Abdelsayed, M. S. El-Shall and B. F. Gupton. 2011. Microwave-assisted synthesis of palladium nanoparticles supported on graphene: A highly active and recyclable catalyst for carbon–carbon cross-coupling reactions. J. Catal. 279(1): 1–11.

Stankovich, S., D. A. Dikin, R. D. Piner, K. A. Kohlhaas, A. Kleinhammes, Y. Jia et al. 2007. Synthesis of graphene-based nanosheets via chemical reduction of exfoliated graphite oxide. Carbon 45: 1558–1565.

Suk, J. W., R. D. Piner, J. An and R. S. Ruoff. 2010. Mechanical properties of monolayer graphene oxide. ACS Nano 4: 6557–6564.

Telesio, F., E. Passaglia, F. Cicogna, F. Costantino, M. Serrano-Ruiz, M. Peruzzini et al. 2018. Hybrid nanocomposites of 2D black phosphorus nanosheets encapsulated in PMMA polymer material: new platforms for advanced device fabrication. Nanotech. 29: 295601. 1–8.

Uberman, P. M., L. A. Pérez, S. E. Martín and G. I. Lacconi. 2014. Electrochemical synthesis of palladium nanoparticles in PVP solutions and their catalytic activity in Suzuki and Heck reactions in aqueous medium. RSC Adv. 4: 12330–12341.

Wang, C., X. Zou and L. Liao. 2019. Recent advances in optoelectronic devices based on 2D materials and their heterostructures. Adv. Optical Mater. 7: 1800441. 1–15.

Wang, Z., J. Dong, L. Li, G. Dong, Y. Cui, Y. Yang et al. 2020. The coalescence behavior of two-dimensional materials revealed by multiscale in situ imaging during chemical vapor deposition growth. ACS Nano 14: 1902–1918.

Wen, W., Y. Song, X. Yan, C. Zhu and Y. Lin. 2018. Recent advances in emerging 2D nanomaterials for biosensing and bioimaging applications. Mater. Today 21: 164–177.

Yin, H., D. Hu, X. Geng, H. Liu, Y. Wan, Z. Guo et al. 2019. 2D gold supercrystal-MoS_2 hybrids: Photoluminescence quenching. Mat. Lett. 255: 126531. 1–3.

Yin, P. T., S. Shah, M. Chhowalla and K. Lee. 2015. Design, synthesis, and characterization of graphene–nanoparticle hybrid materials for bioapplications. Chem. Rev. 115(7): 2483–2531.

Zhang, H. 2015. Ultrathin two-dimensional nanomaterials. ACS Nano 9: 9451–9469.

Zhang, Y., F. M. Heim, J. L. Bartlett, N. Song, D. Isheim and X. Li. 2019. Bioinspired, graphene-enabled Ni composites with high strength and toughness. Sci. Adv. 5: 5577. 1–9.

Zhao, Y., S. Zhang, R. Shi, G. I. N. Waterhouse and T. Zhang. 2020. Two-dimensional photocatalyst design: A critical review of recent experimental and computational advances. Mater. Today 3: 78–91.

Zhong, L., S. Gan, X. Fu, F. Li, D. Han, L. Guo et al. 2013. Electrochemically controlled growth of silver nanocrystals on graphene thin film and applications for efficient nonenzymatic H_2O_2 biosensor. Electroch. Acta 89: 222–228.

CHAPTER 11

Graphene-based Materials as Highly Promising Catalysts for Energy Storage and Conversion Applications

*Maximina Luis-Sunga, Stephanie J. Martínez, José Luis Rodríguez,
Gonzalo García* and *Elena Pastor**

Introduction

The depletion of conventional energy sources motivated by the increasing demand of global energy due to the growth of the world population and rapid economic expansion makes the development of renewable and clean energy sources imperative (Wengenmayr and Bührke 2013). The combustion of fossil fuels for obtaining electricity generates gases emissions that harm the planet: carbon dioxide, carbon monoxide, among others, which contribute to generate and enhance the air, soil and water pollution (García et al. 2020). For this reason, the development of energy-related materials and devices for providing clean energy is vital to meet the global energy demand.

New technology devices for energy conversion and storage are currently under intensive investigation (Azadmanjiri et al. 2018, Dasari et al. 2017). There has been a great worldwide endeavor towards the development of electrochemical energy conversion (such as electrolyzers, fuel cells and photovoltaic cells) and storage devices (like lithium-ion batteries and supercapacitors) (Wang and Liu 2015). The performance of these devices largely depends on the materials used and continues to be one of the great challenges of scientific and technological research to find systems with good stability, durability, efficiency and low cost in order to allow their large-scale production and commercialization (Lv et al. 2016). In this respect, Graphene-based Materials (GMs) have attracted considerable research interest in recent years due to their exceptional properties for use in applications to achieve the energy challenge (Ali Tahir et al. 2016).

This chapter is a brief introduction of the main properties that make GMs extremely attractive to build components that are used in devices for the storage and conversion of energy and, also summarizes the most recent developments of these materials, indicating the relationship of their properties with the function they perform in each device.

Graphene and Graphene-based Materials

Graphene has remarkable properties such as high theoretical surface area (2630 m^2 g^{-1}), great electrical conductivity (10^6 S cm^{-1}), high room-temperature charge carrier mobility

Instituto de Materiales y Nanotecnología, Departamento de Química, Universidad deLa Laguna, PO Box 456, 38200, La Laguna, Santa Cruz de Tenerife, Spain.
* Corresponding author: epastor@ull.edu.es

($\sim 10^6$ cm^2 V^{-1} s^{-1}), elevated thermal conductivity (~ 2000 to 5000 Wm^{-1} K^{-1}), and is capable to support large electrical current density (10^8 A cm^{-2}) (Ali Tahir et al. 2016, Brownson and Banks 2014). It also has a unique lightweight characteristic, mechanical strength and flexibility, high optical transparency and light absorption making graphene excellent for many promising electrochemical applications (Kawrani et al. 2020, Wang and Liu 2015).

However, a pure graphene sheet reveals low catalytic activity for many electrochemical reactions and, consequently, catalysts based on Reduced Graphene Oxide (rGO) are usually used (Azadmanjiri et al. 2018, Dasari et al. 2017). Earlier defects were believed to reduce the performance of graphene, but it has been shown that such imperfections in graphene sheets can enhance their electrochemical performance beyond the predicted limits if they are properly synthesized (Azadmanjiri et al. 2018). The electrical conductivity of graphene sheets largely depends on the defects they contain, since these act as active sites that can inhibit and/or even favor charge transport by limiting the flow of electrons and the adsorption strength of certain substances, respectively (Azadmanjiri et al. 2018, Dasari et al. 2017, García et al. 2020).

Chemical and physical approaches are the most common ways to produce rGO and heteroatom-modified rGO from the reduction of Graphene Oxide (GO), which is usually created from graphite by the popular Hummers' method (García et al. 2020). The doping process may produce electronic and geometric distortions of the carbonaceous grid and accordingly, the activity toward several electrochemical reactions can be altered (García et al. 2020, López-Urías et al. 2013). Two types of doping that depend on the interatomic distance between carbon-heteroatom and C-C may occur in the graphene network (Duan et al. 2015, García et al. 2020, López-Urías et al. 2013, Lim et al. 2016, Rivera et al. 2017):

i) B-C, N-C and C-C have similar interatomic distances and therefore, the heteroatom-modification will follow a similar modus operandi. In this case, the improvement toward a specific reaction can be attributed to a positive charge and to the spin-polarization effects produced on carbon atoms near the heteroatom.

ii) P, Si and S show higher interatomic distance than C-C and their modification will display an exohedral doping, i.e., the heteroatom will be located above and/or below the graphene sheet. The graphene lattice undergoes significant distortions in the presence of the heteroatom, resulting in a position displacement of the surrounding carbon atoms out of the plane. The last, together with the mismatch of the outermost orbitals of the heteroatom and C, may induce a non-uniform spin density distribution, which could modify the surface activity toward a specific reaction.

Furthermore, doping graphene with both types of heteroatoms can effectively create more catalytically active sites and develop a combined effect of both dopants on the reaction of interest, affecting different electrochemical parameters such as the onset potential, exchange current, current density and the number of electrons transferred (Rahmani and Habibi 2019).

Another factor to take into account in the catalyst activity is the surface local acidity, which may be altered by the introduction of functional groups in the graphene network (García et al. 2020, Pérez-Rodríguez et al. 2018). Figure 11.1 summarizes the most common graphene modifications.

Thus, the outstanding properties, facile synthesis and easy functionalization qualify graphene-based materials as promising candidates for a wide range of electrochemical devices such as lithium-ion batteries, capacitors, fuel cells and photocatalytic devices (Hu et al. 2015).

Graphene-based Materials for Energy Devices

In recent years, great efforts have been made to take advantage of GMs properties looking for improvement of the efficiency, stability and cost-effectiveness of energy-related devices in order to achieve extensive commercialization of electrolyzers, fuel cells, lithium-ion batteries, capacitors and photovoltaic systems. Numerous research advances have been reported and therefore, it is

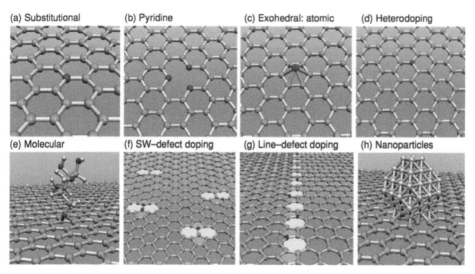

Figure 11.1: Molecular models representing different types of doped graphenes. (a) Substitutional doping, in which a carbon atom is removed and replaced by the dopant; (b) pyridine-like doping, in which a single vacancy is generated and the low coordinated carbon atoms (three carbons around the vacancy) are replaced by dopants; (c) exohedral atomic doping where a dopant is hosted on the surface of graphene; (d) heterodoping which consists in simultaneously doping the graphene with two or more types atoms different from carbon; (e) molecular doping or chemical functionalization of the graphene sheet; (f) 55–77 Thrower–Stone–Wales defects in which 5–7–5–7 rings are generated by rotating a C–C bond by 90 degrees (dopants are also included in some defects); (g) 5–7 line-defect in combination with substitutional doping; and (h) nanoparticles or clusters anchored on the graphene surface. Adapted from (López-Urías et al. 2013) with permission from John Wiley and Sons.

necessary to highlight the latest progress in the use of graphene for energetic applications offering a perspective of the progress achieved in this field.

Fuel Cells

Fuel Cells (FCs) are devices able to convert chemical energy into electricity and heat with higher efficiency than systems operating with internal combustion since no combustion occurs and, consequently, there is no limitation by the Carnot cycle (García et al. 2020). They can operate continuously as long as fuel and an oxidant are supplied and have great advantages such as a high energy conversion efficiency, low pollution and independence from the depletion of fossil fuels (Koper 2013). Figure 11.2 shows a basic scheme of a generic fuel cell with the common components in most fuel cells. These components are an anode and a cathode separated by an electrolyte. The separated fuel (gas- or liquid-phase) and oxidant streams enter through flow channels. The reagents are transported by diffusion and/or convection to the surface of the electrodes where the electrochemical reactions that generate the current take place (Mench 2008).

Despite the significant advantages of FCs as power sources, a real FC has several drawbacks, which decrease their working performance and increase their cost (García and Koper 2011). In this regard, one of the most important issues to solve is the one related to the anode and cathode electrodes, which usually use platinum-based materials that are very active catalysts although their elevated cost, low abundance, facile poisoning and poor durability are the main obstacles in the development of these devices (Heidary et al. 2016). Therefore researchers have been looking for innovative catalysts with excellent properties for fuel cell applications. In this regard, graphene-based materials have been used as electrodes additives, bipolar plates and proton-conducting electrolyte membranes (Iqbal et al. 2019).

FCs can be classified according to the electrolyte used and its operating temperature. The predominant types are the Alkaline Fuel Cell (AFC), Phosphoric Acid Fuel Cell (PAFC), Molten

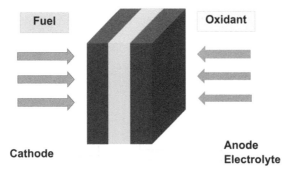

Figure 11.2: The basic scheme of a generic fuel cell.

Carbonate Fuel Cell (MCFC), Solid Oxide Fuel Cell (SOFC) and Ionic Exchange Membrane Fuel Cell (IEMF), which includes Proton Exchange Membrane Fuel Cell (PEMFC), Anion Exchange Membrane Fuel Cell (AEMFC), Direct Methanol Fuel Cell (DMFC) and Direct Ethanol Fuel Cell (DEFC), among others (Mench 2008).

Ionic Exchange Membrane Fuel Cell (IEMF)

This section deals with IEMFCs. The sluggish kinetics of the Oxygen Reduction Reaction (ORR) and the catalyst poisoning at the cathode of these systems are the main challenges to overcome in order to introduce this technology in the energy market (García et al. 2020, Martínez-Huerta and García 2015). Therefore, the basic principles of the ORR will be examined next.

Oxygen Reduction Reaction (ORR)

The ORR is five times slower than the Hydrogen Oxidation Reaction (HOR) in acid media and an elevated amount of expensive materials, such as platinum and palladium, are generally used to solve this issue (Gasteiger et al. 2005). A typical PEMFC using platinum as the catalyst at both the electrodes, i.e., cathode and anode, usually includes about 80–90% of the precious metal in the cathode side (Gasteiger et al. 2005, Sui et al. 2017). Additionally, mixed potentials at the cathode electrode are very important in direct alcohol fuel cells (DMFCs and DEFCs) because the membranes are not impermeable to the fuel (Aricò et al. 2001, García and Koper 2011). Hence, in such systems, the mixed potential appears due to the crossover of the alcohol through the electrolyte from the anode to the cathode (Aricò et al. 2001, García and Koper 2011).

Fortunately, GMs can be catalytically active toward the ORR and tolerant to alcohol poisoning, particularly in alkaline media (García et al. 2020). Nevertheless, it becomes clear that a high amount of scarce noble metals is not economically viable, and a drastic reduction is mandatory. In order to achieve these goals, the improvement and optimization of the cathode are crucial, and to accomplish these requirements, a full understanding of the ORR mechanism is necessary (Duca and Koper 2020, García et al. 2020, Roca-Ayats et al. 2018). The complete ORR involves four electrons:

$$O_2 + 4H^+ + 4e^- \rightarrow 2H_2O \qquad E^0 = 1.229 \; V \; vs \; RHE \qquad (1)$$

This reaction consists of several elementary steps, pathways and intermediates, which depend on the pH, dissolved ions in the electrolyte, catalyst and support materials, surface atomic structure, surface energy, oxygen coverage and potential, among other factors (García et al. 2020). Briefly, two main mechanistic pathways will be followed according to the adsorption strength of O_2 on the catalyst surface.

i) Strong O_2 adsorption (O_2 is activated for bond breaking and water is formed):

$$O_2 + H^+ + e^- \rightarrow OOH_{ads} \qquad (2)$$

$$OOH_{ads} + H^+ + e^- \rightarrow O_{ads} + H_2O \qquad (3)$$

and/or

$$OOH_{ads} \rightarrow OH_{ads} + O_{ads} \qquad (4)$$

$$O_{ads} + H^+ + e^- \rightarrow OH_{ads} \qquad (5)$$

$$OH_{ads} + H^+ + e^- \rightarrow H_2O \qquad (6)$$

ii) Weak O_2 adsorption (O-O bond is preserved and hydrogen peroxide is produced):

$$O_2 + e^- \rightarrow O_{2,ads}^{\cdot-} \qquad (7)$$

$$O_{2,ads}^{\cdot-} + H_2O + e^- \rightarrow HO_{2,ads}^- + OH^- \qquad (8)$$

$$HO_{2,ads}^- + H^+ \rightarrow H_2O_2 \qquad (9)$$

Four-electron reduction also occurs on weakly binding catalysts under certain conditions producing O_{ads} and OH_{ads}:

$$HO_{2,ads}^- \rightarrow O_{ads} + OH^- \qquad (10)$$

$$HO_{2,ads}^- + H_2O + e^- \rightarrow OH_{ads} + 2OH^- \qquad (11)$$

that finally follows reactions (5) and (6).

All reactions involving strong O_2 adsorption implicate concerted proton-coupled electron transfer (ORR at Pt-based catalysts follow this reaction mechanism). However, the adsorption strength of O_2 on GMs is usually weak, and the reaction mechanism is rather complex and comprises some reaction steps with decoupled proton-electron transfer (Koper 2013). Nevertheless, the prevalence of one pathway among the others depends on the properties of the electrocatalyst and the surrounding media.

On the other hand, the oxygen reduction to hydrogen peroxide leads to a low energy conversion efficiency and produces intermediates that can convert to harmful free radicals and degrade both the polymeric membrane and the catalyst causing the deterioration of the components of the fuel cell especially in acidic media (Koper 2013, Mench 2008, Jorissen and Garche 2015). Thus, research work in alkaline media has attracted more attention due to the following advantages (García et al. 2020, García and Koper 2011, Koper 2013, Mench 2008): (i) the ORR kinetics is faster and more efficient than in acidic media; (ii) it confers reduced adsorption energies of anions; (iii) it is not so corrosive; and (iv) a wide range of non-precious materials such as GMs catalysts can be used. However, several drawbacks such as the carbonation of the electrolyte by the CO_2 present in the air, that severely blocks active sites, and the durability of the catalyst need to be solved (García et al. 2020, García and Koper 2011, Koper 2013, Mench 2008).

The recent advances of ORR at GM electrocatalysts will now be illustrated and summarized.

Nitrogen-doped Graphene-based Catalysts

Numerous transition metal and metal-free catalysts based on graphene have been developed through the past years (Lim et al. 2016, Tsang et al. 2020). For instance, graphene-based catalysts with iron have been extensively investigated. Thus, iron- and nitrogen-doped graphene-based catalyst (Fe-N-Gra) was tested in a polymer-electrolyte fuel cell to investigate its suitability for AEMFC and PEMFC applications exhibiting better performance in alkaline solution with an ORR onset potential of 0.94 V (vs RHE). The number of electrons (n) transferred per O_2 molecule was calculated by the Koutecky-Levich equation obtaining a value for Fe-N-Gra close to 4 indicating that the ORR follows the direct 4-electron reduction pathway in alkaline medium (Sibul et al. 2020). Also a CoFe carbide/N-doped graphene composite has been developed by a refluxing strategy before a post-annealing procedure. The results show similar electrochemical performance as Pt/C catalyst for oxygen electro-reduction in the alkaline medium due to the huge surface area, the mesoporous structure and the synergistic effect of both metals (Gautam et al. 2018).

On the other hand, bimetallic iron/nickel metal-organic frameworks (Fe/Ni-MOFs) catalyst was combined with Nitrogen-doped Graphene (NG) obtaining bimetallic Fe/Ni-MOFs/NG nanocomposites (Qin, Huang, Wang et al. 2019). This material exhibits an excellent ORR performance with an onset potential of 1.09 V (vs RHE), a limiting current density of 8.56 mA cm^{-2} at a rotation rate of 1600 rpm and a Tafel slope value of 58.17 mV dec^{-1} in alkaline media. Porous frameworks and carbon-heteroatom bonds may cause agglomeration of catalyst particles, drastically reducing active sites. Moreover, the migration of ions and electrons are blocked causing a great impact on the ORR activity. In the present case, the metal ligands in the MOF constitute the potential binding active sites that also coordinate with other functional groups of the graphene oxide structures (-OH, -COC-, and COOH) or with other metal ions and groups making bridges. This causes the Fe/Ni-MOFs to grow *in situ* and become trapped between the graphene layers and also provides synergistic effects between the metal ions and the π–π conjugated structures of graphene jointly improving the catalytic activity (Qin, Huang, Wang et al. 2019).

Finally, remarkable results have been obtained by Qin et al. with 3D Flower-Like Bi_2SiO_5@ Nitrogen doped Graphene nanomaterials (FLBNG) as a cathodic catalyst (Qin, Huang, Shen et al. 2019). The catalyst exhibited good methanol immunity, high durability and high activity for oxygen reduction in acidic and basic solutions. The onset potential value for the ORR reaches to 1.09 V (vs RHE), the Tafel slope is 45.637 mV dec^{-1} and its limiting current density is 6.49 mA cm^{-2} in alkaline electrolyte. In this composite, nitrogen-doped 3D graphene decreases aggregations between graphene nanolayer structures in addition to improving the electrochemical reduction of oxygen because nitrogen doping could intrinsically adjust electron extraction or contributed properties of graphene leading to regulate the chemical adsorption energy of oxygen. Furthermore, this material has superior durability and better immunity to poisoning during the ORR compared to platinum-based catalysts (Qin, Huang, Shen et al. 2019).

Sulfur-doped Graphene-based Catalysts

Sulfur doped graphene-based catalysts were prepared by Morales-Acosta et al. Graphene oxide was synthesized, reduced and further sulfonated (Morales-Acosta et al. 2019). The reduction was carried out through a chemical and thermal process, performing the functionalization using sulfuric acid or aryl diazonium salt of sulfanilic acid as sulfonating agents, obtaining $rGO-SO_3H$. The $rGO-SO_3H$ catalyst thermally reduced and doped with sulfuric acid showed better electron transfer activity compared to that chemically reduced and sulfonated with sulfanilic acid (Morales-Acosta et al. 2019).

Nitrogen and Sulfur Dual-doping Graphene-based Catalysts

Heteroatom-doped graphene quantum dots have been developed through a hydrothermal treatment using ammonium hydroxide and sodium sulfide as precursors (Fan et al. 2019). For comparison, undoped pristine Graphene Quantum Dots (GQDs), Sulfur-doped Graphene Quantum Dots (SGQDs), Nitrogen-doped Graphene Quantum Dots (N-GQDs) and Nitrogen and Sulfur co-doped Graphene Quantum Dots (N,S-GQDs) have been synthesized. It is difficult to directly apply pristine GQDs, N-GQDs, S-GQDs and N,S-GQDs for electrochemical applications because of its poor electrical conductivity. In order to overcome this drawback, these materials were anchored on rGO. The electrochemical results of N,S-GQDs composite (vs Ag/AgCl, KCl (3M) reference electrode) displayed an ORR onset potential of –0.11 V (Fig. 11.3a) and a limiting current density of 4.82 mA cm^{-2} in alkaline solution (0.1 M KOH), close to those of Pt/C (–0.08 V and 5.08 mA cm^{-2}). Notably the different types of N-C or C-S-C bonding in N,S-GQDs can significantly improve the electrocatalytic activity for the ORR obtaining a value of 3.82 for the electron transfer number (Fig. 11.3b) which confirms the direct 4-electron reduction of oxygen and the effectivity of its facile synthesis method to fabricate metal-free electrocatalysts based on graphene for ORR (Fan et al. 2019).

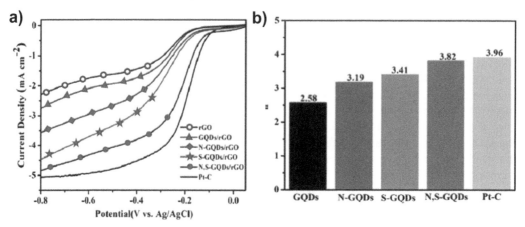

Figure 11.3: (a) Linear sweep voltammetry of rGO, GQDs/rGO, N-GQDs/rGO, S-GQDs/rGO, N,S-GQDs/rGO and Pt/C at 1600 rpm in O_2-saturated 0.1 M KOH electrolyte; and (b) electron transfer number calculated from the slope of Koutecky-Levich plots of GQDs, N-GQDs, S-GQDs and N,S-GQDs supported on rGO and Pt/C. Adapted from (Fan et al. 2019) with permission from Elsevier.

All these investigations indicate that the structure and electronic parameters can adjust the material properties resulting in an improvement of the catalytic properties for the ORR at the cathode of fuel cells.

Hydrogen Oxidation Reaction (HOR)

The electrochemical oxidation of hydrogen (HOR) takes places at the anode side of both proton exchange (PEMFC) and anion exchange (AEMFC) fuel cells. In these electrochemical devices, H_2 is used as fuel that is efficiently transformed into electrical energy without emission of contaminants. The overall reaction for HOR occurs subsequently:

$$H_2 \rightarrow 2H^+ + 2e^- \qquad E^0 = 0.00\ V\ vs\ RHE \qquad (12)$$

HOR does not occur at carbon materials but is extremely facile on Pt-based catalysts in acidic media with an exchange current density of 20–80 mA cm^{-2}, which is about 10^5 to 10^7 times faster than ORR (García et al. 2020). In this regard, literature about HOR on Platinum-Group Metals (PGM)-free catalysts in acidic media is scarce and employing graphene is uncommon, although tuning the catalyst surface acidity and/or the amount of intercalated acid into the graphene structure may help to fabricate PGM-free catalysts for the HOR in acidic media (Das et al. 2014, García et al. 2020, Pérez-Rodríguez et al. 2018). Ni, Mo, Co and their alloys are the most used PGM-free catalysts for the HOR in basic solutions and their exchange current density were reasonable (0.4–28 µA cm^{-2}), although there is still room to improve it (Lim et al. 2016, Sheng et al. 2014, Zhuang et al. 2016).

Direct Alcohol Fuel Cell (DAFC)

Among all fuels that can be used in fuel cells, the most investigated are hydrogen, methanol and ethanol. The use of ethanol and methanol (liquid biofuels) is a beneficial choice due to the convenience of production, storage, transport and handling (Santos et al. 2019).

PGMs are the most used electrocatalysts not only for Methanol Oxidation Reaction (MOR) but also for the Ethanol Oxidation Reaction (EOR) (Roca-Ayats et al. 2018). Nevertheless, PGMs have the great disadvantage that they can be easily poisoned by reaction intermediates, even in alkaline media. Commercial applications are severely limited due to their high cost (Martínez-Huerta and García 2015).

On the other hand, it has been proven that the surface chemistry of the catalytic support improves the behavior of the catalyst for the alcohol oxidation reaction (Iqbal et al. 2019, Rocha et al. 2015). Carbon nanotubes, carbon black and Graphene Nanosheets (GNS) stand out among the most frequently used carbon materials as catalyst supports due to their excellent electrical conductivity and high chemical stability (Iqbal et al. 2019, Rocha et al. 2015). Thus, graphene has lately been proposed as a candidate for electrochemical devices such as fuel cells, due to its high active surface area (Iqbal et al. 2019).

Methanol Oxidation Reaction (MOR)

Methanol electro-oxidation is a complex reaction involving the transfer of six electrons and several catalytic steps (Guillén-Villafuerte et al. 2013). Although the thermodynamic potential for the full electro-oxidation of methanol is close to that of hydrogen oxidation, the overall reaction is much more demanding due to the multi-electron transfer steps to form carbon dioxide (Martínez-Huerta and García 2015):

$$CH_3OH + H_2O \rightarrow CO_2 + 6H^+ + 6e^- \qquad E^0 = 0.04 \ V \ vs \ RHE \qquad (13)$$

It involves different intermediate species such as CO, formaldehyde, and/or formic acid. The sluggish kinetics of MOR and low efficiencies are attributed to the formation of these species (Guillén-Villafuerte et al. 2013, Roca-Ayats et al. 2019). For this reaction, Pt is well known as the most suitable electrocatalyst due to its superior electrocatalytic activity, however, its scarceness and high price have still restricted the development of DMFCs (Martínez-Huerta and García 2015, Wang and Ding 2016).

Graphene-based Catalysts

Platinum-based bimetallic nanoparticles have been reported to have shown extraordinary catalytic activity compared to those of total Pt counterparts (Hanifah et al. 2019). In this regard the authors fabricated platinum-palladium nanoparticles on rGO sheets (PtPd/rGO) through one-pot of hydrothermal-assisted chemical reduction method by using formic acid and demonstrated higher electrochemical surface area (0.91 cm^2) and electrocatalytic activity toward MOR compared to Pt/rGO (0.18 cm^2) and Pd/rGO (0.11 cm^2) catalysts.

On the other hand, it was proven that nickel and nickel-based compounds are efficient electrocatalysts for the oxidation of small alcohol molecules such as methanol and ethanol (Rahmani and Habibi 2019, Sarwar et al. 2018). Thus, research of nickel-based electrocatalysts, like NiCo alloys deposited on graphene, exhibited great efficiencies and high current density for methanol oxidation (Rahmani and Habibi 2019, Sarwar et al. 2018). Furthermore, bimetallic alloys of NiCu supported on rGO and then embedded with Pt nanoparticles (Pt-NiCu/rGO) were synthesized (Yousaf et al. 2018), combining wet-chemical and dry-chemical methods, constituting overall low-cost anode catalyst for DMFCs. In addition, the electrocatalytic performance of binary Ni@Pt nanodisk supported on rGO was studied for the methanol reaction by (Flórez-Montaño, Calderón-Cárdenas et al. 2016), proving good CO tolerance and high current efficiency (specific mass activity) in alkaline medium.

Nitrogen-doped Graphene-based Catalysts

Several reports have shown that by introducing heteroatoms into the graphenic support, it was possible to increase the electrocatalytic performance of the oxidation reaction of both methanol and ethanol (Arteaga et al. 2019, Kiyani et al. 2016, Koper 2013).

For instance (Kiyani et al. 2016), Pd and Pd-Co nanoparticles supported on Nitrogen-doped Graphene sheets (NG) were prepared using ethylene glycol as a reducing agent and studied the MOR in alkaline medium. The report revealed that PdCo/NG has better electrocatalytic activity than Pd/NG, and that both electrocatalysts are more stable than commercial Pt/C under potentiotatic condition (−0.2 V vs. Ag/AgCl) due to an enhanced tolerance toward residual carbon species (Kiyani et al. 2016).

In another interesting study, platinum nanoparticles supported on carbon black Vulcan XC-72R (C), N-doped rGO (N-rGO) and thermally treated rGO (rGO-TT) were synthesized by the formic acid method (Arteaga et al. 2019). The main results indicated that the support determines the state of oxidation of Pt nanoparticles at the surface of the catalysts, which control the electrocatalytic methanol oxidation activity. A small addition of nitrogen into the graphene structure induces a change of the Rate-Determining Step (RDS) from a chemical reaction after an electrochemical path to the first electron transfer. Thus, Pt/N-rGO displayed the worst performance toward the MOR, while Pt/rGO-TT not only enhances the catalytic activity toward the methanol oxidation but also toward the CO oxidation reaction (Arteaga et al. 2019).

Ethanol Oxidation Reaction (EOR)

Compared with methanol, ethanol is even more environmentally friendly, and simple methods could be used to produce large amounts of ethanol directly favorable from agricultural products or biomass (Luo et al. 2020). That is one of the reasons why DEFCs have recently received considerable attention. Among the challenges of these fuel cells, the most are frequently related to the slow electrode kinetics of the EOR at the anode (Flórez-Montaño, García et al. 2016). The full oxidation of ethanol to carbon dioxide has a favorable thermodynamic potential of 0.08 V (vs. RHE), although the efficiency of DEFC is drastically limited by the formation of partial oxidation products (maintaining intact the carbon-carbon bond, such as acetaldehyde and acetic acid) and strongly adsorbed intermediates (CO_{ads} and $CH_{x,ads}$) (Flórez-Montaño, García et al. 2016). The overall reaction for EOR is the following:

$$CH_3CH_2OH + 3H_2O \rightarrow 2CO_2 + 12H^+ + 12e^- \qquad E^0 = 0.08 \ V \ vs \ RHE \qquad (14)$$

Graphene-based Catalysts

In this context, graphene has emerged as an outstanding catalyst support due to its high surface area and excellent electrical conductivity (Boulaghi et al. 2018). For example, the electrooxidation of ethanol by the nickel (II)-Bis (1,10-Phenanthroline) complex was reported by (Santos et al. 2019). The authors found that Ni(II)(Phen)$_2$/rGO displayed better surface area than Ni(II)(Phen)$_2$ due to the presence of rGO (0.755 cm^2 compared to 0.522 cm^2, respectively).

Abdel Hameed et al. 2020 synthesized core-shell structured Cu@Pd/SnO$_2$-rGO electrocatalysts in two-steps protocol using ethylene glycol during the reduction step, containing 20% wt SnO$_2$. For its fabrication, the tin oxide was supported on rGO and then copper ions were chemically reduced on SnO$_2$-rGO. Later a palladium layer was deposited as a shell above the formed electrocatalyst. The researchers explained the enhanced ECSA value of Cu@Pd/SnO$_2$-rGO (584.36 m^2 g^{-1}), compared to Pd/rGO (54.65 m$^2 \cdot$g^{-1}), due to the synergistic effect of manifold components in the material as well as the good dispersion of Pd nanoparticles and their small nanosize. Thus, this electrocatalyst might be applied as anode materials in alkaline alcohol fuel cells.

It is possible to enhance the alcohol adsorption on Pd and its binding strength to the catalyst surface using a sensitization-activation method with a tin layer (Subramani and Gantigiah 2019). The authors successfully synthesized Co-Pd/Sn/rGO electrocatalysts by electroless deposition of Co nanoparticles on rGO, which exhibits significant electrocatalytic activity for methanol and ethanol oxidation in acidic medium (Subramani and Gantigiah 2019).

Finally, (Wang et al. 2019) synthesized highly active MoS$_2$ nanoflowers/graphene nanosheets (GNS) composites through a hydrothermal method and used it as Pt supports. Compared to the commercial Pt/C (0.6 mA cm^{-2}) and Pt/MoS$_2$ (1.0 mA cm^{-2}), Pt/MoS$_2$/GNS (1.3 mA cm^{-2}) had a significantly higher electrocatalytic activity and stability for ethanol electrooxidation. Accordingly, the introduction of GNS not only increases the ECSA value but also prevents the aggregation of MoS$_2$ (Wang et al. 2019).

Nitrogen-doped Graphene-based Catalysts

As mentioned earlier, it is possible to incorporate metal-organic frameworks as a catalyst support to increase stability at high overpotentials. Thus, (Yao et al. 2019) used ZIF (Zeolitic Imidazolate Framework)-GO composite microspheres as precursors for hollow N-doped graphene microspheres

(ZGC) through a spray-drying technique followed by calcination. The structure of the ZIF-GO composite microspheres has GO wrapped around and connected to ZIF-8 (Zeolitic Imidazolate Framework-8). The optimized weight percentage of GO is 20% and it is known as ZG20. Later, Pd nanoparticles were loaded on the ZG20C to obtain Pd/ZG20C. They have proven that Pd/ZG20C exhibits the best catalytic properties among the synthesized materials and ZG20C favors the dispersion of palladium nanoparticles and the EOR response in alkaline medium (Yao et al. 2019).

Nitrogen and Phosphorus Dual-doping Graphene-based Catalysts

Phosphorous low-doping palladium nanoalloys anchored on Three-Dimensional Nitrogen-Doped Graphene (Pd-P/3DNGS) were synthesized by a facile one-pot surfactant-free hydrothermal method, developing an enhanced electrochemical behavior for the EOR (Yang et al. 2020). Compared with Pd/rGO (40.8 m^2 g^{-1}), the novel electrocatalyst Pd-P/3DNGS displayed an increase in the electrochemically active surface area (67.9 m^2·g^{-1}) when P is introduced to the Pd/3DNGS material.

Electrolyzers

Increasing global warming caused by the combustion of fossil fuels is leading to a search for alternatives that are environmentally friendly, accessible and economically attractive. Hydrogen is considered a clean fuel for the future because it acts as a green energy carrier and provides a method for the storage and transport of energy. A variety of processes are available for H$_2$ production, based on conventional or renewable technologies (Chanda et al. 2015, Isao 2009). The latter includes those technologies that utilize renewable resources. Water electrolysis offers a practical route for sustainable hydrogen production by using a renewable electrical energy source for water splitting (Chanda et al. 2015, García et al. 2020, Kong et al. 2017).

The overall water splitting involves the Hydrogen Evolution Reaction (HER) at the cathode and the Oxygen Evolution Reaction (OER) at the anode (Li et al. 2019). When a potential difference (E > 1.23 V) or electrical current is applied, water splits and H$_2$ is evolved at the cathode and O$_2$ on the anode side, according to the following reaction:

$$H_2O_{(l/g)} \rightarrow H_{2(g)} + \frac{1}{2}O_{2(g)} \qquad \Delta E = 1.23\ V \qquad (15)$$

It is a difficult process and the high overpotential for the whole reaction requires the use of catalysts (Yuan et al. 2019). Currently, the best catalysts are based on noble metals such as Ir/Ru for the OER and Pt for the HER, limiting their large-scale applications because of their shortage and high-cost (Jia et al. 2017). The progress of novel electrode materials contributes to overcoming these problems to ensure the future deployment of this technology. In this regard, graphene-based electrocatalysts have gained special research interest in order to decrease the cost of these devices owing to their natural abundance, tunable electronic properties, high electrical conductivity and high tolerance to extremely acidic and alkaline conditions (García et al. 2020).

Hydrogen Evolution Reaction (HER)

The electrochemical Hydrogen Evolution Reaction (HER) in acidic (Eq. 16) and alkaline (Eq. 17) solutions are the subsequent (Luis-Sunga et al. 2020):

$$2H^+ + 2e^- \rightarrow H_2 \qquad E^0 = 0.00\ V\ vs\ RHE \qquad (16)$$

$$2H_2O + 2e^- \rightarrow H_2 + 2OH^- \qquad E^0 = 0.00\ V\ vs\ RHE \qquad (17)$$

The HER in the whole pH range requires advanced electrocatalysts with a high current density at low overpotentials. The most effective for HER in acidic media are those based on Pt (García et al. 2020). However, their large-scale application is restricted because they are limited and expensive. On the other hand, transition metal-based materials can be used for HER in alkaline media, but their activity is much lower than the catalytic performance of Pt (Chanda et al. 2015).

Recently, non-precious metal supported on doped-graphene materials have been studied (Huang et al. 2019, Luis-Sunga et al. 2020). A strong coupling between the metal centers and doped graphene were observed which facilitated the electronic transfer increasing the electrocatalytic activity. Furthermore, the durability of the catalyst improved due to the presence of covalent bonds between the doping and metal atoms (Duan et al. 2015).

Nitrogen and Boron-doped Graphene-based Catalysts

Graphene materials doped with nitrogen have exhibited high HER performance as can be observed in several synthesized materials (Yang et al. 2018). Deng et al. reported that increasing the nitrogen content as a dopant and reducing the number of graphene layers that encapsulate CoNi nanoparticles significantly enhance the electron density, which increases the catalytic activity towards HER in acidic medium (Deng et al. 2015).

Furthermore, heteroatom-doped metal-free graphene catalysts have been developed (i.e., nitrogen, sulfur, boron, phosphorus, etc.) to tune the electronic properties and improve HER activity (Ito et al. 2015, Sathe et al. 2014, Yan et al. 2015). Three-dimensional interconnected arms of carbon nitride backbone wrapped with nitrogen-doped graphene sheets (CNx@N-rGO) have been synthesized by high-temperature annealing of graphene oxide-coated melamine foam. It was reported that a HER overpotential of 193 mV to achieve a current density of 10 mAcm^{-2} which is far superior to the previously reported Pt-free systems (Gangadharan et al. 2017). Besides, boron doping creates surface defect sites and plenty of surface-active reduction centers (electron-rich) as it can be seen on B-substituted graphene developed by (Sathe et al. 2014).

Sulfur-doped Graphene-based Catalysts

The conductivity of graphene may be improved by S-doping because it causes a more effective reduction and a superior electron donor capacity (Rahmani and Habibi 2019, Tian et al. 2017). In this regard, (Tian et al. 2017) reported a strategy to use a plasma etching method to create topological defects on S-doped graphene. Results exhibited an enhanced HER activity with an overpotential of 178 mV at a current density of 10 mA cm^{-2}. The coupling between S-doping and the topological defects contributed to the HER catalytic performance (Tian et al. 2017).

Heteroatom Dual-doping Graphene-based Catalysts

Nitrogen and phosphorus dual-doped graphene material have also exhibited a similar onset overpotential, exchange current density and Tafel slope to Pt/C (Jiang et al. 2015, Zhang et al. 2018). Furthermore, a bottom-up approach was introduced to synthesize nitrogen and phosphorus dual-doped multilayer graphene (Guruprasad et al. 2019). The optimized NP-doped graphene shows a comparable onset potential (–0.12 V vs RHE), exchange current density (0.0243 mA cm^{-2}) and Tafel slope (79 mV per decade) to some of the traditional metallic catalysts (Jiang et al. 2015).

Oxygen Evolution Reaction (OER)

The Oxygen Evolution Reaction (OER) in acidic (Eq. 18) and alkaline (Eq. 19) solutions is the following:

$$2H_2O \rightarrow O_2 + 4H^+ + 4e^- \qquad E^0 = 1.23\ V\ vs\ RHE \qquad (18)$$

$$4OH^- \rightarrow O_2 + 2H_2O + 4e^- \qquad E^0 = 1.23\ V\ vs\ RHE \qquad (19)$$

The OER is a kinetically slow process that involves a 4-electron transport process along with the formation of the O-O bond, and a high overpotential is required (Jamesh and Sun 2018, Li et al. 2018). In this context, graphene oxide and graphenic materials have been characterized as highly active catalysts towards OER in alkaline medium, and numerous graphene-based electrocatalysts have been extensively investigated (Chanda et al. 2015).

A good example of the OER at GMs is the study reported by (Chakrabarty et al. 2019), in which flower-like $ZnCo_2O_4$ attached to a rGO sheet was synthesized by urea-assisted solvothermal method. The OER results reveal that an overpotential of 0.30 V is required for overcoming a current density of 10 mA cm^{-2}. The study indicates that porous $ZnCo_2O_4$ increases the catalytic surface area and that the high conductivity of rGO improves the charge transport process (Chakrabarty et al. 2019).

Nitrogen-doped Graphene-based Catalysts

Edge/defect-rich 3D nitrogen-doped and oxygen-functionalized self-supported graphene nanosheets on the surface of a carbon cloth (N,O-VAGNs/CC) was synthesized through the micro-wave plasma-enhanced chemical vapor deposition (MPECVD) system (Li et al. 2018). In contrast to undoped graphene, the N,O-VAGNs/CC has many functional groups, a great number of defects and active-edges that increase the catalytic performance toward OER. This graphene design enhances the OER catalytic activity, showing results (onset potential of 1.47 V vs RHE and a Tafel slope of 38 mV dec^{-1} in alkaline solution) that indicate an even higher performance than state-of-the-art RuO_2 catalysts, also highlighting its durability and stability in alkaline media (Li et al. 2018).

A composite material of $CuCo_2O_4$ as oxygen evolution electrocatalyst was also reported. It developed efficient OER activity with a small overpotential of 0.36 V in 1 M KOH at the current density of 10 mA cm^{-2}, when compared to $CuCo_2O_4$ nanoparticles (Bikkarolla and Papakonstantinou 2015).

Phosphorus and Sulfur-doped Graphene-based Catalysts

Xiao et al. investigated metal-free phosphorus-doped graphene (G-P) prepared by hydrothermal treatment of red phosphorous followed by ball milling of phosphorus and graphite (Xiao et al. 2016). The onset potential for OER on G-P is 1.48 V vs. RHE and the Tafel slope value is 62 mV dec^{-1} in alkaline solution, which are better results than those of undoped graphene (about 1.54 V vs. RHE and Tafel slope of 91 mV dec^{-1}). This material contains edge-selectively phosphorus-doped sheets and has few graphene layers with many defects. Its aggregated morphology and high surface area enhance the catalytic activity and durability for the OER (Xiao et al. 2016).

Additionally, a sequential two-step strategy to dope sulfur into carbon nanotube–graphene nanolobes was developed (El-Sawy et al. 2016). It was reported that increasing the incorporation of heterocyclic sulfur into the carbon ring of carbon nanotubes not only enhances OER activity with an overpotential of 350 mV at a current density of 10 mA cm^{-2} but also retains 100% stability after 75 hours. Furthermore, the bidoped sulfur carbon nanotube–graphene nanolobes behave like the state-of-the-art catalysts for OER but outperform those systems in terms of turnover frequency (TOF) which is two orders of magnitude greater than 20 wt.% Ir/C at 400 mV overpotential with very high mass activity 1000 mA g^{-1} at 570 mV (El-Sawy et al. 2016).

Nitrogen and Sulfur Dual-doping Graphene-based Catalysts

A hybrid of $CoFe_2O_4$ supported on nitrogen/sulfur dual-doped reduced graphene oxide networks (CFO/NS-rGO) was fabricated by (Yan et al. 2016). The nitrogen/sulfur dual-doped graphene (NS-rGO) was synthesized by a hydrothermal process and then the catalysts was obtained by the covalent coupling between NS-rGO and $CoFe_2O_4$. The porous structure and 3D networks give this material an OER activity comparable to RuO_2/C catalyst and high durability (Yan et al. 2016).

Besides, cobalt sulfide (Co_9S_8) nanoparticles embedded in nitrogen/sulfur co-doped graphene (Co_9S_8/NSG) hybrid material were synthesized (Zhong et al. 2019). It has a small overpotential of 0.26 V with a Tafel slope of 55 mV dec^{-1} at a current density of 10 mA cm^{-2} exhibiting an enhanced performance toward the OER and good stability in alkaline electrolytes (Zhong et al. 2019).

Bifunctional Electrocatalysts for OER and HER

It should be considered that producing two different types of single-function catalysts requires the need to use different equipment, processes and characterizations leading to an increase in the cost (Jia et al. 2017). However, it is difficult to pair two electrode reactions in an integrated electrolyzer to achieve realistic applications because the stability of the catalysts for each reaction usually corresponds to different pH ranges (Jia et al. 2017). Most catalysts cannot provide good bifunctionality at the same pH and electrolyte. For example, some transition-metal-based layered double hydroxides are efficient for OER in alkaline medium but are inert for HER in the same electrolyte (Yuan et al. 2019).

Nevertheless, bifunctional graphene-based materials for OER and HER were developed. In this context, several bifunctional catalysts such as coral-like $NiCo_2O_4$ nanostructure supported on reduced graphene oxide ($NiCo_2O_4$-rGO) were prepared by a controlled and facile hydrothermal reaction procedure (Debata et al. 2018). These materials presented a low onset potential, high current density, lower Tafel slope and long-term stability in both HER and OER, revealing a high catalytic performance due to the high specific surface area, large number of active sites and better anchoring of the $NiCo_2O_4$ nanorods on the graphene sheets (Debata et al. 2018). The behavior towards the overall water-splitting was tested in a two-electrode electrolyzer in alkaline medium using the $NiCo_2O_4$-rGO as the catalyst in both anode and cathode electrodes (Fig. 11.4) (Debata et al. 2018).

Additionally, a Ni-Fe LDH-NS@DG10 hybrid catalyst (formed by exfoliated Ni-Fe Layered Double Hydroxide (LDH) nanosheet (NS) and Defective Graphene (DG)) demonstrated high electrocatalytic activity for OER in an alkaline solution (overpotential of 0.21 V at a current density of 10 mA cm^{-2} and Tafel slope of 52 mV dec^{-1}) and elevated HER performance in an alkaline solution (overpotential of 115 mV by 2 mg cm^{-2} loading at a current density of 20 mA cm^{-2}) (Jia et al. 2017).

In order to investigate the charge distribution, Density Functional Theory calculations (DFT) were carried out revealing that the synergetic effects of highly exposed 3D transition metal atoms and carbon defects are essential for the bifunctional activity for the OER and HER. The authors concluded that the catalyst activity may be attributed to: (i) the direct interfacial contact between 3D transition metal atoms and defects on carbon enhancing the electron transfer and shortening the diffusion distance; (ii) the highly active defective sites on the layered graphenic structure with a large specific surface area and high conductivity; and (iii) the design of the layered Ni–Fe LDH-NS on DG sheets endows distinct thickness, dispersion and structure of the catalyst (Jia et al. 2017).

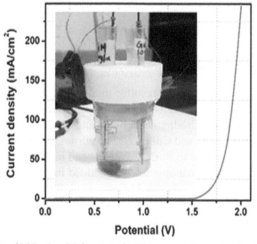

Figure 11.4: Polarization curve of $NiCo_2O_4$-rGO in a two-electrode system (inset showing H_2 and O_2 formation). Adapted from (Debata et al. 2018) with permission from Elsevier.

Furthermore, the outstanding OER and HER activity of Ni-Fe LDH-NS@DG10 hybrid catalyst demonstrated by a solar power-assisted water-splitting device, achieved a record of 20 mA cm^{-2} at a voltage of 1.5 V (Jia et al. 2017).

Nitrogen-doped Graphene-based Catalysts

Ultrathin exfoliated black phosphorus nanosheets on N-doped graphene (EBP@NG) was synthesized and used as a bifunctional catalyst for HER and OER in alkaline medium (KOH 1M) (Yuan et al. 2019). It showed an OER overpotential of 350 mV and a Tafel slope of 82 mV dec^{-1} whereas 125 mV at 10 mA cm^{-2} were revealed in HER studies. This material was tested in an electrolyzer using EBP@NG as an anode and cathode with a low cell voltage obtaining a result of 1.54 V vs RHE at 10 mA cm^{-2}, which is superior to the expensive integrated Pt/C@RuO$_2$ catalyst (1.60 V vs RHE at 10 mA cm^{-2}) (Yuan et al. 2019).

Additionally, the material was tested as cathodic and anodic catalysts in a H-type cell and Faraday efficiencies higher than 97% were acquired for H$_2$ and O$_2$ production (Fig. 11.5a). Furthermore, a solar energy-assisted electrolyzer was constructed and the evolved H$_2$ and O$_2$ bubbles were observed, making it a promising system for distributed energy storage and conversion (Fig. 11.5b) (Yuan et al. 2019).

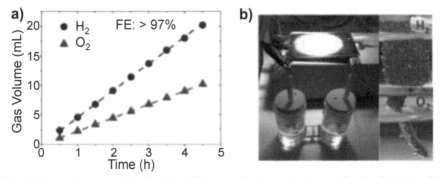

Figure 11.5: (a) Plots on the amount of evolved H$_2$ and O$_2$ versus time at a constant current density of 10 mA cm^{-2} for EBP@ NG and (b) photograph of a solar power-driven water splitting device. Reprinted from (Yuan et al. 2019) with permission from American Chemical Society.

Material for fuel cells and electrolizers and their principal characteristics are summarized in Table 11.1.

Batteries

Energy consumption has increased rapidly, thus it is essential to find energy storage devices that are low-cost, environmentally friendly and with a long service life (Bulbula et al. 2018). Electrochemical energy storage (such as Li and Na-ion batteries) has been widely studied and is considered one of the most attractive and environmentally friendly energy storage systems that can be used due to its versatility, high efficiency and flexibility (Chao 2019).

The components of an ion battery consist of the anode, separator, cathode, electrolyte and a current collector. Lithium-ion and sodium-ion batteries work in a similar manner. The materials allow the displacement of the ions (Li$^+$ or Na$^+$) between the cathode and the anode during the charge/discharge process. During discharge, the cation migrates from the anode to the cathode through the separator and electrolyte, and the electrons from the anode to the cathode through the external circuit. In this process, the potential of the cathode decreases and the potential of the anode increases, leading to a low cell voltage and the release of electrochemical energy. During charging, the migration of ions and electrons is reversed. These processes are known as the rocking-chair mechanism (Chao 2019).

Table 11.1: Electrochemical properties of graphene-based catalysts for fuel cells and electrolyzers.

Hydrogen Fuel Cell

Catalyst	Reaction	Medium	Onset (V vs RHE)	Limiting current density (mA cm^2) 1600 rpm	n° e$^-$	Ref.
Iron- and nitrogen-doped graphene-based catalyst (Fe-N-Gra)	ORR	0.1 M KOH	0.94	~ 6.0	~ 4	Sibul et al. 2020
		0.5 M H$_2$SO$_4$	0.89	4.8 (1900 rpm)	~ 4	
CoFe carbide/N-doped graphene composite		0.1 M KOH	0.93	~ 6.0	3.65	Gautam et al. 2018
Bimetallic iron/nickel metal-organic frameworks on nitrogen-doped graphene (Fe/Ni-MOFs/NG)		0.1 M KOH	1.09	8.56	3.9	Qin, Huang, Wang et al. 2019
3D flower-like Bi$_2$SiO$_5$@ nitrogen doped graphene nanomaterials (FLBNG)		0.1 M KOH	1.09	6.49	3.88	Qin, Huang, Shen et al. 2019
		0.1 M HClO$_4$	1.19	5.86	3.87	
N,S-doped graphene quantum dots (N,S-GQDs)		0.1 M KOH	0.90	4.82	3.82	Fan et al. 2019

Direct Alcohol Fuel Cell

Catalyst	Reaction	Medium	Onset (V vs RHE)	Ref.
Platinum-palladium nanoparticles on rGO sheets (PtPd/rGO)	**MOR**	0.5 M H$_2$SO$_4$	0.60	Hanifah et al. 2019
NiCo alloys deposited on graphene (NiCoErN-GO/CCE)		0.1 M KOH	0.44	Rahmani and Habibi 2019
Bimetallic alloys of NiCu supported on rGO decorated with Pt nanoparticles (Pt-NiCu/rGO)		0.1 M KOH	~ 0.37	Yousaf et al. 2018
Binary Ni@Pt nanodisk supported on rGO (Ni@Pt/rGO)		0.1 M NaOH	0.30	Flórez-Montaño et al. 2016
Pt supported on thermally treated rGO (Pt/rGO-TT)		0.5 M H$_2$SO$_4$	0.70	Arteaga et al. 2019
Cobalt nanoparticles on activated rGO (Co-Pd/Sn/rGO)		0.5 M H$_2$SO$_4$	0.70	Subramani and Gantigiah 2019
NiCo alloys deposited on graphene (NiCoErNGO/CCE)	**EOR**	0.1 M KOH	0.44	Rahmani and Habibi 2019
Nickel (II)-Bis(1,10-Phenanthroline) complex supported on rGO (Ni(II)(Phen)$_2$/rGO)		1 M NaOH	~ 0.50	Santos et al. 2019

Table 11.1 Contd. ...

...Table 11.1 Contd.

Direct Alcohol Fuel Cell

Catalyst	Reaction	Medium	Onset (V vs RHE)	Ref.
Tin oxide as a promoter for copper@palladium nanoparticles on graphene sheets ($Cu@Pd/SnO_2$-rGO)		0.5 M NaOH	0.39	Abdel Hameed et al. 2020
Phosphorous low-doping palladium nanoalloys anchored on three-dimensional nitrogen-doped graphene (Pd-P/3DNGS)		1 M KOH	0.32	Yang et al. 2020

Electrolyzers

Catalyst	Reaction	Medium	Onset (V vs RHE)	Tafel Slope (mV dec^{-1})	Ref.
Three-dimensional interconnected arms of carbon nitride backbone wrapped with nitrogen-doped graphene sheets (CNx@N-rGO)	**HER**	0.5 M H_2SO_4	−0.04	54	Gangadharan et al. 2017
B-substituted graphene		0.5 M H_2SO_4	−0.20	~ 99	Sathe et al. 2014
S-doped graphene		0.5 M H_2SO_4	~ −0.10	86	Tian et al. 2017
Nitrogen and phosphorus dual-doped graphene		0.5 M H_2SO_4	−0.12	79	Jiang et al. 2015
Nitrogen and phosphorus dual-doped multilayer graphene		0.5 MH_2SO_4	−0.09	47	Guruprasad et al. 2019
		1 M KOH	−0.25	110	
Coral-like $NiCo_2O_4$ nanostructure supported on rGO ($NiCo_2O_4$-rGO)		2 M KOH	−0.17	91	Debata et al. 2018
NiFe LDH-NS@DG10 hybrid catalyst		1 M KOH	~ −0.15	110	Jia et al. 2017
Ultrathin exfoliated black phosphorus nanosheets on N-doped graphene (EBP@NG)		1 M KOH	~ −0.15	109	Yuan et al. 2019
Flower-like $ZnCo_2O_4$ attached to a rGO sheet	**OER**	1 M KOH	1.51	59	Chakrabarty et al. 2019
Edge/defect-rich 3D nitrogen-doped and oxygen-functionalized self-supported graphene nanosheets on the surface of a carbon cloth (N,O-VAGNs/CC)		1 M KOH	1.47	38	Li et al. 2018

Table 11.1 Contd. ...

...Table 11.1 Contd.

Electrolyzers					
Catalyst	**Reaction**	**Medium**	**Onset (V vs RHE)**	**Tafel Slope (mV dec^{-1})**	**Ref.**
$CuCo_2O_4$ nanoparticleson nitrogenated graphene ($CuCo_2O_4$/NrGO)		0.1 M KOH	1.52	57	Bikkarolla and Papakonstantinou 2015
Metal-free phosphorus-doped graphene (G-P)		1 MKOH	1.48	62	Xiao et al. 2016
Sulfur into carbon nanotube–graphene nanolobes		1 M KOH	1.55	95	El-Sawy et al. 2016
Hybrid of $CoFe_2O_4$ supported on nitrogen/ sulfur dual-doped reduced graphene oxide networks (CFO/NS-rGO)		0.1 M KOH	1.74	88	Yan et al. 2016
Cobalt sulfide (Co_9S_8) nanoparticles embedded in nitrogen/sulfur co-doped graphene (Co_9S_8/ NSG)		1 M KOH	~ 1.45	55	Zhong et al. 2019
Coral-like $NiCo_2O_4$ nanostructure supported on reduced graphene oxide ($NiCo_2O_4$-rGO)		2M KOH	1.55	47	Debata et al. 2018
NiFe LDH-NS@DG10 hybrid catalyst		1 M KOH	1.41	52	Jia et al. 2017
Ultrathin exfoliated black phosphorus nanosheets on N-doped graphene (EBP@NG)		1 M KOH	~ 1.25	82	Yuan et al. 2019

These Lithium-Ion Batteries (LIBs) are considered as one of the most useful energy storage/ conversion systems for micro-electro-mechanical devices, electric vehicles and smart grids due to their high energy density and environmental friendliness. LIBs present limited charge/discharge rate, safety, temperature range and stability that impede their development in many sectors (Chen et al. 2017).

Moreover, current LIBs cannot meet the increasing performance requirements for these applications due to different troubleshooting needs. For instance, the severe fading capacity of Li-alloying anode materials due to the large volumetric contraction and expansion which cause cracking and pulverization of the active catalysts during the lithium extraction/insertion processes (Chen et al. 2017). The cathode materials also show much lower capacities than anode components and inorganic intercalation in the cathode has drawbacks due to the lack of mineral resources, high cost and limited reversible capacity. Therefore, it is necessary to develop new electrocatalysts with good stability and large capacity (Xiong et al. 2017).

In this context, graphene appears as a promising material for batteries due to its mechanical, optical and thermal properties. Furthermore, graphene does not undergo a significant volume change when it forms a composite. Thus, it can be used as a conductive current collector, an effective buffer network to support electron transport and maintain the structural integrity of the entire electrode. Catalysts supported on graphenic materials can be developed for the cathode or anode of the batteries in order to obtain high energy and power densities. For this reason, one of the most important applications of graphene in energy storage is to form composites with active electrode species and improve its performance for lithium- or sodium-ion batteries (Chao 2019).

Keeping in mind the descriptions above, several cathode graphene-based materials have been developed, such as a high-performance composite cathode comprised of nano-Lithium Manganese Oxide (nano-LMO) particles and Graphene Nano Flakes (GNFs) with a significantly enhanced packing density (Chen et al. 2017). They were fabricated using an ethyl cellulose-stabilized dispersion of primary nano-LMO particles and pristine GNF. The charge transfer was enhanced showing an improved rate capability (~ 75% capacity retention at a 20 C cycling current rate at room temperature) and extraordinary electrochemical performance at low temperatures close to full capacity retention at −20°C. GNFs act as a conductive additive compared to carbon black improving its electrical properties and mechanical resistance. Furthermore, it appears to stabilize the LMO/electrolyte interface by forming a thin and stable solid-electrolyte interface layer on the graphene (Chen et al. 2017).

On the other hand, liquid-exfoliated few-layer graphene flakes have been synthesized through jet cavitation and used as a conductive additive for commercially available cathode materials ($LiFePO_4$ (LFP), nano LFP, $LiCoO_2$ (LCO) and $Li(Ni_{1/3}Mn_{1/3}Co_{1/3})O_2$ (NCM)) in lithium batteries (Wang et al. 2019). Results indicate that a graphene loading of ~ 3 wt% can enhance the specific capacity from 150 to 178 mAh g^{-1} in LFP. For LCO and NCM, the addition of graphene results in a maximum specific capacity of 156 and 168 mAh g^{-1}, respectively, which is a remarkable improvement compared to commercial materials without graphene (140.6 and 152.9 mAh g^{-1}, respectively). Modified graphene is used as a conductive additive because the conductive network reduces the internal resistance of the cathodes by avoiding a voltage polarization or blocking the diffusion path of lithium ions (Wang et al. 2019).

Besides, compounds that are combined with an rGO matrix tend to be well-distributed, helping to reduce particle agglomeration due to the different volume variations experienced during the charge and discharge processes (Jia et al. 2019).

In this manner, the use of rGO has also been proposed as an eco-friendly and multifunctional conductive binder for the manufacture of flexible Hard Carbon (HC-rGO) anodes with high performance for Sodium-Ion Batteries (SIBs) (Sun et al. 2018). It was observed that the amount of oxygenated functional groups largely influences the electrochemical performance of the graphene cathode material. Figure 11.6 depicts the performance of HC-rGO films synthesized with a different mass ratio of HC:rGO (2:1, 1:1 and 1:2). It can be recognized that HC-rGO films show promising results and offer a high reversible electrode capacity of up to 372.4 mAh g^{-1} with a loss of capacity of ca. 9% after 200 charge/discharge cycles.

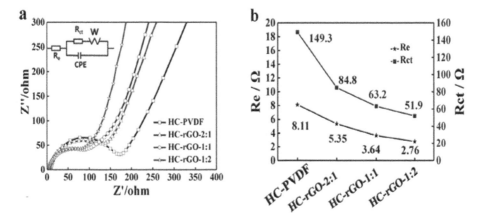

Figure 11.6: (a) Nyquist plots of the HC-rGO electrode as well as the conventional PVDF-bonded hard carbon electrode. The inset in (a) shows the equivalent circuit diagram and (b) the comparison of resistance values for different anodes. Adapted from (Sun et al. 2018) with permission from Elsevier.

In order to study the improved rate capabilities of rGO-bonded films, Nyquist plots were measured for the HC-rGO films and HC-fluorine-containing polymer binder (HC-PVDF) electrodes (Fig. 11.6a). Each plot contains, in the high to middle-frequency region, a depressed semicircle that corresponds to the charge transfer resistance (R_{ct}) and, in the low-frequency region, a straight line which represents the Warburg impedance (W) related to the sodium diffusion in the electrodes. The ohmic resistance (R_e) corresponds to the intercept with the X-axis in the high-frequency region and it represents the totality of electrical resistances, such as the inherent resistance of the active material and the electrolyte; and the contact resistance at the interface between the current collector and the active material. Figure 11.6b shows that the R_e of the HC-PVDF, HC-rGO-2:1, HC-rGO-1:1 and HC-rGO-1:2 are 8.11 Ω, 5.35 Ω, 3.63 Ω and 2.76 Ω, respectively. The R_e of the HC-rGO film progressively decreases with the rise of the mass ratio of rGO confirming the conductive agent purpose of rGO in the HC-rGO film. This conductivity enhancement and the disorder of the structure permit a faster charge/discharge. It is observed that the charge transfer resistance falls intensely with the increment of rGO mass ratio obtaining 149.3 Ω for PVDF-bonded HC electrode to 63.2 Ω for HC-rGO-1:1 and 51.9 Ω for HC-rGO-1:2. The two-dimensional assembly allows rGO to have a flexible main structure and act as a strong binder resulting in a flexible hard carbon film that can be used as an anode for sodium-ion batteries without any current collector. In this way, rGO produces a 3D conductive network that increases the interface between the electrode and electrolyte (Sun et al. 2018).

ZnSnS$_3$ nanosheets have also been synthesized, anchored on highly conductive rGO (ZnSnS$_3$@ rGO) sheets as a novel anode material for SIBs and LIBs (Jia et al. 2019). They were prepared through the aqueous reaction of Na$_2$SnO$_3$ and Zn(CH$_3$COO)$_2$, followed by solvothermal and annealing processes. It delivered an excellent Na- and Li-ion-storage performance with a large specific capacity, ultralong cycle life and high rate capability. In Na-ion cells, the ZnSnS$_3$@rGO nanocomposite displayed the capacity of 472.2 mA h g^{-1} at 100 mA g^{-1} and retained a specific capacity of 401.2 mA h g^{-1} after 200 cycles. And when used in Li-ion cells, it exhibited a capacity of 959.2 mA h g^{-1} at a current density of 100 mA g^{-1} and maintained a specific capacity of 551.3 mA h g^{-1} at a high current density of 1 A g^{-1} upon 500 cycles. These electrochemical results reveal that the uniform dispersion of the metal elements and the interconnected carbonaceous matrix can enhance the adaptability for large volume changes and provide a rapid ion/electron transport (Jia et al. 2019).

Nitrogen-doped Graphene-based Catalysts

The greater electronegativity of nitrogen (N = 3.5 and C = 2.55) and a smaller diameter than C, allow N atoms to be incorporated into the carbon network, promoting stronger interactions between the graphene structure doped with N and Li-ions, which significantly favors the Li-insertion. Hence, a general hydrothermal strategy for fabricating N-doped functionalized graphene (known as NGNS) for the cathode in LIBs was reported (Xiong et al. 2017). These doped-graphene materials presented high specific capacity/capacitance and better cycling stability in comparison to non-doped-graphene materials. NGNS was synthesized using urea as a nitrogen source and reducing agent. The results showed that NGNS-II with 9.26% at nitrogen content has a reversible capacity of 146 mAh g^{-1} after 1000 cycles at a current density of 1 A g^{-1}. Its high electrocatalytic performance is attributed to their unique 3D porous structure, high nitrogen doping and active oxygen-containing functional groups. The 3D porous structure act as transmission channels for enough electrolyte infiltration and facile ion diffusion. Furthermore, it was also demonstrated that a raised concentration of nitrogen doping can dramatically enhance the catalytic properties (Xiong et al. 2017).

The type and density of the heteroatoms in doped graphene must be carefully selected to achieve an optimal balance between the ion storage and the diffusion to obtain graphene electrode materials with good electrochemical activity. The encouraging results can accelerate the development of Li-/Na-ion batteries by improving these materials.

Capacitors and Supercapacitors

There is a growing interest in the use of capacitors for different applications, from portable devices to large-scale technology, due to their high power supply, fast charge/discharge performance and stable cycle life. Depending on the mechanism of energy storage (e.g., non-Faradaic or Faradaic reactions) and cell configurations (e.g., symmetric or asymmetric systems), Electrochemical Capacitors (ECs) can be classified as (i) Electric Double-Layer Capacitors (EDLCs), (ii) Pseudo-Capacitors (PCs), and (iii) hybrid energy storage systems (e.g., Hybrid Super-Capacitors (HSCs)) (Lim et al. 2016, Pérez del Pino et al. 2019, Yadav and Devi 2020).

Traditional ECs offer a high power density (> 10 kW kg^{-1}) and an excellent cycle life ($> 10,000$ cycles in EDLCs). Nevertheless, they expose a lower energy density (< 30 W h kg^{-1}) than Li-ion batteries (~ 150 W h kg^{-1}) (Lim et al. 2016). These capacitors require new constituent nanomaterials for reaching an upper power and longer cycling life than batteries (Pérez del Pino et al. 2019). Otherwise, supercapacitors (HSCs) combine the high power density of the capacitors and the superior energy density of a battery (Yadav and Devi 2020).

Although significant advances have been made, these devices present deficiencies such as low specific capacitance, low rate capability (the accompanying decrease of the electrical conductivity decay) in conductive polymers and poor electrical conductivity in metal oxides which are limiting their practical applications (Lim et al. 2016).

Carbonaceous materials have managed to provide a breakthrough for supercapacitors. Among them, graphene-based electrode materials outshine the others because of properties such as large theoretical capacitance, theoretical surface area, good conductivity, mechanical stability and flexibility (Rahmani and Habibi 2019).

In this regard, rGO can efficiently enhance storage capacity (Hota et al. 2020) and also displays the ability to act synergistically with the catalyst particles improving its electrochemical performance (Ates et al. 2018). Kumar Jha et al. produced a nanocomposite of Mn_3O_4 and rGO from GO by an *in situ* approach using $MnCl_2 \cdot 4H_2O$ as a reducing agent in an aqueous medium. The rGO was then extracted by dispersing the nanocomposite in HF. A rGO all-solid-state supercapacitor was then fabricated in a gel polymer electrolyte (Poly Vinyl Alcohol-sulfuric acid: PVA-H$_2$SO$_4$) showing a specific capacitance value of ~ 310 F g^{-1} at a current density of 1 A g^{-1} along with long durability (10,000 charge-discharge cycles). It exhibited remarkable mechanical stress (1,000 bending cycles with $> 80\%$ specific capacitance retention) as well as durability (30,000 cycles with 100% specific capacitance retention) (Jha et al. 2019).

Additionally, a nanoparticle/Poly Carbazole (rGO/Zn/PCz) nanocomposite has also been developed by an *in situ* polymerization method and reducing graphene oxide to rGO with sodium borohydride (Ates et al. 2018). This nanocomposite exhibited an increased capacitance (33.88 F g^{-1}) compared to Zn/PCz (19.05 F g^{-1}), PCz (12.57 F g^{-1}) and rGO (20.78 F g^{-1}) at the scan rate of 10 mV/s by a Cyclic Voltammetry (CV) technique. The improved capacitance results in high-power (P = 442.5 W kg^{-1}) and energy-storage (E = 1.66 W h kg^{-1}) capabilities of the rGO/Zn/PCz. It also retains 96.53% of the initial capacitance of rGO/Zn/PCz at a scan rate of 100 mV/s after 1,000 cycles due to the higher mechanical strength and larger surface area obtained by rGO composite (Ates et al. 2018).

Besides, it successfully synthesized amorphous MoS_2 on rGO (MoS$_2$-RGO) with two different sizes (50 nm and 5–7 nm) for supercapacitors (Hota et al. 2020). The capacitance retention of the material is 90% after 5,000 cycles. It was found that the specific capacitance was largely increased with decreasing size of the amorphous nanoparticle. MoS$_2$-RGO composite containing 50 nm MoS$_2$ has a value of specific capacitance of 270 F g^{-1}, while that of 5–7 nm exhibits a specific capacitance of 460 F g^{-1}. This difference can be attributed to the presence of a larger number of active sulfur edges for the ultra-small MoS$_2$ which improves the charge transport increasing the storage capacity of the system (Hota et al. 2020).

Nitrogen-doped Graphene-based Catalysts

N-doped rGO has been investigated as solid-state symmetric supercapacitor showing a high specific capacity (141.1 mA h g^{-1}) and high energy density (28.2 Wh kg^{-1}) with remarkable stability of 95,4% after 10,000 Galvanostatic Charging-Discharging (GCD) cycles (Mishra et al. 2019).

Figure 11.7a depicts the average specific capacity and average specific capacity retention evaluated by a symmetric supercapacitor offering outstanding stability of 93.2% after 8 hours voltage holding tests. Figure 11.7b shows the Electrochemical Impedance Spectra (EIS) of the supercapacitor before and after 10,000 GCD cycles and 8 hours voltage holding tests. The addition of nitrogen on the surface or the bulk of rGO can deliver an approach to enhance the pair of electrons in the host system. This effect is believed to harmonize the electron donor/acceptor behavior improving the electrochemical performance of the supercapacitor (Mishra et al. 2019).

Moreover, N-doped rGO in combination with NiO nanostructures were studied for energy storage devices (Pérez del Pino et al. 2019). It was fabricated through a Reactive Inverse Matrix-Assisted Pulsed Laser Evaporation technique (RIMAPLE), which is a promising strategy for controlling the deposition of flexible N-doped rGO. The last material presents high volumetric capacitance (up to 114 F cm^{-3} using imidazole as the precursor). The N-doped rGO films coated with NiO nanostructures expose less volumetric capacitance (up to 19 F cm^{-3} using imidazole as the precursor) but it presents higher capacitive contribution to charge storage and more stability on charge-discharge cycles (100% of capacitance retention and coulombic efficiency after 10,000 cycles) (Pérez del Pino et al. 2019).

At the end, nickel sulfide nanoflakes were anchored at N-doped graphene for supercapacitors using a hydrothermal method (Rahmani and Habibi 2019). It demonstrated a notable electrochemical performance with a specific capacitance of 1120 F g^{-1} at 1 A g^{-1} and cyclic stability of 82% for 3,000 charge/discharge cycles. This nanocomposite is benefited, on the one hand, by the high specific capacity of nickel sulfide that increases the energy density and, on the other, by the conductive network and mechanical cushion offered by graphene. Therefore, N-doped graphene can provide good channels for charge-transport and improve the electrochemical activity of NiS nanoflakes (Rahmani Habibi 2019).

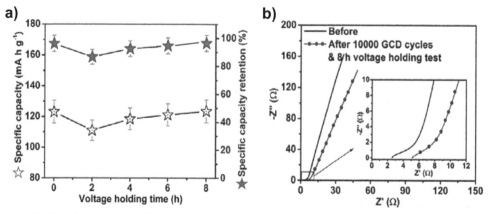

Figure 11.7: (a) N-doped RGO specific capacity and retention at 3.1 A g^{-1} and (b) EIS spectra before/after 10,000 GCD cycles and 8 hours voltage holding test. Adapted from (Mishra et al. 2019) with permission from Elsevier.

Photocatalytic Devices

Solar-driven water splitting is an attractive process to convert solar light into hydrogen for generating clean energy without contamination or by-products (Pérez del Pino et al. 2018). The design of photoactive electrodes capable of absorbing visible light is one of the important challenges for the development of Photo Electrochemical Cells (PECs) (Pan et al. 2019).

To accomplish solar-water-splitting, the PEC system uses photoactive semiconductors that are capable of absorbing sunlight to produce electrons and holes. The photoelectrochemical behavior depends on the presence of active sites on the surface of photocatalysts with a bandgap higher than 1.23 eV in order to photogenerate current at the lowest possible potential for water splitting (Eq. 22) (Kawrani et al. 2020). When the photocatalyst is under an adequate photonic excitation it performs electronic transitions and creates e⁻/H⁺ pairs. The charges are then separated before the electrons are excited from the valence band to the conduction band. The oxidation reaction involves the separation of the water molecule into H^+ (Eq. 20) and the reduction occurs when H^+ gains an electron giving rise to H_2 (Eq. 21) (Fig. 11.8). The reaction on the surface of the photocatalyst takes place when the reduction and oxidation potentials are above and below the levels of the conduction band and the valence band, respectively (Marlinda et al. 2020).

$$\text{Oxidation reaction:} \quad H_2O \xrightarrow{hv} \frac{1}{2}O_2 + 2H^+ \tag{20}$$

$$\text{Reduction reaction:} \quad 2H^+ + 2e^- \rightarrow H_2 + \frac{1}{2}O_2 \tag{21}$$

$$\text{Global reaction:} \quad H_2O \xrightarrow{hv} H_2 + \frac{1}{2}O_2 \tag{22}$$

The main problem for H_2 generation through water-splitting is to accomplish oxidation of the water with photogenerated holes (Pérez del Pino et al. 2018). Rapid recombination of electron-hole pairs photogenerated was achieved through the combination of semiconductor heterostructures and noble metal co-catalysts (e.g., Ru, Rh, Pd, Pt, Au and Ag). However, the limitation of noble metals due to their cost and unavailability prompted researchers to develop inexpensive metal-free co-catalysts to improve the photocatalytic performance (Elbakkay et al. 2018). The use of organic compounds whose oxidation is thermodynamically more favorable than water prevents the recombination of electron holes in the semiconductor favoring the water-splitting process (Pérez del Pino et al. 2018).

In regard to this, a promising alternative to transition metals and rare earth metals are metal-free carbon nanomaterials, which offer excellent properties such as high electrical conductivity, large surface area and adjustable morphology. Furthermore, they can be generated in abundant quantities by cheap, robust and non-toxic processes.

In this context, graphene-based materials have attractive qualities due to their extraordinary physicochemical properties as photo- and co-catalysts for the development of low-cost and effective photocatalysts (Pérez del Pino et al. 2018).

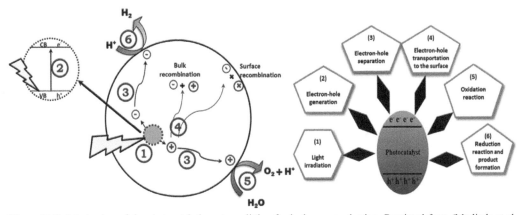

Figure 11.8: Mechanism of the photocatalytic water splitting for hydrogen production. Reprinted from (Marlinda et al. 2020) with permission from Elsevier.

Nevertheless, graphene sheets are unsuitable for immediate use in water-splitting due to their hydrophobicity. However, Graphene Oxide (GO) is hydrophilic, but to initiate charge transfer and create active sites for catalytic activity, it is necessary to restore the sp^2 structure and reduce GO to rGO (Marlinda et al. 2020).

rGO is also a semiconductor covalently bonded to oxygenated functional groups (e.g., carbonyl, hydroxyl, epoxy, carboxyl groups depending on the reduction degree of GO) and has a large surface area with a wide bandgap. By tailoring the degree of reduction, the bandgap can be modified to exploit the photocatalytic region of visible light (Akyüz et al. 2019).

In this regard, rGO has been widely investigated for the creation of innovative photocatalysts such as TiO_2-rGO and other composites (Akyüz et al. 2019, Lettieri et al. 2018, Subramanyam et al. 2019). Among the numerous semiconductors that can be found, TiO_2 has outstanding properties that make it an ideal photoanode for water splitting due to its chemical stability, large surface area, excellent charge transfer, resistance to photo-corrosion, abundance, low toxicity and low cost (Akyüz et al. 2019, Lettieri et al. 2018, Subramanyam et al. 2019). Nevertheless, TiO_2 presents a bandgap that limits its photoactivity to the UV region, in addition to undergoing rapid recombination of photogenerated electron-hole pairs (Elbakkay et al. 2018). In this case, it has been widely reported that graphene and rGO improve the efficiency of many photocatalytic semiconductors by rapidly capturing photogenerated electrons in the semiconductor conduction band (Akyüz et al. 2019, Lettieri et al. 2018, Subramanyam et al. 2019).

Therefore, different methods using rGO for enhancing the photocatalytic response of TiO_2 have been proposed. For instance, an oxygen vacancy TiO_{2-x}/reduced graphene oxide nanocomposite (TiO_{2-x}/rGO) was synthesized by facile hydrothermal reactions (Gao et al. 2018). This composite exhibited an improved photocatalytic activity compared to TiO_{2-x}, due to the lattice oxygen deficiency, which extends the TiO_{2-x} absorption to the visible light region; and the high mobility of electrons in graphene sheets which suppresses the recombination of the electron-hole pairs (Gao et al. 2018).

Noble metal-free core-shell nanoparticles of Graphene (G)-wrapped CdS and TiO_2 (CdS@G@TiO_2) were also fabricated by a simple hydrothermal method (Zubair et al. 2020). With the addition of graphene between the CdS and TiO_2 an increment in the absorption and intensity in the UV-Vis region was observed compared to CdS@TiO_2 as shown in Fig. 11.9a. The graphene layer in CdS@xG@TiO_2 was optimized by changing the content of Graphene Quantum Dots (GQDs) (where x = 20, 30, 40, 50, 60, 80 μL correspond to 100 mg mL^{-1} concentration of GQD solution). The most optimized sample (i.e., CdS@50G@TiO_2) produced 1510 μmol g^{-1} h^{-1} of H_2 from water splitting

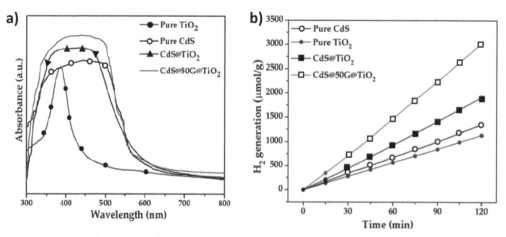

Figure 11.9: (a) UV-Vis Diffuse Reflectance Spectroscopy (DRS) spectra of pure TiO_2, pure CdS, CdS@TiO_2 and CdS@50G@TiO_2; and (b) photocatalytic activity for hydrogen production from water splitting of pure CdS, pure TiO_2, CdS@TiO_2 and CdS@50G@TiO_2. Adapted from (Zubair et al. 2020) with permission from MDPI.

under simulated solar light with air mass 1.5 global (AM 1.5 G) condition, which is ~ 2.7 times superior to pure TiO_2 and ~ 2.2 times superior to pure CdS (Fig. 11.9b). The increased catalytic activity is attributed to the improvement of visible light absorption and efficient charge separation and transference due to the incorporation of graphene between CdS and TiO_2. Graphene acts as an electron reservoir and a charging transmitter, in addition to providing a protective effect against CdS photo-corrosion by efficiently transferring holes from CdS to TiO_2 (Zubair et al. 2020).

Furthermore, heteroatom-doping of carbon materials is an effective strategy with great interest due to its improvement in the electrical conductivity and electronic structure of the photocatalyst. Doping atoms with n-type electron-donating or p-type electron-withdrawing groups modify the electron transfer properties of rGO leading to changes in its conductivity, photocatalytic activity and stability (Akyüz et al. 2019). For instance, replacing oxygenated functional groups by nitrogenous chemical groups on the edge of the GO sheet transforms GO into an n-type semiconductor (Pérez del Pino et al. 2018). The replacement of a carbon atom in the graphene network with a nitrogen atom induces the alteration of the optical and electronic properties. Thus, modifying graphene sheets to exhibit p- or n-type properties can lead to the photocatalytic activity for water splitting with enhanced efficacy at visible wavelengths (Pérez del Pino et al. 2018).

Nitrogen-doped Graphene-based Catalysts

A series of new core-shell cobalt particles loaded on a nitrogen-doped graphene catalyst (Co@CoO/NG) was prepared for H_2 and O_2 production through the water splitting process (Qiao et al. 2019). Co@CoO/NG was obtained by calcining ZIF-67 (Zeolitic Imidazolate Framework-67) using an *in situ* preparation process. It revealed an excellent performance (water oxidation apparent quantum efficiency of 10.22% at λ = 450 nm, O_2 production rate of 543,198 μmol g^{-1} h^{-1} and H_2 production rate of 330 μmol g^{-1} h^{-1}). N-doped graphene produces a synergistic effect with the Co@CoO core-shell structure since N-doping may increase light absorption, promote electron transfer and reduce the electron/hole recombination, which enhances the photocatalytic activity (Qiao et al. 2019).

Additionally, heteroatom (N, B and P) doped rGO-metal chalcogenide nanocomposites ($Cd_{0.60}Zn_{0.40}S$-rGO) have been synthesized using the solvothermal method (Akyüz et al. 2019). Doping with N, B and P increased the charge transfer capacity of the nanocomposites leading to improved photocatalytic and photoelectrochemical activity. N-doped $Cd_{0.60}Zn_{0.40}S$-rGO photocatalyst exhibited the best photocatalytic hydrogen production rate (1114 μmol h^{-1} g^{-1}) with 17.8% of apparent quantum efficiency in the photocatalytic system and it was 1.5 times higher than that of the undoped photocatalyst (Akyüz et al. 2019). It was also tested on a PEC system showing a current density of 0.92 mA cm^{-2} and exhibiting a PEC hydrogen production three times higher than $Cd_{0.60}Zn_{0.40}S$-rGO. The obtained results exposed that heteroatom doping of rGO leads to promising materials for renewable hydrogen production (Akyüz et al. 2019).

Sulfur Doped Graphene-based Catalysts

Sulfur-doped titanium oxide on the surface of sulfur-doped reduced graphene oxide nanocomposites (S-TiO_2/S-rGO) was developed as an efficient photoanode for PEC water splitting using dimethyl sulfoxide (DMSO), GO and tetrabutyl orthotitanate (TBOT) as starting materials (Elbakkay et al. 2018). In order to investigate the influence of rGO, the weight of GO was varied (0, 50, 100, 200 mg) and reduced with DMSO. The obtained samples were known as S-TiO_2, S-TiO_2/S-rGO(0.05), S-TiO_2/S-rGO(0.1) and S-TiO_2/S-rGO(0.2), respectively (Elbakkay et al. 2018).

The nanocomposite displayed improved photoelectrochemical performance and long-term stability. The UV-Vis diffuse reflectance spectra of different samples (TiO_2, S-TiO_2, TiO_2/rGO(0.1), S-TiO_2/S-rGO(0.05), S-TiO_2/S-rGO(0.1) and sulfur-doped titanium oxide on the surface of rGO (S-TiO_2/rGO(0.1)) are given in Fig. 11.10. It can be observed that TiO_2 and S-TiO_2 nanoparticles suffer increased reflectance in the visible range while TiO_2/rGO(0.1), S-TiO_2/S-rGO(0.05), S-TiO_2/S-rGO(0.1) and S-TiO_2/S-rGO(0.2) nanocomposites revealed a diminution in reflectance, which

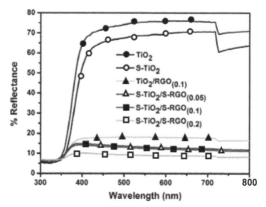

Figure 11.10: UV–vis diffuse reflectance spectra of TiO_2, S-TiO_2 nanoparticles, TiO_2/rGO(0.1), S-TiO_2/S-rGO(0.05), S-TiO_2/S-rGO(0.1) and S-TiO_2/S-rGO(0.2) nanocomposites. Adapted from (Elbakkay et al. 2018) with permission from Elsevier.

may be attributed to the formation of Ti-O-C bonds and facilitated charge transfer from TiO_2 to the rGO or the S-rGO (Elbakkay et al. 2018).

Finally, S-TiO_2/S-rGO(0.2) nanocomposite displayed a photocurrent density of 3.36 mA cm^{-2} at 1 V (vs Ag/AgCl) which is three times higher than bare synthesized TiO_2. The observed improvement in photocurrent density is attributed to the higher rate of separation of photogenerated electrons/holes, in addition to the efficient collection of visible light resulting in a successful combination of S-TiO_2 and S-rGO for its use as a photoanode for PEC water splitting (Elbakkay et al. 2018).

Conclusion

In this chapter, the general concept of (photo)electrochemical energy storage and conversion devices was described, with special attention to graphene-based catalysts. To solve the principal catalytic, durability and cost problems at the electrodes of these technologies, a fundamental and applied study of all involved (photo)electrochemical reactions on graphene-based materials in a wide pH range was considered. All information reviewed and scrutinized here may help to improve the fabrication of novel catalysts in order to decrease the cost and enhance the performance of (photo)electrochemical energy storage and conversion systems.

Acknowledgements

This work has been supported by the Spanish Ministry of Science (MICINN) under project ENE2017-83976-C2-2-R (co-founded FEDER). S.J.M. thanks the ACIISI for her PhD fellowships and G.G. acknowledges the 'Viera y Clavijo' program (ACIISI & ULL) for financial support.

References

Abdel Hameed, R. M., A. E. Fahim and N. K. Allam. 2020. Tin oxide as a promoter for copper@palladium nanoparticles on graphene sheets during ethanol electro-oxidation in NaOH solution. J. Mol. Liq. 297: 111816. https://doi.org/10.1016/j.molliq.2019.111816.

Akyüz, D., R. M. Zunain Ayaz, S. Yılmaz, Ö. Uğuz, C. Sarıoğlu, F. Karaca et al. 2019. Metal chalcogenide based photocatalysts decorated with heteroatom doped reduced graphene oxide for photocatalytic and photoelectrochemical hydrogen production. Int. J. Hydrogen Energy 44(34): 18836–18847. https://doi.org/10.1016/j.ijhydene.2019.04.049.

Ali Tahir, A., H. Ullah, P. Sudhagar, M. Asri Mat Teridi, A. Devadoss and S. Sundaram. 2016. The application of graphene and its derivatives to energy conversion, storage, and environmental and biosensing devices. Chem. Rec. 16: 1591–1634. https://doi.org/10.1002/tcr.201500279.

Aricò, A. S., S. Srinivasan and V. Antonucci. 2001. DMFCs: From fundamental aspects to technology development. Fuel Cells 1(2): 133–161. https://doi.org/10.1002/1615-6854(200107)1:2<133::AID-FUCE133>3.0.CO;2-5.

Arteaga, G., L. M. Rivera-Gavidia, S. J. Martínez, R. Rizo, E. Pastor and G. García. 2019. Methanol oxidation on graphenic-supported platinum catalysts. Surfaces 2(1): 16–31. https://doi.org/10.3390/surfaces2010002.

Ates, M., S. Caliskan and E. Özten. 2018. Supercapacitor study of reduced graphene oxide/Zn nanoparticle/ polycarbazole electrode active materials and equivalent circuit models. J. Solid State Electrochem. 22(10): 3261–3271. https://doi.org/10.1007/s10008-018-4039-3.

Azadmanjiri, J., V. K. Srivastava, P. Kumar, M. Nikzad, J. Wang and A. Yu. 2018. Two- and three-dimensional graphene-based hybrid composites for advanced energy storage and conversion devices. J. Mater. Chem. A 6(3): 702–734. https://doi.org/10.1039/c7ta08748a.

Bikkarolla, S. K. and P. Papakonstantinou. 2015. $CuCo_2O_4$ nanoparticles on nitrogenated graphene as highly efficient oxygen evolution catalyst. J. Power Sources 281: 243–251. https://doi.org/10.1016/j.jpowsour.2015.01.192.

Boulaghi, M., H. Ghafouri Taleghani, M. Soleimani Lashkenari and M. Ghorbani. 2018. Platinum-palladium nanoparticles-loaded on N-doped graphene oxide/polypyrrole framework as a high performance electrode in ethanol oxidation reaction. Int. J. Hydrogen Energy 43(32): 15164–15175. https://doi.org/10.1016/j. ijhydene.2018.06.092.

Brownson, D. A. C. and C. E. Banks. 2014. The Handbook of Graphene Electrochemistry. Edited by Springer. https:// doi.org/10.1007/978-1-4471-6428-9.

Bulbula, S. T., Y. Lu, Y. Dong and X. Y. Yang. 2018. Hierarchically porous graphene for batteries and supercapacitors. New J. Chem. 42(8): 5634–5655. https://doi.org/10.1039/c8nj00652k.

Chakrabarty, S., A. Mukherjee, W. N. Su and S. Basu. 2019. Improved bi-functional ORR and OER catalytic activity of reduced graphene oxide supported $ZnCo_2O_4$ microsphere. Int. J. Hydrogen Energy 44(3): 1565–1578. https:// doi.org/10.1016/j.ijhydene.2018.11.163.

Chanda, D., J. Hnát, A. S. Dobrota, I. A. Pašti, M. Paidar and K. Bouzek. 2015. The effect of surface modification by reduced graphene oxide on the electrocatalytic activity of nickel towards the hydrogen evolution reaction. Phys. Chem. Chem. Phys. 17(40): 26864–26874. https://doi.org/10.1039/c5cp04238k.

Chao, D. 2019. Graphene network scaffolded flexible electrodes: from lithium to sodium ion batteries. Edited by Springer. https://doi.org/10.1007/978-981-13-3080-3.

Chen, K. S., R. Xu, N. S. Luu, E. B. Secor, K. Hamamoto, Q. Li et al. 2017. Comprehensive enhancement of nanostructured lithium-ion battery cathode materials via conformal graphene dispersion. Nano Lett. 17(4): 2539–2546. https://doi.org/10.1021/acs.nanolett.7b00274.

Das, R. K., Y. Wang, S. V. Vasilyeva, E. Donoghue, I. Pucher, G. Kamenov et al. 2014. Extraordinary hydrogen evolution and oxidation reaction activity from carbon nanotubes and graphitic carbons. ACS Nano 8(8): 8447–8456. https://doi.org/10.1021/nn5030225.

Dasari, B. L., J. M. Nouri, D. Brabazon and S. Naher. 2017. Graphene and derivatives—Synthesis techniques, properties and their energy applications. Energy 140: 766–778. https://doi.org/10.1016/j.energy.2017.08.048.

Debata, S., S. Patra, S. Banerjee, R. Madhuri and P. K. Sharma. 2018. Controlled hydrothermal synthesis of graphene supported $NiCo_2O_4$ coral-like nanostructures: An efficient electrocatalyst for overall water splitting. Appl. Surf. Sci. 449: 203–212. https://doi.org/10.1016/j.apsusc.2018.01.302.

Deng, J., P. Ren, D. Deng and X. Bao. 2015. Enhanced electron penetration through an ultrathin graphene layer for highly efficient catalysis of the hydrogen evolution reaction. Angew. Chem. Int. Ed. 54(7): 2100–2104. https:// doi.org/10.1002/anie.201409524.

Duan, J., S. Chen, M. Jaroniec and S. Z. Qiao. 2015. Heteroatom-doped graphene-based materials for energy-relevant electrocatalytic processes. ACS Catal. 5(9): 5207–5234. https://doi.org/10.1021/acscatal.5b00991.

Duca, M. and M. T. M. Koper. 2020. Fundamental Aspects of Electrocatalysis. InK. Wandelt, Surface and Interface Science (pp. 773–890). Edited by Wiley. https://doi.org/10.1002/9783527680603.ch59.

El-Sawy, A. M., I. M. Mosa, D. Su, C. J. Guild, S. Khalid, R. Joesten et al. 2016. Controlling the active sites of sulfur-doped carbon nanotube-graphene nanolobes for highly efficient oxygen evolution and reduction catalysis. Adv. Energy Mater. 6(5): 1501966. https://doi.org/10.1002/aenm.201501966.

Elbakkay, M. H., W. M. A. El Rouby, S. I. El-Dek and A. A. Farghali. 2018. S-TiO_2/S-reduced graphene oxide for enhanced photoelectrochemical water splitting. Appl. Surf. Sci. 439: 1088–1102. https://doi.org/10.1016/j. apsusc.2018.01.070.

Fan, T., G. Zhang, L. Jian, I. Murtaza, H. Meng, Y. Liu et al. 2019. Facile synthesis of defect-rich nitrogen and sulfur Co-doped graphene quantum dots as metal-free electrocatalyst for the oxygen reduction reaction. J. Alloy Compd. 792: 844–850. https://doi.org/10.1016/j.jallcom.2019.04.097.

Flórez-Montaño, J., A. Calderón-Cárdenas, W. Lizcano-Valbuena, J. L. Rodríguez and E. Pastor. 2016. Ni@Pt nanodisks with low Pt content supported on reduced graphene oxide for methanol electrooxidation in alkaline media. Int. J. Hydrogen Energy 41(43): 19799–19809. https://doi.org/10.1016/j.ijhydene.2016.06.166.

Flórez-Montaño, J., G. García, O. Guillén-Villafuerte, J. L. Rodríguez, G. A. Planes and E. Pastor. 2016. Mechanism of ethanol electrooxidation on mesoporous Pt electrode in acidic medium studied by a novel electrochemical mass spectrometry set-up. Electrochim. Acta 209: 121–131. https://doi.org/10.1016/j.electacta.2016.05.070.

Gangadharan, P. K., S. M. Unni, N. Kumar, P. Ghosh and S. Kurungot. 2017. Nitrogen-doped graphene with a three-dimensional architecture assisted by carbon nitride tetrapods as an efficient metal-free electrocatalyst for hydrogen evolution. ChemElectroChem. 4(10): 2643–2652. https://doi.org/10.1002/celc.201700479.

Gao, G., Q. Zhu, H. Chong, J. Zheng, C. Fan and G. Li. 2018. Synthesis of biphasic defective TiO_2–x/reduced graphene oxide nanocomposites with highly enhanced photocatalytic activity. Chem. Res. Chinese U. 34(2): 158–163. https://doi.org/10.1007/s40242-018-7369-x.

García, G. and M. T. M. Koper. 2011. Carbon monoxide oxidation on Pt single crystal electrodes: Understanding the catalysis for low temperature fuel cells. ChemPhysChem. 12(11): 2064–2072. https://doi.org/10.1002/cphc.201100247.

García, G., F. Alcaide and E. Pastor. 2020. Graphene for the electrocatalysts used for fuel cells and electrolyzers. *In*: S. Sadjadi [ed.]. Emerging Carbon Materials for Cataysis. Edited by Elsevier.

Gasteiger, H. A., S. S. Kocha, B. Sompalli and F. T. Wagner. 2005. Activity benchmarks and requirements for Pt, Pt-alloy, and non-Pt oxygen reduction catalysts for PEMFCs. Appl. Catal. B-Environ. 56(1-2 SPEC. ISS.): 9–35. https://doi.org/10.1016/j.apcatb.2004.06.021.

Gautam, J., T. D. Thanh, K. Maiti, N. H. Kim and J. H. Lee. 2018. Highly efficient electrocatalyst of N-doped graphene-encapsulated cobalt-iron carbides towards oxygen reduction reaction. Carbon 137: 358–367. https://doi.org/10.1016/j.carbon.2018.05.042.

Guillén-Villafuerte, O., G. García, R. Guil-López, E. Nieto, J. L. Rodríguez, J. L. G. Fierro et al. 2013. Carbon monoxide and methanol oxidations on $Pt/X@MoO_3/C$ (X = Mo_2C, MoO_2, MoO) electrodes at different temperatures. J. Power Sources 231: 163–172. https://doi.org/10.1016/j.jpowsour.2012.12.099.

Guruprasad, K., T. Maiyalagan and S. Shanmugam. 2019. Phosphorus doped MoS_2 nanosheet promoted with nitrogen, sulfur dual doped reduced graphene oxide as an effective electrocatalyst for hydrogen evolution reaction. ACS Appl. Energy Mater. 2(9): 6184–6194. https://doi.org/10.1021/acsaem.9b00629.

Hanifah, M. F. R., J. Jaafar, M. H. D. Othman, A. F. Ismail, M. A. Rahman, N. Yusof et al. 2019. One-pot synthesis of efficient reduced graphene oxide supported binary Pt-Pd alloy nanoparticles as superior electro-catalyst and its electro-catalytic performance toward methanol electro-oxidation reaction in direct methanol fuel cell. J. Alloys Compd. 793: 232–246. https://doi.org/10.1016/j.jallcom.2019.04.114.

Heidary, H., M. Jafar Kermani and N. Khajeh-Hosseini-Dalasm. 2016. Performance analysis of PEM fuel cells cathode catalyst layer at various operating conditions. Int. J. Hydrogen Energy 41(47): 22274–22284. https://doi.org/10.1016/j.ijhydene.2016.08.178.

Hota, P., M. Miah, S. Bose, D. Dinda, U. K. Ghorai, Y. K. Su et al. 2020. Ultra-small amorphous MoS_2 decorated reduced graphene oxide for supercapacitor application. J. Mater. Sci. Technol. 40: 196–203. https://doi.org/10.1016/j.jmst.2019.08.032.

Hu, C., L. Song, Z. Zhang, N. Chen, Z. Feng and L. Qu. 2015. Tailored graphene systems for unconventional applications in energy conversion and storage devices. Energy Environ. Sci. 8(1): 31–54. https://doi.org/10.1039/c4ee02594f.

Huang, H., M. Yan, C. Yang, H. He, Q. Jiang, L. Yang et al. 2019. Graphene nanoarchitectonics: recent advances in graphene-based electrocatalysts for hydrogen evolution reaction. Adv. Mater. 31(48): 1903415. https://doi.org/10.1002/adma.201903415.

Martínez Huerta, M. V. and G. García. 2017. Fabrication of electro-catalytic nano-particles and applications to proton exchange membrane fuel cells. *In*: D. Y. C. Leung and J. Xuan [eds.]. Micro & Nano-Engineering of Fuel Cells. Edited by CRC Press.

Iqbal, M. Z., A. U. Rehman and S. Siddique. 2019. Prospects and challenges of graphene based fuel cells. J. Energy Chem. 39: 217–234. https://doi.org/10.1016/j.jechem.2019.02.009.

Isao, A. 2009. Alkaline water-electrolysis. *In*: T. Otha [ed.]. Energy Carriers and Conversion Systems with Emphasis On Hydrogen - Vol I. Edited by Encyclopedia of Life Support Systems (EOLSS).

Ito, Y., W. Cong, T. Fujita, Z. Tang and M. Chen. 2015. High catalytic activity of nitrogen and sulfur co-doped nanoporous graphene in the hydrogen evolution reaction. Angew. Chem. Int. Ed. 54(7): 2131–2136. https://doi.org/10.1002/anie.201410050.

Jamesh, M. I. and X. Sun. 2018. Recent progress on earth abundant electrocatalysts for oxygen evolution reaction (OER) in alkaline medium to achieve efficient water splitting—A review. J. Power Sources 400(July): 31–68. https://doi.org/10.1016/j.jpowsour.2018.07.125.

Jha, P. K., V. Kashyap, K. Gupta, V. Kumar, A. K. Debnath, D. Roy et al. 2019. *In situ* generated Mn_3O_4-reduced graphene oxide nanocomposite for oxygen reduction reaction and isolated reduced graphene oxide for supercapacitor applications. Carbon 154: 285–291. https://doi.org/10.1016/j.carbon.2019.08.012.

Jia, H., M. Dirican, N. Sun, C. Chen, C. Yan, P. Zhu et al. 2019. Advanced ZnSnS$_3$@rGO anode material for superior sodium-ion and lithium-ion storage with ultralong cycle life. ChemElectroChem. 6(4): 1183–1191. https://doi.org/10.1002/celc.201801333.

Jia, Y., L. Zhang, G. Gao, H. Chen, B. Wang, J. Zhou et al. 2017. A heterostructure coupling of exfoliated Ni–Fe hydroxide nanosheet and defective graphene as a bifunctional electrocatalyst for overall water splitting. Adv. Mater. 29(17): 1700017. https://doi.org/10.1002/adma.201700017.

Jiang, H., Y. Zhu, Y. Su, Y. Yao, Y. Liu, X. Yang et al. 2015. Highly dual-doped multilayer nanoporous graphene: Efficient metal-free electrocatalysts for the hydrogen evolution reaction. J. Mater. Chem. A 3(24): 12642–12645. https://doi.org/10.1039/c5ta02792f.

Jorissen, L. and J. Garche. 2015. Polymer electrolyte membrane fuel cells. *In*: J. Töpler and J. Lehmann [eds.]. Hydrogen and Fuel Cell: Technologies and Market Perspectives. Edited by Springer. https://doi.org/10.1007/978-3-662-44972-1.

Kawrani, S., M. Boulos, M. F. Bekheet, R. Viter, A. A. Nada, W. Riedel et al. 2020. Segregation of copper oxide on calcium copper titanate surface induced by graphene oxide for water splitting applications. Appl. Surf. Sci. 516: 146051–146062. https://doi.org/10.1016/j.apsusc.2020.146051.

Kiyani, R., S. Rowshanzamir and M. J. Parnian. 2016. Nitrogen doped graphene supported palladium-cobalt as a promising catalyst for methanol oxidation reaction: Synthesis, characterization and electrocatalytic performance. Energy 113: 1162–1173. https://doi.org/10.1016/j.energy.2016.07.143.

Kong, X., Q. Liu, D. Chen and G. Chen. 2017. Identifying the active sites on N-doped graphene toward oxygen evolution reaction. ChemCatChem. 9(5): 846–852. https://doi.org/10.1002/cctc.201601268.

Koper, M. T. M. 2013. Theory of multiple proton-electron transfer reactions and its implications for electrocatalysis. Chem. Sci. 4(7): 2710–2723. https://doi.org/10.1039/c3sc50205h.

Lettieri, S., V. Gargiulo, D. K. Pallotti, G. Vitiello, P. Maddalena, M. Alfè et al. 2018. Evidencing opposite charge-transfer processes at TiO$_2$/graphene-related materials interface through combined EPR, photoluminescence and photocatalysis assessment. Catal. Today 315: 19–30. https://doi.org/10.1016/j.cattod.2018.01.022.

Li, D., B. Ren, Q. Jin, H. Cui and C. Wang. 2018. Nitrogen-doped, oxygen-functionalized, edge- and defect-rich vertically aligned graphene for highly enhanced oxygen evolution reaction. J. Mater. Chem. A 6(5): 2176–2183. https://doi.org/10.1039/c7ta07896j.

Li, X., X. Duan, C. Han, X. Fan, Y. Li, F. Zhang et al. 2019. Chemical activation of nitrogen and sulfur co-doped graphene as defect-rich carbocatalyst for electrochemical water splitting. Carbon 148: 540–549. https://doi.org/10.1016/j.carbon.2019.04.021.

Lim, E., C. Jo and J. Lee. 2016. A mini review of designed mesoporous materials for energy-storage applications: From electric double-layer capacitors to hybrid supercapacitors. Nanoscale 8(15): 7827–7833. https://doi.org/10.1039/c6nr00796a.

López-Urías, F., R. Lv, H. Terrones and M. Terrones. 2013. Doped graphene: theory, synthesis, characterization, and applications. *In*: D. E. Jiang and Z. Chen [eds.]. Graphene Chemistry: Theoretical Perspectives. Edited by John Wiley & Sons, Ltd. https://doi.org/10.1002/9781118691281.

Luis-Sunga, M., L. Regent, E. Pastor and G. García. 2020. Non-precious metal graphene-based catalysts for hydrogen evolution reaction. Electrochem. 1(2): 75–86. https://doi.org/10.3390/electrochem1020008.

Luo, L., C. Fu, F. Yang, X. Li, F. Jiang, Y. Guo et al. 2020. Composition-graded Cu-Pd nanospheres with Ir-doped surfaces on N-doped porous graphene for highly efficient ethanol electro-oxidation in alkaline media. ACS Catal. 10(2): 1171–1184. https://doi.org/10.1021/acscatal.9b05292.

Lv, W., Z. Li, Y. Deng, Q. H. Yang and F. Kang. 2016. Graphene-based materials for electrochemical energy storage devices: Opportunities and challenges. Energy Storage Mater. 2: 107–138. https://doi.org/10.1016/j.ensm.2015.10.002.

Marlinda, A. R., N. Yusoff, S. Sagadevan and M. R. Johan. 2020. Recent developments in reduced graphene oxide nanocomposites for photoelectrochemical water-splitting applications. Int. J. Hydrogen Energy 45(21): 11976–11994. https://doi.org/10.1016/j.ijhydene.2020.02.096.

Mench, M. M. 2008. Fuel Cell Engines. Edited by John Wiley & Sons, Inc. https://doi.org/10.1002/9780470209769.

Mishra, R. K., G. J. Choi, Y. Sohn, S. H. Lee and J. S. Gwag. 2019. Nitrogen-doped reduced graphene oxide as excellent electrode materials for high performance energy storage device applications. Mater. Lett. 245: 192–195. https://doi.org/10.1016/j.matlet.2019.03.010.

Morales-Acosta, D., J. D. Flores-Oyervides, J. A. Rodríguez-González, N. M. Sánchez-Padilla, R. Benavides, S. Fernández-Tavizón et al. 2019. Comparative methods for reduction and sulfonation of graphene oxide for fuel cell electrode applications. Int. J. Hydrogen Energy 44(24): 12356–12364. https://doi.org/10.1016/j.ijhydene.2019.02.091.

Pan, Z., G. Zhang and X. Wang. 2019. Polymeric carbon nitride/reduced graphene oxide/Fe$_2$O$_3$: All-solid-state Z-scheme system for photocatalytic overall water splitting. Angew. Chem. Int. Ed. 58(21): 7102–7106. https://doi.org/10.1002/anie.201902634.

Pérez-Rodríguez, S., G. García, M. J. Lázaro and E. Pastor. 2018. DEMS strategy for the determination of the difference in surface acidity of carbon materials. Electrochem. Commun. 90: 87–90. https://doi.org/10.1016/j.elecom.2018.04.014.

Pérez del Pino, A., A. González-Campo, S. Giraldo, J. Peral, E. György, C. Logofatu et al. 2018. Synthesis of graphene-based photocatalysts for water splitting by laser-induced doping with ionic liquids. Carbon 130: 48–58. https://doi.org/10.1016/j.carbon.2017.12.116.

Pérez del Pino, Á., M. A. Ramadan, P. Garcia Lebière, R. Ivan, C. Logofatu, I. Yousef et al. 2019. Fabrication of graphene-based electrochemical capacitors through reactive inverse matrix assisted pulsed laser evaporation. Appl. Surf. Sci. 484: 245–256. https://doi.org/10.1016/j.apsusc.2019.04.127.

Qiao, S., J. Guo, D. Wang, L. Zhang, A. Hassan, T. Chen et al. 2020. Core-shell cobalt particles Co@CoO loaded on nitrogen-doped graphene for photocatalytic water-splitting. Int. J. Hydrogen Energy 45(3): 1629–1639. https://doi.org/10.1016/j.ijhydene.2019.10.157.

Qin, X., Y. Huang, Y. Shen, M. Zhao and X. Gao. 2019. Porous 3D flower-like bismuth silicate@nitrogen-doped graphene nanomaterial as high-efficient catalyst for fuel cell cathode. Ceram. Int. 45(18): 24515–24527. https://doi.org/10.1016/j.ceramint.2019.08.179.

Qin, X., Y. Huang, K. Wang, T. Xu, Y. Wang, M. Wang et al. 2019. Highly efficient oxygen reduction reaction catalyst derived from Fe/Ni mixed-metal-organic frameworks for application of fuel cell cathode. Ind. Eng. Chem. Res. 58(24): 10224–10237. https://doi.org/10.1021/acs.iecr.9b01412.

Rahmani, K. and B. Habibi. 2019. NiCo alloy nanoparticles electrodeposited on an electrochemically reduced nitrogen-doped graphene oxide/carbon-ceramic electrode: A low cost electrocatalyst towards methanol and ethanol oxidation. RSC Adv. 9(58): 34050–34064. https://doi.org/10.1039/c9ra06290d.

Rivera, L. M., S. Fajardo, M. Arévalo, C. del, G. García and E. Pastor. 2017. S- and N-doped graphene nanomaterials for the oxygen reduction reaction. Catalysts 7(9): 278. https://doi.org/10.3390/catal7090278.

Roca-Ayats, M., S. Pérez-Rodríguez, G. García and E. Pastor. 2018. Recent advances on electrocatalysts for PEM and AEM fuel cells. *In*: F. J. Rodríguez-Varela and T. W. Napporn [eds.]. Advanced Electrocatalysts for Low-Temperature Fuel Cells. Edited by Springer International Publishing. https://doi.org/10.1007/978-3-319-99019-4_2.

Roca-Ayats, M., K. L. Yeung, M. Hernández-Caricol, W. Y. Chen, R. Deng, J. L. G. Fierro et al. 2019. Titanium carbonitride–graphene composites assembled with organic linkers as electrocatalytic supports for methanol oxidation reaction. Catal. Today. https://doi.org/10.1016/j.cattod.2019.06.079.

Rocha, R. P., A. G. Gonçalves, L. M. Pastrana-Martínez, B. C. Bordoni, O. S. G. P. Soares, J. J. M. Órfão et al. 2015. Nitrogen-doped graphene-based materials for advanced oxidation processes. Catal. Today 249: 192–198. https://doi.org/10.1016/j.cattod.2014.10.046.

Santos, J. R. N., D. S. S. Viégas, I. C. B. Alves, A. D. Rabelo, W. M. Costa, E. P. Marques et al. 2019. Reduced graphene oxide-supported nickel(ii)-bis(1,10-phenanthroline) complex as a highly active electrocatalyst for ethanol oxidation reaction. Electrocatalysis 10(5): 560–572. https://doi.org/10.1007/s12678-019-00539-0.

Sarwar, E., T. Noor, N. Iqbal, Y. Mehmood, S. Ahmed and R. Mehek. 2018. Effect of Co-Ni ratio in graphene based bimetallic electro-catalyst for methanol oxidation. Fuel Cells 18(2): 189–194. https://doi.org/10.1002/fuce.201700143.

Sathe, B. R., X. Zou and T. Asefa. 2014. Metal-free B-doped graphene with efficient electrocatalytic activity for hydrogen evolution reaction. Catal. Sci Technol. 4(7): 2023–2030. https://doi.org/10.1039/c4cy00075g.

Sheng, W., A. P. Bivens, M. Myint, Z. Zhuang, R. V. Forest, Q. Fang et al. 2014. Non-precious metal electrocatalysts with high activity for hydrogen oxidation reaction in alkaline electrolytes. Energy Environ. Sci. 7(5): 1719–1724. https://doi.org/10.1039/c3ee43899f.

Sibul, R., E. Kibena-Põldsepp, S. Ratso, M. Kook, M. T. Sougrati, M. Käärik et al. 2020. Iron- and nitrogen-doped graphene-based catalysts for fuel cell applications. ChemElectroChem. 7(7): 1739–1747. https://doi.org/10.1002/celc.202000011.

Subramani, S. M. and K. Gantigiah. 2019. Deposition of cobalt nanoparticles on reduced graphene oxide and the electrocatalytic activity for methanol and ethanol oxidation. Mater. Res. Express 6(12): 124001. https://doi.org/10.1088/2053-1591/ab5320.

Subramanyam, P., T. Vinodkumar, D. Nepak, M. Deepa and C. Subrahmanyam. 2019. Mo-doped BiVO$_4$@reduced graphene oxide composite as an efficient photoanode for photoelectrochemical water splitting. Catal. Today 325: 73–80. https://doi.org/10.1016/j.cattod.2018.07.006.

Sui, S., X. Wang, X. Zhou, Y. Su, S. Riffat and C. Liu. Jun. 2017. A comprehensive review of Pt electrocatalysts for the oxygen reduction reaction: Nanostructure, activity, mechanism and carbon support in PEM fuel cells. J. Mater. Chem. A 5(5): 1808–1825. https://doi.org/10.1039/C6TA08580F.

Sun, N., Y. Guan, Y. T. Liu, Q. Zhu, J. Shen, H. Liu et al. 2018. Facile synthesis of free-standing, flexible hard carbon anode for high-performance sodium ion batteries using graphene as a multi-functional binder. Carbon 137: 475–483. https://doi.org/10.1016/j.carbon.2018.05.056.

Tian, Y., Z. Wei, X. Wang, S. Peng, X. Zhang and W. Liu. 2017. Plasma-etched, S-doped graphene for effective hydrogen evolution reaction. Int. J Hydrogen Energy 42(7): 4184–4192. https://doi.org/10.1016/j.ijhydene.2016.09.142.

Tsang, C. H. A., H. Huang, J. Xuan, H. Wang and D. Y. C. Leung. 2020. Graphene materials in green energy applications: Recent development and future perspective. Renew. Sust. Energy Rev. 120: 109656. https://doi.org/10.1016/j.rser.2019.109656.

Wang, Jiangli, X. You, C. Xiao, X. Zhang, S. Cai, W. Jiang et al. 2019. Small-sized Pt nanoparticles supported on hybrid structures of MoS_2 nanoflowers/graphene nanosheets: Highly active composite catalyst toward efficient ethanol oxidation reaction studied by in situ electrochemical NMR spectroscopy. Appl. Catal. B: Environ. 259(March): 118060. https://doi.org/10.1016/j.apcatb.2019.118060.

Wang, Jingshi, Z. Shen and M. Yi. 2019. Liquid-exfoliated graphene as highly efficient conductive additives for cathodes in lithium ion batteries. Carbon 153: 156–163. https://doi.org/10.1016/j.carbon.2019.07.008.

Wang, Q. Y. and Y. H. Ding. 2016. Mechanism of methanol oxidation on graphene-supported Pt: Defect is better or not? Electrochim. Acta 216: 140–146. https://doi.org/10.1016/j.electacta.2016.08.052.

Wang, Z. and C. J. Liu. 2015. Preparation and application of iron oxide/graphene based composites for electrochemical energy storage and energy conversion devices: Current status and perspective. Nano Energy 11: 277–293. https://doi.org/10.1016/j.nanoen.2014.10.022.

Wengenmayr, R. and T. Bührke. 2013. *In*: R. Wengenmayr and T. Bührke [eds.]. Renewable energy: Sustainable energy concepts for the energy change. Edited by John Wiley & Sons. https://www.wiley.com/en-us/Renewable+Energy%3A+Sustainable+Energy+Concepts+for+the+Energy+Change%2C+2nd+Edition-p-9783527671366.

Xiao, Z., X. Huang, L. Xu, D. Yan, J. Huo and S. Wang. 2016. Edge-selectively phosphorus-doped few-layer graphene as an efficient metal-free electrocatalyst for the oxygen evolution reaction. Chem. Commun. 52(88): 13008–13011. https://doi.org/10.1039/c6cc07217h.

Xiong, D., X. Li, Z. Bai, H. Shan, L. Fan, C. Wu et al. 2017. Superior cathode performance of nitrogen-doped graphene frameworks for lithium ion batteries. ACS Appl. Mater. Inter. 9(12): 10643–10651. https://doi.org/10.1021/acsami.6b15872.

Yadav, S. and A. Devi. 2020. Recent advancements of metal oxides/Nitrogen-doped graphene nanocomposites for supercapacitor electrode materials. J. Energy Storage 30: 101486. https://doi.org/10.1016/j.est.2020.101486.

Yan, H., C. Tian, L. Wang, A. Wu, M. Meng, L. Zhao et al. 2015. Phosphorus-modified tungsten nitride/reduced graphene oxide as a high-performance, non-noble-metal electrocatalyst for the hydrogen evolution reaction. Angew. Chem. Int. Ed. 54(21): 6325–6329. https://doi.org/10.1002/anie.201501419.

Yan, W., X. Cao, J. Tian, C. Jin, K. Ke and R. Yang. 2016. Nitrogen/sulfur dual-doped 3D reduced graphene oxide networks-supported $CoFe_2O_4$ with enhanced electrocatalytic activities for oxygen reduction and evolution reactions. Carbon 99: 195–202. https://doi.org/10.1016/j.carbon.2015.12.011.

Yang, H., S. Li, R. Jin, Z. Yu, G. Yang and J. Ma. 2020. Surface engineering of phosphorus low-doping palladium nanoalloys anchored on the three-dimensional nitrogen-doped graphene for enhancing ethanol electroxidation. Chem. Eng. J. 389: 124487. https://doi.org/10.1016/j.cej.2020.124487.

Yang, L., Y. Lv and D. Cao. 2018. Co,N-codoped nanotube/graphene 1D/2D heterostructure for efficient oxygen reduction and hydrogen evolution reactions. J. Mater. Chem. A 6(9): 3926–3932. https://doi.org/10.1039/c7ta11140a.

Yao, C., Q. Zhang, Y. Su, L. Xu, H. Wang, J. Liu et al. 2019. Palladium nanoparticles encapsulated into hollow N-doped graphene microspheres as electrocatalyst for ethanol oxidation reaction. ACS Appl. Nano Mater. 2(4): 1898–1908. https://doi.org/10.1021/acsanm.8b02294.

Yousaf, A. Bin, S. A. M. Alsaydeh, F. S. Zavahir, P. Kasak and S. J. Zaidi. 2018. Ultra-low Pt-decorated NiCu bimetallic alloys nanoparticles supported on reduced graphene oxide for electro-oxidation of methanol. MRS Commun. 8(3): 1050–1057. https://doi.org/10.1557/mrc.2018.140.

Yuan, Z., J. Li, M. Yang, Z. Fang, J. Jian, D. Yu et al. 2019. Ultrathin black phosphorus-on-nitrogen doped graphene for efficient overall water splitting: dual modulation roles of directional interfacial charge transfer. J. Am. Chem. Soc. 141(12): 4972–4979. https://doi.org/10.1021/jacs.9b00154.

Zhang, J., X. Wang, Y. Xue, Z. Xu, J. Pei and Z. Zhuang. 2018. Self-assembly precursor-derived MoP supported on N,P-codoped reduced graphene oxides as efficient catalysts for hydrogen evolution reaction. Inorg. Chem. 57(21): 13859–13865. https://doi.org/10.1021/acs.inorgchem.8b02359.

Zhong, J., T. Wu, Q. Wu, S. Du, D. Chen, B. Chen et al. 2019. N- and S-co-doped graphene sheet-encapsulated Co_9S_8 nanomaterials as excellent electrocatalysts for the oxygen evolution reaction. J. Power Sources 417: 90–98. https://doi.org/10.1016/j.jpowsour.2019.02.024.

Zhuang, Z., S. A. Giles, J. Zheng, G. R. Jenness, S. Caratzoulas, D. G. Vlachos et al. 2016. Nickel supported on nitrogen-doped carbon nanotubes as hydrogen oxidation reaction catalyst in alkaline electrolyte. Nat. Commun. 7(1): 10141. https://doi.org/10.1038/ncomms10141.

Zubair, M., I. H. Svenum, M. Rønning and J. Yang. 2020. Core-shell nanostructures of graphene-wrapped CdS nanoparticles and TiO_2 ($CdS@g@TiO_2$): The role of graphene in enhanced photocatalytic H_2 generation. Catalysts 10(4): 358. https://doi.org/10.3390/catal10040358.

Chapter 12

From Bulk to Nano
Understanding the Transition through Computer Simulations

Jimena A. Olmos-Asar

Introduction

When going from the infinite bulk to the nano regime, one finds that several properties of the material change. A large surface/volume ratio appears, increasing as the size of the system decreases, and with it, a higher density of low-coordinated atoms on the surface become potential reactive sites. Furthermore, electronic confinement starts to play an important role in the material's behaviour. These changes, among other novel properties, give the nanomaterial unique characteristics that usually differ significantly from those of their bulk counterpart.

Despite the incredible progress of the experimental techniques in the last decades, an atomistic description of these complex nanoscale systems remains a big challenge. The recent development of new theoretical methodologies, numerical techniques and software, together with the ever-increasing computational power, has been responsible for the rise of computer simulations, a powerful and reliable tool that provides an in-depth atomic/electronic description of systems, from bulk to nanostructures and surfaces.

Computer simulations can be described as a set of techniques associated with the use of a computer to simulate a system with a program or code. It involves a mathematical model composed by equations, which try to mimic the real system, and the results are obtained in the form of data to be later interpreted. The 'experiments' are performed inside a computer, and for this reason, there is no limit in treating dangerous or expensive systems. Moreover, materials which have not been yet synthesized can be idealized in the computer. The possibility to replicate real-world events allows researchers in all types of academic disciplines and commercial industries to figure out how things would function or act in certain environments without having to physically replicate those conditions.

Once a level of theory is chosen, according to the system to simulate and the expected results, the main limitation of a simulation is the computational capabilities. In the last few years, with the appearance and development of high-performance supercomputers, larger systems could be handled. On the other hand, the enhancement of resolution in experimental techniques allows scientists to measure matter at really low size levels (even individual atoms). The combination of

INFIQC-CONICET – Departamento de Química Teórica y Computacional, Facultad de Ciencias Químicas, Universidad Nacional de Córdoba.
Email: jimena.olmos.asar@unc.edu.ar

these capabilities gets the experimental and theoretical studies to the same size level, so the results can be directly compared, and a global study of a material can be reached.

In this chapter, the predictive power of computer simulations is discussed, evolving from merely descriptive tools to methodologies that have revolutionized the research in the development of new materials. Different levels of theory and computational tools to characterize a material are also briefly described. Finally, some properties, which are characteristic of low dimensional materials, are presented.

Prediction vs Description

Traditionally, computer simulations have been useful to describe the characteristics and the behaviour of matter. When some property of a known system, synthesized in the laboratory, are needed to be explained or better understood, computer simulations are a reliable method to provide a description at an atomistic level.

Technological eras have always been related to some particular material (stone age, steel age and more recently, silicon age, with an outstanding boosting of information and communication technologies). Once a material is found and proved to be efficient and to satisfy human needs, it is exploited during a considerable period of time. This is mainly due to the investments associated with its development and massive production. For this reason, the initial guess and choice of a material are key factors for the long-lasting success of a technological sector. This chosen material may have a series of desirable properties. It may be, for example, compatible with other technologies, cheap, non-toxic and resistant (Curtarolo et al. 2013).

Whereas for most of humanity materials discovery was mainly done in a trial and error approach, in the modern era, material designs can be facilitated by the implementation of computer simulations that can rapidly explore different candidates *in silico*. The field of computational chemistry has become increasingly predictive in the 21st century, with applications in several areas such as drug design, catalysis, and materials for energy storage, just to mention a few (Walsh et al. 2013). The simulations allow the anticipation of some properties of a compound (with reasonable accuracy) before it has been achieved in the laboratory. Some examples of materials first predicted theoretically and synthesized later are graphene (Wallace 1947, McClure 1956, Slonczewski and Weiss 1958) and several (other)[1] topological insulators (Pankratov et al. 1987). This ambitious strategy allows saving time and money for industries since all the efforts would only be put towards the fabrication of an already predicted well-performed material.

The first step when studying some new material is to know its structure. The Potential Energy Surface (PES) is the energy landscape that depends on the atomic positions relative to each other (internal coordinates). It contains wells that are local minimum values, corresponding to particularly stable configurations. There is one of these minima which will be the lowest value of the PES: the global minimum, which corresponds to the thermodynamically most stable structure.

Predicting the most stable conformation of a system can be a real challenge. Since materials at the nanoscale may have a different structure than the bulk counterpart, particular approaches have to be used in the search. This process implies exploring the PES of the system to find its global minimum, and for this reason, is named Global Optimization. Several methods follow this goal. The Genetic Algorithms (Holland 1975), the Basin-hopping (Wales and Doye 1997), and the Simulated Annealing (Kirkpatrick et al. 1983) are perhaps the most used these days. A few of them are more efficient for some systems, but the main idea in all the cases is very similar.

Global Optimizations start from a first guess structure (certain point of the configurational space), and the calculation of its energy. The next step consists of proposing a new structure, usually obtained from the transformation of the previous one. The energy of both structures is compared, and the most stable is chosen as a probable candidate for the global minimum.

---

[1] The Quantum Spin Hall effect was first proposed in graphene. However, in this material the band gap opening by the intrinsic spin-orbit coupling is very small, making experimental observations extremely difficult.

When the system is relatively small and simple, Global Optimization techniques are usually able to obtain the global minimum in acceptable simulation time. However, the more complicated the PES, the more difficult to find the most stable conformation of the material. There is an exponential increase in the search space with the size of the system (Doye and Wales 1998). The situation is even more complicated when dealing with multi-element materials, due to the existence of isomers and homotops (Rossi and Ferrando 2007).

The increase in computational power and storage capacity and the Internet connection gave another advantage to scientists: the appearance of public material databases, built with data coming from computer simulations. The systematization of the different materials is usually done by atomic composition, and they assure constant update, fast access, easy searching and the possibility to filter information to select the materials.

Today, there are a number of these databases which condense a lot of information on several materials. A few that can be mentioned, the AFLOW database (Curtarolo et al. 2012, Rose et al. 2017), the Materials Project (Jain et al. 2011), Novel Materials Discovery (NoMaD) (Scheffler et al. 2014), Open Quantum Materials Database (OQMD) (Saal et al. 2013), the Computational Materials Repository (Landis et al. 2012) and Automated Interactive Infrastructure and Database (AIIDA) (Pizzi et al. 2016). These repositories can be used for accelerating the synthesis optimization process using knowledge of similar systems, for training models to automatically classify structures and defects, and for identifying materials with similar behaviours. Databases are helping in saving a lot of computational costs, ensuring that there are no repeated calculations already performed.

The 'high-throughput' (HT) computational materials design is based on computational database construction and intelligent data mining to find promissory materials (Curtarolo et al. 2003). The method is used in three steps: (i) thermodynamic and electronic structure calculations of existing and hypothetical materials; (ii) systematic storage of the information in databases; (iii) materials characterization and intelligent selection of those with the desired properties. The last step is the most challenging: the researcher should search these data and look for materials of interest. When the search involves all the three steps, the expansion of the database is achieved.

With the rapid development of supercomputers and high-performance computations, high-throughput materials screening is a growing new methodology in materials science for the discovery of novel kinds of materials with desired functionality. HT techniques are created to search in depth inside the databases (combining the available experimental and theoretical libraries in a physics-based framework) to identify candidate materials that may have optimized properties for subsequent experimental evaluation. The approach is transforming the way materials science research is performed worldwide. The screening of a database is performed by means of 'descriptors'. These are empirical quantities, which not necessarily represent observables, connecting the calculated microscopic parameters to macroscopic properties of the materials (Curtarolo et al. 2013).

The emergence of Artificial Intelligence (AI) methods has the potential to substantially enhance the role of computers in science and engineering (Butler et al. 2018). The combination of large data and AI has been described as the 'fourth paradigm of science' (Agrawal and Choudhary 2016) and the number of applications in materials chemistry is growing at a striking rate. An area of AI that is constantly evolving in recent years is Machine Learning (ML).

Machine Learning is based on models that learn from existing (training) data, without human input. The idea of these techniques is to extract insight from this data by identifying patterns to create knowledge. In traditional computational approaches, the computer uses an algorithm provided by a scientist and acts just as a calculator. On the other hand, ML approaches learn the characteristics of a dataset and build a model to make predictions.

Taking into account the development of computational materials science, it is possible to classify the methodologies into three generations (Butler et al. 2018). The first generation involves using local optimization techniques to find some particular property of a material. The second generation is related to structure prediction for some fixed composition, using Global Optimization methods. The third generation is based on statistical learning, allowing the discovery of novel compositions, as well as faster prediction of properties and crystalline structures, using the large amount of data available and using ML algorithms (Schleder et al. 2019).

The accessibility of ML technology relies on three factors: open data, open software and open education (Butler et al. 2018). Finally, there are four main components in successfully applying ML to materials: (i) acquiring large enough datasets, (ii) designing feature vectors that can appropriately describe the material (descriptors), (iii) implementing a validation strategy for the models, and (iv) interpreting the ML model where applicable.

Advantages and Limitations of Computational Techniques

As mentioned earlier, computer simulations are based on mathematical models to mimic the real world. Different models have different levels of accuracy. The more complex the mathematical representation, the greater the computational resources needed to perform the calculations. The computational cost also grows rapidly as the size of the system increases. For this reason, there is always a compromise between the computational effort, the human time one is willing to invest and the degree of accuracy one wants to represent the material.

The most accurate is the quantum mechanics description, in which the electrons are explicitly included in the equations used to model the system.

To define an atom's position, one needs only to define where its nucleus is. An efficient way to apply quantum mechanics is to remember the fact that the nucleus is much heavier than the electrons, and for this reason, it moves much slower. The lowest energy configuration for the electrons under the influence of the electric field generated by the nuclei is searched. This is known as the ground state of the electrons. The separation of the nuclei and electrons into separate unconstrained mathematical equations is known as the Born-Oppenheimer approximation (Born and Oppenheimer 1927).

The Density Functional Theory (DFT) is based on two theorems (Hohenberg and Kohn 1964) which proved that, in a system with N electrons, the external potential felt by the electrons is a unique function of the electronic density and that the ground state energy is minimal for the exact density. It means that knowing the electron density, it is possible to obtain the precise energy of the system. In other words, there exists a one-to-one mapping between the ground-state wave function and the ground-state electron density. Therefore, all the ground state properties of the system depend directly on this density. Solving the Schrodinger equation for a system of N electrons is a rather difficult task. However, the electronic density is a quantity which is related to the space coordinates of the system and its calculation is much easier. In this way, one can reduce the dimensions of many-body calculation from 3N (three times the number of electrons) to only three (the cartesian coordinates). Although the Hohenberg and Kohn Theorems prove the existence of a universal functional, they do not give any hint about its nature or how to calculate the density of the ground state. To do this, it is necessary to implement the Kohn-Sham formulation (Kohn and Sham 1965), which maps the system of interacting electrons on a fictitious auxiliary system of non-interacting electrons.

The formalism of the DFT is, in principle, exact. However, in practice, a series of approximations are required in order to solve the equations. The mathematical formulation of the electron-electron interactions contained in the exchange-correlation functional is unknown. The search for an 'exact' functional is still a subject of intensive research (Perdew 2001). Since, to date, there is no such thing as the best correct functional form for the exchange and correlation terms, different functionals might end up being chosen depending on the problem to be solved. A very large variety of functionals can be found in the literature; some are parameter-free and others are semi-empirical, i.e., contain parameters fitted from experimental data.

On the other hand, sometimes it is convenient to treat valence and core electrons differently. All-electron methods take explicitly into account all the electrons of the system. However, many properties of the materials and the chemistry involved in the reactions are mostly determined by the valence electrons alone, making the computational effort of all-electron approaches sometimes excessive. The pseudopotential method (Phillips and Kleinman 1959) simplifies the atomic description by implementing an effective potential to represent the core electrons and the nuclei, and only the valence electrons, attached to this potential, are explicitly described. Some popular approaches are the norm-conserving (Hamann et al. 1979) and ultrasoft (Vanderbilt 1990) pseudopotentials and the Projector Augmented Waves (PAW) (Kresse and Joubert 1999).

The main limitation of DFT is inherent to its fundamental definition: it only accurately describes the ground state. Thus, the study of excited states and other second-order properties are not in principle provided by the methodology. For this, the perturbation theory and some other techniques, included in what is known as 'beyond DFT methods', have been developed in the last decades. The GW approximation (Aryasetiawan and Gunnarsson 1998), which solves the Green's function for carrying out self-energy calculations and obtain excited-states properties can be mentioned. Additionally, some auxiliary theories need to be added in the description. For example, strongly correlated systems, such as d-band electrons in some transition metals, need to be handled with the Hubbard U parameter (Liechtenstein et al. 1995, Dudarev et al. 1998). When long-range dispersion, which is underestimated in the DFT formalism, is significant, some van der Waals (vdW) correction has to be added in the calculation (Du and Smith 2005).

Quantum-mechanics based methods are the most accurate to describe a material, and for this reason, they could be the first choice in a simulation. However, the computational power that is required imposes a relevant limitation. With the resources available nowadays, only hundreds to a thousand atoms can be handled.

Based on the Born-Oppenheimer approximation, as it was mentioned earlier an atom can be imagined as a heavy nucleus surrounded by much lighter electrons, which move so fast that they can adapt instantaneously to the atomic position. Taking this approach to a step forward, a way to further simplify the complexity of calculations is to describe the motion of these heavy nuclei according to Newton's laws of dynamics. The electrons, on their side, are not explicitly taken into account; they are included in the parameterization of some interatomic classical potentials, which represent the forces between different nuclei. According to Newton's laws, when two atoms are close, a force acts on both of them (third Newton law) and is transformed into an acceleration (second Newton law) that makes them move.

The quality of a classical simulation depends almost exclusively on the accuracy of the interatomic potentials, and if they are able to properly represent the system in study. A classical potential is a mathematical expression that describes the energy as a function of the spatial configuration of the interacting particles. In other words, the interatomic potentials sample the energy landscape (PES) of the system. Forces are obtained as the gradient of these energy functions (Frenkel 2007). The more complex the potential expression, the more cumbersome the simulation will be (Voter 1996).

These mathematical functions have parameters that depend on the particular system to be simulated. There are several methods to obtain their value. One of the most used techniques is to fit the potential to some data obtained with a higher accuracy method, such as quantum-mechanics calculations. Some experimental data can also be useful to parameterize a potential. These days the use of automation or semi-automation, such as machine learning techniques, have a great impact on the parameterization of different classical potentials.

Quantitatively, to obtain accurate results it is required that a model is able to describe a database of physical properties of the system. Transferability is also desirable, i.e., the potential could describe structures and processes beyond those used to fit it. This is particularly important if the simulation is to be used as a predictive tool.

All the potential functions and the corresponding parameters used in the computer simulation constitute the force field, which samples the complete Potential Energy Surface of the system. *All-atom* force fields describe the system at the atomistic level, providing parameters for every species (including hydrogen). The *united-atom* method treats a small group of atoms (such as the hydrogen and carbon atoms in methyl and methylene groups) as one identity, usually called a pseudo-atom. It is particularly useful when the dynamics of the group as a whole is much more relevant than the motion inside the pseudo-atom (Leach 2001). *Coarse-grained* potentials, which are often used for simulating macromolecules (such as proteins), group several atoms in only one centre, and although chemical details are lost, they allow achieving high computational efficiency (Levitt and Warshel 1975, Levitt 2014).

The simplest family of classical potentials is the one of pair potentials. As it can be deduced from its name, they represent the interatomic bond (or non-bonding interaction) between two particles. They have usually been constructed in analogy with binding curves of diatomic molecules, or with the variation of atomic configurations at a constant average density of the material. A harmonic function is usually used when the (covalent) bond breaking is not allowed during the simulation. A more realistic description is obtained using other forms like the Lennard-Jones (Lennard-Jones 1931) and Morse (Morse 1929) potentials. They adequately describe, for example, the energetics of systems in which van der Waals-type interactions dominate, and for some metallic systems (Vitek 1996).

Pair potentials can be augmented by a new term, which is related to the electronic density of the material. This simple modification had a great impact on the simulation of metals (Voter 1996). The cohesive energy of an atom is mostly determined by the local electron density around it, due to the presence of its neighbours. The embedding energy is associated with placing an atom in that electronic environment, and is a function only of the identity of the atom and the electron density; it can be calculated and tabulated. The remaining pair-potential term results from electrostatic contributions (Raeker and Depristo 2008). The main advantages of these many-body potentials, which can also be interpreted as based on a tight-binding model of the energetics, are that they can describe inhomogeneous systems (like surfaces) and also that the bond strength depends on the coordination (Foiles 1996). Some of these potentials include the Embedded Atom Method (EAM) (Daw et al. 1993, Voter 1994, Foiles 1992), the Effective Medium Theory (Norskov 1990), the Finnis-Sinclair N-body potentials (Finnis and Sinclair 1984), the Second-Moment Approximation of the Tight-Binding (Ducastelle and Cryot.Lackmann 1971) and the Glue model (Ercolessi et al. 1986).

Bond-order potentials, such as Brenner (Brenner 1990), Tersoff (Tersoff 1988a, b), AIREBO (Stuart 2000) and COMB (Phillpot and Sinnott 2009, Shan et al. 2010) formulations, depend on the atomic local coordination, i.e., on the number of the nearest neighbours, which determine the chemical environment of a particle and indicates the strength of a chemical bond (Abell 1985). Usually the greater the coordination, the weaker the individual interaction. These potentials have as a strength, the fact that they take into account the angularity of the system, being able to simulate different allotropes. Among the limitations of these formulations, one can mention issues related to charge transfer and ionic bonding in solids, for example. These types of interactions are not explicitly addressed. Nonetheless, they provide a simple but impressive framework from which, in general terms, both structure and chemical dynamics can begin to be understood (Brenner 1996).

Most potentials mentioned earlier are inadequate for modelling changes in atomic connectivity. In other words, they are not efficient for simulating chemical reactions as bonds break and form. ReaxFF, a reactive force field, describes reactive events through a bond-order formalism, where the bond order is empirically calculated from interatomic distances. The formulation is divided into bond-order dependent and independent contributions. Additionally, ReaxFF uses polarizable charge descriptions to represent both reactive and non-reactive interactions between atoms. For this reason, this force field is able to accurately model both covalent and electrostatic interactions for a diverse range of materials (Senftle et al. 2016).

The main advantage of representing a material using a classical approach is that it becomes possible to simulate much larger systems (in the order of the micrometre) and to obtain a dynamic description for longer periods of time (microseconds). In this way, more realistic materials, as well as slower processes, can be observed within this approximation. However, the chemistry (i.e., the electronic description) of the system is lost, and the formation and breaking of bonds are typically not well represented. As mentioned earlier, there are some classical potentials which attempt to simulate the formation and breaking of bonds, as well as the change in hybridization according to the atomic environment, such as the ReaxFF. However, a classical approach will always depend on the parameterization, and is not a good technique to predict unknown scenarios. Finally, classical simulations are not a good choice when quantum and relativistic effects are determinants (Johanson et al. 2004). This is a common situation when the size of the system is small, on the order of a few atoms.

QM-MM (Quantum Mechanics – Molecular Mechanics) method, first proposed by Warshel and Levitt (Warshel and Levitt 1976), combine the quantum and classical approximations in a convenient manner: a relatively small part of the system (the chemically active site) is represented by quantum mechanics, and the rest of the material is approximated by classical methods. Both pieces interact through some quantum-classical potential. Basically, the energetics of the system is approximated as the addition of the energy of the active site, calculated with a QM method, the energy of the classical part, calculated with MM, and an interaction term between both systems, containing bonding and non-bonding terms (Lin and Truhlar 2007). In this way, the chemical representation of the system is not lost, while conserving the efficiency in the calculation. This technique is particularly useful for the simulation of some biological or catalytic systems or to simulate chemical reactions in an explicit solvent.

Finally, it is possible to simulate large scale systems, in the order of macroscopic samples, losing the atomistic description. Continuum models are based on differential equations to simulate processes that occur in matter. They are very useful to predict the behaviour of some material on a real scale when the microscopic description is already known. However, these larger-scale methods, such as the finite element method, depend on critical inputs from experimental data and atomistic simulations.

Low Dimensional Materials

The materials one deals with in our daily life, those which can be seen, touched and manipulated with our hands, are bulk materials, typically in the macroscopic scale. They exist in a three-dimensional world and are massive in each direction.

Low dimensional materials are those which, at least in one of the three cartesian dimensions, are so small that their properties lay at some point between the properties of the bulk and individual atoms.

Two-dimensional materials are extensive in two of the three cartesian directions and very thin in the remaining one. Examples of 2D materials are sheets and nanofilms. One-dimensional materials are those which can be thought as existing in only one dimension, and in the other two, they have nanometric size. Examples of 1D materials are nanorods, nanotubes and nanowires. Finally, one can think of materials with no dimensionality. They can be imagined as really small spheres containing only a few atoms or molecules. Nanoparticles and fullerene are the greatest examples of 0D materials.

A picture of low dimensional materials is shown in Fig. 12.1.

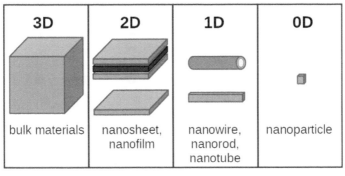

Figure 12.1: Bulk and low dimensional materials.

Periodic Boundary Conditions

As mentioned earlier, the complexity and computational cost of a simulation depend, mostly, on the size of the system one wants to represent. In the case of atomistic classical simulations, the computational cost scales linearly with the number of atoms. Within the DFT approximation, the scaling is much worse: it increases with the cube of the number of states (N^3), although some algorithms are able to significantly reduce this scaling, but always worse than linear.

One trick to limit the number of particles to be simulated is to take into account the periodicity of the system and to apply Periodic Boundary Conditions (PBC). The idea of this implementation is to choose a simulation cell which contains all the atoms that are necessary to fully represent the material. This is called the primary cell, which can be thought as a small portion of the massive system, and whose limits are given by the dimensions of the simulation box. This cell, in turn, is surrounded in all the directions of periodicity by exact replicas of itself, called image cells. In other words, the primary cell is periodically replicated in all the directions to mimic an infinite system. The velocities of the image particles (in the case of a Molecular Dynamics simulation) are the same as those of the original particles, and their positions are those of the primary cell, plus some multiple of the vector cell.

For example when simulating magnesium oxide (MgO), which has a Face-Centred Cubic (FCC) structure, one could think of a simple cubic simulation box containing only eight atoms and then replicate it in the three cartesian axes to have a perfect bulk. This is shown in Fig. 12.2a. Another alternative is to choose an FCC simulation box containing one atom of each kind.

The individual cells are separated by imaginary borders. For this reason, the particles do not collide with the walls of the simulation box; they are free to leave one cell and enter the next one. Due to the fact that the particles within different cells are periodic images, when one particle leaves a cell, another one enters at the opposite edge with the same velocity, keeping the number of particles in each cell always constant, as shown in Fig. 12.2b.

With this methodology, the size of the simulated systems is reduced significantly. However, some details have to be taken into account. First of all, it is important to note that some collective properties may be missed. This is because all the atoms in the periodic images will behave exactly the same way as the original ones. Another limitation of using PBC is that one is representing a periodic system. It means that the simulated material is regular and perfect, and in this representation, it is not possible to simulate most defects, such as dopants, vacancies or other singularities in the system. When it is necessary to introduce some of these characteristics, then a larger cell is needed, called a supercell and the PBC have to be adapted accordingly.

When simulating bulk materials, the PBC are applied in the three cartesian directions to mimic a 3D infinite system. However, when the size of the material is reduced and one moves to the nanometric scale in some dimension, i.e., when low dimensional materials are simulated, the PBC are different.

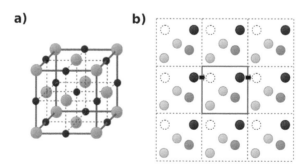

Figure 12.2: (a) Primary simulation cell of magnesium oxide (MgO). Mg: large grey spheres; O: small black spheres (b) primary simulation cell and periodic images surrounding it in a bidimensional system.

To simulate surfaces or 2D materials, PBC are applied in only two of the cartesian directions, leaving the other one free. In this way, the system is represented as infinite in the periodic plane, but finite in the perpendicular direction. On the other hand, to simulate 1D materials, PBC are applied only in a single direction, while for 0D materials there is no need to implement PBC at all due to the non-periodicity.

Some methodologies work only under periodic boundary conditions by definition. The most common example is density functional calculations using plane waves to represent the electronic wave functions. Is it possible, however, to use these methodologies to simulate low dimensional materials. The trick is to add enough empty space (a large simulation box) in the finite direction. For example, to simulate a 2D material, the simulation cell in the plane should be left unchanged, while in the perpendicular direction a large quantity of vacuum is added in order to minimize the interaction between periodic images.

Material's Properties at the Nanoscale

At the nanoscale, the properties of materials are different from those of the bulk. For example, the melting point at the macroscopic scale is an extensive property, independent on the quantity of matter that is being heated. However, at the mesoscale, the melting point decreases as the size of the system is reduced. Even more, at the nanoscale, when size and shape effects become very important, clusters formed by a few hundred or fewer atoms show important deviations from this law, sometimes melting at even higher temperatures than the bulk (Shvartsburg and Jarrold 2000, Breaux et al. 2004).

The main reason for the appearance of these novel properties is the large area/volume ratio that these materials present. The low coordination of atoms makes them more reactive and with a particular rearrangement of the electronic levels, and for this reason nanomaterials encounter several applications in different areas such as catalysis, optoelectronics, solar cells, LEDs, transistors, displays, quantum computing and medical imaging (Kumar et al. 2018).

The dependence of the properties on the size of the nanomaterial allows the tuning of some particular characteristic even by keeping the composition constant and just varying its size (Alivisatos 1996).

Nanostructures may be differentiated by their size, morphology, composition and architecture. Size and morphology are directly related to the coordination of the atoms, while the composition and architecture determine the surface ensemble (Alayoglu 2017). All these properties can now be controlled experimentally at a very accurate level, and for this reason, nanostructures produced at large scales exhibit high uniformity. This is particularly important in catalysis, which is a surface phenomenon and at the nanometric level is governed by factors of atomic dimensions. In the case of monometallic systems, size, shape and coordination are basically the characteristics that determine the electronic structure of the surfaces, which is different from the bulk. For multi elements nanoparticles, the atomic arrangement is another key factor determining its catalytic properties.

It is then, of great importance to find the best structure of a catalyst. With this purpose, global minimization techniques can be applied to explore the PES and to find the thermodynamically most stable configurations for the nanoparticles. Depending on the size of the system, *ab-initio* methods could be applied or, if the catalyst is large enough, classical potentials might be enough to accurately describe the PES.

When reducing the size of a material to the nanometre scale, one of the most immediate observations is the restriction in the movement of electrons due to the quantum confinement effect. This leads to a discretization of the electron energy levels, which are separated by 'forbidden' energy gaps (Nenadovic et al. 1985, Zhirnov and Cavin III 2015). The quantum confinement effect is observed when the size of the particle is so small that it is comparable to the wavelength of the electron. The confinement of an electron and hole in nanocrystals significantly depends on the material properties and will be observed when the spacing between the states exceeds the temperature (Neikov and Yefimov 2019, Alivisatos 1996).

Ultrasmall particles, which are in the molecular limit and designated as clusters, are attractive in electronics because of their quantum confinement effects. These particles have been demonstrated to have unique electronic properties that are different from those of their molecular building blocks and larger nanoparticles (Alivisatos 1996). As an example, a metal-to-insulator transition was reported for Au_{55} clusters, also known as Schmid's cluster, as early as the 2000s (Boyen et al. 2001). In Fig. 12.3, the Density of States (DOS) for copper bulk and nanoparticles containing 13 and 147 atoms is shown. As can be observed, discretization of the states occurs when the size of the system is reduced. Cu bulk is a conductor, while several energy gaps appear at the DOS of Cu_{13}.

Quantum Dots (QDs) are materials in which quantum confinement effects can be observed. They are very small semiconductor crystals, in the order of nanometre size, containing only a few hundred to a thousand atoms. As a result, they tightly confine electrons or electron-hole pairs called 'excitons'. QDs exhibit electronic properties that are intermediate between those of bulk and isolated molecules. Their optoelectronic properties are determined by their size and shape. For example, when QDs are excited by a photon of energy hv (where v is the frequency of the incident photon and h is the Planck's constant), they will emit energy in a particular wavelength that depends directly on their size. This is due to the variation in their band gaps. As a consequence, QDs' properties can be specifically tuned to have the desired output by altering the dot size and shape, making them significant for optoelectronic applications (Kumar et al. 2018).

Finally, it is of interest to mention that many nanomaterials present enhanced mechanical properties with respect to their bulk counterpart. At the macroscopic level, many metals have an important strength, but their ductility is very poor or vice-versa. However, due to nanoscopic and interface effects, nanostructured metallic materials can present high strength, hardness, ductility and wear resistance (Ovid'ko et al. 2018 and references therein), making them very promising in structural and functional applications.

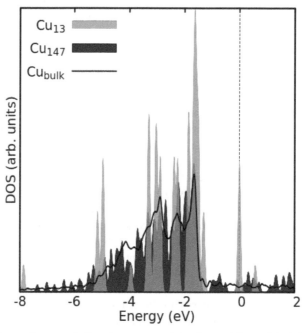

Figure 12.3: Density of states for copper bulk and two nanoparticles: Cu_{13} and Cu_{147}. Calculations were done within the DFT approach, using ultra soft pseudo potentials and the LDA functional for the exchange and correlation term.

Conclusions

In this Chapter we delved into the power of computer simulations to mimic the real world. We reviewed some of the computational and theoretical techniques that provide not only an accurate description of systems, but also allows one to imagine, create and predict novel materials before they can be experimentally synthesized. The increasing rise in the computational power is nowadays allowing, on one side, the chance to emulate real-sized systems and, on the other hand, to test several atomistic combinations to span lots of data and to find the best behaved materials. Finally, we walked through some characteristic properties of low dimensional materials.

Acknowledgements

J.A.O.-A. is a research member of Consejo Nacional de Investigaciones en Ciencia y Tecnología (CONICET). The author thanks Dr. Fábio R. Negreiros for reading and giving valuable insights on this chapter.

References

Abell, G. C. 1985. Empirical chemical pseudopotential theory of molecular and metallic bonding. Phys. Rev. B. 31: 6184.

Agrawal, A. and A. Choudhary. 2016. Perspective: Materials informatics and big data: realization of the 'fourth paradigm' of science in materials science. APL Mater. 4: 053208.

Alayoglu, S. 2017. Achievements, present status, and grand challenges of controlled model nanocatalysts. Stud. Surf. Sci. Catal. 177: 85–121.

Alivisatos, A. P. 1996. Semiconductor clusters, nanocrystals, and quantum dots. Science 271: 933–937.

Aryasetiawan, F. and O. Gunnarsson. 1998. The GW method. Rep. Prog. Phys. 61: 237.

Born, M. and J. R. Oppenheimer. 1927. On the quantum theory of molecules. Annalen der Physik (in German) 389: 457–484.

Boyen, H. -G., G. Kästle, F. Weigl, P. Ziemann, G. Schmid, M. G. Garnier et al. 2001. Chemically induced metal-to-insulator transition in Au 55 clusters: effect of stabilizing ligands on the electronic properties of nanoparticles. Phys. Rev. Lett. 87: 276401.

Breaux, G. A., D. A. Hillman, C. M. Neal, R. C. Benirschke and M. F. Jarrold. 2004. Gallium cluster "magic melters". J. Am. Chem. Soc. 126: 8628–8629.

Brenner, D. W. 1990. Empirical potential for hydrocarbons for use in simulating the chemical vapor deposition of diamond films. Phys. Rev. B. 42: 9458–9471.

Brenner, D. W. 1996. Chemical dynamics and bond-order potentials. MRS Bull. 21: 36–41.

Butler, K. T., D. W. Davies, H. Cartwright, O. Isayev and A. Walsh. 2018. Machine learning for molecular and materials science. Nature 559: 547–555.

Curtarolo, S., D. Morgan, K. Persson, J. Rodgers and G. Ceder. 2003. Predicting crystal structures with data mining of quantum calculations. Phys. Rev. Lett. 91: 135503.

Curtarolo, S., W. Setyawan, S. Wang, J. Xue, K. Yang, R. H. Taylor et al. 2012. AFLOWLIB.ORG: A distributed materials properties repository from high-throughput *ab initio* calculations. Comput. Mater. Sci. 58: 227–235.

Curtarolo, S., G. L. W. Hart, M. Buongiorno Nardelli, N. Mingo, S. Sanvito and O. Levy. 2013. The high-throughput highway to computational materials design. Nat. Mat. 12: 191–201.

Daw, M. S., S. M. Foiles and M. I. Baskes. 1993. The embedded-atom method: a review of theory and applications. Mater. Sci. Rep. 9: 251.

Doye, J. P. K. and D. J. Wales. 1998. Thermodynamics of global optimizations. Phys. Rev. Lett. 80: 1357–1360.

Du, A. J. and S. C. Smith. 2005. Van der Waals-corrected density functional theory: benchmarking for hydrogen–nanotube and nanotube–nanotube interactions. Nanotechnology 16: 2118.

Ducastelle, F. and F. Cyrot-Lackmann. 1971. Moments developments: II. Application to the crystalline structures and the stacking fault energies of transition metals. Phys. Chem. Solids 32: 285.

Dudarev, S. L., G. A. Botton, S. Y. Savrasov, C. J. Humphreys and A. P. Sutton. 1998. Electron-energy-loss spectra and the structural stability of nickel oxide: an LSDA+U study. Phys. Rev. B. 57: 1505.

Ercolessi, F., M. Parrinello and E. Tosatti. 1986. Au(100) reconstruction in the glue model. Surf. Sci. 177: 314.

Finnis, M. W. and J. E. Sinclair. 1984. A simple empirical N-body potential for transition metals. Philos. Mag. A 50: 45.

Foiles, S. M. 1992. Atomistic simulations of surfaces and interfaces. Equilibrium Structure and Properties of Surfaces and Interfaces. Springer US, 89.

Foiles, S. M. 1996. Embedded-Atom and related methods for modeling metallic systems. MRS Bull. 21: 24–28.

Frenkel, D. 2007. Understanding Molecular Simulation: From Algorithms to Applications. Academic Press.

Hamann, D. R., M. Schlüter and C. Chiang. 1979. Norm-conserving pseudopotentials. Phys. Rev. Lett. 43: 1494–1497.

Hohenberg, P. and W. Kohn. 1964. Inhomogeneous electron gas. Phys. Rev. 136: B864.

Holland, J. 1975. Adaptation in Natural and Artificial Systems. University of Michigan Press, Ann. Anbor, MI.

Jain, A., G. Hautier, C. J. Moore, S. P. Ong, C. C. Fischer, T. Mueller et al. 2011. A high-throughput infrastructure for density functional theory calculations. Comput. Mater. Sci. 50: 2295–2310.

Johanson, M. P., D. Sundholm and J. Vaara. 2004. Au_{32}: A 24-carat golden fullerene. Angew. Chem. Int. Ed. 43: 2678–2681.

Kirkpatrick, S., C. D. Gellat and M. P. Vecchi. 1983. Optimization by simulated annealing. Science 220: 671680.

Kohn, W. and L. J. Sham. 1965. Self-consistent equations including exchange and correlation effects. Phys. Rev. 140: A1133.

Kresse, G. and D. Joubert. 1999. From ultrasoft pseudopotentials to the projector augmented-wave method. Phys. Rev. B 59: 1758–1775.

Kumar, D. S., B. J. Kumar and H. M. Mahesh. 2018. Quantum nanostructures (QDs): An overview. Ssynthesis of inorganic nanomaterials: Advances and key technologies. Micro and Nano Technologies. Woodhead Publishing, 59–88.

Landis, D. D., J. S. Hummelshøj, S. Nestorov, J. Greeley, M. Dułak, T. Bligaard et al. 2012. The computational materials repository. Comput. Sci. Eng. 14: 51–57.

Leach, A. 2001. Molecular Modeling: Principles and Applications. Harlow: Prentice Hall.

Lennard-Jones, J. E. 1931. Cohesion. Proc. Phys. Soc. 43: 461–482.

Levitt, M. and A. Warshel. 1975. Computer simulation of protein folding. Nature 253: 694–698.

Levitt, M. 2014. Birth and future of multiscale modeling for macromolecular systems (Nobel lecture). Angew. Chem. Int. Ed. 53: 10006–10018.

Liechtenstein, A. I., V. I. Anisimov and J. Zaane. 1995. Density-function theory and strong interactions: Orbital ordering in Mott-Hubbard insulators. Phys. Rev. B. 52: R5467.

Lin, H. and D. G. Truhlar. 2007. QM/MM: what have we learned, where are we, and where do we go from here? Theor. Chem. Acc. 117: 185–199.

McClure, J. W. 1956. Diamagnetism of graphite. Phys. Rev. 104: 666–671.

Morse, P. M. 1929. Diatomic molecules according to the wave mechanics. II. Vibrational levels. Phys. Rev. 34: 57.

Neikov, O. D. and N. A. Yefimov. 2019. Nanopowders. Handbook of non-ferrous metal powders. Technologies and Applications. Elsevier, 271–311.

Nenadovic, M. T., T. Rajh and O. I. Micic. 1985. Size quantization in small semiconductor particles. J. Phys. Chem. 89: 397–399.

Norskov, J. K. 1990. Chemisorption on metal surfaces. Rep. Prog. Phys. 53: 1253.

Ovid'ko, I. A., R. Z. Valiev and Y. T. Zhu. 2018. Review on superior strength and enhanced ductility of metallic nanomaterials. Prog. Mater. Sci. 94: 462–540.

Pankratov, O. A., S. V. Pakhomov and B. A. Volkov. 1987. Supersymmetry in heterojunctions: Band-inverting contact on the basis of Pb1-xSnxTe and Hg1-xCdxTe. Solid State Commun. 61: 93–96.

Perdew, J. P. 2001. AIP Conference Proceedings 577: 1.

Phillips, J. C. and L. Kleinman. 1959. New method for calculating wave functions in crystals and molecules. Phys. Rev. 116: 287.

Phillpot, S. R. and S. B. Sinnott. 2009. Simulating multifunctional structures. Science 325: 1634–1635.

Pizzi, G., A. Cepellotti, R. Sabatini, N. Marzari and B. Kozinsky. 2016. AiiDA: automated interactive infrastructure and database for computational science. Comput. Mater. Sci. 111: 218–230.

Raeker, T. J. and A. E. Depristo. 1991. Theory of chemical bonding based on the atom–homogeneous electron gas system. Intern. Rev. Phys. Chem. 93: 1.

Rose, F., C. Toher, E. Gossett, C. Oses, M. Buongiorno Nardelli, M. Fornari et al. 2017. AFLUX: The LUX materials search API for the AFLOW data repositories. Comput. Mater. Sci. 137: 362–370.

Rossi, G. and R. Ferrando. 2007. Structural Properties of Pure and Binary Nanoclusters Investigated by Computer Simulations. Nanomaterials: Design and Simulation. Elsevier. 35–58.

Saal, J. E., S. Kirklin, M. Aykol, B. Meredig and C. Wolverton. 2013. Materials design and discovery with high-throughput density functional theory: the open quantum materials database (OQMD). JOM 65: 1501–1509.

Scheffler, M., C. Draxl and Computer Center of the Max-Planck Society, Garching. 2014. The NoMaD Repository, http://nomad-repository.eu.

Schleder, G. R., A. C. M. Padilla, C. Mera Acosta, M. Costa and A. Fazzio. 2019. From DFT to machine learning: recent approaches to materials science—a review. J. Phys. Mater. 2: 032001.

Senftle, Thomas, Hong Sungwook, Islam Md Mahbubul, Kylasa Sudhir, Zheng Yuanzia, Shin Yun Kyung et al. 2016. The ReaxFF reactive force-field: development, applications and future directions. Comput. Mater. 2: 15011.

Shan, T. -R., Bryce D. Devine, Jeffery M. Hawkins, Aravind Asthagiri, Simon R. Phillpot, Susan B. Sinnott et al. 2010. Second generation charge optimized many-body (COMB) Potential for Si/SiO_2 and amorphous silica. Phys. Rev. B. 82: 235302.

Shvartsburg, A. A. and M. F. Jarrold. 2000. Solid clusters above the bulk melting point. Phys. Rev. Lett. 85: 2530.

Slonczewski, J. C. and P. R. Weiss. 1958. Band structure of graphite. Phys. Rev. 109: 272–279.

Stuart, S. J., A. B. Tutein and J. A. Harrison. 2000. A reactive potential for hydrocarbons with intermolecular interactions. J. Chem. Phys. 112: 6472.

Tersoff, J. 1988a. New empirical approach for the structure and energy of covalent systems. Phys. Rev. B 37: 6991–7000.

Tersoff, J. 1988b. Empirical interatomic potential for carbon, with applications to amorphous carbon. Phys. Rev. Lett. 61: 2879–2882.

Vanderbilt, D. 1990. Soft self-consistent pseudopotentials in a generalized eigenvalue formalism. Phys. Rev. B 41: 7892–7895.

Vitek, V. 1996. Pair potentials in atomistic computer simulations. MRS Bull. 21: 20–23.

Voter, A. F. 1994. The embedded atom method. Intermetallic Compounds. John Wiley & Sons, 1: 77.

Voter, A. F. 1996. Interatomic potentials for atomistic simulations. MRS Bull. 21: 17–19.

Wales, D. J. and J. P. K. Doye. 1997. Global optimization by basin-hopping and the lowest energy structures of lennard-jones clusters containing up to 110 atoms. J. Phys. Chem. A 101: 5111–5116.

Wallace, P. R. 1947. The band theory of graphite. Phys. Rev. 71: 622–634.

Walsh, A., A. A. Sokol and C. R. A. Catlow. 2013. Computational Approaches to Energy Materials. Wiley-Blackwell, New York.

Warshel, A. and M. Levitt. 1976. Theoretical studies of enzymic reactions: dielectric, electrostatic and steric stabilization of the carbonium ion in the reaction of lysozyme. J. Mol. Biol. 103: 227–249.

Zhirnov, V. V. and R. K. Cavin III. 2015. Basic physics of ICT. Microsystems for Bioelectronics. Scaling and Performance Limits. Micro and Nano Technologies, William Andrew, 19–49.

CHAPTER 13

Computer Simulations of Ultra-thin Materials
From 2D to Heterostructures

Fabio R. Negreiros

Introduction

Since the extraction of a single graphene layer from graphite in 2004, 2D materials have gained increasing attention. A significant number of works have been dedicated to their fabrication, experimental and theoretical characterization. These materials can be seen as an atomically-thin slab with an infinitely large surface area, being stable enough to present little or no defects in some chosen environment. In principle, a 2D layer can be cut from any material, but in practice, one should always consider the stability of the 2D film created and the energetic cost to create it. There are several materials in which a layer cannot be easily characterized or other cases such as metals, that an ultra-thin layer can only be stable if adsorbed on specific substrates, given their tendency to form clusters or islands to optimize the atomic coordination. Therefore, the actual size of the database containing all possible 2D layers that exist in nature is considerably lower than the total number of existing 3D materials.

Many works and reviews have been written in the last few years about 2D and derived materials (Geim and Grigorieva 2013, Novoselov et al. 2016, Khan et al. 2020, Heard et al. 2019, Barcaro and Fortunelli 2019, Zhang et al. 2019, Sankar et al. 2019, to cite a few). Although many experimental techniques have been applied to their fabrication and characterization, an even larger number of works have focused on computer simulation. The reason for this lies in the very nature of the system, which is atomically thin in one of its directions, making them particularly difficult to experimentally synthesize and analyze, but comparatively straightforward to simulate. 2D films can be simulated with rather small cells, containing only a few atoms in the surface normal direction. Furthermore, since they are infinite in the plane, they are benefited from the periodic boundary conditions usually adopted in most numerical simulations. For example, graphene can be represented with only two atoms, while 2D materials of the most common chalcogenides and oxides can be represented with only a few atoms.

From the simulation side, Density Functional Theory or DFT, presented in the earlier chapter of this book, has been the most used method to determine the properties of the 2D layers. DFT gives not only the most stable structures and their formation energy but also the ground state electronic

INFIQC, CONICET, Departamento de Química Teórica y Computacional, Facultad de Ciencias Químicas, Universidad Nacional de Córdoba, Argentina.
Email: fabio.ribeiro@unc.edu.ar

and magnetic structure. Second-order properties, such as phonon and electronic excitations can also be determined with the perturbation theory, even though the accuracy obtained can be rather poor for some systems, especially when optical properties are desired.

Although the theoretical characterization of 2D films can present some additional challenges with respect to other classes of systems, such as bulk and surfaces, the standard numerical methodologies that have been used in condensed matter can be safely adopted, in quite a straightforward manner. Given that they are very well established and understood, in the last decade, a lot of effort has been focused in developing more automated searches by using high-throughput techniques in order to predict new unknown stable 2D systems, that could in principle be fabricated experimentally. The overall idea consists of creating an automatic procedure to generate and characterize 2D layers, giving special focus on their stability, since it will ultimately determine if a new material should exist or not. The main approach would require little to no human interference, and also allows a complete exploration of the landscape of existing structures. The difficulties that could arise are related to the algorithm chosen to create the material, which should be automatic and ideally explore all the possible cases; the accuracy of the theory used to describe its properties and stability, which in general lie completely or to some extent in DFT calculations; the computational power available.

The 'Computational 2D Materials Database' (Haastrup et al. 2018) is an example of such an approach. It is a growing, free and open database that contains the structural, thermodynamic, elastic, electronic, magnetic and optical properties of thousands of different 2D materials, evaluated at the DFT level. For hundreds of them, state-of-the-art many-body perturbation theory within the G_0W_0 approximation (Hedin 1965) and the Bethe-Salpeter equation (Salpeter and Bethe 1951) were also used to more accurately characterize their electronic and optical properties. In another example (Nicolas et al. 2017), a large database of 2D systems was built taking as a starting point more than 100000 stable bulk compounds, from which potential 2D layers are extracted and characterized at the DFT level in order to check their stability and overall structural properties. A further study of the energy required to extract the layer, evaluated with respect to the interlayer interaction energy in the bulk case, was also performed, so as to determine which layers were the most easily exfoliable. Given the experimental difficulties that arise from fabricating these materials, this type of analysis should offer valuable information, giving a complete theoretical description of both the stability and ease of fabrication of new materials. Other databases, such as the ones mentioned in Chapter 12, usually also contain particular data about 2D materials. But more important, it is necessary to highlight that the creation of a single complete database of 2D films, characterized in the richest and most accurate possible way, whether by using high-throughput techniques or even by collecting the several already existing works on different materials, should be the ultimate goal.

Most 2D structures studied and reported in the literature were shown to exhibit interesting and unique properties, different from bulk, surface, isolated atoms and also nanoparticles, is therefore defined as a new class of materials. However, to be multifunctional, which is the main topic of this book, a single 2D film will not likely be the only element present in the real operative system. For example, support might be required for practical reasons. One might also seek to optimize several different properties, being forced to combine more than one 2D layer: due to the very ultra-thin nature of 2D films, they have the potential to be used as building blocks to form a new material, called a heterostructure, that can potentially present all the individual properties of each layer that compose it, possibly even enhancing them. Therefore, a proper characterization of these materials has to take into account not only the properties of its constituent pieces but also how they interact with each other and with other elements.

The characterization of 2D materials is briefly presented and discussed, starting with the isolated atomically-thin cases. Two types of system are then considered: heterostructures, that are built from the stacking of two or more 2D films, weakly interacting with each other, called Van Der Waals heterostructures; interacting TMO-2D films, that can be adsorbed in a substrate or another 2D material. The focus will be on the computer simulation techniques that have been typically used, and the type of properties studied. A few recent works where computer simulations provided a significant and valuable contribution to the characterization and understanding of these materials will also be highlighted.

The 2D Limit

Graphene-like 2D Layers

In 2004, a single graphene layer was successfully extracted from graphite using a mechanical exfoliation technique, and was shown to have unique and remarkable electronic properties (Novoselov et al. 2004). As of 2020, a great deal of work has been devoted to completely characterize graphene in its pristine or defected forms, experimentally and theoretically (Castro Neto et al. 2009, Sarma et al. 2011, Cooper et al. 2012). It is most likely the most studied 2D material to date, but more important than that, it opened the doors to the study of several other graphene-like structures.

Graphene is an atomically-thin 2D material, with strong planar covalent bonds that make it very stable at standard conditions, even when isolated. It has a honeycomb lattice and can be simulated with a simple hexagonal cell with two carbon atoms inside, as illustrated in Fig. 13.1a. In its pristine form, since it contains a small number of electrons, it can, and has been, simulated at nearly all levels of theory. Indeed, several different methodologies have been used to simulate it, from analytical studies to numerical ones, such as classical potentials, tight-binding approaches, DFT and also many-body perturbation. Its impressive structural and mechanical properties, such as high hardness as well as elasticity, have been successfully reproduced and accurately predicted in numerous works, performing static and molecular dynamics calculations (Akinwande et al. 2017, Kumar et al. 2019). The fundamental theory behind its particular electronic properties, such as semi-metal electronic character with conduction and valence bands almost touching at Dirac points, high electron and hole mobility (acting as massless relativistic particles), among many others, have been reviewed several times in the literature already (Castro Neto et al. 2009, Abergel et al. 2010). Herein only highlighted was that nearly all electronic/optical properties were predicted or reproduced by numerical simulations, at several degrees of accuracy by different methodologies. As an example of what can be extracted from these calculations, about a decade ago state-of-the-art *ab initio* GW many-body calculations were performed on top of standard DFT calculations in undoped graphene (Trevisanutto 2008) to accurately take into account electron-electron self-energy corrections for this material. The calculations confirmed that there is no opening of the energy gap at the reciprocal space point K. Furthermore, a small change of the linear dispersion close to the Dirac point was

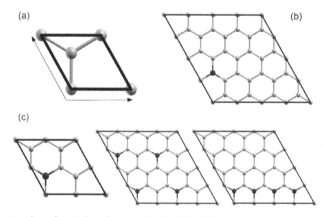

Figure 13.1: In (a) the top view of a graphene hexagonal unit cell with 2 carbon atoms is shown. In (b) where one of the carbon atoms is replaced by a dopant, giving a 3.125% doping content is depicted as a 4 × 4 supercell. In (c), three possible ways of simulating a 12.5% doping content are shown: using a 2 × 2 supercell with one dopant atom; a 4 × 4 supercell with 4 dopant atoms, sorted in a similar way as in the 2 × 2 case; a 4 × 4 supercell with 4 dopant atoms forming a line in the (100) direction, which may or not be more stable than the earlier configuration. The carbon atoms are in light gray, and the doping is in dark gray.

obtained, with a slightly large Fermi velocity close to K and the appearance of a kink, which was connected to low-energy single-particle excitation, a kink that was also observed by angle-resolved photoemission spectroscopy measurements. Furthermore, the band plot overall changed little with respect to standard DFT-LDA calculations, with a gap opening of around 0.6 eV at Gamma and M special points. Later a 17% increase in the Fermi velocity was obtained, which is in better agreement with known experimental results.

In realistic operative conditions, all materials present defects that might perturb or even change some of its properties, and studying them is an important step for a more complete understanding of a system. For graphene, this is even more relevant, since its properties are very dependent on its very flat and pristine structure. Furthermore, there might be a practical interest in changing some of its properties, such as increasing its nearly zero electronic gaps for nano-electronic applications (Sahu and Rout 2017, Han et al. 2007). One way to achieve this is by doping the surface or adding adatoms to it. In this case, many different defects have been computationally simulated and characterized in graphene, and among the things that one has to keep in mind to properly simulate these defects, probably the most important one is building of the atomic cell. The 2-atoms unit cell shown in Fig. 13.1a cannot represent any except the simpler cases, such as pristine graphene or a 50% mixed structure. In most cases, multiple unit cells are required, so one has to build what is called a supercell. In Fig. 13.1b an illustrative case of substitutional doping of a random element in the graphene layer is shown. For the 3.125% doping content, a specific 4 × 4 supercell size has to be used to accurately take into account this amount of doping. Furthermore, for one specific content, an even larger supercell could be required to simulate different configurations. For 12.5% doping, as an example, one could use a 2 × 2 supercell with one C atom replaced, or a 4 × 4 cell with 4 C atoms replaced, as illustrated in Fig. 13.1c. In this last case, the distribution of the doping atoms is not unique, and it could be found that a (100) line of doping atoms is more stable than putting them far from each other, minimizing their interaction. Although the example given in Fig. 13.1c is about substitutional doping, many other defects, such as adatoms and strained sheets, also require the use of supercells. Thus, their choice is an important step in characterizing nearly all structural and chemical defects.

In general, the use of supercells results in a larger computational power required and longer simulation time, and more often than not, this requires a compromise between accuracy and computationally feasibility. This ultimately limits the accuracy and/or type of property that can be extracted from the simulation. As in the illustrative example of this case, the change on the interband π plasmon frequency of graphene was evaluated for the pristine case and substitutionally doped with 1% of B and N (Hage et al. 2018). Using scanning transmission electron microscopy, a decrease/increase of this frequency was obtained through high-resolution electron energy loss spectroscopy for B/N doping, respectively. Typically, accurate theoretical simulations of excitonic effects involve beyond-DFT approaches, which are computationally expensive, and in addition, an accurate representation of about 1% doping would require a large 7 × 7 supercell, which could make such type of calculations very expensive. Therefore, the theoretical simulations were done in a much smaller 3 × 3 supercell, a bit far from the real experimental system. Nevertheless, a good qualitative and acceptable quantitative agreement was obtained between theory and experiment.

Transition Metal Dichalcogenides

Due to the unique properties presented by graphene, several other atomically-thin 2D layers have recently started to be considered. Inspired by the process of exfoliating graphite to obtain graphene, Transition Metal Dichalcogenides (TMD) were usually considered. Their bulk structure consists of a stacking of 2D layers 3-atoms wide with strong in-plane covalent bonds and weak interlayer Van Der Waals interactions, which also makes them ideal for exfoliation. 2D-TMD can be similarly fabricated and characterized, and present a hexagonal structure similar to graphene although out of the plane, as illustrated in Fig. 13.2.

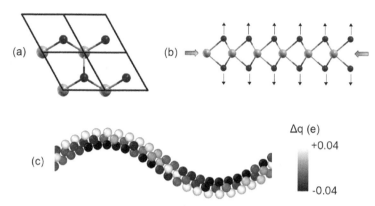

Figure 13.2: (a) Top view of a unit cell (limited by a solid black line) of a MoS$_2$, one typical example of a Transition Metal Dichalcogenides (TMD). Mo atoms are in light gray, while S atoms are in dark gray. (b) Side view of the structure shown in (a), where a lateral compression (represented by solid arrows) giving rise to a perpendicular displacement of the sulfur atoms (represented by thin black arrows) were illustrated. In (c), a charge density difference between the Bader charge of the curved shown monolayer with respect to the pristine MoS$_2$ flat monolayer. Loss/gain of electrons is represented by negative/positive Δq values, in black/white respectively.

Even though 2D TMD might be structurally similar to graphene in some aspects, some notable differences can be highlighted (Akinwande et al. 2017). Since its width typically consists of three atoms, the additional degree of freedom in the perpendicular direction help the system ease the strain when submitted to in-plane stress, as illustrated in Fig. 13.2b. Furthermore, due to the charge transference from the transition metal to sulfur, a charge rearrangement occurs when the material is submitted to defects such as wrinkling and folding. For a MoS$_2$ monolayer isotropically stressed laterally, for example, one gets a natural charge transference from the sulfur atoms inside the curve to those outside it, which creates an extra dipole field that lowers the energy of the strained slab, as illustrated in Fig. 13.2c (Negreiros et al. 2019a). Later sulfur vacancy formation energies are considerably lower in energy than carbon vacancies in graphene, and it is usually a much more common defect in 2D TMD, which makes them interesting candidates for catalysis, especially in reactions involving desulfurization (Zhang et al. 2020). 2D-TMD also differ from graphene in the considerably large electronic bandgap, around a few eV, and a spin-orbit coupling that causes valence-band splitting, which makes them potential candidates to be used in spintronics (Latzke et al. 2015).

Transition Metal Oxides

Another family of 2D materials are Transition Metal Oxides (TMO) (Yang et al. 2019, Netzer and Fortunelli 2016). In many TMO bulk structures, a clear layer-by-layer stacking can be usually identified in at least one spatial direction, which would suggest that 2D-TMO should be similar to 2D-TMD and graphene. There are, however, at least two important differences to be noted. First, 2D-TMO might not present such weak inter-layer interactions, since the interaction in all directions is of ionic nature and typically quite strong. This also means that the 2D layers could be less easily exfoliable, and other methods of fabrication are more typically used to overcome these issues. In addition, from a theoretical point of view, TMO are more challenging than the earlier two families. In TMO the inter-atomic interaction is ionic, so there are large charge transferences between the atoms, with the formation of cations and anions with a strong oxidation and reduction states. The electronic structure is, therefore, significantly changed with respect to the neutral atoms that compose it, and on top of that, defects such as vacancies might add electrons or holes to the system that are not trivially described by most theoretical methods that exist today. When focusing on DFT, for example, one finds that most standard functions fail to describe correctly the electronic structure of these materials, due to a self-interaction error inherent in the theory that is not error compensated in the case of TMO. Therefore, additional corrections (typically called "beyond-DFT methods") are often require for a more accurate characterization of TMO (Li et al. 2013, Mandal et al. 2019), which results in larger computational power and time required that can vary from 10% with DFT+U Hubbard corrections to orders of magnitude more, such as hybrid functionals and many-body perturbation theory.

Apart from the 2D-TMD and graphene-like monolayers, 2D-TMO can have a structure significantly different from the layers extracted from the bulk host. As an example of such a case, free-standing WO_3 bilayers were shown to have a different structural arrangement compared to a bilayer directly extracted from its monoclinic bulk structure, as illustrated in Fig. 13.3 (Negreiros et al. 2019b). At the DFT+U level, this alternative bilayer model proposed is '0.3 eV/WO_3 unit' more stable than the one derived from the monoclinic bulk WO_3, and structurally resembles the α-phase of the MoO_3 bulk crystal. Its stability was shown to be connected to symmetric rearrangement, which eliminates any residual dipole in the system, resulting in electronic shell closure and a large opening in the electronic gap. In general, given the more polar character of transition oxides, one can expect that the ultra-thin layers might often show some sort of reconstruction, especially if adsorbed to substrates that could strongly interact with them.

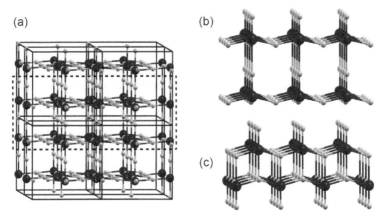

Figure 13.3: (a) Monoclinic bulk structure of WO_3. The unit cell is doubled in each direction for visualization purposes. In (b), the optimized free-standing bilayer directly extracted from the monoclinic bulk shown in (a). In (c), a different bilayer structure is proposed, which is 0.3 eV/WO_3 unit more stable than (b) according to DFT+U calculations, as detailed in the main text. Tungsten and oxygen atoms are in black and white, respectively.

Additional Families and Further Considerations

Even though graphene-like, 2D-TMD and 2D-TMO layers presented so far are composed of an important and considerable part of all known 2D materials, they are far from being the only known cases. 2D boronphosphide, hexagonal boron nitride, ultra-thin metallic films (only a few-atoms wide) are other examples of typical materials. Furthermore, new materials have been recently considered that do not exactly fit in any of these mentioned families. As an example, the new family of 2D materials was derived from the perovskite structures discovered only a decade ago (Kojima et al. 2009). These materials have a general formula ABC_3, where A and B are two distinct types of cations located in the corners (atom A) and center (atom B) of a cubic or near-cubic structure, while the anions C (typically oxygen) are in face-centerd positions. Ultra-thin perovskite films can be extracted from the layered bulk material, being only a few-atoms wide. They have been fabricated by deposition, either by pulsed laser deposition or molecular-beam epitaxy techniques, on top of a substrate that can be either another oxide or metallic (Netzer and Fortunelli 2016). As it typically occurs for TMO, simulating these 2D materials also include the substrate in which they are adsorbed, since the interface is relevant for the properties of the overall material.

Among many computational works that have been done in these materials in the last years (Even et al. 2014, Jayan and Sebastian 2019), most work done today are performed at the DFT level, which generally gives accurate structural and mechanical properties and some basic electronic properties. More accurate second-order features, such as optical properties, which are especially important since perovskites are commonly used in photovoltaic applications, are also evaluated with more accurate functionals, such as hybrid functionals or within the GW Green-function many-body formalism. A study on the stability of distinct 2D films compared to their bulk hosts is highlighted (Yang et al. 2018), which were found to be extremely dependent on the perpendicular termination of the extracted layers and on the material itself. In fact, for the $CH_3NH_3PbI_3$ perovskite, the extracted 2D-layer CH_3NH_3I-terminated was found to be surprisingly even more stable than the 3D bulk, so this layer could be grown non-stoichiometric under CH_3NH_3I-rich conditions. This surprisingly reinforces the uniqueness of the 2D materials, and how different they are with respect to their bulk counterparts.

Van Der Waals Heterostructures

In these types of materials, the in-plane bonds are strong, of ionic or covalent nature, while the inter-plane interaction is weak, having a Van der Waals character. This allows the building of a nanosheet composed of two or more 2D layers, and since the interaction between them is weak, one can potentially combine the intrinsic properties of each layer into a single material or even enhance them.

Typical methods of experimental fabrication of heterostructures are mechanical or chemical exfoliation, chemical vapor deposition and molecular beam epitaxy. Even though applying these techniques to generate high-quality 2D materials is generally a challenging task, simulating them in the computer is usually no more complicated than separately simulating the 2D layers that compose it, since the interaction between each layer is weak and the main properties of the heterostructure are given by a combination of its constituent parts. However, it should be noted that there are elements that can make the characterization of these materials far from trivial, such as the possibility of distinct stacking modes between the films, even different rotations with respect to each other, that can give rise to Moiré patterns, for example; the potentially large charge rearrangements that might occur due to the atomically thin nature of the interacting layers; synergic effects, that could enhance some property due to a more complex interaction between layers.

In graphene, for example, the interaction between distinct monolayers already shows a surprising complexity depending on the angle they make with each other (Cao et al. 2018, Lucignano et al. 2019). For this twisted bilayer graphene structure, the existence of magic angles where a Moiré pattern appears and the superlattice presents regions with a high and low density of electrons, depending on the stacking between the monolayers were discovered. When the layers are exactly on top of each other, the stacking is of AA type, and when half of the carbon atoms are on top of the center of the other layer's hexagon, the type is AB, as illustrated in Fig. 13.4a. Charge accumulation and depletion were found to occur in the AA and AB stackings, respectively. Due to this strong interlayer coupling, a flattening of the bands near the Fermi energy occurs, where insulating states at half-filling are present, which were related to the strong electron correlation that exists in the regions with high electron density. This very important result showed the sensitivity of the electronic properties of a somewhat simple system such as bilayer graphene, that could be changed only by a simple rotation of one layer with respect to each other. It also opened the door to an extra functional application of this material, that can have its electronic properties changed with an on/off key, a back and forth rotation of a layer, that keeps the overall system intact. Moreover, this rotation can be thought as a thermodynamically feasible process, due to the van der Waals nature of the interlayer interactions.

In another experimental and theoretical work, superlattices with Moiré patterns were shown to modulate not only the electronic properties of a system but also the optical ones (Tran et al. 2019). Interlayer exciton resonances were found in hBN-encapsulated $MoSe_2/WSe_2$ VDW

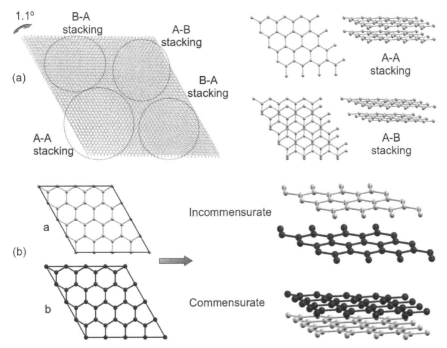

Figure 13.4: (a) Two 40 × 40 graphene sheets are shown, where one is rotated with respect to the other at the magic angle of 1.1°. Two types of stacking that appear from the rotation are highlighted in the figures on the right. (b) Two hypothetical cases illustrating the interface between two hexagonal 2D layers with a size mismatch (b > a). An incommensurate interface, where each layer has its own lattice size, and a commensurate interaction, when both layers have the same lattice size 'x', with 'a < x < b' were illustrated. Therefore, in the commensurate case, both layers are strained, one is compressed while the other is expanded in order to match their lattices and maximize the interlayer interaction.

heterostructures when, a twist angle is similarly applied between them. As it occurred for graphene, different local alignments appear due to the rotation of one of the layers, and three clear regions with hexagonal symmetry can be identified, each with a different stacking mode between the interacting 2D $MoSe_2$/WSe_2 materials. DFT and GW-BSE calculations performed on these three stacking types revealed a small exciton binding energy variation, less than 5% within the moiré supercell. Furthermore, samples with different stackings showed distinct quasiparticle gaps, which confirmed the experimental predictions. The results suggest that VDW heterostructures can be engineered to generate the so-called excitonic crystals for nano-optics applications.

Another unexpected example of the weakly interacting interface was found when an ultra-thin WO_3 bilayer was successfully grown on an Ag(001) substrate (Negreiros et al. 2019b). When $(WO_3)_3$ clusters were deposited on the Ag substrate by molecular beam epitaxy, the growth that initially started in a non-regular way eventually forms a stable flat 2D WO_3 bilayer, that weakly interacts with the Ag substrate and is also incommensurate, presenting various rotation orientations with different Moiré patterns. DFT calculations further revealed that the obtained behavior was connected to the formation of the non-polar bilayer WO_3 structure, mentioned earlier, which is stable even in the free-standing scenario and is representative of how the 2D-TMO materials can change with respect to the bulk-TMO analogue. So the free-standing 2D-WO_3 bilayer was shown to weakly interact with the Ag(001) surface, forming a type of VDW heterostructure with this surface.

As a final example, a possible interface-induced strain could occur when the growth of one 2D material into another is commensurate. This can happen when, for example, the lateral size of the interacting 2D films are different, and it might turn out to be thermodynamically favorable to strain one or both films so as to maximize the inter-layer interaction. This is shown in Fig. 13.4b, where an illustrative example of commensurate and incommensurate interactions is pictured. The final most stable case depends, thus, on the delicate balance between the energy gain by favoring the inter-layer interaction and the energy cost to strain one or both films. Since for the Van der Waals heterostructures, the first term is, by definition, weak, this balance is typically in favor of the incommensurate case. When, however, it is not, additional analysis on the property changes of each strained film has to be performed for a complete characterization of the structure. The strain is known to cause significant changes in the electronic properties, such as the absolute position of the Fermi level and the electronic bandgap. An example of such a case can be found in the work of Rahman et al. 2018, where the authors performed first-principles calculations at the DFT level to study two heterostructures, GaS/GaSe and GaSe/InS, which have a lattice mismatch of 4.7 and 2.1%, respectively. They found that in the larger mismatch GaS/GaSe heterostructure, an incommensurate mode is energetically favorable, while for the lower GaSe/InS case a commensurate mode is favorable. This was found to be not only connected to the mismatch, but also to the sensitivity of the electronic structure of each material due to lateral positive or negative strain. For the three GaS, GaSe and InS monolayers considered, the strain was ultimately responsible for changing the band alignment between them, since the lowest unoccupied molecular orbitals were found to be much more sensitive to the applied strain than the highest occupied molecular orbitals.

Interacting 2D-TMO Films: More General Heterostructures

Apart from the earlier case, when a 2D material interacts more strongly with a substrate or with another 2D layer some or all its properties might be significantly changed. In this case, computer simulations have to accurately characterize the interface, since the knowledge of the isolated components might not be able to provide a good representation of the combined system anymore.

Of the relevant elements to be taken into account to characterize the interface between two interacting materials, two can be highlighted. First, the size and symmetry of the combined film have to be correctly simulated, for there is typically a strong energy gain by maximizing the interaction between the materials, and more often than not the structure and even the symmetry of the 2D film(s) are considerably changed. Second, the accuracy of the theoretical approach adopted should be specially taken into account: given that the interacting components might have quite different electronic natures, for example, an oxide with a strong electron localization, and a metal where the electrons are very delocalized, the potential/functional to be used has to be equally accurate for all cases. As an example, a standard GGA functional, that generally gives good structural parameters and work function predictions for metals, can often incorrectly describe most oxides, especially electronically. So, the structure of the interface between an oxide and a metal, for example, can be inaccuratedly described due to a poor band alignment between the different materials, which in turn predicts a wrong charge rearrangement, and therefore most interface properties. Thus, the use of uncorrected GGA functionals to characterize a metal/oxide interface is likely to give erroneous results, even if the 'metal part' of the system is perfectly described.

As mentioned earlier, one typical example of strongly reactive 2D layers are ultra-thin TMO. Since 2D-TMO show a strong surface polarization and the surface atoms are less coordinated than the bulk ones, there are typically two possible outcomes when interacting with another material: or they are reconstructed to reduce or eliminate its strong polarization or they are strongly bounded, and a large charge rearrangement typically occurs in order to stabilize the under-coordinated polarized 2D-oxide (Yang et al. 2019, Netzer and Fortunelli 2016).

As an example of how the 2D-TMO typically bond to a particular material, $(WO_3)_3$ clusters were adsorbed on the Pd(001) surface by vapor phase deposition (Doudin et al. 2016), and LEED and STM experimental techniques complemented with DFT+U calculations provided a detailed description of the structure formed, a 'WO_2+O' 2-atoms-thin layer in a well-ordered c (2×2) structure that resembles a (001) bilayer taken from the cubic bulk WO_3 structure. Even though a small charge transfer from Pd to the oxide was obtained, a strong charge rearrangement occurred so as to compensate the polarization of the 2D-WO_3 film, which changed the capability of the oxide to release oxygen. This result contrasts the growth of 2D-WO_3 bilayers on Ag(001), detailed earlier, where the oxide reconstructed to eliminate its polarization, and ended up interacting only weakly with the metal surface, showing Moiré patterns typical of VDW heterostructures.

In a more practical example, graphene-supported 2D-TMO heterostructures have the potential of combining the photocatalyst, gas-sensing and supercapacitor properties of different TMO material with the structural stability, thermal and electronic high conductivity and optical transparency of graphene (Azadmanjiri et al. 2018). Overall, the most important elements to be taken into account when simulating these heterostructures is related to the amount of electron transfer and the type of bonding formed at the interface. Given the high electronegativity of oxygen, even when it is in a formal 2^- oxidation state, it was noted that the charge is usually donated by graphene to the 2D-oxide. However, it is also possible to form C-O bonds which share electrons, minimizing this charge donation. The typical graphene/2D-TMO heterostructure formed was shown to sometimes not only combine the properties of each constituent material but also enhance them, such as graphene/SnO_2 as a supercapacitor, reduced graphene oxide/SnO_2 for gas sensing; graphene oxide/TiO_2 for water splitting catalysts; Graphene/TiO_2 as potentially both anode and electrode for Lithium-Ion Batteries (LIB), to cite a few examples.

Heterolayered 2D nanohybrids stacking ultra-thin TMD (MoS_2) and TMO (MnO_2) layers were also recently produced and characterized (Jin et al. 2018). These heterostructures were first experimentally synthesized by self-assembly of exfoliated nanosheets. They were then characterized by *in situ* spectroscopic techniques together with DFT calculations, and the properties of the heterostructure formed were shown to be significantly improved as electrocatalysts and electrodes for LIBs and supercapacitors. The reason for such improvement was connected to the porosity enhancement of the TMD at the interface. Furthermore, it was also found that the trigonal 1T phase of the 2D-TMD layer was stabilized by the interface with the oxide, due to the strong neighbor interaction. Since the MoS_2-1T phase is known to exhibit a metallic character, as opposed to the large-gap semiconductor nature of the hexagonal phase, the charge transfer kinetics and reactivity of the TMD was significantly improved by this phase change. Thus, this combined experimental/theoretical work is another example of heterostructures presenting enhanced properties, clearly showing a way to produce multifunctional heterostructures from stacking TMO and TMD nanosheets.

Perspectives/Conclusions

In this brief chapter, we reviewed the most common types of 2D films, some of their unique features and properties which make them a separate new class of material. As of 2021, the total number of exfoliated/synthesized or theoretically predicted 2D layers is ever growing, and the different families in which they are grouped has also been constantly revised. But more important than their individual properties, is their potential to be combined to form a new system, that could merge or even enhance the properties of its counterparts, which is often achievable. Therefore, they show great promise as potential candidates for the new state-of-the-art multifunctional materials.

Acknowledgements

Negreiros F.R. thanks the Consejo Nacional de Investigacionesen Ciencia y Tecnología (CONICET) and the National University of Cordoba for the support. The author also thanks Dr. Jimena A. Olmos Asar for reading, correcting and providing valuable feedback.

References

Abergel, D., V. Apalkov, J. Berashevich, K. Ziegler and T. Chakraborty. 2010. Properties of graphene: a theoretical perspective. Adv. Phys. 59: 262–482.
Akinwande, D., C. J. Brennan, J. S. Bunch, P. Egberts, J. R. Felts and H. Gao. 2017. A review on mechanics and mechanical properties of 2D materials—Graphene and beyond. Extreme Mech. Lett. 13: 42–77.

Azadmanjiri, J., V. K. Srivastava, P. Kumar, J. Wang and A. Yu. 2018. Graphene-supported 2D transition metal oxide heterostructures. J. Mater. Chem. A 6: 13509–13537.

Barcaro, G. and A. Fortunelli. 2019. 2D oxides on metal materials: concepts, status, and perspectives. Phys. Chem. Chem. Phys. 21: 11510.

Cao, Y., V. Fatemi, A. Demir, S. Fang, S. L. Tomarken, J. Y. Luo et al. 2018. Correlated insulator behaviour at half-filling in magic-angle graphene superlattices. Nature 556(7699): 80–84.

Castro Neto, A. H., F. Guinea, N. M. R. Peres, K. S. Novoselov and A. K. Geim. 2009. The electronic properties of graphene. Rev. Mod. Phys. 81: 109–162.

Cooper, D. R., B. D'Anjou, N. Ghattamaneni, B. Harack, M. Hilke, A. Horth et al. 2012. Experimental review of graphene. ISRN Cond. Mat. Phys. 501686.

Doudin, N., D. Kuhness, M. Blatnik, G. Barcaro, F. R. Negreiros, L. Sementa et al. 2016. Nanoscale domain structure and defects in a 2-D WO_3 layer on Pd(100). J. Phys. Chem. C 120: 28682–28693.

Even, J., L. Pedesseau, E. Tea, S. Almosni, A. Rolland, C. Robert et al. 2014. Density functional theory simulations of semiconductors for photovoltaic applications: hybrid organic-inorganic perovskites and III/V heterostructures. Int. J. Photoenergy 2014: 649408.

Geim, A. K. and I. V. Grigorieva. 2013. Van der Waals heterostructures. Nature 499: 419–425.

Haastrup, S., M. Strange, M. Pandey, T. Deilmann, Per S. Schmidt, N.i F. Hinsche et al. 2018. The computational 2D materials database: High-throughput modeling and discovery of atomically thin crystals. 2D Mater. 5: 042002.

Hage, F. S., T. P. Hardcastle, M. N. Gjerding, D. M. Kepaptsoglou, Che R. Seabourne, K. T. Winther et al. 2018. Local plasmon engineering in doped graphene. ACS Nano 12: 1837–1848.

Han, M. Y., B. Özyilmaz, Y. Zhang and P. Kim. 2007. Energy band-gap engineering of graphene nanoribbons. Phys. Rev. Lett. 98: 206805.

Heard, C. J., J. Čejka, M. Opanasenko, P. Nachtigall, G. Centi and S. Perathoner. 2019. 2D oxide nanomaterials to address the energy transition and catalysis. Adv. Mater. 31: 1801712.

Hedin, L. 1965. New method for calculating the one-particle green's function with application to the electron-gas problem. Phys. Rev. 139: A796–A823.

Jayan, K. D. and V. Sebastian. 2019. A review on computational modelling of individual device components and interfaces of perovskite solar cells using DFT. AIP Conference Proceedings 2162: 020036.

Jin, X., S. Shin, J. Kim, N. Lee, H. Kimb and S. Hwang. 2018. Heterolayered 2D nanohybrids of uniformly stacked transition metal dichalcogenide–transition metal oxide monolayers with improved energy-related functionalities. J. Mater. Chem. A 6: 15237–15244.

Khan, K., A. K. Tareen, M. Aslam, R. Wang, Y. Zhang, A. Mahmood et al. 2020. Recent developments in emerging two-dimensional materials and their applications. J. Mater. Chem. C. 8: 387.

Kojima, A., K. Teshima, Y. Shirai and T. Miyasaka. 2009. Organometal halide perovskites as visible-light sensitizers for photovoltaic cells. J. Am. Chem. Soc. 131: 6050–6051.

Kumar, A., K. Sharma and A. R. Dixit. 2019. A review on the mechanical and thermal properties of graphene and graphene-based polymer nanocomposites: understanding of modelling and MD simulation. Molecular Simulation. DOI:10.1080/08927022.2019.1680844.

Latzke, D. W., W. Zhang, A. Suslu, T. Chang, H. Lin, H. Jeng et al. 2015. Electronic structure, spin-orbit coupling, and interlayer interaction in bulk MoS_2 and WS_2. Phys. Rev. B 91: 235202.

Li, W., C. F. J. Walther, A. Kuc and T. Heine. 2013. Density functional theory and beyond for band-gap screening: performance for transition-metal oxides and dichalcogenides. J. Chem. Theory Comput. 9: 2950–2958.

Lucignano, P., D. Alfè, V. Cataudella, D. Ninno and G. Cantele. 2019. Crucial role of atomic corrugation on the flat bands and energy gaps of twisted bilayer graphene at the magic angle θ ~ 1.08°. Phys. Rev. B 99: 195419.

Mandal, S., K. Haule, K. M. Rabe and D. Vanderbilt. 2019. Systematic beyond-DFT study of binary transition metal oxides. NPJ Comp. Mat. 5: 115.

Negreiros, F. R., G. J. Soldano, S. Fuentes, T. Zepeda, M. José-Yacamáncan and M. M. Mariscal. 2019a. The unexpected effect of vacancies and wrinkling on the electronic properties of MoS_2 layers. Phys. Chem. Chem. Phys. 21: 24731–24739.

Negreiros, F. R., T. Obermüller, M. Blatnik, M. Mohammadi, A. Fortunelli, F. P. Netzer et al. 2019b. Ultrathin WO_3 bilayer on Ag(100): A model for the structure of 2D WO_3 nanosheets. The J. of Phys. Chem. C 123: 27584–27593.

Netzer, F. and A. Fortunelli. 2016. Oxide materials at the two-dimensional limit. DOI: 10.1007/978-3-319-28332-6.

Nicolas, M., M. Gibertini, P. Schwaller, D. Campi, A. Merkys, A. Marrazzo et al. 2017. Two-dimensional materials from high-throughput computational exfoliation of experimentally known compounds (Data download). Materials Cloud Archive 2017.0008/v1, doi: 10.24435/materialscloud:2017.0008/v1.

Novoselov, K. S., A. K. Geim, S. V. Morozov, D. Jiang, Y. Zhang, S. V. Dubonos et al. 2004. Electric field effect in atomically thin carbon films. Science 22: 666–669.

Novoselov, K. S., A. Mishchenko, A. Carvalho and A. H. Castro Neto. 2016. 2D materials and van der Waals heterostructures. Science 353: 9439.

Rahman, A. U., J. M. Morbec, G. Rahman and P. Kratzer. 2018. Commensurate versus incommensurate heterostructures of group-III monochalcogenides. Phys. Rev. Mater. 2: 094002.

Sahu, S. and G. C. Rout. 2017. Band gap opening in graphene: a short theoretical study. Int. Nano Lett. 7: 81–89.

Salpeter, E. E. and H. A. Bethe. 1951. A relativistic equation for bound-state problems. Phys. Rev. 84: 1232–1242.

Sankar, I. V., J. Jeon, S. K. Jang, J. H. Cho, E. Hwang and S. Lee. 2019. Heterogeneous integration of 2D materials: recent advances in fabrication and functional device applications. Nano 14: 12.

Sarma, S. Das, S. Adam, E. H. Hwang and E. Rossi. 2011. Electronic transport in two-dimensional graphene. Rev. Mod. Phys. 83: 407–470.

Tran, K., G. Moody, F. Wu, X. Lu, J. Choi, K. Kim et al. 2019. Evidence for moiré excitons in van der Waals heterostructures. Nature 567: 71–75.

Trevisanutto, P. E., C. Giorgetti, L. Reining, M. Ladisa and V. Olevano. 2008. Ab initio GW many-body effects in graphene. Phys. Rev. Lett. 101: 226405.

Yang, T., T. T. Song, M. Callsen, J. Zhou, J. W. Chai, Y. P. Feng et al. 2019. Atomically thin 2D transition metal oxides: structural reconstruction, interaction with substrates, and potential applications. Adv. Mater. Interfaces 6: 1801160.

Yang, Y., F. Gao, S. Gao and S. Wei. 2018. Origin of the stability of two-dimensional perovskites: a first-principles study. J. Mater. Chem. A 6: 14949–14955.

Zhang, X., A. Chen and Z. Zhou. 2019. High-throughput computational screening of layered and two-dimensional materials. Comput. Mol. Sci. 9: 1385.

Zhang, X., S. Y. Teng, A. C. M. Loy, B. S. How, W. D. Leong and X. Tao. 2020. Transition metal dichalcogenides for the application of pollution reduction: a review. Nanomaterials 10: 1012.

CHAPTER 14

Exploring Water Nanoconfinement in Mesoporous Oxides through Molecular Simulations

Matias H. Factorovich,[1] Damian A. Scherlis[1] and Estefania Gonzalez Solveyra[2,]*

Introduction

In the last couple of decades, water and TiO_2/SiO_2 interactions have attracted great attention and have been a topic of extensive research, both experimental and theoretical (Henderson 2002, Diebold 2003, Verdaguer et al. 2006, Sun et al. 2010, Xu et al. 2019, Renou et al. 2014). Given the fundamental interest in both the physical-chemical properties in confinement and their applications, great precision in the control of size and morphology of nanoscopic cavities in inorganic solid materials was achieved, particularly for mesoporous oxides (Soler-Illia et al. 2002, Lee et al. 2008, Rozes and Sánchez 2011, Contreras et al. 2019). The behaviour of water confined inside hydrophilic pores has been studied in depth by techniques such as NMR, Raman dispersion and scanning differential calorimetry, focusing on its structural, transport and dynamic properties (Takahara et al. 1999, Grünberg et al. 2004, Takahara et al. 2005, Schoch et al. 2008, Steiner et al. 2011, Pajzderska et al. 2014), on adsorption and capillary condensation (Llewellyn et al. 1995, Branton et al. 1995, Inagaki and Fukushima 1998, Oh et al. 2003, Ng and Mintova 2008, Kocherbitov and Alfredsson 2011), and on the solid-liquid transition (Morishige and Uematsu 2005, Chen et al. 2006, Jähnert et al. 2008, Erko et al. 2011, Morishige 2018).

However, one of the central and most complex tasks consists of the precise characterization of the physical-chemical conditions inside the pores, where two main aspects come into play: surface phenomenon and confinement effects. The difficulty to obtain a microscopic vision in the interior of such nanoscopic cavities constitutes an invitation to molecular simulations. In the last decades, simulations methods based on classical and quantum mechanics have contributed greatly to the understanding and rationalization of the physical-chemical properties of materials and surfaces, including structure resolution, study of electronic properties, adsorption-desorption phenomenon and transport properties. The current theoretical and methodological developments combined with the increase in computation capabilities have made it possible for simulations to deal with more

[1] Departamento de Química Inorgánica, Analítica y Química Física, Facultad de Ciencias Exactas y Naturales, Universidad de Buenos Aires, Ciudad Universitaria, Buenos Aires, Pab II, C1428EGA, Argentina.

[2] Department of Biomedical Engineering, Department of Chemistry, and Chemistry of Life Processes Institute, Northwestern University, Evanston, Illinois 60208, United States.

* Corresponding author: estefania.solveyra@northwestern.edu

complex and realistic problems (Challenges in theoretical chemistry 2008). There are numerous examples of Density Functional Theory (DFT) studies on nanosciences and materials science (Zhang and Lindan 2003, Leszczynki 2004, Hafner et al. 2011, Neugebauer and Hickel 2013). On the other hand, classical Molecular Dynamics (MD) and Monte Carlo (MC) methods allow for extending the analysis to dynamical and structural properties of mesoscopic systems for tenths of nanoseconds, even microseconds (Phillpot 2000, Frenkel and Smit 2002, Gubbins et al. 2011, Prokhorenko et al. 2018). In this way, these methods constitute a natural complement to DFT methods. Whereas the latter techniques result to appropriately study reactive problems, classical simulations allow exploring the equilibrium and dynamical properties of systems of hundreds of thousands of atoms. The next step is to further extend the spatial-temporal scope of the simulations involving coarse-grained models, capable of increasing the simulations by a factor of a thousand with respect to full atomistic models (Noid 2013, Tang et al. 2017, Gkeka et al. 2020). Hence, simulations in materials chemistry count on a series of methods that allow going from the molecular to the macroscopic, spanning from the role of electrons and chemical reactivity to the collective behaviour of molecules in the mesoscale (Schleder et al. 2019).

Given the relevance of water in the majority of mesoporous materials applications, next the relation between nanoconfinement and properties of water in mesoporous materials will be explored, with a focus on phase transitions phenomenon (solid-liquid and liquid-vapour equilibria), and structure and transport properties, resorting to molecular simulation techniques at different scales (atomistic and coarse-grained MD).

Water Confined in Mesoporous Materials

Understanding the behaviour of water in porous matrices is a transversal objective to the basic interest of physical chemistry under confinement and current and potential applications of these materials, ranging from fields like physics, chemistry and biology, to geology and engineering. How is water distributed inside the pores? How does confinement and the surface phenomenon affect the structure, dynamics and phase equilibria of water in these materials? These questions are of great interest, but answering them poses serious challenges to current experimental methods.

Given the extensive development of silica (SiO_2) mesoporous materials (from both the synthetic and characterization points of view), a great deal of the studies on confined water has been done in these materials, with pore sizes between 1–10 nm. Regarding the water structure inside MCM-41 materials, results from neutron diffraction experiments along with numerical modelling and atomistic simulations showed a non-homogeneous distribution of water molecules inside the pores, as well as a strong distortion of the radial distribution function compared to bulk water (Soper 2012, Mancinelli et al. 2009). The inhomogeneous occupation of the available volume and occurrence of depleted zones inside the pores stress the ambiguous meaning of the term density when the fluid is confined in the nanoscale (Lombardo et al. 2009, Mancinelli et al. 2011).

NMR relaxometry experiments on water dynamics in silica matrices showed different diffusivities and relaxation mechanisms, ascribed to different types of water molecules inside the pores: a first layer of strongly adsorbed and practically immobile molecules, a second layer with higher mobility and correlation times five times faster (0.1 ns) and a third group of basically free water molecules, with correlation times in the order of the picoseconds (Steiner et al. 2011). In the same line, quasi-elastic neutron scattering experiments on water in mesoporous TiO_2 nanocrystals revealed the existence of three components for water diffusion inside the pores, associated with molecules in the first, second and third hydration layers (Mamontov et al. 2007).

Regarding the filling and adsorption mechanisms of the pores, NMR experiments of water inside silica pores of 3.3 and 8 nm suggested two mechanisms in place: in narrow pores, capillary condensation is preceded by the formation of a single monolayer, while for larger pores a multilayer radial growth is observed (Grünberg et al. 2004). Analogue results were found for water inside TiO_2 aerosols (Velasco et al. 2017).

Monte Carlo and Molecular Dynamics simulations provide a powerful tool to explore the behaviour of water in nanoconfined environments. Resorting to grand canonical MC (GCMC) and MD simulations, Milischuk and Ladanyi studied the structure and dynamics of water inside amorphous silica pores of 2–4 nm in diameter identifying two distinct layers of interfacial water, and a homogeneous region where water density reaches a value of 90% relative to that of bulk water (Milischuk and Ladanyi 2011). The diffusion of water between planar silica surfaces was exhaustively studied by Debenedetti and coworkers as a function of temperature, hydrophilicity, and confinement (Giovambattista et al. 2009, Castrillón et al. 2009a, b). Meanwhile, Wei et al. compared the diffusion of water molecules confined in slit pores of TiO_2 and graphite and it was observed that water mobility is highly restrained, with diffusion coefficients strongly dependent on pore size (Wei et al. 2011).

Despite the extended use of titania mesoporous materials, water behaviour in these systems has not been as thoroughly studied as their silica counterparts. Titania is far more hydrophilic, with water-TiO_2 interactions being approximately three times stronger than with silica. This is expected to largely affect the behaviour of water in these materials. Density Functional Theory (DFT) calculations have shown that for water multilayers, molecular adsorption dominates in surfaces without defects of anatase (101) and rutile (110) (Tilocca and Selloni 2004, Bandura et al. 2008, Sun et al. 2010, Sánchez et al. 2011). Based on this, several atomistic force fields for water and TiO_2 systems have been developed in the last decade (Bandura and Kubicki 2003, Alimohammadi and Fichthorn 2011), allowing to perform MD simulations of water on planar titania surfaces as well as on TiO_2 nanoparticles in aqueous solutions (Předota et al. 2004, Koparde and Cummings 2007). However, examples of water confined in TiO_2 nanopores are still scarce. In this context, the results discussed next address structural and dynamical properties of water confined in TiO_2 cylindrical nanopores of different sizes and varying degrees of filling, explored through classical MD simulations (Gonzalez Solveyra et al. 2013, Velasco et al. 2017).

Water Structure and Dynamics inside TiO_2 Nanopores

Atomistic MD simulations of water confined in TiO_2 nanopores of 1.3, 2.8 and 5.1 nm were performed in the canonical ensemble (constant number of molecules, volume and temperature, NVT). Water molecules were represented with the SPC/E model (Berendsen et al. 1987) and the mesoporous matrix and the interactions between TiO_2 and water were represented with the force field proposed by Bandura and Kubicki (Bandura and Kubicki 2003). Pores were filled with different amounts of water molecules to explore filling mechanisms. Further details can be found in (Gonzalez Solveyra et al. 2013, Velasco et al. 2017). The results on filled pores will be described first.

The strong hydrophilicity of the TiO_2 surface along with the crystal nature of the rutile phase results in a very well-defined multilayer water structure, as shown in Fig. 14.1. The first layer is tightly bound to the surface, showing almost no exchange with more external layers during the simulations and very low diffusion coefficients (Fig. 14.2, left axis). There is a second layer strongly bound as well but exhibiting more mobility than the first one, a third less structured layer and finally a central region in which water density becomes homogeneous. In the smallest 1.3 nm nanopore, the remaining space beyond the bilayer allows only for a quasi 1D arrangement of water molecules, originating the peak in the central region of Fig. 14.1, upper panel. For the 2.8 nm pore and the 5.1 nm pore, the strongly adsorbed bilayer is followed by a third less structured water layer, in which water molecules already regained around 75% of the maximum mobility (Fig. 14.2). The central portion of the larger pores corresponds to water exhibiting an almost homogeneous density, with mobility comparable to that of bulk water. The profiles obtained for water confined in TiO_2 pores resemble those of water on planar rutile TiO_2 surfaces (Wei et al. 2011, Předota et al. 2004, Předota et al. 2007).

When comparing the density profiles obtained for water inside titania nanopores with those of silica pores of similar sizes, water density in the proximity of the SiO_2 pores ranges from 0.8 to

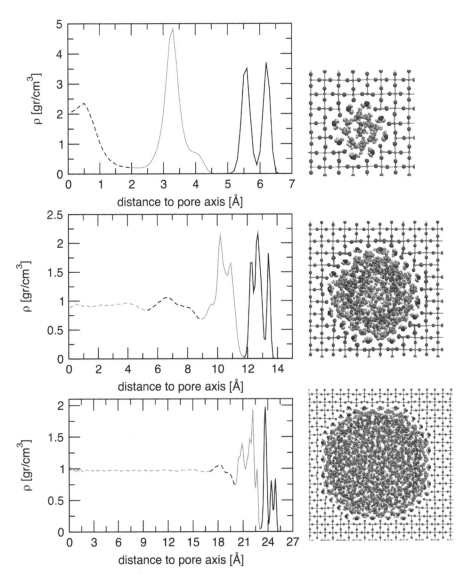

Figure 14.1: Radial density profiles for the 1.3 (upper panel), 2.8 nm (middle panel) and the 5.1 nm (lower panel) pores at 100% filling. Black and grey full lines correspond to the first and second adsorbed layers respectively. Black dashed lines correspond to the third region, and grey dashed lines to the central portion of water regaining bulk-like properties (Adapted with permission from (Gonzalez Solveyra et al. 2013). Copyright 2013 American Chemical Society).

1.4 g/cm³ (Milischuk and Ladanyi 2011, Shirono and Daiguji 2007, Mancinelli et al. 2009) whereas in the TiO_2 nanopores these values get close to 2 g/cm³ and go to zero between layers. The first two layers are more compact in TiO_2 nanopores and much more structured, following the much higher hydrophilicity of the surface.

Given the highly structured arrangement of water molecules inside the TiO_2 nanopores, as reflected in the radial density profiles, it is not surprising that the mobility of water largely depends on the distance to the surface (Fig. 14.2). The first water monolayer exhibits almost no translational mobility. In this region, water movements are the result of molecular vibrations more than diffusional behaviour. For the smallest pore, diffusion in the second layer is approximately 30% of the value at the centre of the pore, where it still only reaches 1/3 of the bulk value. Confinement seems to have little impact when comparing results between the bigger pores. The obtained results are rather

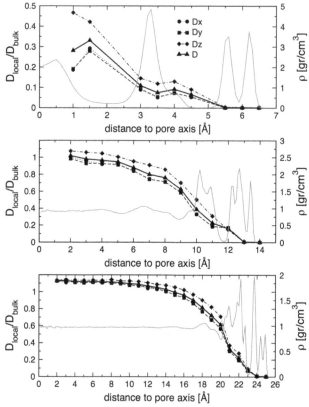

Figure 14.2: Diffusion coefficients as a function of the distance to the pore axis for the 1.3 (upper panel), 2.8 (middle panel), and 5.1 nm (lower panel) pores. Black full lines correspond to the total diffusion coefficient (D), while dotted, dashed and dot-dashed lines correspond to its x, y, z component respectively (D_x, D_y, D_z). Grey superimposed lines correspond to the radial density profile for each system (Adapted with permission from (Gonzalez Solveyra et al. 2013). Copyright 2013 American Chemical Society).

similar for the 2.8 and the 5.1 nm pores, and in turn, they are in agreement with earlier results for water and titania planar interfaces (Předota et al. 2004). In the larger pores, the diffusivity increases rapidly moving away from the surface, reaching a plateau at around 10 Å from the wall. Simulations for water inside silica nanopores have also shown that the mobility of water molecules increases moving away from the surface, but reaching a value closer to that of bulk water at a smaller distance from the wall in comparison with titania (typically between 6–8 Å) (Milischuk and Ladanyi 2011, Shirono and Daiguji 2007). When examining the central region of the 2.8 and 5.1 nm pores, the diffusivity of confined water is slightly higher than that of bulk water. This is related to the fact the water density in this region is smaller than that of bulk water, allowing for faster mobility (0.93 and 0.96 g/cm³ for the 2.8 and 5.1 nm pores respectively). Similar results were obtained for confined water in silica nanopores of diameters between 2 and 4 nm (Milischuk and Ladanyi 2011, Jähnert et al. 2008).

With regards to diffusion anisotropy, coefficients are larger in the Z axial direction, along which there is no confinement, but the effect is significant only for the smaller 1.3 nm pore. A somewhat surprising increase in the perpendicular components of the diffusion coefficients (D_x and D_y) can be observed in the interlayer spaces, where there is a low probability of finding a water molecule: while water is highly stabilized within each layer, permanence between them is rather unstable, which implies that a molecule would move faster from one layer to the adjacent one.

Water Filling inside TiO$_2$ Nanopores

To gain insights on the liquid-vapour transition and to obtain details on the filling mechanisms in mesoporous TiO$_2$ matrices and how they depend on pore size, the three pores were equilibrated with different content of water molecules.

For the smaller 1.3 nm pore, increasing water content is acquired in the gradual formation of the multilayer structure as shown in Fig. 14.1, without any signs of capillary condensation or the development of a second homogeneous phase. Water tends to aggregate in small clusters on completion of the first layer. This was ascribed to the pronounced curvature of the 1.3 nm pore, which encourages the formation of these small clusters in order to maximize water-water interactions. The lack of capillary condensation may seem contradictory to experimental and computational results for water inside less hydrophilic pores of similar sizes (Branton et al. 1995, Inagaki and Fukushima 1998, Llewellyn et al. 1995, de la Llave et al. 2010, de la Llave et al. 2012). However, this discrepancy can be rationalized keeping in mind that the quasi-solid water layer at the TiO$_2$ interface diminishes the effective radius of the pore. In this way, water in TiO$_2$ nanopores of 1.3 nm might be expected to behave in a similar way to less hydrophilic pores of less than 1 nm, for which capillary condensation is not observed (Naono and Hakuman 1993).

For the larger pores, capillary condensation is observed for water contents of 71 and 65% for the 2.8 and 5.1 nm pores respectively (Fig. 14.3a–b). At this point, the system exhibits two phases: a liquid plug corresponding to the high-density phase that connects the space between pore walls, and a low-density phase, adsorbed at the pore wall (Fig. 14.3c). Beyond this critical filling value, the density of each phase remains constant; both are in equilibrium, and adding more water molecules results in the growth of the liquid plug. Moreover, once capillary condensation occurs, the surface density of the adsorbed phase is similar for the 2.8 and 5.1 nm pore, in accordance with previous MD results (de la Llave et al. 2012).

Figure 14.3 panel d shows the density of the adsorbed phase (Γ) as a function of pore-filling for the three systems studied. For the 2.8 nm pore, the progressive addition of water molecules results in a homogeneous growth of the adsorbed bilayer. The phase transition occurs in equilibrium: once reached the equilibrium surface density ($\Gamma_{eq} \sim 14$ nm^{-2}), the condensed phase forms and a water plug appears. As mentioned earlier, the first layer of adsorbed water decreases the effective pore size, such that the 2.8 nm pore can be compared to a less hydrophilic pore of ~ 2 nm, for which capillary condensation is observed, but without hysteresis (de la Llave et al. 2012).

On the other hand, for the 5.1 nm pore, a strong supersaturation is needed for capillary condensation to occur. The density of the adsorbed phase before capillary condensation reaches a value close to 25 nm^{-2}, and it drops to ~ 15 nm^{-2} once the condensed phase appears. The transition occurs out of equilibrium, probably accompanied by a hysteresis loop (as will be discussed next). After completing the strongly adsorbed bilayer (filling > 40%), the subsequent growth of the adsorbed phase proceeds in an inhomogeneous manner: water molecules begin to accumulate in a region of the pore, resulting in an axially localized increase of surface density eventually triggering capillary condensation and the filling of the pore. This radial thickening of a water aggregate resembles the hydrophobic transition characterized in 3.0 and 4.0 nm nanopores (de la Llave et al. 2012).

Figure 14.3 panels a–b show the radial density corresponding to the adsorbed phase in equilibrium with the liquid phase. The grey shaded areas indicate the difference between the profiles just before and after capillary condensation, highlighting that the vapour-liquid transition occurs in equilibrium for the 2.8 nm nanopore, whereas a significant supersaturation is observed for the 5.1 nm pore. These filling mechanisms were also observed by NMR experiments of mesoporous TiO$_2$ aerosols with different water filling (Velasco et al. 2017).

Molecular dynamics simulations discussed so far corresponded to the canonical ensemble, constant NVT. This allowed studying the structure and dynamics of water inside the pores and also rationalizing the filling mechanism of the pores when the water content is varied in separate simulations. However, this method does not allow one to link the observed behaviour to the vapour

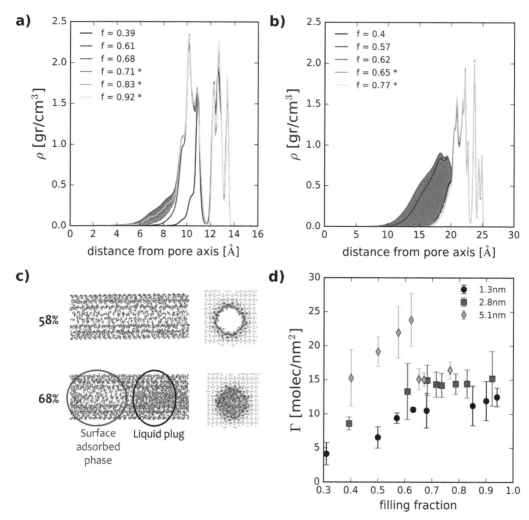

Figure 14.3: (a–b) Radial density profiles as a function of the distance from the pore axis for 2.8 and 5.1 nm, nanopores. Filling degree of each curve is given in the figure, and the asterisk refers to filling for which capillary condensation is observed. In these cases, the radial density plotted corresponds to the adsorbed phase in equilibrium with the liquid phase (whose density is not shown). Grey shaded areas indicate the difference between the profiles just before and after capillary condensation, highlighting that the vapour-liquid transition occurs in equilibrium for the 2.8 nm nanopore, whereas a significant supersaturation is observed for the 5.1 nm pore. (c) Capillary condensation phenomenon in the 2.8 nm pore, indicating the formation of the liquid plug-in contact with the adsorbed phase. (d) The density of the surface-adsorbed water phase (Γ) as a function of filling percentage in the 1.3 (black circles), 2.8 (grey squares), and 5.1 nm (light grey diamonds) diameter pores. For fillings exhibiting phase coexistence, the density was calculated excluding the region occupied by the condensed phase (Adapted with permission from (Gonzalez Solveyra et al. 2013). Copyright 2013 American Chemical Society, and from (Velasco et al. 2017). Copyright 2017, American Chemical Society).

pressure inside the pores, as could be measured in adsorption-desorption experiments. For that, Grand Canonical MD simulations (GCMD) of water inside the pores is needed, which will be discussed next.

Liquid-Vapour Equilibrium: Adsorption Mechanisms

Sorption studies of water through molecular simulations are widespread in the literature. In general Monte Carlo simulations in the Grand Canonical ensemble have shown that up to four liquid-vapour phase transitions are possible during the adsorption process depending on the pore size and hydrophilicity of the pore (Brovchenko et al. 2004, Shirono and Daiguji 2007). In pores ranging from 1 to 3 nm diameter, these transitions were identified as monolayer transition, bilayer transition, pre-wetting and a capillary condensation and, in particular, the first three are present in pores with high hydrophilicity. Thus adjusting the hydrophilicity of the pore changes the behaviour of the adsorption, producing a significant shift of the capillary condensation pressure towards higher values (Schreiber et al. 2002). Molecular dynamics simulation in the canonical ensemble with the mW coarse-grained model of water has also been used to study the adsorption phenomenon (Molinero and Moore 2009). In these simulations, pores with different diameter and different hydrophilicity were partially filled with different amounts of water molecules (de la Llave et al. 2010, de la Llave et al. 2012). These simulations revealed two filling mechanisms for pores in the range 3.0–4.0 nm, depending on the water-surface affinity: (i) a localized growth of a water droplet for surfaces of moderate hydrophobicity, and (ii) an homogeneous filling leading to water densities above equilibrium for hydrophilic nanopores, for which this situation is considered as supersaturation, just before capillary condensation. Hybrid Monte Carlo-Molecular dynamic simulations in the grand canonical (GCMD) ensemble have also been made. This method allows obtaining equilibrium, non-equilibrium, and dynamical information on the liquid-vapour transition (Factorovich et al. 2014a).

This section will be divided into two parts. First, the Grand Canonical Screening (GCS) method is presented. GCS allows retrieving the vapour pressure of condensed bulk phases or aggregates (Factorovich et al. 2014b, Pickering et al. 2018). Second, results from GCMD simulation regarding desorption, adsorption pressure, filling and emptying dynamics are presented (Factorovich et al. 2014a).

Grand Canonical Screening

Adsorption experiments are usually presented in plots of the amount of adsorbate adsorbed as a function of the pressure relative to the vapour pressure at the working temperature (P/P_{vap}). Several methods allow access to the coexistence properties of a liquid-vapour system, including the vapour pressure. Some of these methods are the Gibbs ensemble Monte Carlo simulations developed by Panagiotopoulos (Panagiotopoulos et al. 1988, Panagiotopoulos 2000, Frenkel and Smit 2001) the largely sophisticated and accurate Transition Matrix Monte Carlo (TMMC), (Errington 2003) and a simple method called Grand Canonical Screening (GCS) (Factorovich et al. 2014b). Here the focus will be on the latter, which despite its simplicity can give vapour pressures with accuracy in between the Gibbs ensemble and the TMMC method.

The GCS method consists in carrying out a series of simulations with a conventional grand canonical scheme, but each of these at a different chemical potential (fugacity or pressure), including the equilibrium pressure in the screening range. During the simulation time, the evolution of the number of particles is recorded; it can increase, decrease or fluctuate around a value. From the analysis of the results produced by these series of computational experiments, it is possible to obtain the equilibrium pressure by averaging the highest pressure at which the simulation lose particles with the lowest pressure at which the simulation gain particles.

The system's layout for the GCS method is a simulation box that is partially filled with the fluid under study. This liquid is deployed in the form of a slab, i.e., a system that is discontinuous in one of the dimensions of the box but is homogeneous in the other two, thus exposing a pair of surfaces. If the slab has the proper dimensions, it allows retrieving bulk properties of the fluid.

Figure 14.4 shows results of the GCS procedure for computing the vapour pressure for two water models, SPC/E (Berendsen et al. 1987) (A) and the coarse-grained mW (Molinero and Moore 2009) (B). The system consists of water molecules represented with these models arranged in slab geometry. The SPC/E simulations are pure Monte Carlo runs performed with the open-source software TOWHEE (Martin et al. 2013), whereas for the mW a modified version of the code LAMMPS (Plimpton 1995) was used to perform molecular dynamics in the Grand Canonical ensemble.

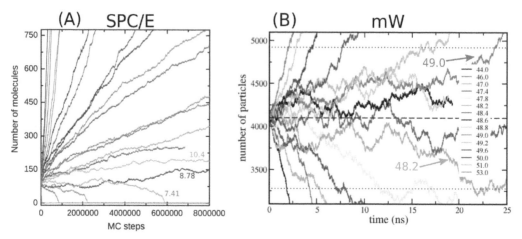

Figure 14.4: Number of water molecules versus GCMC simulation step (A) or molecular dynamic time (B) of a planar vapour-liquid interface using the SPC/E(A) or mW (B) model. Each curve corresponds to a different pressure, which is indicated in mbar (A) or mbar x 1000 (B) (Adapted with permission from (Factorovich et al. 2014b). Copyright 2014, American Chemical Society).

As can be seen for SPC/E water, the system passes from evaporation to condensation when the pressure jumps from 7.4 mbar to 8.8 mbar: this allows placing the equilibrium vapour pressure at 8.1 ± 0.7 mbar. This outcome is in excellent agreement with the value of 7.65 ± 0.38 mbar reported for SPC/E water at 300 K by Liu and Monson based on simulations in the Gibbs ensemble (Liu and Monson 2005). For mW the value is 0.486 ± 0.004 mbar (Molinero and Moore 2009).

Two statements are worth mentioning for the GCS method. The first is that, as this method is based on the explicit simulation of interphase, it is not limited to compute vapour pressures of bulk liquids, as is the case for the Gibbs ensemble, but can also be used to compute the vapour pressure of clusters regardless of the geometry of the vapour-liquid interphase. For example, this method has been used to compute the vapour pressure of nanodroplets as a function of their sizes and also the solid-vapour equilibrium and water's quasi-liquid layer (Factorovich et al. 2014c, Pickering et al. 2018). Second, regards the precision of the method, as can be seen from Fig. 14.4B, the uncertainty of the equilibrium vapour pressure can be decreased by further exploration of the pressure around the unknown equilibrium value. However, as the pressure approaches the vapour pressure, the ratio between effective particle insertions and deletions points to 1, such that the sampling needs to be extended over longer simulation times to explore the final bias to evaporation or condensation. In the example case for the mW water with 25 ns of simulation time, a 1% error was obtained. In these conditions, the precision of the GCS approach is higher than that achieved with the Gibbs ensemble method but lower than that corresponding to TMMC, which has been reported to be below 0.2% (Errington 2003).

Adsorption Isotherms

Here the results from GCMD simulations of water adsorption are presented (Factorovich et al. 2014a). The water is modelled with the mW potential. The pore is formed by the same mW particles, where a channel of 2.8 nm diameter was built by removing a cylindrical block from a box of water molecules, to produce cylindrical nanopores with amorphous walls. The interactions between the water molecules and the particles of the pore were adjusted by tuning the parameters of the Stillinger Weber force field. The hydrophobicity of the surface was characterized in terms of the contact angle (θ), by simulating a droplet of mW water over a planar surface of the same particles making the pore. For a detailed description of the pore construction or the contact angle assessment, readers are referred (Giovambattista et al. 2007, Factorovich et al. 2014a).

Equilibrium Desorption Pressure and the Kelvin Equation

Figure 14.5A–B shows the isotherms for 2.8 nm diameter pores for two different hydrophilicities. Both hysteresis loops are of type H1, as expected for homogeneous cylindrical pores of this size (Thommes 2010). For a contact angle equal to zero (Fig. 14.5A), capillary condensation takes place at $P/P_0 = 0.59$, with a water surface density at the condensation of $\Gamma_c = 6.4$ nm^{-2}, while desorption occurs at $P/P_0 \approx 0.44$, with a surface coverage at the equilibrium of $\Gamma_{eq} = 3.4$ nm^{-2}.

Isotherm (A) is in good agreement with experimental measurements performed in hydroxylated silica mesopores of comparable sizes (Zhao and Lu 1998, Inagaki and Fukushima 1998, Kocherbitov and Alfredsson 2011, Ng and Mintova 2008). For these silica pores the adsorption branch is generally found somewhere in the P/P_0 range between 0.5 and 0.7 (Inagaki and Fukushima 1998, Ng and Mintova 2008), with desorption typically shifted 0.1 to lower pressures. The adsorption-desorption hysteresis from our simulations are slightly larger than usually seen in experiments, which can be attributed to two major reasons: (i) the distribution of pore sizes present in synthesized mesoporous materials—rather than a uniform pore diameter, which makes the adsorption branch sweep up two tenths in units of relative pressure instead of being a vertical; and (ii) capillary condensation in these pores is an activated phenomenon, which may require higher chemical potentials in simulations than in experiments, because of the different accessible sampling times.

Figure 14.5B illustrates the water sorption isotherm for a pore with $\theta = 82.9°$, which roughly represents the interaction of water with a graphitic material. Our simulations show that condensation of water in these systems should occur in conditions of vapour supersaturation. Comparison of

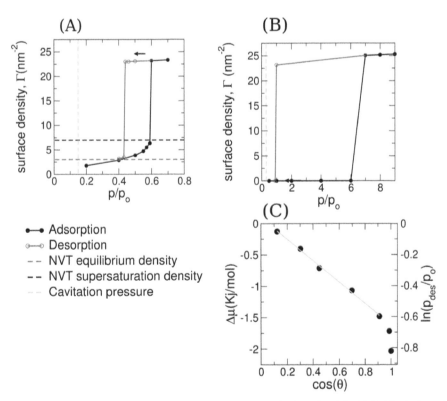

Figure 14.5: Water sorption isotherms obtained in pores with contact angles (A) 0° and (B) 82.9°. The horizontal dashed lines indicate the results from canonical molecular dynamics simulations for the density of the surface adsorbed phase in equilibrium (light grey) and at the point of condensation (black). (C) The logarithm of the relative desorption pressure as a function of the cosine of the contact angle θ. The linear dependence indicates that the Kelvin equation is satisfied for contact angles above 20°. The light grey line guides the eye; from its slope, the surface tension of the mW model can be recovered (Adapted with permission from (Factorovich et al. 2014a). Copyright 2014, American Chemical Society).

Fig. 14.5A–B indicates that an increase in θ affects the condensation pressure (P_c) to a higher extent than the desorption pressure (P_d), enlarging the hysteresis loop. Experimental evidence presented by several authors is consistent with these observations (Easton and Machin 2000, Liu and Monson 2005, Monson 2008). Also Γ_{eq} and Γ_c exhibit a dramatic decrease with respect to the more hydrophilic pores.

The light grey lines of Fig. 14.5A–B show the desorption branches of the isotherms for the hydrophilic and hydrophobic pores. These desorption isotherms were constructed by performing simulations that started with partially filled pores, which expose a liquid-vapour interface (in the form of a condensed plug). With this layout, evaporation may proceed as in open-ended pores, at the liquid-gas interface, and this process is assumed to take place in equilibrium (Fan et al. 2011, Thommes 2010). Figure 14.5 shows that the surface density, Γ_{eq}, changes from a value of 3.4 nm^{-2} for θ = 0, to nearly zero for θ = 82.9°, evidencing its strong dependence with the hydrophilicity of the pore. Increasing pore hydrophobicity favours the stabilization of thicker adsorbed layers against the formation of a condensed phase at a given pressure. The values Γ_{eq} obtained through GCMD simulations are in good agreement with those reported in the canonical ensemble (de la Llave et al. 2010, de la Llave et al. 2012).

The equilibrium desorption pressures P_d obtained for the pores of different hydrophilicity are accurately predicted by the Kelvin equation:

$$\ln\left(\frac{P_d}{P_{vap}}\right) = -\frac{2\sigma V_m \cos(\theta)}{RT}\frac{1}{r_{pore}}$$

where σ is the liquid-vapour surface tension, V_m is the molar volume of the liquid, θ is the contact angle between the water and the pore surface, R is the universal gas constant and T is the system's temperature.

As shown in Fig. 14.5C, the logarithm of the desorption pressure resulting from our GCMD simulations is well represented as a linear function of the cosine of the contact angle for values of θ above 20°. The liquid-vapour surface tension calculated from the slope of this curve is σ = 0.066 N/m at 298 K, in perfect agreement with the surface tension reported for the mW model and close to the experimental value of 0.071 N/m (Molinero and Moore 2009). As the desorption pressures follow the Kelvin equation, it confirms that the process is taking place at equilibrium, while this relation is not influenced for the condensation pressures, since it involves an activated process, as mentioned above. It may be surprising to find that the validity of the Kelvin equation extends to water in pores of 2.8 nm diameter. This behaviour has been seen, for example, in water nanodroplets down to diameters of 2 nm or in Lennard-Jones fluids (Factorovich et al. 2014c, Tarazona et al. 1987, Bruno et al. 1987). However the reason why this macroscopic theory holds down to this nanoscopic scale is not clear. Evans noted that the Kelvin equation is valid as long as the fluid-wall interaction is weak enough to prevent wetting (Evans 1990). This is consistent with the deviations observed in Fig. 14.5C for low contact angles.

Qualitative Description of Activated Mechanisms, Adsorption and Cavitation

The GCMD approach delivers a dynamical description of the condensation process in real-time. Figure 14.6A–B displays a temporal sequence of images extracted from the GCMD simulations illustrating these mechanisms of condensation for θ = 0° and θ = 9.3°.

Figure 14.6 reveals different mechanisms for the nucleation of the liquid phase, depending on the water-surface interactions. In the hydrophilic case (Fig. 14.6A), the filling of the pore is heterogeneous. Above a certain surface density (of around 7 nm^{-2} for this pore diameter), a fluctuation closes the gap across the centre of the pore and the liquid is formed. For higher contact angles (Fig. 14.6B) the mechanism changes and involves the formation of a droplet that grows in a certain spot on the surface. It follows that water molecules are more stable as part of aggregate than adsorbed on the surface. Finally, in a highly hydrophobic environment, the nucleation process

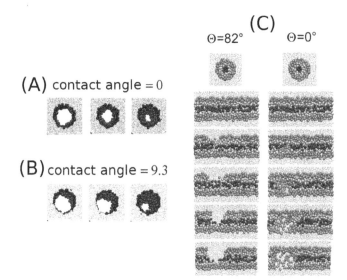

Figure 14.6: (A) and (B) Front views of the pore model depicting two different nucleation mechanisms. From left to right, the temporal sequence of snapshots taken from the GCMD simulations at the condensation pressure. Light grey points: particles of the pore. Dark spheres: water molecules inside the pore. (C) Front and side cross-section views of the pore model depicting two different cavitation mechanisms: heterogeneous (left) and homogeneous (right). From top to bottom, the temporal sequence of snapshots taken from the GCMD simulations at the cavitation pressure. For clarity, water molecules are displayed in different grey tones according to their distance to the wall. The particles of the pore wall are represented as light grey points. Adapted with permission from (Factorovich et al. 2014a). Copyright 2014, American Chemical Society.

follows a homogeneous pathway not different from the nucleation of liquid from the vapour phase and so the transition takes place close to the vapour spinodal. These observations are in agreement with earlier findings from canonical molecular dynamics simulations (de la Llave et al. 2012). The two simulations frameworks (NVT and μVT) are complementary: the canonical simulations allow for a thorough characterization of configurations along with the adsorption jump that are not easily stabilized in the grand canonical ensemble, where only one phase may be observed at a time. In turn, the present simulations in the μVT ensemble allow for establishing the pressure that corresponds to the different fillings in the NVT molecular dynamics calculations.

In the case of blocked-end pores, desorption takes place by cavitation, which is an activated process that involves the nucleation of a bubble inside the condensed phase (Thommes 2010). Cavitation can be seen in simulations by studying the desorption in the absence of a liquid-vapour interface. This type of desorption is indicated with vertical green lines in Fig. 14.5A–B, and occurs at very low pressure, $P/P_0 \approx 0.2$. Depending on the hydrophilicity, cavitation follows different mechanisms, as shown in Fig. 14.6C, for contact angles of 0° and 82.9°. In the hydrophilic pore, the nucleation of a bubble is determined by the intermolecular forces in the liquid phase and occurs homogeneously, whereas, in the hydrophobic pore, it tends to form heterogeneously, in contact with the wall. Still, these different mechanisms translate only in a minor shift in the cavitation relative pressures.

In summary, in this section we mentioned several simulation methods to evaluate the vapour pressure of fluids, which can be compared to experiments. Results of the equilibrium and dynamic behaviour of the sorption phenomenon showed that the desorption branch of the isotherm in open-ended pores is a non-activated process in which pressure is accurately described by the Kelvin equation, with deviations occurring for contact angles below 20°. The effect of hydrophobicity is to separate the adsorption and the desorption branches, enlarging the hysteresis. The dynamics of equilibrium desorption (cavitation) is largely influenced by the liquid-wall interaction. When the pore is hydrophilic the cavitation bubble tends to form in the centre of the pore, contrary to the hydrophobic case in which the bubble forms at the walls. Finally, our simulations showed that canonical molecular dynamics (NVT ensemble) can describe the liquid-vapour transition and the equilibrium states with the same accuracy as grand canonical simulations (μVT).

Next, attention is focused on the solid-liquid equilibrium of confined water, resorting to NVT molecular simulations.

Solid-Liquid Equilibrium of Confined Water

Under nanometric confinement, liquid water can be thermodynamically stable at temperatures below the homogeneous nucleation transition of bulk water ($T_H = 235$ K). The confinement effects on the solid-liquid equilibrium derive from the balance between the surface and the bulk contributions to the free energy of the system, generally favouring the liquid phase over the solid one. This is known as capillary fusion. As one is dealing with condensed phases, the relevant thermodynamic variable is the temperature, and the physical manifestation of the phenomenon is the decrease in the melting point with respect to that of the bulk.

Melting and freezing of water in silica nanopores have been thoroughly studied by Differential Scanning Calorimetry (DSC), Neutron Diffraction (ND), X-ray Diffraction (DRX), Nuclear Magnetic Resonance (NMR) and dielectric measurements (Pearson and Derbyshire 1974, Hall et al. 1985, Schmidt et al. 1995, Morishige and Kawano 1999, Schreiber et al. 2001, Kittaka et al. 2006, Liu et al. 2006, Webber et al. 2007, Jähnert et al. 2008, Seyed-Yazdi et al. 2008, Jelassi et al. 2010, Morishige 2018). These studies have shown a marked decrease of the melting temperature of confined ice with respect to that of the bulk, and the presence of non-crystallized water in the pores below the melting temperature of confined ice.

Confinement affects not only the thermodynamics of the system but also the structure of water and ice inside of the pores. Unlike the macroscopic world, where bulk ice consists mainly in the polymorph of hexagonal ice, I_h, results from neutron and X-ray diffraction experiments of ice in silica nanopores have shown that confined ice consists in a hybrid of cubic, I_c, and hexagonal ice, with profuse stacking faults (Morishige and Iwasaki 2003, Morishige and Uematsu 2005, Baker et al. 1997). In nanopores with diameters smaller than 50 nm, ice forms as a defective form of cubic ice and its stability with respect to the most common hexagonal ice increases as the pore size decreases. In these pores, confined ice consists of alternating layers of hexagonal and cubic ice (Baker et al. 1997, Moore et al. 2012). NMR and diffraction experiments have also shown the presence of a disordered component for water in nanopores below T_m (Morishige and Iwasaki 2003, Liu et al. 2006, Webber et al. 2007, Seyed-Yazdi et al. 2008, Jelassi et al. 2010). While there is an experimental agreement on the existence of non crystallizable water in the pores, the current experimental methods do not allow resolving the spatial distribution of water and ice in the pores.

Regarding the effect of pore hydrophilicity and degree of filing, DSC experiments have shown that water T_m in silica nanopores of 3 nm diameter to be rather insensitive to these parameters (Morishige and Iwasaki 2003, Jähnert et al. 2008, Findenegg et al. 2008, Jelassi et al. 2010, Deschamps et al. 2010). However, later DRX studies showed important differences in the structure of water at 298 K in hydrophilic and hydrophobic nanopores, while the melting temperatures were essentially the same (Jelassi et al. 2011, Morishige 2018).

The molecular resolution provided by computational simulations makes them an optimal tool to characterize the solid-liquid transition and the structure and properties of confined water and ice and to address the different experimental observations and unanswered questions. However, using MD to characterize the solid-liquid equilibrium represents a challenge given the elevated computational cost and the insufficient accuracy of the existing water models to describe the mentioned processes. A reliable analysis of the phenomenon of nucleation and crystalline growth involves extensive sampling times in systems with thousands of molecules, which translates into long simulation times. The mW coarse-grained water model mentioned earlier to study water sorption isotherms allows overcoming these limitations. It also reproduces the structures of different water phases (I_h, I_c, high and low-density amorphous ice), and the thermodynamic anomalies of water and the enthalpy of phase transitions (liquid-vapour and solid-liquid), with a precision higher than or comparable to popular atomistic water models.

Here the results from MD simulations with the mW coarse-grained model for water will be presented, studying the melting, crystallization and the structure of water and ice in confined environments. Particularly, the focus will be on the effect of the wall-water interactions and the filling of the pores on the solid-liquid equilibrium of water in cylindrical silica nanopores (Gonzalez Solveyra et al. 2011). For a detailed description of filled pores, readers can refer to (Moore et al. 2010).

Water and Ice in Amorphous Silica Nanopores

Molecular dynamic simulations of water confined in cylindrical silica-like nanopores were performed in the canonical ensemble (NVT). Both pore walls and water molecules were represented by the coarse-grained model mW (Molinero and Moore 2009). The nanopores were constructed from an instantaneous configuration of mW bulk water molecules equilibrated at 298 K and 1 atm. The pore walls were made of particles outside a cylinder of radius R = 3 nm from the centre of the block, and their motion was restrained to vibrations around their initial position. Pores were filled with different amount of water molecules and their hydrophilicity varied. MD simulations were performed with the LAMMPS package. All the systems were equilibrated at 298 K and were then subjected to different treatments at low temperatures (quenching, cooling and heating ramps, thermalization at constant T). In order to identify liquid water from ice, and further characterize and quantify ice as hexagonal and cubic ice during crystallization and melting, the CHILL algorithm was used. Further details can be found in (Gonzalez Solveyra et al. 2011, Moore et al. 2010).

Water Crystallization in Partially Filled Nanopores

While melting corresponds to an equilibrium process, crystallization is a phenomenon governed by kinetics and as such, it depends on the speed of the cooling ramp. Initially, water was crystallized inside hydrophilic nanopores with filling fractions f = 0.4, 0.6, 0.9, and 1 (filled pore) (Gonzalez Solveyra et al. 2011). Crystallization was achieved in two steps: first, the systems were cooled down from 230 K to 180 K at a cooling ramp of 1 K/ns. This resulted in the crystallization for all the systems studied, except the one with less water filling (f = 0.4), for which an additional equilibration at 180 K for 2 μs was needed (Fig. 14.7). Given the stochastic nature of the nucleation process, water crystallization occurs more easily (meaning at higher temperatures for the same cooling ramp) for pores with greater water content, in accordance with experimental results in partially filled silica nanopores of similar sizes (Morishige and Iwasaki 2003). The second step consisted of equilibrating the crystallized systems at 190 and 200 K, to favour molecular reorganization to form one crystallite inside the pores (annealing process). In all cases, this two-step procedure yielded an ice plug that occluded the pore.

As discussed earlier, the initial configurations at 298 K contained two phases of water in the pore: a liquid plug and a surface-adsorbed phase. On cooling, water crystallized exclusively in the liquid phase, forming a crystallite whose size depends largely on the water content of the pore. The adsorbed phase of the pore walls does not form ice. For all systems studied, the crystallization process originated homogeneously, that is, without assistance from the pore wall, irrespective of their water content. An initial ice nucleus in the centre of the liquid plug continued growing until forming only one crystallite, in accordance with previous MD simulations of filled nanopores (Moore et al. 2010).

An in depth analysis of the crystallization at 180 K for the 40% filled nanopore showed two very distinct regimes, as shown in Fig. 14.7. First, an induction period during which sub-critical ice nuclei form and dissolve. From the size of the nuclei that formed and dissolved during this period, the size of the critical nucleus can be estimated in around 20 water molecules, which is in agreement with that of bulk water at the same temperature (Moore and Molinero 2010). The induction period is followed by a growth period, in which the critical-sized nucleus increases in size, giving rise to a

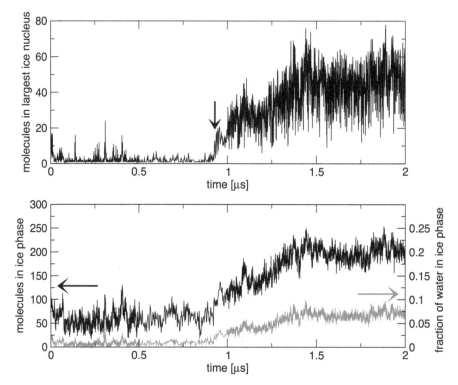

Figure 14.7: Crystallization of ice in a 40% filled hydrophilic pore at 180 K. Upper panel: number of water molecules in the largest ice nucleus as a function of time, counting molecules in the core as cubic + hexagonal ice. The black arrow indicates the formation of the critical nucleus, ending the induction period and starting the period of growth of the crystallite. Lower panel: The black curve (label in the left axis) corresponds to the temporal evolution of the number of water molecules in the core of the ice nuclei (cubic + hexagonal). The grey curve (label in right axis) shows the total fraction of water in the ice phase, including cubic, hexagonal, and intermediate ice. Reprinted with permission from (Gonzalez Solveyra et al. 2011). Copyright 2011, American Chemical Society.

unique massive crystallite. Reaching the critical-sized nucleus took around 900 ns to occur, which is three orders of magnitude larger than the induction time for ice nucleation of bulk mW water at the same temperature (Moore and Molinero 2010). This difference can be ascribed to a smaller driving force to crystallize, given that the melting temperature of confined ice is approximately 60 K lower than that of bulk. Secondly, nucleation is prevented inside the 40% filled pores given the low availability of water molecules to form a crystal (see Table 14.1), as also described by Cooper and coworkers (Liu et al. 2007, Cooper et al. 2008).

Structure of Water and Ice in Silica-like Nanopores

The structure of confined ice showed stacked layers of cubic and hexagonal ice, for all filling degrees. Neutron diffraction studies of water in silica nanopores of 7 nm suggest that I_h forms only for $f > 0.9$, disappearing for $f < 0.5$. However in our simulations, both I_h and I_c were observed in the pores, with a 1:2 ratio, similar to filled pores (Moore et al. 2010). This higher proportion of cubic ice has also been observed experimentally for confined water systems (Morishige and Uematsu 2005). Hexagonal ice was observed even for the pores with $f = 0.4$, suggesting that this polymorph may exist in confined ice, even if its presence cannot be detected experimentally.

In line with experimental studies evidencing the presence of noncrystallizable water inside silica pores (Schreiber et al. 2001, Engemann et al. 2004, Petrov and Furó 2011), our simulations show that this component is in the form of small aggregates of adsorbed water on the surface (only for hydrophilic pores) and around the ice crystal, as a result of interfacial melting. In Fig. 14.8 it

Hydrophilic nanopore   **Hydrophobic nanopore**

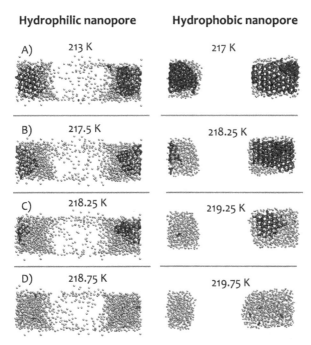

Figure 14.8: Representative snapshots showing the progression of melting of confined ice in 60% filled hydrophilic and hydrophobic nanopores (left and right panels). The pore walls are omitted for clarity. Panels A–D: water molecules in liquid water (light grey) and total ice ($I_c + I_h$ + interfacial black), where total ice includes cubic, hexagonal and interfacial ice. The corresponding temperatures for each snapshot are indicated on top of each panel. Adapted with permission from (Gonzalez Solveyra et al. 2011). Copyright 2011, American Chemical Society.

can be seen that this component is present both for hydrophilic and hydrophobic pores, even at 180 K. In the latter, an adsorbed phase on the pore walls is not observed (Fig. 14.8 right panel), and the thickness and density of water molecules in the quasi-liquid layer around the ice crystal are smaller. This translates into a higher fraction of water in the solid phase when compared with the hydrophilic pore (Table 14.1). Simulations on filled pores also showed the density and structure of the quasi-liquid layer to be affected by pore hydrophilicity (Moore et al. 2012). The development of this quasi-liquid layer obtained from the fact that the liquid phase can better accommodate the pore structure than the solid phase (Parry et al. 2006). This phenomenon has also been observed in the ice-vacuum interface (Bluhm et al. 2002, Dash et al. 2006, Stewart and Evans 2005, Pickering et al. 2018).

The exchange between molecules in the quasi-liquid phase and the ice and their translational mobility was characterized by computing the survival probability in the liquid phase, as shown in Fig. 14.9 for the hydrophilic pore with f = 0.6 at 190 and 215K (both below the corresponding melting temperature, T_m = 216K, see Table 14.1).

Our simulations suggest that a reversible and dynamic conversion exists between water in the liquid or quasi-liquid phase and ice. The abrupt decrease at short times is due to the molecules in the interface. At 190 K, the exchange is not fast enough and one can see that molecules in the centre of the ice do not exchange with the liquid phase (Fig. 14.9, right panel). However, at 215K a complete exchange of molecules between both phases is achieved, accompanied by significant translational mobility in both phases. It is worth noting, that at 215K (1 K below T_m), the ice crystal retains its structure, even though molecules have exchanged almost completely with the liquid phase. This indicates that the component of the water that does not crystallize in the pores at $T < T_m$ does not adopt a form of plastic ice, as suggested in the literature-based in NMR relaxation times (Webber et al. 2007), but rather it retains the behaviour of a disordered liquid or glass (Schranz and Soprunyuk 2019).

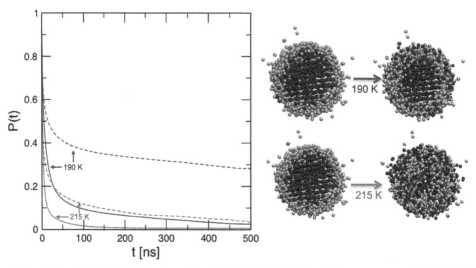

Figure 14.9: (left panel) Probability that water molecules in the 60% filled hydrophilic nanopores remain in the liquid (full line) or crystal (dashed line) states at 190 K (dark grey) and 215 K (light grey) (right panel). Snapshots perpendicular to the pore axis for water in the pore at time t = 0 (left snapshots) and after 500 ns (right snapshots) at these two temperatures. Molecules that originally were classified as part of the liquid are shown in grey and molecules that at t = 0 were classified as ice are shown in black. Adapted with permission from (Gonzalez Solveyra et al. 2011). Copyright 2011, American Chemical Society.

Melting Temperature and Enthalpy of Confined Water

Figure 14.8 shows the melting progression of hydrophilic and hydrophobic pores with f = 0.6. The melting process was abrupt for all pore fillings studied, irrespective of the hydrophilicity of the pores, proceeding from the outer part of the crystal towards the centre of the pore until the ice nucleus is small enough to melt completely. This can be observed in Fig. 14.8 for the hydrophobic pore, where a continuous increase of the thickness of the liquid interfacial layer is observed, followed by an abrupt transition at $T \approx T_m$.

Ice melting inside the pores was monitored through the fraction of water in the solid phase. The computed values for the melting temperatures of all systems studied are collected in Table 14.1.

Our results show that T_m is rather insensitive to pore filling, although a slight decrease is observed when going from f = 0.4 to 1. This is comparable to experimental results (Jelassi et al. 2010), and can be attributed to an increase in the surface/volume ratio of the ice crystal, as it becomes smaller with smaller water content inside the pores. It is also worth noting that the observed differences are

Table 14.1: Temperature and enthalpy of melting of ice in partially filled pores. The suffix *w* corresponds to the enthalpy of melting calculated by considering all the water molecules in the system, while the suffix *ice* refers to the value normalized with the water molecules in the ice phase (including cubic, hexagonal, and interfacial ice). Bulk values for the mW taken from (Molinero and Moore 2009).

f	Pore-wall interactions	N_{Ice}	N_{Total}	T_f [K]	ΔH_m^w [kcal/mol water]	ΔH_m^{ice} [kcal/mol ice]
0.4	hydrophilic	211	948	212.0 ± 1	0.23 ± 0.03	1.04 ± 0.14
0.6	hydrophilic	475	1399	216.5 ± 1	0.32 ± 0.05	0.95 ± 0.14
0.6	hydrophobic	591	1378	216.0 ± 1	0.41 ± 0.06	0.95 ± 0.14
0.9	hydrophilic	754	2123	219.0 ± 1	0.33 ± 0.05	0.92 ± 0.14
1.0	hydrophilic	2295	4429	219.0 ± 1	0.55 ± 0.83	1.06 ± 0.16
bulk	-	-	-	274.0 ± 1	1.26 ± 0.03	1.26 ± 0.01

Adapted from (Gonzalez Solveyra et al. 2011) with permission from the American Chemical Society.

very small in comparison to the depression with respect to bulk water $T_m = 274$ K for the mW model (Molinero and Moore 2009).

With regards to pore hydrophobicity, our simulations showed that even if the fraction of ice inside the hydrophobic pores is much higher than in hydrophilic pores with the same degree of filling, the melting temperature is essentially the same (Table 14.1), in agreement with experiments for water inside silica pores (Findenegg et al. 2008, Jelassi et al. 2010, Morishige 2018). However, the structure of water inside the hydrophobic pores is different, as shown in Fig. 14.8 and discussed above. Thus, the similarity in T_m should not be associated with a comparable structure of water and ice in the pores.

To complete the analysis of the melting process, the enthalpy of melting (ΔH_m) was computed for all the systems studied, resorting to two normalization schemes: (i) by the total number of water molecules inside the pores (ΔH_m^w), which allows a more direct comparison with experiments, and (ii) by the number of molecules in the ice (ΔH_m^{ice}), which allows a more precise characterization of the energetics of the phase transition. Results are shown in Table 14.1. It can be seen that ΔH_m^w increases with pore filling, following the increase of the fraction of water inside the pores. Meanwhile, ΔH_m^{ice} is practically the same for all systems studied, irrespective of pore-filling or hydrophilicity, but is in turn smaller than ΔH_m^{bulk} at the melting temperature for bulk water, in agreement with experimental results (Jähnert et al. 2008, Findenegg et al. 2008). This comes from the fact that the structure of the liquid, with decreasing temperature, resembles that of ice.

In summary, the results discussed here showed that the solid-liquid equilibrium in confined environments differs for the macroscopic counterpart in at least two aspects. First, melting temperatures of confined ice are significantly depressed as compared to bulk ice. Second, the structure of confined ice shows important differences. The existence of a quasi-liquid layer of water between the ice crystal and the pore wall was observed, even for $T \ll T_m$, showing a dynamic and reversible exchange with ice. This non-crystallized component acts as a buffer for the wall-water interactions, such that the hydrophilicity of the pores does not affect T_m significantly. However, pore hydrophilicity does impact on the density of the adsorbed phase and the fraction of ice inside of the pores.

General Conclusions

The thermodynamic properties of confined water differ significantly from the bulk counterpart and depend not only on parameters of the fluid but also on the conditions imposed by the confining matrix (size, morphology, surface-water interactions). This provides a tool for characterizing porous systems in different scales. For this, it is crucial to count with a precise description of the physical-chemical properties inside of the pores, which combine confinement and surface effects. Importantly, the works reviewed in this chapter show that these two effects cannot, in general, be studied separately, but they are intimately intertwined. Confinement and surface interactions operate jointly to determine the properties of water in nanospaces. To think of the effects of confinement alone is unrealistic. Even in hydrophobic cavities, water molecules in the outermost layers of the nanophase align to form an ice-like structure that determines the properties of the whole fluid.

These studies demonstrate that the structure and mobility of confined water are only affected in the vicinity of the interface, typically not much further than the second monolayer, or 1 nm from the walls. Water molecules lying beyond this distance exhibit behaviour that, from a microscopic standpoint, cannot be discerned from that of a molecule in the bulk phase. Then, as a general message, these studies convey that the incidence of confinement in most physical-chemical properties of water will be only seen in pores below 5 or 4 nm because in larger pores most of the fluid will retain the behaviour of the bulk liquid.

Appropriately chosen simulation techniques may bridge the gap between the microscopic world and the insights provided by current experimental techniques. In line with this, the results discussed earlier illustrate the potential of molecular simulation at different scales as a tool to explore the characteristics of confined water in mesoporous oxides, from the structural and transport properties to the phase transitions phenomenon (liquid-vapour and solid-liquid equilibria).

Acknowledgements

E.G.S. and M.H.F. acknowledge CONICET for the doctoral fellowships when the research projects presented here were conducted. We also acknowledge the Center of High Performance Computing of the University of Utah for the technical support and the allocation of computing time.

References

Alimohammadi, M. and K. A. Fichthorn. 2011. A force field for the interaction of water with TiO_2 surfaces. The J. Phys. Chem. C 115(49): 24206–24214.

Baker, J. M., J. C. Dore and P. Behrens. 1997. Nucleation of ice in confined geometry. J. Phys. Chem. B 101(32): 6226–6229.

Bandura, A. V. and J. D. Kubicki. 2003. Derivation of force field parameters for TiO_2-H_2O systems from *ab Initio* calculations. J. Phys. Chem. B 107(40): 11072–11081.

Bandura, A. V., J. D. Kubicki and J. O. Sofo. 2008. Comparisons of multilayer H_2O adsorption onto the (110) surfaces of α-TiO_2 and SnO_2 as calculated with density functional theory. J. Phys. Chem. B 112(37): 11616–11624.

Berendsen, H. J. C., J. R. Grigera and T. P. Straatsma. 1987. The missing term in effective pair potentials. J. Phys. Chem. 91(24): 6269–6271.

Bluhm, H., D. F. Ogletree, C. S. Fadley, Z. Hussain and M. Salmeron. 2002. The premelting of ice studied with photoelectron spectroscopy. J. Phys.: Condens. Matter 14: 227–233.

Branton, P. J., P. G. Hall, M. Treguer and K. S. W. Sing. 1995. Adsorption of carbon dioxide, sulfur dioxide and water vapour by MCM-41, a model mesoporous adsorbent. J. Chem. Soc., Faraday Trans. 91: 2041–2043.

Brovchenko, I., A. Geiger and A. Oleinikova. 2004. Water in nanopores. I. Coexistence curves from gibbs ensemble Monte Carlo simulations. J. Chem. Phys. 120: 1958–1972.

Bruno, E., U. Marini Bettolo Marconi and R. Evans. 1987. Phase transitions in a confined lattice gas: Prewetting and capillary condensation. Physica A 141: 187–210.

Castrillón, S. -V., N. Giovambattista, I. A. Aksay and P. G. Debenedetti. 2009a. Effect of surface polarity on the structure and dynamics of water in nanoscale confinement. J. Phys. Chem. B 113(5): 1438–1446.

Castrillón, S. -V., N. Giovambattista, U. A. Aksay and P. G. Debenedetti. 2009b. Evolution from surface-influenced to bulk-like dynamics in nanoscopically confined water. J. Phys. Chem. B 113(23): 7973–7976.

Chen, S. -H., F. Mallamace, C. -Y. Mou, M. Broccio, C. Corsaro, A. Faraone et al. 2006. The violation of the Stokes-Einstein relation in supercooled water. Proc. Natl. Acad. Sci. 103(35): 12974–12978.

Contreras, C. B., O. Azzaroni and G. J. A. A. Soler-Illia. 2019. 1.16 – Use of confinement effects in mesoporous materials to build tailored nanoarchitectures. *In*: D. L. Andrews, R. H. Lipson and T. Nann [eds.]. Comprehensive Nanoscience and Nanotechnology (Second Edition). Oxford: Academic Press.

Cooper, S. J., C. E. Nicholson and J. Liu. 2008. A simple classical model for predicting onset crystallization temperatures on curved substrates and its implications for phase transitions in confined volumes. J. Chem. Phys. 129: 124715–124727.

CRC Handbook of Chemistry and Physics, 81th ed.; CRC: Boca Raton, 2000–2001.

Dash, J. G., A. W. Rempel and J. S. Wettlaufer. 2006. The physics of premelted ice and its geophysical consequences. Rev. Mod. Phys. 78(3): 695–741.

De la Llave, E., V. Molinero and D. A. Scherlis. 2010. Water filling of hydrophilic nanopores. J. Chem. Phys. 133(3): 034513–034523.

de la Llave, E., V. Molinero and D. A. Scherlis. 2012. Role of confinement and surface affinity on filling mechanisms and sorption hysteresis of water in nanopores. J. Phys. Chem. C 116(2): 1833–1840.

Deschamps, J., F. Audonnet, N. Brodie-Linder, M. Schoeffel and C. Alba-Simionesco. 2010. A thermodynamic limit of the melting/freezing processes of water under strongly hydrophobic nanoscopic confinement. Phys. Chem. Chem. Phys. 12(7): 1440–1443.

Diebold, U. 2003. The surface science of titanium dioxide. Surf. Sci. Reports 48(5-8): 53–229.

Easton, E. B. and W. D. Machin. 2000. Adsorption of water vapor on a graphitized carbon black. J. Colloid Interface Sci. 231: 204–206.

Engemann, S., H. Reichert, H. Dosch, J. Bilgram, V. Honkimäki and A. Snigirev. 2004. Interfacial melting of ice in contact with SiO_2. Phys. Rev. Lett. 92(20): 205701–205701.

Erko, M., G. Findenegg, N. Cade, A. Michette and O. Paris. 2011. Confinement-induced structural changes of water studied by Raman scattering. Phys. Rev. B 84(10): 104205.

Errington, J. R. 2003. Direct calculation of liquid–vapor phase equilibria from transition matrix Monte Carlo simulation. J. Chem. Phys. 118: 9915–9925.

Evans, R. 1990. Fluids adsorbed in narrow pores: phase equilibria and structure. J. Phys.: Condens. Matter 2(46): 8989–9007.

Factorovich, M. H., E. Gonzalez Solveyra, V. Molinero and D. A. Scherlis. 2014a. Sorption isotherms of water in nanopores: relationship between hydropohobicity, adsorption pressure, and hysteresis. J. Phys. Chem. C 118(29): 16290–16300.

Factorovich, M. H., V. Molinero and D. A. Scherlis. 2014b. A simple grand canonical approach to compute the vapor pressure of bulk and finite size systems. J. Chem. Phys. 140(6): 64111–64119.

Factorovich, M. H., V. Molinero and D. A. Scherlis. 2014c. Vapor pressure of water nanodroplets. J. Am. Chem. Soc. 136(12): 4508–4514.

Fahrenkamp-Uppenbrink, J., P. Szuromi, J. Yeston and R. Coontz. 2008. Challenges in theoretical chemistry. Science 321(5890): 783.

Fan, C., D. D. Do and D. Nicholson. 2011. On the cavitation and pore blocking in slit-shaped ink-bottle pores. Langmuir 27(7): 3511–3526.

Findenegg, G. H., S. Jähnert, D. Akcakayiran and A. Schreiber. 2008. Freezing and melting of water confined in silica nanopores. Chem. Phys. Chem. 9(18): 2651–2659.

Frenkel, D. and B. Smit. 2001. Understanding Molecular Simulation: From Algorithms to Applications. Computational Science Series: Elsevier Science.

Frenkel, D. and B. Smit. 2002. Understanding Molecular Simulation: From Algorithms to Applications. Vol. 1: Elsevier (previamente editado por Academic Press).

Giovambattista, N., P. G. Debenedetti and P. J. Rosky. 2007. Effect of surface polarity on water contact angle and interfacial hydration structure. J. Phys. Chem. B 111: 9581–9587.

Giovambattista, N., P. J. Rossky and P. G. Debenedetti. 2009. Effect of temperature on the structure and phase behavior of water confined by hydrophobic, hydrophilic, and heterogeneous surfaces. J. Phys. Chem. B 113(42): 13723–13734.

Gkeka, P., G. Stoltz, A. Barati Farimani, Z. Belkacemi, M. Ceriotti, J. D. Chodera et al. 2020. Machine learning force fields and coarse-grained variables in molecular dynamics: application to materials and biological systems. J. Chem. Theory Comput.

Gonzalez Solveyra, E., E. de la Llave, D. A. Scherlis and V. Molinero. 2011. Melting and crystallization of ice in partially filled nanopores. J. Phys. Chem. B 115(48): 14196–14204.

Gonzalez Solveyra, E., E. De la Llave, V. Molinero, G. J. A. A. Soler-Illia and D. A. Scherlis. 2013. Structure, dynamics, and phase behavior of water in TiO_2 nanopores. J. Phys. Chem. C 117(17): 3330–3342.

Grünberg, B., T. Emmler, E. Gedat, I. Shenderovich, G. H. Findenegg, H. -H. Limbach et al. 2004. Hydrogen bonding of water confined in mesoporous silica MCM-41 and SBA-15 studied by 1H Solid-State NMR. Chem-Eur. J. 10(22): 5689–5696.

Gubbins, K. E., Y. -C. Liu, J. D. Moore and J. C. Palmer. 2011. The role of molecular modeling in confined systems: impact and prospects. Phys. Chem. Chem. Phys. 13(1): 58–85.

Hafner, J., C. Wolverton and G. Ceder. 2011. Toward computational materials design: the impact of density functional theory on materials research. MRS Bulletin 31(9): 659–668.

Hall, P. G., R. T. Williams and R. C. T. Slade. 1985. Nuclear magnetic resonance and dielectric relaxation investigations of water sorbed by Spherisorb silica. J. Chem. Soc., Faraday Trans. 181: 847–855.

Henderson, M. A. 2002. The interaction of water with solid surfaces: Fundamental aspects revisited. Surf. Sci. Reports 46(1-8): 1–308.

Inagaki, S. and Y. Fukushima. 1998. Adsorption of water vapor and hydrophobicity of ordered mesoporous silica, FSM-16. Micropor. Mesopor. Mat. 21(4): 667–672.

Jähnert, S., F. Vaca Chávez, G. E. Schaumann, A. Schreiber, M. Schönhoff and G. H. Findenegg. 2008. Melting and freezing of water in cylindrical silica nanopores. Phys. Chem. Chem. Phys. 10(39): 6039–6051.

Jelassi, J., H. L. Castricum, M. -C. Bellissent-Funel, J. Dore, J. B. W. Webber and R. Sridi-Dorbez. 2010. Studies of water and ice in hydrophilic and hydrophobic mesoporous silicas: pore characterisation and phase transformations. Phys. Chem. Chem. Phys. 12(12): 2838–2849.

Jelassi, J., T. Grosz, I. Bako, M. C. Bellissent-Funel, J. C. Dore, H. L. Castricum et al. 2011. Structural studies of water in hydrophilic and hydrophobic mesoporous silicas: An x-ray and neutron diffraction study at 297 K. J. Chem. Phys. 134(6): 64509–64518.

Kittaka, S., S. Ishimaru, M. Kuranishi and T. Matsuda. 2006. Enthalpy and interfacial free energy changes of water capillary condensed in mesoporous silica, MCM-41 and SBA-15. Phys. Chem. Chem. Phys. 8: 3223–3231.

Kocherbitov, V. and V. Alfredsson. 2011. Assessment of porosities of SBA-15 and MCM-41 using water sorption calorimetry. Langmuir 27(7): 3889–3897.

Koparde, V. N. and P. T. Cummings. 2007. Molecular dynamics study of water adsorption on TiO_2 nanoparticles. The J. Phys. Chem. C 111(19): 6920–6926.

Lee, J., M. Christopher Orilall, S. C. Warren, M. Kamperman, F. J. DiSalvo and U. Wiesner. 2008. Direct access to thermally stable and highly crystalline mesoporous transition-metal oxides with uniform pores. Nat. Mater. 7(3): 222–228.

Leszczynki, J. 2004. Computational Materials Science: Theoretical and Computational Chemistry. Amsterdam: Ed. Elsevier Science.

Liu, E., J. C. Dore, J. B. W. Webber, D. Khushalani, S. Jähnert, G. H. Findenegg et al. 2006. Neutron diffraction and NMR relaxation studies of structural variation and phase transformations for water/ice in SBA-15 silica: I. The over-filled case. J. Phys.: Condens. Matter 18(44): 10009–10028.

Liu, J., C. E. Nicholson and S. J. Cooper. 2007. Direct measurement of critical nucleus size in confined volumes. Langmuir 23(13): 7286–7292.

Liu, J. C. and P. A. Monson. 2005. Does water condense in carbon pores? Langmuir 21: 10219–10225.

Llewellyn, P. L., F. Schüth, Y. Grillet, F. Rouquerol, J. Rouquerol and K. K. Unger. 1995. Water sorption on mesoporous aluminosilicate MCM-41. Langmuir 11(2): 574–577.

Lombardo, T. G., N. Giovambattista and P. G. Debenedetti. 2009. Structural and mechanical properties of glassy water in nanoscale confinement. Faraday Discuss 141: 359–376.

Mamontov, E., L. Vlcek, D. J. Wesolowski, P. T. Cummings, W. Wang, L. M. Anovitz et al. 2007. Dynamics and structure of hydration water on rutile and cassiterite nanopowders studied by quasielastic neutron scattering and molecular dynamics simulations. J. Phys. Chem. C 111(11): 4328–4341.

Mancinelli, R., S. Imberti, A. K. Soper, K. H. Liu, C. Y. Mou, F. Bruni et al. 2009. Multiscale approach to the structural study of water confined in MCM41. JJ. Phys. Chem. B 113(50): 16169–16177.

Mancinelli, R., F. Bruni and M. A. Ricci. 2011. Structural studies of confined liquids: The case of water confined in MCM-41. J. Mol. Liq. 159(1): 42–46.

Martin, M. G., B. Chen, J. J. P. J. J. C. D. Wick, J. M. Stubbs and J. I. Siepmann. 2013. MCCCS Towhee, Version 7.0.6, see http://towhee.sourceforge.net.

Milischuk, A. A. and B. M. Ladanyi. 2011. Structure and dynamics of water confined in silica nanopores. J. Chem. Phys. 135(17): 174709.

Molinero, V. and E. B. Moore. 2009. Water modeled as an intermediate element between carbon and silicon. J. Phys. Chem. B 113(13): 4008–4016.

Monson, P. A. 2008. Contact angles, pore condensation, and hysteresis: insights from a simple molecular model. Langmuir 24: 12295–12302.

Moore, E. B., E. De la Llave, K. Welke, D. A. Scherlis and V. Molinero. 2010. Freezing, melting and structure of ice in a hydrophilic nanopore. Phys. Chem. Chem. Phys. 12(16): 4124–4134.

Moore, E. B. and V. Molinero. 2010. Ice crystallization in water's "no-man's land". J. Chem. Phys. 132(24): 244504–244513.

Moore, E. B., J. T. Allen and V. Molinero. 2012. Liquid-ice coexistence below the melting temperature for water confined in hydrophilic and hydrophobic nanopores. J. Phys. Chem. C 116: 7507–7514.

Morishige, K. and K. Kawano. 1999. Freezing and melting of water in a single cylindrical pore: The pore-size dependence of freezing and melting behavior. J. Chem. Phys. 110(10): 4867–4872.

Morishige, K. and H. Iwasaki. 2003. X-ray study of freezing and melting of water confined within SBA-15. Langmuir 19(7): 2808–2811.

Morishige, K. and H. Uematsu. 2005. The proper structure of cubic ice confined in mesopores. J. Chem. Phys. 122(4): 1–4.

Morishige, K. 2018. Influence of pore wall hydrophobicity on freezing and melting of confined water. J. Phys. Chem. C 122(9): 5013–5019.

Naono, H. and M. Hakuman. 1993. Analysis of porous texture by means of water vapor adsorption isotherm with particular attention to lower limit of hysteresis loop. J. Colloid Interface Sci. 158(1): 19–26.

Neugebauer, J. and T. Hickel. 2013. Density functional theory in materials science. WIREs Computational Molecular Science 3(5): 438–448.

Ng, E. -P. and S. Mintova. 2008. Nanoporous materials with enhanced hydrophilicity and high water sorption capacity. Micropor. Mesopor. Mat. 114(1-3): 1–26.

Noid, W. G. 2013. Perspective: Coarse-grained models for biomolecular systems. J. Phys. Chem. 139(9): 090901.

Oh, J. S., W. G. Shim, J. W. Lee, J. H. Kim, H. Moon and G. Seo. 2003. Adsorption equilibrium of water vapor on mesoporous materials. J. Chem. Eng. Data 48(6): 1458–1462.

Pajzderska, A., M. A. Gonzalez, J. Mielcarek and J. Wąsicki. 2014. Water behavior in MCM-41 as a function of pore filling and temperature studied by NMR and molecular dynamics simulations. J. Phys. Chem. C 118(41): 23701–23710.

Panagiotopoulos, A. Z., N. Quirke, M. Stapleton and D. J. Tildesley. 1988. Phase equilibria by simulation in the Gibbs ensemble. Mol. Phys. 63: 527–545.

Panagiotopoulos, A. Z. 2000. Monte Carlo methods for phase equilibria of fluids. J. Phys.: Condens. Matter 12: R25.

Parry, A. O., C. Rascón and L. Morgan. 2006. Signatures of non locality for short-ranged wetting at curved substrates. J. Chem. Phys. 124(15): 151101–151104.

Pearson, R. T. and W. Derbyshire. 1974. NMR studies of water adsorbed on a number of silica surfaces. J. Colloid Interface Sci. 46(2): 232–248.

Petrov, O. and I. Furó. 2011. A study of freezing–melting hysteresis of water in different porous materials. Part I: Porous silica glasses. Micropor. Mesopor. Mat. 138(1): 221–227.

Phillpot, S. R. 2000. An introduction to the molecular-dynamics simulation of materials. *In*: J. Lépinoux, D. Mazière, V. Pontikis and G. Saada [eds.]. Multiscale Phenomena in Plasticity: From Experiments to Phenomenology, Modelling and Materials Engineering. Dordrecht: Springer Netherlands.

Pickering, I., M. Paleico, Y. A. P. Sirkin, D. A. Scherlis and M. H. Factorovich. 2018. Grand canonical investigation of the quasi liquid layer of ice: is it liquid? J. Phys. Chem. B 122(18): 4880–4890.

Plimpton, S. 1995. Fast parallel algorithms for short-range molecular dynamics. J. Comput. Phys. 117(1): 1–19.

Předota, M., A. V. Bandura, P. T. Cummings, J. D. Kubicki, D. J. Wesolowski, A. A. Chialvo et al. 2004. Electric double layer at the Rutile (110) surface. 1. Structure of surfaces and interfacial water from molecular dynamics by use of *ab initio* potentials. J. Phys. Chem. B 108(32): 12049–12060.

Předota, M., P. T. Cummings and D. J. Wesolowski. 2007. Electric double layer at the Rutile (110) surface. 3. Inhomogeneous viscosity and diffusivity measurement by computer simulations. J. Phys. Chem. C 111(7): 3071–3079.

Prokhorenko, S., K. Kalke, Y. Nahas and L. Bellaiche. 2018. Large scale hybrid Monte Carlo simulations for structure and property prediction. NPJ Computational Mater. 4(1): 80.

Renou, R., A. Szymczyk and A. Ghoufi. 2014. Water confinement in nanoporous silica materials. J. Chem. Phys. 140(4): 044704.

Rozes, L. and C. Sánchez. 2011. Titanium oxo-clusters: precursors for a Lego-like construction of nanostructured hybrid materials. Chem. Soc. Rev. 40: 1006–1030.

Sánchez, V. M., E. de la Llave and D. A. Scherlis. 2011. Adsorption of R--OH molecules on TiO_2 surfaces at the solid-liquid interface. Langmuir 27(6): 2411–2419.

Schleder, G. R., A. C. M. Padilha, C. M. Acosta, M. Costa and A. Fazzio. 2019. From DFT to machine learning: recent approaches to materials science—a review. J. Phys. Mater. 2(3): 032001.

Schmidt, R., E. W. Hansen, M. Stoecker, D. Akporiaye and O. H. Ellestad. 1995. Pore size determination of MCM-51 mesoporous materials by means of 1H NMR spectroscopy, N2 adsorption, and HREM. A preliminary study. J. Am. Chem. Soc. 117(14): 4049–4056.

Schoch, R., J. Han and P. Renaud. 2008. Transport phenomena in nanofluidics. Rev. Mod. Phys. 80(3): 839–883.

Schranz, W. and V. Soprunyuk. 2019. Water in mesoporous confinement: Glass-to-liquid transition or freezing of molecular reorientation dynamics? Molecules 24(19).

Schreiber, A., I. Ketelsen and G. H. Findenegg. 2001. Melting and freezing of water in ordered mesoporous silica materials. Phys. Chem. Chem. Phys. 3(7): 1185–1195.

Schreiber, A., H. Bock, M. Schoen and G. H. Findenegg. 2002. Effect of surface modification on the pore condensation of fluids: experimental results and density functional theory. Mol. Phys. 100(13): 2097–2107.

Seyed-Yazdi, J., H. Farman, J. C. Dore, J. B. W. Webber and G. H. Findenegg. 2008. Structural characterization of water/ice formation in SBA-15 silicas: III. The triplet profile for 86 Å pore diameter. J. Phys.: Condens. Matter 20: 205108–205120.

Shirono, K. and H. Daiguji. 2007. Molecular simulation of the phase behavior of water confined in silica nanopores. J. Phys. Chem. C 111(22): 7938–7946.

Soler-Illia, G. J. d. A. A., C. Sánchez, B. Lebeau and J. Patarin. 2002. Chemical strategies to design textured materials: from microporous and mesoporous oxides to nanonetworks and hierarchical structures. Chem. Rev. 102(11): 4093–4138.

Soper, A. K. 2012. Density profile of water confined in cylindrical pores in MCM-41 silica. J. Phys.: Condens. Matter 24(6): 64107–64118.

Steiner, E., S. Bouguet-Bonnet, J. -L. Blin and D. Canet. 2011. Water behavior in mesoporous materials as studied by NMR relaxometry. J. Phys. Chem. A 115(35): 9941–9946.

Stewart, M. C. and R. Evans. 2005. Wetting and drying at a curved substrate: Long-ranged forces. Phys. Rev. E 71: 011602–011615.

Sun, C., L. Liu, A. Selloni, G. Lu and S. Smith. 2010. Titania-water interactions: a review of theoretical studies. J. Mater. Chem.

Takahara, S., M. Nakano, S. Kittaka, Y. Kuroda, T. Mori, H. Hamano et al. 1999. Neutron scattering study on dynamics of water molecules in MCM-41. J. Phys. Chem. B 103(28): 5814–5819.

Takahara, S., N. Sumiyama, S. Kittaka, T. Yamaguchi and M. -C. Bellissent-Funel. 2005. Neutron scattering study on dynamics of water molecules in MCM-41. 2. Determination of translational diffusion coefficient. J. Phys. Chem. B 109(22): 11231–11239.

Tang, Q., P. C. Angelomé, G. J. A. A. Soler-Illia and M. Müller. 2017. Formation of ordered mesostructured TiO_2 thin films: a soft coarse-grained simulation study. Phys. Chem. Chem. Phys. 19(41): 28249–28262.

Tarazona, P., U. M. B. Marconi and R. Evans. 1987. Phase equilibria of fluid interfaces and confined fluids. Mol. Phys. 60(3): 573–595.

Thommes, M. 2010. Physical adsorption characterization of nanoporous materials. Chemie Ingenieur Technik 82(7): 1059–1073.

Tilocca, A. and A. Selloni. 2004. Vertical and lateral order in adsorbed water layers on anatase TiO_2 (101). Langmuir 20(19): 8379–8384.

Velasco, M. I., M. B. Franzoni, E. A. Franceschini, E. Gonzalez Solveyra, D. Scherlis, R. H. Acosta et al. 2017. Water confined in mesoporous TiO_2 aerosols: Insights from NMR experiments and molecular dynamics simulations. J. Phys. Chem. C 121(13): 7533–7541.

Verdaguer, A., G. M. Sacha, H. Bluhm and M. Salmeron. 2006. Molecular structure of water at interfaces: Wetting at the nanometer scale. Chem. Rev. 106(4): 1478–1510.

Webber, J. B. W., J. C. Dore, J. H. Strange, R. Anderson and B. Tohidi. 2007. Plastic ice in confined geometry: the evidence from neutron diffraction and NMR relaxation. J. Phys.: Condens. Matter 19: 415117–415129.

Wei, M. J., J. Zhou, X. Lu, Y. Zhu, W. Liu, L. Lu et al. 2011. Diffusion of water molecules confined in slits of rutile TiO_2 (110) and graphite (0001). Fluid Phase Equilib. 302(1): 316–320.

Xu, Z., L. Zhang, L. Wang, J. Zuo and M. Yang. 2019. Computational characterization of the structural and mechanical properties of nanoporous titania. RSC Adv. 9(27): 15298–15306.

Zhang, C. and P. J. D. Lindan. 2003. Multilayer water adsorption on rutile TiO[sub 2](110): A first-principles study. J. Chem. Phys. 118(10): 4620–4630.

Zhao, X. S. and G. Q. Lu. 1998. Modification of MCM-41 by surface silylation with trimethylchlorosilane and adsorption study. J. Phys. Chem. B 102: 1556–1561.

Index